## Space and Sculpture in the Classic Maya City

In this book, Alexander Parmington combines an examination of space, access control, and sculptural themes and placement to propose how images and texts controlled movement in Classic Maya cities. Using Palenque as a case study, this book analyzes specific building groups and corresponding sculptures to provide insight into the hierarchical distribution and use of ritual and administrative space in temple and palace architecture. Identifying which spaces were the most accessible and therefore more public and which spaces were more segregated and consequently more private, Dr. Alexander Parmington demonstrates how sculptural, iconographic, and hieroglyphic content varies considerably when found in public/common or private/elite space. Drawing on specific examples from the Classic Maya and other early civilizations, he demonstrates that, by examining the intent in the distribution of architecture and art, the variation and function of the artistic themes represented in sculpture and other monumental works of art can be better understood.

Alexander Parmington is an Archaeologist and Heritage Consultant at the Wurundjeri Tribe Land and Compensation Cultural Heritage Council Inc. in Melbourne, Australia. He has worked extensively in Australia, Mexico, Guatemala, and Honduras. He has contributed articles to several journals and organizations, including *Mexicon*, the Ministerio de Cultura y Deportes de Guatemala, the Instituto de Antropología e Historia de Guatemala, and the Foundation for the Advancement of Mesoamerican Studies.

# Space and Sculpture in the Classic Maya City

Alexander Parmington

CAMBRIDGE
UNIVERSITY PRESS

CAMBRIDGE UNIVERSITY PRESS
Cambridge, New York, Melbourne, Madrid, Cape Town,
Singapore, São Paulo, Delhi, Tokyo, Mexico City

Cambridge University Press
32 Avenue of the Americas, New York, NY 10013-2473, USA

www.cambridge.org
Information on this title: www.cambridge.org/9781107002340



First published 2011

Printed in the United States of America

*A catalog record for this publication is available from the British Library.*

*Library of Congress Cataloging in Publication data*

Parmington, Alexander, 1971–
Space and sculpture in the Classic Maya City / Alexander Parmington.
p. cm.
Includes bibliographical references and index.
ISBN 978-1-107-00234-0 (hardback)
1. Spatial analysis (Statistics) in archaeology – Mexico – Palenque (Chiapas)
2. Maya sculpture – Mexico – Palenque (Chiapas) 3. Maya architecture – Mexico –
Palenque (Chiapas). 4. Palenque Site (Mexico) 5. Palenque (Chiapas, Mexico) –
Antiquities. I. Title.
F1435.1.P2P37 2011
972′.75–dc22 2010042795

ISBN 978-1-107-00234-0 Hardback

# Contents

# List of illustrations

# List of tables

# Summary


Spatial analysis of specific building types, combined with the thematic inquiry of corresponding sculpture, has proven to be a helpful method in gaining more developed insights into the hierarchical distribution and use of ritual and administrative "space" in Classic Maya (A.D. 250–900) city centres. Evidence suggests that monumental art, its subject matter, and its placement were exploited by the Maya elite as instruments of communication and control at important Maya sites. Drawing on specific examples from Palenque, in this study I demonstrate how "access analysis" of building group "archetypes" (Andrews 1975) can be used to detect shifts in the thematic content of monumental art, subject to differences in accessibility. I argue that sculpture and other artistic media assigned to distinct types of spaces (differentiated by progressive enclosure, channeled movement, and changes in elevation) may, like the spaces themselves, be sorted from most "public" to most "private," and the scenes, activities, and symbolism represented can be characterised similarly. Underpinning my investigation is the premise that "public space" and "private space" correspond to notions of "common space" and "elite space" and, furthermore, that monumental art contained within common and elite categories of space was accessed by corresponding common and elite audiences.

The presence of divergent sculptural themes at varying levels of accessibility in both the Cross Group and the early phases of the Palace at Palenque indicates that different types of imagery were directed at different audiences at Palenque. Although the Palace Group investigation detected some correspondence between spatial context and differences in the themes represented at the earlier stages of its development, analysis of subsequent phases implies that the importance of sculptural embellishment gradually diminished over time. The declining significance of sculpture and other decoration in the Palenque Palace is signaled by the relatively minor artistic contributions at later phases and by changes in the accessibility of artwork that existed prior to the structural modifications undertaken by later rulers.

# Acknowledgements

I would like to thank all those who assisted me directly and indirectly with my doctoral research, which forms the basis of this book. First and foremost, I am grateful to Peter Mathews for his guidance and support; he at all times made himself available for advice and direction, without which this work would not have been possible. I also thank the members of the Department of Archaeology, La Trobe University, for their assistance over the years, especially my friends and colleagues, Ernesto Arredondo, Péter Bíró, Armando Anaya Hernández, Tim Murray, Eduardo Pérez de Heredia, and Claudia Solis. In addition to these people, I will be forever grateful to La Trobe University for awarding me an APA scholarship and the Faculty of Humanities and Social Sciences for the provision of a grant that funded the field component of this study. I also thank my brother Jason for volunteering his time to assist me in the field and helping me with some of the imaging I have included in this book.

I am also thankful to those individuals, institutions, and organisations that have kindly permitted the reproduction of the many drawings that appear in this publication; each has been acknowledged separately, in text, with its respective illustration. A special thanks goes to Merle Greene Robertson for the contribution she has made to Maya studies at Palenque and the insights that her work has provided during the years that I undertook my own research. I am also appreciative of the advice and direction provided by Beatrice Rehl and Amanda Smith of Cambridge University Press and by Brigitte Coulton of Aptara during the production phase of the book.

I especially thank my wife Jo and my family for the love, patience, and support they have provided over the years; they have been a constant source of inspiration and motivation. I dedicate this work to my beautiful daughters, Charlotte and Jazmine.

# INTRODUCTION

The objectives of this book are to describe the methods and present the results of an investigation into the utilisation of sculpture by the Classic Maya as "signposting," as a means of signaling the function and the hierarchical division of ritual and administrative "spaces" in a city centre. Due to limitations in the breadth of literacy among the greater Maya (A.D. 250–900) population, imagery (as displayed on monuments and architecture) was used by the Maya elite as a supplementary communicative tool. Monumental art was used to signal, among other things, the "identities of sacred places and [their] function through sculptural composition . . . to people using them or coming into the spaces they addressed" (Schele and Mathews 1998: 27). Most often, attempts to understand the function and exclusivity of space within Classic Maya city centres have focused on the interpretation of architectural variation (i.e., points of restricted access implying public and private domains; see Harrison 1970; Andrews 1975; Pendergast 1992). Additional methods of inquiry have included epigraphic decipherment, ethnohistoric investigation, ethnographic analogy, and analysis of artefact residues and their distribution (e.g., Fash 1983; Schele and Mathews 1998; McAnany and Plank 2001).

A primary objective of the present inquiry was to determine whether an analysis of the spatial distribution of sculpture within a Maya site would influence current proposals concerning elite-civic demarcation and area function maintained by ethnographic, ethnohistoric, artefactual, and epigraphic assessment. My intention was to apply a social dimension to the inquiry of "space" through the analysis of behaviour and symbolism represented on monumental art and architecture. It is argued that among the Maya "a full range of activities took place in residential compounds, including lineage festivals, administrative overseeing, manufacture, gathering of tribute, adjudications, child rearing, food preparation, and a hundred other enterprises" (Schele and Mathews 1998: 29). As noted by Tilley (1994: 10–11), the "spatial experience [should not be viewed as] innocent and neutral, but invested with power relating to age, gender, social position and relationships with others. Different societies, groups and individuals act out their lives in different spaces." I saw value in establishing whether imagery, as displayed on monumental art and architecture, could further contribute to our understanding of social order and control within important Maya sites. Cross-culturally, the utilisation of visual media to communicate to the broader masses has long been acknowledged. For the "Greeks, mythologic, heroic, or historic sculpture represented something which had a very vivid interest for everyone. Similarly . . . mediaeval monuments [and] statuary had a meaning perfectly understood by all; it was a means of instruction. The iconography of our great northern cathedrals is a veritable encyclopedia instructing the multitude through the eyes" (Viollet-le-Duc 1987: 214).

"Access analysis" of archetypal building groups, combined with the thematic–analytical investigation of fixed sculptural media (positioned within specific spatial contexts), has proven to be an effective method of assessing the role that imagery played in signaling the function and hierarchical division of bounded space. General questions that I sought to address in my inquiry were: (1) What was it that motivated the Maya elite to position sculpture where they did? (2) Is there specific imagery marking specific spaces? (3) Acknowledging the multidimensionality of meaning communicated by Maya sculpture, are there embedded within compositions clues as to how certain space was used and socially demarcated?

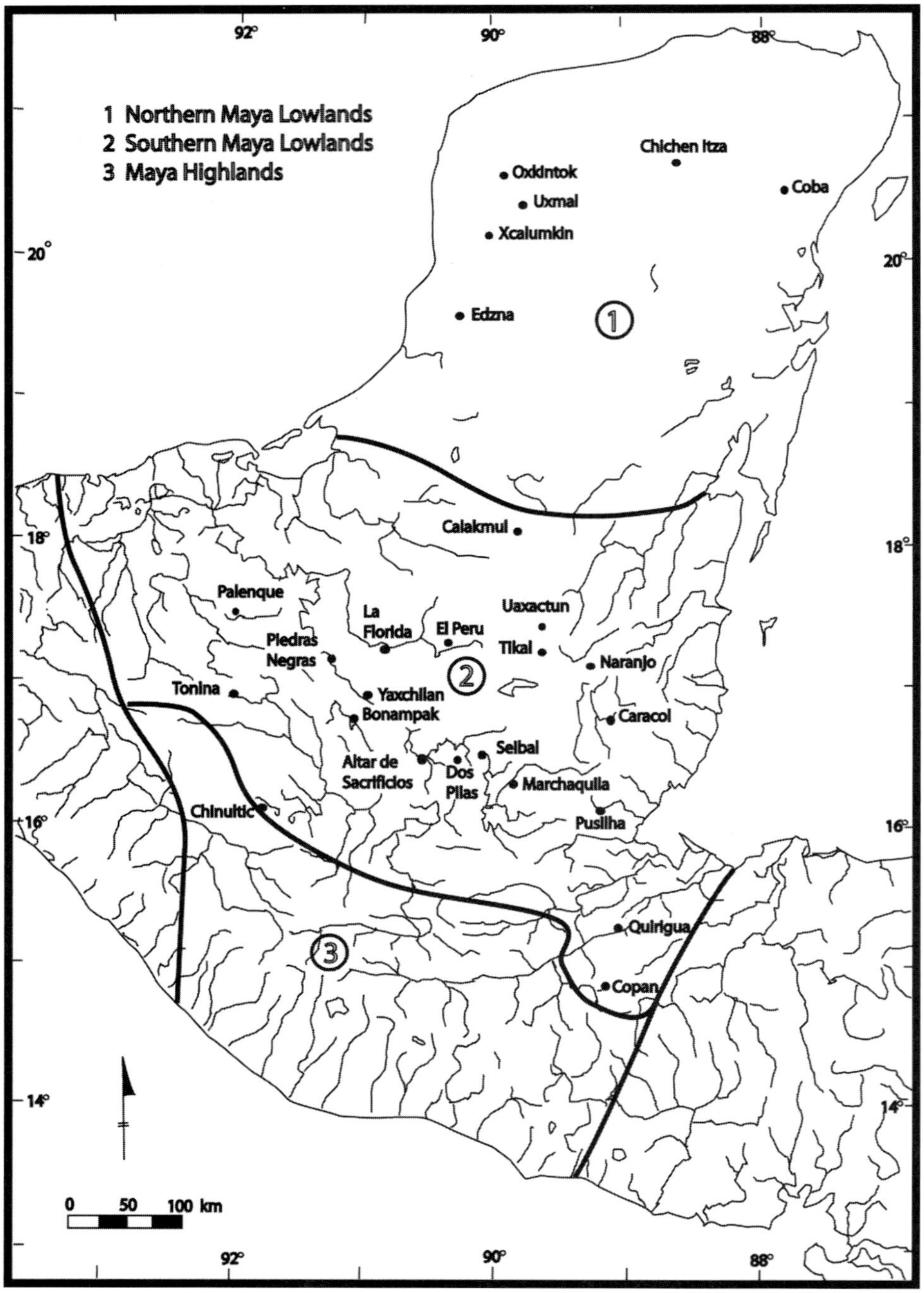


Map showing major sites in the Maya region. Mathews (1996: 3), reprinted with permission.

## Region and Cultural Background

The modern states and countries that collectively comprise the Maya region are the Mexican states of Yucatan, Quintana Roo, Campeche, Tabasco, and Chiapas, as well as Guatemala, Belize, and the westernmost regions of Honduras and El Salvador. With a cultural tradition extending at least as far back as 2,500 years, the ancient Maya were successful farmers, traders, and builders of cities that maintained large populations of inhabitants. In the past, the people of Mesoamerica shared a common cultural tradition – more specifically, "definitions about how to grow and distribute food, what constituted government, and how the world worked, both on the mundane and the cosmological level" (Schele and Mathews 1998: 15). Speaking many of the languages spoken by their ancient predecessors, descendants of the ancient Maya (numbering in the millions) still live in traditional houses in close proximity to the cities once ruled by their ancestors (Mathews 1996: 2).

Shortly after 1,000 B.C. the Maya began to construct their cities in the lowland forests and the highland mountains of Central America. The configuration of basic building groups in Maya centres mimicked, in their general layout and design, family compounds made from pole and thatch. Family compounds generally comprised a square central plaza surrounded on three sides by a platform surmounted by inward facing structures. Today, little survives of these residential structures that once occupied the peripheries of city centres. Their presence, size, and orientation, however, can be surmised from surviving earth mounds. Referred to as patio-groups, hundreds have been excavated. Patio-groups usually comprise two to six buildings that surround a central courtyard. Often arranged in clusters numbering between five and twelve, these architectural assemblages represent the basic building prototype in Maya settlement studies (Willey 1980: 255–256).

The Classic Period (A.D. 250–900) was characterised by the rapid growth of kingdoms, where ongoing conflict after the fifth century A.D. between city-states resulted in two great alliances comprising many of the smaller kingdoms of the Maya region. These great alliances were to be headed by two great "super-powers," Calakmul and Tikal (Schele and Mathews 1998: 17–18). Once the foci of ritual and administrative overseeing, some of the finest examples of major city centres of the period included Palenque, Yaxchilan, Bonampak, Tikal, Copan, and Piedras Negras. In the northern Maya lowlands, other Maya cites thrived towards the end of the Classic Period; included among these were Chichen Itza, Xcalumk'in, and the 'Puuc' cities of Uxmal, Kabah, Sayil, and Labna (see map of Maya region).

Architectural features commonly associated with major city centres include pyramidal temples, vaulted multiroomed palaces supported by single or multiterraced platforms, and the presence of one or more ballcourts. Artistic features commonly appearing in major centres include monolithic stelae and altars, most often positioned around plazas in the city's core. In addition, the larger Maya sites generally comprise several building groups connected by a network of causeways (Bullard 1960: 360–361), the longest of which may have provided access to smaller neighbouring sites. During the Classic Period, the western Maya lowlands also saw the rise of many subordinate centres, some of the largest of which were affiliated with sites like Yaxchilan (e.g., La Pasadita) and Piedras Negras (e.g., El Cayo).

1

# Defining the Maya built environment

## 1.1. Architecture, space, and cognition

Since ancient times, cities and towns have incorporated dominant architectural features as a means of signaling the priorities of civic and/or elite populations, as a way of making apparent the hierarchy of popular or imposed institutions, whether religious, governmental, or technological in nature. Often it has been architectural features, such as the degree of decoration displayed on buildings, combined with the relative size of functionally specific structures, that have been used as a way of signaling this order of priority, where building size has been especially important in communication over distance. Traditionally, the term "skyline" referred to a line in the distance where the earth and sky met. Today, the term has come to represent those buildings of a town or city visible at a distance on the horizon (Kostof 1991: 279). More often the prominent buildings influence how a city is perceived, signaling the prevailing social and/or political order of the time. For instance, in modern societies it is the secular architecture of enterprise that now dominates the landscape of most cities, where corporate skyscrapers now overshadow the religious architecture of churches and cathedrals (Kostof 1991: 280–294). In the ancient past, funerary monuments and temple architecture outshone all other forms, the most prominent being the Ziggurats of Mesopotamia and the temples and pyramids of Egypt and Mesoamerica.

It is argued that "environmental cognition is a vitally important part of the interaction between people and their physical surroundings" (Walmsley 1988: 22). Human beings both seek and create meaning from the landscape as a way of establishing some sense of order and security, raising questions as to what can and cannot be interpreted symbolically. For some time it has been acknowledged that "everything can assume symbolic significance" (Jaffé 1966: 232), whether natural objects like stones, plants, animals, mountains, wind, water, and fire, or things made by human beings such as houses, boats, and cars (Jaffé 1966: 232). It is through symbolic constructs that we as human beings assign meaning to a world that can and cannot be experienced directly. More specifically, "place naming, star naming, maps, myth and tale, the orientation of buildings, the spatial implications in dances and ceremonies, all facilitate the construction and maintenance of spatial patterns of the world in which the individual must live and act" (Hallowell 1977: 133).

In an environment shaped by the human constructs imposed upon it, much of the meaning communicated by the landscape comes from symbols contrived as regulatory mechanisms. For example, symbols may be used to signal function, accessibility, order of importance, acceptable behaviour in a particular environment, the intensity at which a given activity is performed, in addition to the relative significance of proximity (Walmsley 1988: 21–22). One function of symbols and rules encoded within the landscape is to articulate domains, for example, separating public and private localities and ensuring appropriate levels of inclusion, exclusion, interaction, and protection (Rapoport 1977: 289). In this sense, the built environment may be viewed as influencing how, in what context, and under what conditions individuals and/or groups interact and communicate with each other (Rapoport 1976: 19).

Spatial attributes are primarily understood in terms of the static and shifting relationships between objects contained within them, as the ability to perceive space relies on discerning the proximity of things in relation to other things. Spatial perception permits individuals to orient themselves, coordinate behaviour with others, as well as manipulate and transform their environment, whereas culturally constituted views of space permit people to participate in a world that is commonly perceived and understood (Hallowell 1977: 132). The distribution of space in the built environment displays patterns because of the influence of culture; different cultures comprise different rules, which guide behaviour and subsequently affect the design and organisation of architectonic space. Rules affect the placement, distribution, and utilisation of architectural features assigned to the landscape; therefore, the built environment can be understood in terms of the rules encoded within it (Rapoport 1977: 14). The effect that culture has on spatial patterning is discernible when comparing the urban landscapes of different countries. For example, the layout of North American cities such as New York generally conform to a grid pattern, whereas cities such as Paris in France display a more hierarchical layout where streets radiate outward from an architectonic epicentre (Nanda 1994: 81). In both examples, navigating through a city would make more sense to those people who reside in it, as particular cultures determine what emphasis is placed on spatial relationships, in addition to the significance of objects and reference points used for spatial orientation (Hallowell 1977: 131).

Human beings view themselves, in addition to objects, as being "in" space; consequently, they function as points of reference from which the rest of the world is perceived. "What spatial orientation . . . involves is a constant awareness of varying relations between the 'self' and other objects in a spatial schema of traditionally defined points" (Hallowell 1977: 133). Subsequently, to be oriented in a world that extends beyond the senses, the individual must view himself or herself in relation to a greater spatial plan, perceiving the self relative to objects and things conceived rather than seen (Hallowell 1977: 134). Social, psychological, and cultural aspects are represented in spatial terms, in "the intervals, relationships and distances between people and people, people and things, and things and things . . . [that lie] at the heart of the built environment" (Rapoport 1977: 9). Accordingly, features that we assign to space become markers of human action, values, and experiences and it is culture that determines the way in which these markers are ordered and interpreted (Pearson and Richards 1994a: 4).

The arrangements of architectural features in the built environment are affected by how people relate to each other in space, and are therefore subject to variation depending on behavioural context. Space articulated by architecture provides a "physical frame" to behaviour and allows architecture to be interpreted as arenas of "activity and performance of a range of social and cultural roles" (Arnold 2002: 135). Grahame (1997) points out that, in attempting to understand spatial layout in architecture, it becomes important to note that human beings are not simply located in space, but that they also have feelings towards space. Because of this emotional response to the environment, and the fact that social interaction occurs within it, space is constantly altered and transformed. It is through this process that structure is assigned to the world not only in a physical sense but morally, socially, and cosmologically, as it is through constructs derived through classification and categorisation that our sense of morality and social order are manifested and transformed.

It remains important to consider the relationship between space, social structure, and social consciousness; in this way the experience of architecture may be viewed as an interaction between subjective feeling and external influences (Arnold 2002: 134). There are three forms of knowledge that collectively control the way individuals perceive and react to their environments. They are: (1) operational knowledge, (2) responsive knowledge, and (3) inferential knowledge. Combined, these three forms of knowledge do two things: First, they facilitate movement, and, second, they provide a general frame of reference by which individuals understand and relate to their environment (Walmsley 1988: 21–22).

Operational knowledge relates to a person's understanding of the system or mode of operation in a given environment. By committing to memory such things as locations and physical attributes important to the functioning of an environment, one acquires this form of knowledge. In modern-day towns and cities, examples would include buildings, parks, and other features that, in addition to providing spatial contexts for activities (e.g., education, administration, and recreation), provide cues for the successful navigation around structured environments.

An individual's responsive knowledge comes as a result of reactions to distinctive features in the physical environment that stand out to differentiate elements – features such as the colour, size, and elaborateness of buildings – and the presence of communicative media, such as signs and billboards. Like operational knowledge, responsive knowledge provides cues that may assist in the selection of appropriate routes through an environment. To Walmsley (1988), knowledge of a place may also be derived from sounds and smells, which also fall under the category of responsive knowledge.

At the centre of inferential knowledge is the human being's ability to create systems of generalised categories to help organise and understand his or her environment. Inferential knowledge differs from both the operational and responsive forms in the sense that it is not derived from direct experience but rather through the individual's ability to extrapolate beyond what he or she has previously experienced and to probabilistically infer things that have not been experienced directly. Walmsley (1988) uses the modern example to make his point, noting how prior experience with a central business district may allow an individual to survey the skyline of an unvisited city and discern the proximity of its central business district based on the proximity and height of certain buildings.

As a result of the three forms of knowledge just described (operational, responsive, and inferential), cities and towns come to be known in terms of the actions of the occupants, the images projected by the environment, and the systems of symbols and categories selected to order and classify features of the landscape (Walmsley 1988: 21–22). Contrast in architecture is important in providing identity to formal systems leading to mutual reinforcement where a duality of interdependence is attained by the tension of opposing characteristics. Examples of these paired oppositions include solid/void, dominant/subordinate, positive/negative, complex/simple, high/low, large/small, wide/narrow, and also public/private (von Meiss 1990: 44; see also Pearson and Richards 1994a: 24).

## 1.2. Classic Maya architecture and spatial planning

Maya cities were affected by a number of factors, principally social, economic, engineering, historical, and ideational influences (Ashmore and Sabloff 2002: 202; see also Ashmore and Sabloff 2003). In attempting to understand intent in the configuration of Maya architecture, discerning the weight that each influential factor has had on the arrangement of architectural features from site to site remains an ongoing problem, as it was a mix of factors that controlled the size, orientation, and general layout of architectonic space in Maya centres. For example, climate would have affected the choices made in the design and orientation of buildings – factors such as prevailing winds, ventilation, and maximisation of seasonal light (Aveni and Hartung 1986: 3). Furthermore, pressures caused by population growth would have affected the concentration and distribution of architectural features. For example, ancient Maya settlements in the Petén Region resemble modern industrial centres in the sense that increases in populations forced neighbouring cities and towns to coalesce. In such instances, breaks between settlements may completely disappear, whereas civic and political identities are retained (Bullard 1960: 371).

### 1.2.1. Topography and architecture

In most other parts of the world the spread of human settlement has often followed the contours of the landscape where the orientation and alignment of streets, buildings, and settlements follow the course of rivers, coastlines, and other natural features (Kostof 1991: 53–57). Acknowledging the influence that topography has on the alignment and distribution of architecture, it is clear that local topography would have affected the arrangement of architectural features in Maya city centres. The topography that characterises the Maya region is largely diverse, encompassing volcanic mountains bordering the Pacific coast and lowland tropical rainforests (Mathews 1996: 2). As a result, "the flood plains of Copán and Quirigua; the sloping banks of the Pasion River; the hills along the Usumacinta River; the mountainous shelf of Palenque; the alluvial plains of Comalcalco and the flat Northern Yucatan plain all offered different challenges to the ancient builder" (Pollock 1965: 389–390). Terrain that is steep or hilly limits options for the placement, orientation, and design of architectural features, diminishing the level of control that an architect and/or builder has over how a structure or a group of structures may be configured. Alternatively, in flat terrain where topographical constraints are less prevalent, one would expect the control over the design and configuration of architectural features to increase substantially.

The negative effect that topography might have on the distribution of architecture is often minimised by what has been described as "amendments to the landscape" (Kostof 1991: 55), where land was forced, through a process of modification, to abide by the architectural plan imposed on it. These modifications have included the leveling and terracing of hills for either farming or the erection of buildings, the clearing of forests, reclamation of swamps, the diversion of rivers, and the construction of elevated platforms. One expression of higher architectonic control may be that a settlement appears less organic in form and more regimented in its overall configuration, although a direct correlation cannot be drawn between regular and irregular architectural planning and horizontal or undulating terrain. Kostof (1991: 55) reminds us that the banks of the Nile at Thebes "hosted a lively tangle of the residential quarters, while the tossed topography of the western slopes was the setting for formally planned units (the mortuary temples)."

### 1.2.2. Building materials and construction techniques

Because of varied accessibility to raw materials, construction differed from region to region, affecting the general appearance of surviving architecture. For example, in the southeastern lowlands, sandstone, rhyolite, and marble were quarried and used at sites such as Quirigua, whereas trachyte was used at Copán (Sharer 1994: 631). In the

Maya highlands, pyramids were built with rammed earth and in the lowland region with stone rubble (Schele and Mathews 1998: 17). At Comalcalco in Tabasco, limestone was not readily available, so consequently fired brick and shell were used as an effective alternative (Hammond 1982: 249). Other locally acquired building materials included tuff, sandstone, slate, and dolomite. By mixing lime mortar and rock, an effective adhesive was concocted.

Transportation of quarried stone was undertaken using manpower; beasts of burden were not exploited by the Maya. It is probable that the Maya used a combination of ropes, rollers, and ramps to transport and maneuver heavier stone blocks into position (Pollock 1965: 396–397). Most often dry stone rubble was used as a fill to support superstructures; soil or dressed stone was then placed over the mass. The exploitation of rich limestone beds in the northern, southern, and central lowlands by ancient builders has provided archaeologists with some of the best extant examples of Maya masonry architecture and art. Plaster produced from the burning of the limestone was also used to embellish architectural features with sculptured elements, in addition to smoothing the interior and exterior of buildings as well as surfacing plazas and courtyards (Pollock 1965: 386–389; Sharer 1994: 631).

Many of the stone structures present in Maya city centres, whilst they varied geographically, were surveyed to a relatively high degree of accuracy. The Maya considered any structural angle between 80 and 100 degrees to be a right angle and a slope of 3 degrees or less to be horizontal or flat. Some of the devices used in the planning and construction of buildings and associated sculpture were the lever, the plumb bob, and the water level (Hammond 1982: 249). The symmetry of buildings, as well as the sculpture that adorns them, is one of the first things striking the observer when viewing Maya architecture. Symmetry in architecture and sculptural design was controlled by using a cord as a measuring device. First, a section of cord was cut to a particular body length (e.g., shoulder to outstretched hand); this cord was then used as a unit of length to mark out a square to the required dimensions. Through a process of halving and stretching this cord to determine the portions of the square, this basic shape could be accurately divided or expanded to derive shapes with "golden mean" proportions (Schele and Mathews 1998: 35). To ensure the balanced and harmonious appearance of buildings and their sculptured surfaces, the Maya used this simple but elegant method of measurement.

A common characteristic of Classic Maya masonry architecture is the confined nature of interior spaces (Sharer 1994: 634). One reason for this was technological – the corbelled vault was a common component in Maya architecture that provided no structural support to the structures in which they appeared (Pollock 1965: 402). The corbelled vault was constructed by stacking successive courses of stone inward until they met at the top where a final capstone was placed. Unlike a "true arch," in a corbelled vault, adjacent stones do not support each other, providing a powerfully incorporated whole. Consequently, the configuration is weaker, relying on massive walls to support the vault. For this reason, the widths of rooms within most Maya masonry buildings rarely exceeded 3 metres, and structures often do not exceed more than one storey (Sharer 1994: 634–637).

### 1.2.3. Regional variations in Maya architecture

Whilst acknowledging similarities in architectural traditions throughout the ancient Maya world, there are some regional differences that should be outlined here briefly. There are four principal styles that characterise Classic Maya architecture; these are the Petén architectural style, the Usumacinta architectural style, the Southeastern Lowland architectural style, and the Puuc architectural style (Sharer 1994: 637–639; see also Andrews 1975: 30–32).

Maya buildings varied regionally because of variations in local trends as well as differences in historical and artistic traditions. Petén-style architecture of the central lowlands is characterised by massive platforms that supported heavy structures with small interior spaces, in addition to single entrances and huge mosaic roofcombs. The geographical extent of this regional style is not well defined, although it probably stretches as far as Calakmul to the north and the modern border of Belize to the east. Characteristics of the Usumacinta architectural style are lighter roofcombs and structures with multiple doorways positioned on natural rises or hills. This architectural style encompasses sites such as Yaxchilan, Piedras Negras, and Bonampak, as well as Palenque; however, Palenque architecture represents a technological development in this style. The inclusion of medial walls shared by parallel vaulted rooms combined with inward sloping mansard roofing permitted the supporting walls of structures at Palenque to be less massive (Sharer 1994: 637–638). A consequence of this design was that it permitted the interior of buildings to be made wider and more open, also allowing the number of entrances to be increased (Miller 1999: 38).

Copán, the most southern of all Maya sites (Miller 1999: 49), characterises the southeastern architectural style with its monumental staircases and elaborate Early Classic stucco sculpture. During the Late Classic Period, less emphasis was placed on stucco modeling at Copán, being replaced by a form of mosaic sculpture that was finished in plaster. Both Quirigua and Lubaantun would probably fall within this style (Sharer 1994: 638; see also Hammond 1975 and Miller 1999: 48–54).

Finally, Late Classic Puuc style architecture of the northern lowlands, which peaked between A.D. 771 and A.D. 790, is characterised by structures consisting of either single or multiple entrances surmounted by a small plain roofcomb. Typically, the exterior of Puuc-style buildings was covered by a veneer of limestone masonry adhered to a concrete mass (Kowalski 1998: 404). Some of the finest examples of Puuc-style architecture are found at sites such as Uxmal, Kabah, and Sayil (Kowalski 1998: 401–402). Little stucco sculpture is associated with Puuc-style architecture; generally the lower halves of Puuc buildings are plain, with the upper portions of the building displaying most of the decoration (Kowalski 1998: 404). The buildings are often decorated with highly elaborate stone mosaics; the designs include masks, serpents, stepped fret patterns, lattice and columnar designs, as well as various other geometric patterns (Greene Robertson 1994: 209). This style of architecture is thought to have been emulated by Post Classic sites such as Tulum and Mayapan (Sharer 1994: 638–640).

The goal of the ancient Maya builders and architects was not solely to articulate interior spaces but also to maximise the effect of the structures when observed externally. The decorated exteriors of buildings functioned as backdrops and stages for ritual processions, dances, and performances. In contrast, the less-decorated interiors of many buildings were domains where the gods resided. Architects decorated buildings in two main ways, either with carved stone or with modeled plaster. The most common surfaces to be decorated were the terraces of pyramids, staircases and associated balustrades, building platforms, bearing walls, door jambs, lintels, entablatures, and roofcombs. Building surfaces that were most often heavily decorated were those that would have faced a prospective audience (Schele and Mathews 1998: 40). At Palenque in Chiapas, public sculpture appears most frequently on piers, eaves, entablatures, and roofcombs, whereas balustrades and substructural terraces were decorated less frequently. Architectural decoration appearing at Tikal in Guatemala is located primarily on entablatures and roofcombs; sculpture is rarely found on terraces and pyramidal substructures. At Copán in Honduras, sculpture appears most commonly on the corners of buildings, entablatures, roofcombs, stairways, and speakers' platforms. At Bonampak and Yaxchilan, sculptors focused primarily on decorating stairways, entablatures, and roofcombs (Schele 1998: 479–480).

### 1.2.4. Sociological and cosmological influences on site plans

The ancient Maya used specific strategies in their site plans, where "the arrangement of ancient Maya buildings and arenas emphatically express[ed] statements about cosmology and political order" (Ashmore and Sabloff 2002: 201). Among the factors that influenced the distribution of architectural features in Maya centres were: (1) the Maya belief in the multilayered heaven and the Underworld, in addition to the gods and ancestors that reside in both realms; (2) the connection between heaven and the Underworld as expressed by the movements of the sun and other celestial bodies such as the moon and Venus; (3) "vertical connectors," such as mountains, "sky bearers" (*bacabs*), and cave portals to the Underworld; and (4) the division of the world into the four cardinal points, including the centre point, and their corresponding colours (Ashmore 1992: 174). Factors such as these were manifested architecturally by the Maya in the arrangement of buildings and associated architectonic space, as expressed in: (1) the north–south axis of city centres; (2) the divergent functions of structures assigned to both ends of the axis, more specifically Underworld associations for southern structures and celestial–supernatural associations for northern buildings; (3) the presence of eastern and western building groups forming a triadic configuration with the northern architecture; (4) the presence of a ballcourt that functioned as a mediator between northern and southern axis architecture; and (5) the use of roads and causeways to connect and integrate all architecture physically and symbolically (Ashmore 1992: 174; see also Ashmore 1986: 36).

It has been proposed that "if anything operated to regulate the arrangement of temples, stairways, platforms, ballcourts, and pyramids [in Maya centres] they were religious in nature" (Fuson 1969: 497). One thing that influenced the distribution and orientation of architecture was astronomical alignment; more specifically, the path of the rising and setting sun as well as acknowledgment of true north, as marked by the celestial pole star (*Xaman Ek*). Solar and celestial observation resulted in the construction of buildings and building groups that were oriented with the four cardinal directions, in addition to buildings that were designed to mark the position of the rising sun during specific times of the year. Structures that collectively comprise Group E at Uaxactun were configured to mark the position of the rising sun during equinox and the summer and winter solstices. To observe these events required standing on the observatory pyramid to the west and facing three minor structures positioned to the east (Fuson 1969: 498–499).

The planet Venus must have also been influential in the design and distribution of Maya architectonic space, given that temples and platforms were dedicated to the planet (Fuson 1969: 502). Referred to as *Noh Ek* (great star) or *Xux Ek* (wasp star), Venus was an important celestial object among the Maya. Thought to be a god, Venus has been identified with *Kukulcan* (also known as Quetzalcoatl), the patron god of rulership (Miller and Taube 1993: 142). Both Venus and the sun were also believed to be manifestations of the Hero Twins as described in the creation mythology of the *Popol Wuj* (Schele and Freidel 1990: 115; Sharer 1994: 288). The Day 1 *Ajaw* was also used in reference to the Venus god, and *Yax* represented the month of the deity (Fuson 1969: 502–503).

Among the contemporary Maya, the perception of their world "embodies an interaction between two kinds of spatialisation" (Hanks 1990: 335), which creates a conceptual division between cardinal directions and cardinal places (Hanks 1990: 300). These two forms of spatial arrangement are reflected in the directional organisation of space (a linear notion of space that is relative to a position or the perspective of an actor) and in locative organisation (an absolute rather than perspectival notion of space that is orthogonal in nature, not linear–successive) (Hanks 1990: 335). In many cultures, particularly important to the orientation of buildings were the cardinal directions; the most important of which was the east – the direction from which the sun rose into the heavens. Even today, structures are oriented or aligned to either optimise or minimise the effect caused by the path of the sun throughout the year (Pearson and Richards 1994b: 15).

Maya buildings were positioned intentionally to align with the cardinal directions (Ashmore 1986: 36). Beyond viewing the cardinal directions as a coordinate system, Maya cosmology views the directions north, south, east, and west as locations, referring to them as *tukantí'itzil ká'an, tukantí'itzil lú'um*, meaning "at the four corners of the sky, at the four corners of the earth" (Hanks 1990: 299). (Note: The words in italics have been changed from Hanks's 1990 orthography to that used in this study). The modern Maya conceive these locations to be joined by lines (east to south to west to north) establishing a perimeter, rather than perceiving them linearly as two overlapping perpendicular lines (joining north to south and east to west). In addition, the Maya also believe that there is a fifth cardinal position referred to as the *chuúmuk* that marks the middle or centre point of the perimeter square. "Whether it is a small domestic space, a municipal one, the top of an altar, a corn field, or the entire world, its four corners plus centre define its schematic totality" (Hanks 1990: 299–300).

The Maya view "domestic space [as] . . . a direct embodiment of cultural order" (Hanks 1990: 315). The use of perimeter lines to connect north, south, east, and west locations allows (1) conceptual divisions to be made between what is perceived as inside/included and outside/excluded and (2) directional distinctions to be made, such as inside/inward and outside/outward (Hanks 1990: 302). Today, the Maya homestead has a specific type of spatial arrangement where the centre, combined with the four corners, forms a unitary whole. This *haál* ("perimeter") functions to separate the inner/private space from the outer/public space. Nonresidents wishing to enter a homestead first acknowledge the outer boundary with an announcement of their presence – this is referred to as "respecting the *nah*" ("house"); there they must wait for permission before crossing the perimeter boundary (Hanks 1990: 324). Activities such as cooking, eating, storage, and sometimes bathing take place in the *k'oób'en* ("kitchen"), whereas sleeping, dressing, and the receiving of guests occur in the *nah* (Hanks 1990: 333–334). In the traditional *xá'anih nah* ("palm roof house"), the rounded ends (*moy*) of the interior are cordoned off with curtains converting them into private spaces where residents may sleep, dress, and bathe unobserved (Hanks 1990: 324–326).

Among the Maya, domestic space is often used for ritual activities where cardinal spaces are formulated diagrammatically on an altar, analogous to that which defines the perimeter of domestic environments (Hanks 1990: 335). In major ceremonies, such as *heétz lú'um* ("fix earth"), *tz'áa tzaák* ("get rain") ceremonies, there are two principal stages, namely (1) *he'ik b'èel* ("opening the road") and/or *k'axik mèesa* ("binding [the] altar") and (2) *wach'ik mèesa* ("untie [the] altar") (Hanks 1990: 336). To open the road is to provide access to spirits who reside at different locations in the vertical and horizontal universe. Summoned by the shaman, these spirits are brought to the place of performance in a specified order, beginning with the east and moving counterclockwise (north is the next direction, followed by west, south, and then, last, the centre). The spirits are then bound to the quadrilateral altar according to their cardinal locations. In the ritual process of "untying," the spirits are returned to their original places of origin in a clockwise order starting with the east, followed by south, west, north, and then the centre (Hanks 1990: 336–338). There is also a linear orientation to such rituals that corresponds to an east–west axis. Observers/participants in domestic and agricultural ceremonies stand behind the shaman on the west side of the altar facing east; the gods positioned to the east face west. "The process of lowering the spirits through prayer combines successive with orthogonal, as well as centred with absolute, orientations . . . just as their placement at the four corners plus the centre anchors the cardinal directions, their simultaneous union defines totality" (Hanks 1990: 338).

Today, the Maya shamans' conception of the universe is as a sphere suspended in a void, where the earth forms a horizontal quadrilateral layer vertically positioned at the midpoint of that sphere. Marking the highest point in the heavens is the sun, the rotation of which, in addition to the moon, defines the edge of a sphere; this edge represents the limits of reality – beyond this boundary is only nothingness (Hanks 1990: 304). Above the quadrilateral earth are seven layers, or atmospheres, which the Maya refer to as *yoók'ol kàab'*, meaning "above earth." Beneath the earth layer (*'lú'um'*) is a body of water on which the earth rests; below this level is the Underworld referred to as *metnal* ("hell"). In current mythology, the Maya believe that spirits are located throughout the multileveled universe (Hanks 1990: 305–306).

## 1.3. Common forms in Maya architecture

While acknowledging the presence of regional variations in architecture at different Maya centres, certain reoccurring forms have been identified at many sites in the northern, southern, and central lowlands. Some of the more common structural forms include palaces, ballcourts, plazas, courtyards, shrines, causeways, temples, and tombs (Andrews 1975: 37–51; Hammond 1982: 43; Sharer 1994: 630–631; Webster 1998: 6; Schele and Mathews 1998: 23–37).

### 1.3.1. Palaces

The term *palacio* ("palace") was first used by eighteenth-century explorers and writers to describe Maya multiroomed structures perceived to be elite-residential in nature. The term *palace* was also used to create a descriptive distinction between multiroomed structures and the more elevated single-roomed structures referred to as "temples" (Christie 2003b: 2–3). A distinction has long been made between Maya palace and temple architecture: Palaces are defined as "large, range-type, vaulted masonry multi-roomed structures, or as architectural groups composed of several such . . . structures surrounding small plazas [courtyards] or patios" (Kowalski 2003: 204). In contrast, temples are defined as "smaller structures with more restricted interior space constructed in more inaccessible locations atop high pyramidal substructures" (Kowalski 2003: 204; see also Pollock 1965; Andrews 1975).

For some time there has been general criticism concerning the assigning of such overarching terms to describe Maya architecture. Oversimplistic terms, such as *palace* and *temple*, effectively group together architectural forms that show considerable variation whilst implying functions that may be incorrect or inappropriate (Webster 1992: 140). For example, Satterthwaite (1935), on excavating the Palace at Piedras Negras, noted that there was no apparent relationship between the term *palace* and the function of the building group. Today, because of advances in settlement studies and other archaeological inquiry, there is consensus among scholars that palace-type structures would have had several functions, primarily religious, political, and residential in nature (Christie 2003b: 5).

An important question remains as to whether elite dignitaries lived in the elaborate centres or whether they occupied less permanent structures outside the core of their cities. "During Classic times, the evidence seems to indicate an association of residential and other functions within palace structures" (Liendo Stuardo 2003: 194). There are several criteria that one must consider in determining the presence of elite-residential architecture in Maya centres; among these are spatial indicators, such as building size and complexity, as well as the presence of items reflecting elite ideology, such as luxury or exotic goods. Other things that should be considered are the proximity of architectural features in relation to the central precinct, as well as relative accessibility of buildings (Guderjan et al. 2003: 19–20).

Whereas the lesser elites probably resided in close proximity to the city centres, the greater civic populations lived on the periphery of city centres in pole-and-thatch houses built on raised platforms around central courtyards (Pollock 1965: 381–382). It has long been argued that "the Maya nobility . . . lived on the outskirts of the city [centres], and [that] the peasant and the working population lived far away in small scattered settlements" (Thompson 1954: 313). Although more recent settlement studies indicate that a wholly concentric model represents an oversimplification of spatial organization in Maya society (see Chase 1992b: 133 and Barnhart 2001), ethnohistorical reference to Chichen Itza does support a proposal of societal demarcation among the Maya expressed architecturally and spatially. Diego de Landa was the first to recognise the hierarchical distribution of space in towns and settlements (Christie 2003b: 3).

> In the middle of the town were their temples with beautiful plazas, and all around the temples stood the houses of the lords and the priests and then (those of) the most

> important people. Then came the houses of the richest and of those who were held in the highest estimation nearest to these, and at the outskirts of the town were the houses of the lower class (de Landa in Tozzer 1941: 62–64).

Further evidence that nobility resided in Maya centres comes from Structure 5D-46 in the Central Acropolis at Tikal, where a cached vessel bearing the title and emblem glyph of one of Tikal's rulers was recovered. The accompanying inscription also records the dedication of the ruler's house (Haviland and Moholy-Nagy 1992: 51).

It is reasonable to propose that those who once inhabited the magnificent stone structures of the Maya held higher status than those who occupied smaller structures of poorer quality. Today, the quality and size of dwellings is considered reflective of social position: The greater the social standing of an individual, the more impressive the house in which he or she resides. Maya palaces are most often positioned in the core of cities, located in the principal ritual administrative precincts. Among other things, one of the differences between elite residences and more common domestic dwellings is that elite buildings generally have more space per living unit than do structures of lower status (Haviland and Moholy-Nagy 1992: 51). As a description of an Aztec palace by Hernan Cortés has revealed:

> There are houses belonging to certain men of rank which are very cool and have many rooms, for we have seen as many as five courtyards in a single house, and the rooms around them very well laid out, each man having a private room. Inside there are also wells and water tanks and rooms for slaves and servants of which they have many. Each of these chieftains has in front of the entrance to his house a large courtyard and some two or three or four of them raised up high with steps up to them and all very well built (Cortés 1986: 30–31).

Palace-type buildings were often the end result of a gradual accumulation of architectural features commissioned by successive rulers over many years, developed through "periodic programs of renovation, burial, and expansion" (McAnany and Plank 2001: 84). For example, the Palace at Palenque, not unlike Group A-5 from Uaxactun, represents an architectural accretion that occurred over many generations (see Chapter 4, Architectural and Sculptural Programs of the Palenque Palace Group). It is argued that generally "palaces are often haphazardly arranged since they grow organically to suit the needs of an expanding elite in residence" (Cohodas 1973: 216). The eight phases of construction that culminated in Group A-5 at Uaxactun illustrate how a building complex can change dramatically over time from what was once a basic triadic temple design to a complex of multichambered galleries (Miller 1999: 30). In addition, at each phase of the building group's construction there would have undoubtedly been a corresponding sculptural program.

Investigations of the Palace at Palenque indicate that Maya architecture may have been becoming increasingly restricted over time. More specifically, buildings were extended and partitioned as a means of segregating elite-ritual and -residential spaces from the more accessible public spaces of the Palace Group (see Chapter 5, Access Analysis of the Architectural and Sculptural Programs of the Palenque Palace Group). Such changes reflect a dramatic shift in the relationship between the ruling class and their subjects during the Classic Period (Christie 2003a: 321). The accumulation of architectural features over time at many city centres is a consequence of free or forced amendments made by successive rulers over the entire history of the site. For example, "the manner in which the Acropolis grew [at Tikal] through addition and accretion is suggestive of an expanding extended family of great power and wealth" (Haviland and Moholy-Nagy 1992: 52), where the installation of secondary controls on pedestrian traffic (i.e., supplementary stairs, masonry screens, and masonry blocks) suggests "a deliberate attempt to preserve privacy" (Harrison 1970: 187).

### 1.3.2. Temple pyramids and shrines

Maya temple pyramids generally comprise a pyramidal platform and an elevated superstructure that contain one or a number of small chambers that have restricted access (Kowalski 2003: 204). Temples appear independently as single structures, in pairs, or in a triadic configuration; the third arrangement establishes an architectural hierarchy and a sense of directionality to any persons who entered the group. The principal structure in this configuration faces the entrance to its central plaza, whereas the remaining two temples are positioned peripherally from the perspective of someone entering the group (Cohodas 1973: 218). In Maya mythology, the triangular arrangement of the Temple Group is argued to be representative of the mythic hearth throne stones that were laid down at the creation of the current world. This triangular formation appears in the constellation of Orion, which the Maya reproduced in the configuration of their buildings (Schele and Mathews 1998: 44).

Most pyramids in Mesoamerica display bilateral mirror symmetry, which is expressed in both their architectural and sculptural arrangements. The "axis of reflection is usually the ramp or staircase" (Kubler 1985: 243), which provided access to a summit temple. One of the few exceptions to this pyramidal form was the "radial pyramid," which has four staircases, one on each side of the structure (Schele and Mathews 1998: 42), corresponding to the four cardinal directions. Most often positioned centrally in open plazas, this type of pyramid could be ascended from several directions (e.g., Structure 4 at Copán). Kubler (1985: 243) argues that "the radial symmetrical pyramid is the most monumental and commanding of pyramidal forms, for all approaches [to such structures] are of equal value." Twin pyramid complexes comprise two closely positioned radial platforms, each of which has no temple superstructure positioned on top; they also generally do not display any architectural sculpture (Schele and Mathews 1998: 42).

Temples are most often located within the core of Maya cities, constructed on either man-made substructures or on natural hills or rises. Temple interiors were principally for the worshipping of gods as well as for ancestor veneration (see Foster 2002: 211–212). They were places where a select few could commune with the dead and make offerings as currency for divine intervention. Often displaying an abundance of imagery, temple exteriors functioned as stages and backdrops for public performances, such as dances, dramas, and ritual processions (Schele and Mathews 1998: 40). In addition, temples were often decorated with mythological imagery and texts; decorated architectural features usually included *alfardas*, piers, entablatures, and roofcombs. In addition, the supporting substructure sometimes portrayed sculptured elements, such as zoomorphic masks that flanked the main stairway.

It is argued that "temples would have been the most restricted space of all" (Schele and Mathews 1998: 29). The purpose of elevating temples was to effectively advertise their presence whilst simultaneously restricting access to their interior via a single stairway. The ancient Maya viewed pyramidal temples as symbolic representations of mountains (Miller and Taube 1993: 160–161; Schele and Mathews 1998: 17, 43). For example, Temple 26 at Copán was constructed to symbolise an "ancestor mountain" (Martin and Grube 2000: 208). The Maya interpreted summit entrances to temples as cave thresholds metaphorically providing access to the interior of the mountain and on into the Underworld. There is intrinsic importance to the thresholds in architecture as "the spatiality of an object reaches its peak when it is pierced, when it can be crossed and thus links the observer to other elements in the environment beyond the object" (von Meiss 1990: 94). Once inside the symbolic mountain, rulers believed that they could commune with gods and ancestors.

In contrast to temples, shrines are enclosed spaces contained within the larger temple structures, often housing mythological and ancestral themes (comprising both text and imagery) as represented on interior friezes or wall panels (see Chapter 3, Access Analysis of the Palenque Cross Group and Its Sculpture). Shrines were sacred spaces that the ancient Maya referred to as either a *pib' nah*, meaning "underground structure," or a *waybill*, meaning "resting place" (Schele and Mathews 1998: 40; Stuart 2006: 109). Often there is an area within the shrine where an offering can be placed. Several of the best examples of these types of inner shrines are found in each of the Cross Group temples at Palenque. Pollock (1965: 409–411) argues that "temples and shrines in their most typical forms, because of their restricted interior arrangements and exulted location . . . seem fairly certain to have housed the esoteric rites of a selected priesthood." Similarly, Hammond argues that "the cramped space inside Maya temples is an argument against any sort of participatory ritual involving more than one or two celebrants" (Hammond 1982: 250–251).

### 1.3.3. Ballcourts

Often located centrally in ancient Maya cities (Miller and Houston 1987: 53), the architecture of ballcourts was highly public, positioned either in an open plaza or in close proximity to a pyramid or acropolis (Cohodas 1991: 254). The highly public nature of ballcourts suggests that the game was an important ritual activity among the general population and the upper echelon in Maya society (see Fash 1991: 126). Generally comprising two parallel rectangular structures (Miller and Taube 1993: 43) oriented north–south, the ballcourt functioned as an arena for a game played with a large rubber ball. Both sides of a ballcourt comprised a vertical or inward sloping wall (Kubler 1985: 244), probably designed as a rebounding surface to keep the ball in play. The main playing surface was the alley in between the parallel structures, which was set at ground level; the end zones of the court were generally left open (Miller and Houston 1987: 47; Cohodas 1991: 254).

Although the precise rules remain unclear, depictions of the ball game in Maya art indicate that the principal objective of the game was to keep the ball airborne or in play, using nothing but the hips or torso. It is probable that advantage was achieved by driving the ball into either end of the court or by sending the ball through small hoops placed high up on the walls. Participants wore impact protection around their waists, padded belts in the Maya lowlands and open-ended "yokes" (heavy belts tied over a softer lining) in the southern highlands

(Ekholm 1991: 241–245). In addition, players wore padding on their forearms and knees to prevent injury when dropping beneath the ball. Tilted or kneeling posture in ball-game imagery is suggestive of an individual about to return a ball into play (Cohodas 1991: 251). There are several ball-game–related glyphs, one of which reads *pitz*, meaning "to play ball"; this glyph was also used as an elite title (see Miller and Houston 1987: 60–63).

The ball game was as a highly ritualised gladiatorial-type conflict, where losers may have been executed or sacrificed. The ballcourt was a "theater showing the reenactment of a cosmic conflict between gods that manifested as sun and the planet Venus" (Miller and Houston 1987: 47). Mesoamerican ballcourts were large and highly decorated (as seen at sites such as Copán and Chichen Itza) or small and relatively plain (as seen at Palenque). In addition to the presence of architectural embellishment, round markers were often placed in the floor of ballcourts. Carved from stone, these markers often displayed Underworld scenes and corresponding hieroglyphic texts. Cohodas argues that "taken together, the axial orientation, absence of end zones, and [the] line of three markers in the Classic Maya courts emphasize architectural symbolism of the world axis which demarcates a cosmological passage through the earth's surface into the Underworld" (Cohodas 1991: 254).

### 1.3.4. Courtyards and plazas

Plazas are generally larger and more open than courtyards, having the space to potentially hold many hundreds, if not many thousands, of people. By comparison, courtyards are much smaller and are typically associated with the interiors of quadrangles and palace-type structures (Andrews 1975: 59–63). Plazas and courtyards functioned as public meeting areas as well as arenas for elite display and conspicuous ritual. More broadly, they have been described as "culturally defined spatial settings for diverse public interactions that may be sacred or mundane – processions, commercial exchanges, executions, games, fiestas or chats on public benches" (Moore 1996: 789–790). The smaller, more restricted nature of the courtyards implies that activities of a more intimate nature took place within their confines; however, this can also be said of more restricted plaza spaces. "In many [Maya] sites there is a clear distinction between plazas surrounded by public buildings, easy to access from several directions, and private residential buildings set around plazas [or courtyards] with restricted entry" (Hammond 1982: 242).

Many plazas and courtyards display a range of sculptural media that may relate to the various activities that once took place within their confines. As seen at sites such as Copán and Tikal, the most prevalent sculptural forms associated with public plaza spaces are monolithic stelae and altars (see Chapter 2, Investigative Considerations and Methodology, this volume), in addition to the decorative elements assigned to buildings that mark plaza boundaries and perimeters (see Chapter 2, Section 2.5.1, Architectural Sculpture). Embellished plazas would include decorative features, such as roofcombs, friezes, and piers, which were designed to face plaza interiors. Examples of this type of architectural embellishment include much of the sculpture that is assigned to the exterior of the Cross Group temples at Palenque, most of which is visible only after entering its plaza.

As with plaza spaces, it is the presence of peripheral structures that articulates the boundaries or limits of a courtyard; courtyards are generally smaller and more restricted than are plazas and are generally rectangular or square in their configuration. Examples of restricted courtyards include those confined to the interior of the Palace at Palenque, more specifically the East Court, West Court, Southeast Court, and the Tower Court. Excluding stelae and altars, many of the sculptural features associated with plazas are also found associated with courtyards; more specifically, decorated friezes, tablets, panels, benches, and thrones also appear within the confines of courtyards.

### 1.3.5. Causeways

The principal objective in the construction of Maya causeways was to demarcate routes of access and communication between buildings within cities (Hanson 1998: 75) and perhaps trade between neighbouring city centres. It has been suggested that Maya causeways were also used for ceremonial purposes (Coe 1965: 111, 2000: 141) – more specifically, as processional routes between ritual and administrative precincts and between neighbouring sites. Referred to as *sacbe* or *sacbeob* meaning "white road" (Coe 2000: 141), Maya causeways range in length from tens of metres to tens of kilometres. Often paved with stones and surfaced with plaster, causeways also functioned as a "form of oriented longitudinal space" (Kubler 1985: 242), providing axial order for the construction of buildings.

## 1.4. Classic Maya social structure

Traditionally, Classic Maya social structure has been defined in terms of a two-tiered system comprising two juxtaposed groups broadly described as "elites" and "non-elites" or "commoners" (Chase and Chase 1992a: 8; Sharer 1994: 490). Although debate persists concerning the descriptive precision of such categories (see Chase and Chase 1992a), "the implication of the term elite is clear; it refers to a segment of a social system that enjoys measurably more prestige, power and/or wealth than the society at large" (Sanders 1992: 278). Determining the level of societal overlap between elites and commoners, however, remains an ongoing problem in Maya studies, inhibiting a more thorough understanding of sociopolitical organisation (Chase and Chase 1992a: 4). In terms of their distribution, elites are thought to have occupied the core of city centres whereas the non-elite lived on the periphery of the sites. An alternative to the concentric model proposes a mix of classes dispersed throughout city centres, "where monumental architecture . . . [combined] both public and elite functions" (Chase and Chase 1992a: 10). In accordance with this view, it has been proposed contentiously that although "the elite may be seen as . . . an important force in the integration and organization of society" (Chase and Chase 1992a: 16), "time and the availability of increased wealth allowed a 'middle class' to emerge that maintained positions of status and power between the elite and non-elite" (Sharer and Traxler 2006: 86; see also Chase and Chase 1992a: 11, Chase and Chase 1992b: 314, and Marcus 1992b: 223, 225). It is argued that this middle class probably comprised preferred specialists such as sculptors, painters, woodcarvers, potters, architects, and costume makers (Chase and Chase 1992a: 12). One of the methods by which the ruling elite sustained their power and prestige was through the control of goods produced or acquired through tribute. Visiting nobles traveled great distances to access commodities and resources under the strategic control of other city centres, presenting themselves as ambassadors from their home kingdoms (Schele and Mathews 1998: 18–19). Clues as to the nature and structure of Maya society during the Classic Period come from archaeological, epigraphic, artefactual, iconographic sources; collectively these sources provide important insights into the Maya social system.

### 1.4.1. Status and epigraphic evidence

During the Late Classic Period (A.D. 700–900), Maya elites were differentiated hierarchically by a number of titles (Webster 1992: 136); these were most prominently *ch'uhul-ajaw, ajaw, ch'ok-ajaw, ba-sajal, sajal, y-ajaw-k'ak'*, and *ah'k'un* (see Houston and Stuart 2001). The most important of these titles was that of *ch'uhul-ajaw*, which functioned to distinguish rulers from the lesser nobility. In reference to the origins of the title:

> the institution of the [*ajaw*] constituted the successful transformation of a mythology focusing on the heroic Ancestor Twins from an ideological affirmation of ethnic brotherhood across a vast segmentry society into a celebration of hierarchical division of that society into living representatives of those ancestors and their worshipers. The consequence . . . was a sudden upsurge in construction in Centres throughout the Southern Lowlands (Freidel and Schele 1988: 549).

As a result of the increasing population in the Maya lowlands and the subsequent increase in social complexity, towards the end of the Early Classic Period, further title ranks were introduced as the *ajaw* title alone could no longer sufficiently differentiate hierarchic variations in Maya society. Introduced during the Late Classic Period, the most important title beneath that of *ajaw* was *sajal*. Both the *ajaw* and the *sajal* titles represented offices into which one was seated or bound. "A *sajal* [for example] may be born into his status but acquires its essence, its sajal-ship (*sajal-il*) only through rituals of enthronement" (Houston and Stuart 2001: 61). Persons who held the office of *sajal* were second in importance to those who held that of *ajaw*; their function was to act as governors appointed by rulers to control subordinate satellite centres under the control of a centralised government (Parmington 2003: 47). Ethnographic as well as ethnohistoric evidence supports this proposal, as noted by Diego de Landa in reference to the founding of Mayapán:

> The lords appointed governors, and if they were acceptable confirmed their sons into the office. They charged them with the kind treatment of poor people, the peace of the town and to occupy themselves in their work of supporting themselves and their lords (de Landa in Tozzer 1941: 27).

### 1.4.2. Status and iconographic evidence

Previous research has indicated that general differences in costume and composition as represented iconographically in monumental art can be used to determine the relative status of the individual, where two or more are portrayed together (see Clarkson 1979 and Parmington 2000, 2003). Compositional elements to consider when attempting to establish the relative status of individuals depicted together are (1) the degree of embellishment in costumes; (2) the degree of embellishment in associated paraphernalia; (3) the number of costume elements; (4) whether persons are seated, standing, or depicted on the same horizontal register; (5) whether any person is seated on a throne; and (6) whether persons are depicted frontally or in profile.

Symbols of royalty that adorned the Maya elite, as represented iconographically, include slipknots and braided mat signs made from cloth or reed, the drum-major headdress, and the Jester God diadem. Feathers, along with jaguar skins, were the privileged items of rulers and high nobility. Other authoritative symbols incorporated in Maya dress included the mantle or cape, decorated hipcloths, the ladies huipil, mufflers or war scarves, shoulder pieces, tunics, Jester God and masked pectorals, skyband belts, dangler belts, backracks, pyrite mirrors, turbans, zoomorphic headdresses, and the God C fret apron. Authoritative objects that were held included hafted axes, spears, incense bags, fans, and battle standards (Parmington 2000: 42–44; Parmington 2003: 52; see also Schele and Miller 1986; Schele and Freidel 1990; Tate 1992).

Costume was important in signaling social status in ancient Mesoamerican cultures. Ethnohistorical references, although less prevalent and explicit than those of Central Mexico, indicated a comparable relationship between status and attire among the ancient Maya. As Anawalt (1981: 27) points out, for the Aztec, "society was sharply stratified . . . the appropriate apparel for the different levels of social hierarchy was precisely controlled by explicit sumptuary laws. The common people wore maguey, yucca, or palm fibre garments. Only the upper classes were allowed cotton clothing, and the decorations, colours, and the amount of featherwork on these costumes was clearly specified." According to Duran:

> The twelve great lords wore certain cloaks, the minor lords others. The common soldier could wear only the simplest style of mantle without any special design or fine embroidery that might set him off from the rest. Even the length of the capes was prescribed: the common man's mantle was not to be worn below the knee; if it reached the ankle, the penalty was death (Duran, in Anawalt 1981: 27).

The presence of comparable statements by colonial period writers supports the view that variations were apparent in the way that the Maya dressed because of differences in their social position (Clarkson 1979: 48). As noted by Diego de Landa:

> [The priest] came forth clothed in a jacket of red feathers . . . and he wore a coroza [headdress] on his head of the same feathers; and under the jacket many cotton ribbons hanging down to the ground like tails. And in his hand he carried an aspergillum, made from a short stick finely worked and for the hairs or bristles of the hyssop there were certain tails of serpents which are like rattlesnakes. And he showed exactly the same gravity as the Pope shows on crowning an emperor; and it was a remarkable thing to see how much nobility this dress gave them (de Landa, in Tozzer 1941: 105).

Polychrome vessels provide some of the strongest pictographic evidence for social stratification among the Maya during the Classic Period (Reents-Budet 1994: 2). Portraying both imagery and text, this type of ceramic is most often found associated with elite burials and dedicatory caches. Scenes painted on pottery provide important insights into courtly activities undertaken by the Maya elite within the confines of buildings. Scenes on polychrome vessels suggest that palace staff and courtiers included, among others, dwarfs, servants, scribes, and musicians. The behaviours recorded on the pottery include both formal and recreational activities, more specifically, the visitation by foreign dignitaries, dancing, the receiving of tribute, and auto-sacrificial rites (Coe 2000: 113–114). The various vessel types that display imagery include cylinder- and jar-shaped pottery used as *uch'ab* ("drinking vessels") for holding liquids, such as cacao, and flat dishes and bowls that were used to hold solid foods. Text accompanying the scenes on painted ceramic often recorded the function of the vessel in addition to its owner and the artist who painted it (Reents-Budet 1994: 72–77). Given the personal nature of Maya polychrome pottery, scenes portrayed on vessels provide an important source of information for understanding both private religious and secular behaviour among the elite classes in Classic Maya society.

### 1.4.3. Status and archaeological evidence

It has been argued that "gradations within archaeological remains suggest that the distinction between the elites and the non-elite is more of a continuum than a well-defined division" (Sharer 1994: 490). Remnants of "monumental earthen constructions, elaborate human burials, diverse arrays of luxury goods, and other remains . . . [exist] as evidence . . . [of the] sociopolitical complexity [among the ancient Maya]" (Sharer 1989: 166). Identifying the context and distribution of specific artefactual types is a useful way of understanding how certain buildings were used by the ancient Maya. For example, analyses of artefact residues from Tikal identified disparities in the types of raw materials available to different status groups at the site, as well as differences in the wealth of persons who once occupied palace-type buildings and those who inhabited the more common domestic architecture (see Haviland and Moholy-Nagy 1992: 54–56). Haviland and Moholy-Nagy (1992) note that discerning differences between upper and lower class refuse remains difficult because of cross-contamination. More specifically, material discarded from common domestic groups in Maya centres was often reused as architectural fill when constructing palaces and other elite architecture. Furthermore, materials that one would expect to be associated with elite residences may appear in more common dwellings because of craft specialisation in which luxury items were manufactured purely for aristocratic consumption. Given the likelihood that servants would have coexisted with the upper echelon in Maya society, one may even see a mix of common and elite refuse associated with elite residential architecture. Given these problems, it is clear that one must rely on both qualitative and quantitative measures to discern elite from common residues.

Haviland and Moholy-Nagy (1992: 55) sorted household material remains from Tikal into three categories and interpreted their distribution. These categories were defined as (1) basic domestic items (i.e., manos, metites, cores, ovate and elongate bifaces, irregular retouched, used and unused flakes, prismatic blades, pointed retouched flakes, and figurines); (2) items commonly but not invariably present (i.e., censers, centrally perforated sherds, bifacial blades, pointed retouched flakes, hammer-stones, and rubbing stones); and (3) items rarely present (all other classifiable artefacts). Comparative analysis of the distribution of these artefact categories revealed that workshop debris, such as debitage, and common items, such as figurines and prismatic blades, were found to be more abundant around small structures. Rare artefacts are present in greater number and diversity inside palace structures. Items that fall under the more common artefact categories found associated with smaller structures were largely made from local materials, such as chert, whereas common artefact types found around palaces were made from more exotic materials, such as obsidian and imported chert. In addition, unaltered animal remains were found in a higher proportion in refuse associated with the Central Acropolis, suggesting that aristocrats ate more meat than the underclasses did. A higher proportion of polychrome pottery was also found in areas associated with Palace Groups, as were carved cache bowl lids and fine blackware sherds found in architectural fill (Haviland and Moholy-Nagy 1992: 55–56).

Things to consider when attempting to discern the relative importance of burials found at Tikal are the size of the grave, the quality of its construction, the number and quality of objects buried with the individual, and the quality and size of the structure assembled above it. The graves of the most important individuals in Maya society were designed to hold more than the body of the deceased; additional space was provided for an abundance of grave goods, which generally included large quantities of pottery, treasured heirlooms, and pieces made especially for the burial. In contrast, lower status burials typically left room only for a few select pieces. Elite tombs were also designed to provide space so that the body of the deceased may be visited (e.g., the Temple of Inscriptions at Palenque). As the head of lower class households were, on occasion, placed in funerary shrines at Tikal, the status of the buried is more accurately discerned in relative terms by assessing the number and quality of burial goods as well as the size and quality of any associated mortuary architecture. Items found in the elite burial sites associated with the North Acropolis at Tikal included, among other things, jade beads, spondylus shell, stingray spines, red pigment, small anthropomorphic sculptures, greenstone face masks, pearl pendants, worked marine shells, stone and shell earplugs, pyrite mosaic mirrors, vases, painted wooden sculptures, bracelets, anklets, fan handles, inscribed bone, jaguar pelts, obsidian debitage, and human sacrifices (Haviland and Moholy-Nagy 1992: 53–54; see also Chase 1992a: 37 and Foster 2002: 208–211).

## 1.5. Text, image, and literacy among the Classic Maya

Maya art was configured to deliver information on two levels – in narrative scenes and through the interpretation of accompanying inscriptions. When combined, text and image provided a narrative complex, functioning to communicate information encompassing entire ritual sequences (Schele and Miller 1986: 33). Within the restrictive social structure of Classic Maya society, literacy in hieroglyphic writing was one of the privileges of an educated minority (Houston and Stuart 1992: 592; Marcus 1992b: 224, 230; Houston 1994: 27–50; see also Brown 1991). Although there may have been varying levels of active literacy from site to site, "there is a general accord that not everyone in Classic Maya society could read" (Fash 1998: 216), and writing and the utilisation of written texts were primarily associated with private custom and ritual. As noted by Diego de Landa:

> They wrote their books on a long sheet doubled in folds, which was then enclosed between two boards finely ornamented; the writing was on one side and the other, according to the folds. The paper they made from the roots of a tree, and gave it a white finish excellent for writing upon. Some of the principal lords were learned in these sciences, from interest, and for the greater esteem they enjoyed thereby; yet they did not make use of them in public (de Landa, 1978: 13).

Scribes used a range of media on which to record texts and depict scenes; these were most prominently limestone panels, tablets, altars, and stelae, in addition to murals and surfaces modeled with stucco. Smaller forms of artistic media that the Maya used to display text and imagery included screenfold books, bones, jade, shells, and polychrome pottery. Archaeologically, the smaller painted, carved, and sculptured pieces most often appear in elite tombs as items either owned by the deceased or placed there as offerings. Wood was also used by the Maya for artistic purposes (e.g., Tikal Temple I and II lintels); however, due to the ephemeral nature of the material, such finds are relatively rare (Coe and Van Stone 2001: 13–14).

Sculptors and scribes would have held considerable status in Maya society, as they were most likely members of royal families. The highest scribal rank was that of *aj k'uhuun*, meaning "the keeper of the holy books." These individuals had many important responsibilities, such as royal librarian, historian, genealogist, tribute recorder, master of ceremonies, astronomer, and mathematician. The scribes who had the greatest reputation were given additional titles, such as *aj its'aat* and *aj miats* (Coe and Van Stone 2001: 13). The primary purpose of hieroglyphic texts was to record private ownership, dynastic titles, events, and religious rites (Marcus 1992a: 81–84; Schele and Mathews 1998: 18; Miller 1999: 107). Designed to be read in double columns, from left to right and from top to bottom, it is the combination of sound symbols and representative symbols that define Maya texts as "logosyllabic" in nature (see Justeson 1989: 25–39). Diego de Landa in *Relación de las Cosas de Yucatán* (Account of the Affairs of Yucatan) attempted to deconstruct the Maya writing system into an alphabet by defining twenty-nine signs that he believed that the Maya combined to create their written language. Substituting these sound symbols with the Spanish letter equivalent, he obtained not a Maya alphabet, but a list of syllabograms (combinations of consonants and vowels) – a syllabification of words being broken up into their sound constituents (syllables). Today, the Maya syllabary comprises a list of some 150 sounds represented by a wide range of symbols or characters. Some of the symbols that comprise the syllabary are variations of similar signs, and others are allograms (alternative spellings) for a single syllable (Coe and Van Stone 2001: 18–20).

Because of limitations in the breadth of literacy it is likely that larger monumental sculpture and iconography functioned, in part, as a form of signage, its purpose to more effectively communicate doctrine concerning religious and sociopolitical order. Maya text and imagery were purposeful in the sense that they were a means by which a ruler could instill obedience and loyalty among his subjects, the art acting as a petition to the gods promoting a ruler's successes and validating his authoritative position (Marcus 1992a: 14). Furthermore, public monumental art functioned as a historical record reinforcing a common standard of knowledge, "providing emotional security . . . and psychological cohesion" (Tate 1992: 31).

Among the ancient Maya it is likely that iconography functioned as a compensatory mechanism exploited by the upper echelon in society to assist in more effective communication to the broader population. "Scholars . . . agree that the populace could understand and appreciate the architectural sculpture, which was designed and built to be admired and understood [by all]" (Fash 1998: 216). In support of this proposal, at sites like Palenque and Chichen Itza, the Maya positioned textual monuments predominantly in areas of limited accessibility, arguably, due to the limited capacity for their broader comprehension. "The not-very-subtle iconographic adornments [at Chichen Itza] suggest that direct communication with illiterate masses may have been one of the functions of [sculpture on] major buildings" (Lincoln 1986: 154). Furthermore, as variations in proximity of text and imagery indicate, "glyphic monuments had a narrow audience whilst pictorial/iconographic monuments had a much broader appeal" (Lincoln 1986: 155). Glyphic texts identified at Chichen Itza occur most frequently on lintels positioned in contexts that are private. In contrast, sculpture found on columns, wall panels, and benches occur in more readily accessible areas, suggesting that "art seemed to have a potentially wider audience than the Maya glyphs" (Lincoln 1986: 153). Parallels with this observation are found when comparing the relative proportions of text and image on monuments found at Palenque, where a correlation was found between areas of restricted access and the increase in the proportion of text found on monuments. For example, the private internal sanctuaries of the Cross Group all contain higher proportions of text than do the more prominent external sculptural media (for more detail, refer to Chapter 3, Access Analysis of the Palenque Cross Group and Its Sculpture). Examples of monuments that display text exclusively are the panels found in the interior of Palenque's Temple of the Inscriptions, and the Palace's Tablet of 96 Glyphs, neither of which has accompanying scenes or iconographic complements. [Note: The terms "text" and "image" are defined here in accordance with Miller (1989), where "image" refers to pictures (figurative or abstract), rendered on flat or three-dimensional surfaces, that do not function as a component of pictographic, ideographic, and logographic writing.

Conversely, the term "text" refers to writing that is ordered linearly on monuments and other media (Miller 1989: 176).]

In a sense, one has to focus on the subject matter of Maya texts to really understand their purposes, whilst avoiding both modern and ethnocentric interpretations as to their function. Maya texts do not cover the multitude of functions that writing provides in modern society. "The subject matter of the most public inscriptions . . . is relatively restricted and there is much repetition and redundancy" (Coe and Van Stone 2001: 7). Conversely, texts in modern cultures function to inform and educate intricately on a broad scale, concerning innumerable subjects. The more limited prevalence of Maya text suggests that literacy was not crucial to everyday life among the Maya, that it was not important that the broader plebeian population understood texts. Instead, the context and content of texts suggest that they functioned principally to legitimise the authority of the ruler among his peers (Marcus 1992b: 230).

Due to the limited capacity of texts to communicate to the broader masses in Maya society, it is reasonable to argue that the ruling class used alternative methods of transmitting information to the population at large. Imagery was the primary method of communicating religious and political dogma during the Classic Period. In addition, the purpose of imagery as displayed on monuments and architecture was to signal the function and exclusivity of space in Maya centres. "The Maya singled [the] identities of sacred places and function through sculptural composition. Important architecture was . . . built from stone, finished with plaster, and decorated with passages of sculpture and paintings that signaled their function to people using them or coming into the spaces they addressed" (Schele and Mathews 1998: 27).

For years, scholars have wrestled with problems of characterising/defining Mesoamerican writing systems; "of primary interest to the art historians [today] are the ways in which ethnocentric ideas of image and text have affected our perception of the Maya and other Mexican groups" (Baddeley 1983: 55). There are three types of texts found in pre-Columbian art; these types may be described as (1) "discrete texts" that do not appear with the image but relate to it; (2) "conjoined text and image," which are the incorporation of text and image as found on stelae; and (3) "embedded texts," which are texts that are integrated with imagery, such as those encoded in costumes (Berlo 1983: 13). Early last century there was a tendency for European scholars to define Middle American art as focused on representation rather than the transmission of meaning. Today, iconographers acknowledge more readily that the association between image and text in Mesoamerica did not necessarily parallel that of Europe (Miller 1989: 177). "Mesoamerica[n] . . . systems [of communication] do not allow for an easy distinction between pure text or pure image" (Baddeley 1983: 56), as suggested by the Nahuatl word *tlacuiloliztli*, which means both "to paint" and "to write" (Boone 1994: 3).

Some scholars have sought a "reformation of the definition of writing that considers both verbal and nonverbal systems of graphic communication" (Boone 1994: 4). More specifically, they argue for a "semasiographic" (meaningful graphic) definition, which sees writing as "the communication of relatively specific ideas in a conventional manner by the means of permanent, visible marks" (Boone 1994: 15). Verging on iconic in nature, in semasiographic modes of communication there is a natural correspondence between the image portrayed and the object to which it refers. The power of imagery as a communicative medium, whether iconic or otherwise, is its capacity to communicate its message without "a detour through speech" (Boone 1994: 15). Furthermore, the strength of such a system lies primarily in the fact that when one understands the logic that drives and orders it, meaning is conveyed irrespective of the language that one speaks (Boone 1994: 15–16) and irrespective of the level of education that one has received.

Maya sculpture and iconography have a semasiographic/iconic ingredient used by Maya rulers as a form of "signposting," used in part as a means of signaling the function and hierarchical division of associated space in Classic Maya city centres. Within this communicative system, standardisation of Maya imagery would have been fundamental to the clear and effective transmission of information to the broader population, as there was little freedom for artisans to explore their artistic versatility. Design in sculpture was predetermined, controlled by education, and standardised by strict traditions (Schele and Miller 1986: 33). For example, even where regional variation in sculptural style is apparent at different centres, depictions of people were "regularised" to fit ideals of human appearance (Tate 1992: 50–51). Human beings portrayed in Maya sculpture were generally depicted proportionately accurate with a standardised quality to their rendering. Artists, when portraying individuals, tended to focus their attention on characteristic attributes of a person's appearance with depictions verging on caricature (Schele and Miller 1986: 66). According to Schele and Miller, because of limitations in artistic freedom, the artist's general creativity was asserted primarily through the quality of sculptural execution (Schele and Miller 1986: 33).

Although text and image both have the capacity to transmit information, they represent different modes of communication that, by their nature, vary in communicative precision. Because of the multivariate forms, Maya imagery is considered more nebulous in the way that it conveys meaning. Subsequently, Maya imagery stands in stark contrast to the more concise linear organisation of written texts that make them less "context dependent" (Miller 1989: 186). While acknowledging this fact, however, it remains that many of the insights that we have today concerning Maya behaviour and custom come from the interpretation of rendered scenes as displayed in palaces, temples, and tombs. Information – such as time, action, place, and individuals portrayed – are said to be the subject of text, whilst scenes primarily provide displays of action and ritual (Schele and Miller 1986: 66).

Maya scribes and sculptors presented narrative imagery in a number of different ways; one method was in "snapshot," in which an important or characteristic moment of a ritual was captured in a single image (Schele and Miller 1986: 38). There were three moments within a ritual sequence of events on which Maya artisans focused: the inceptive, the progressive, and the completive. The term "inceptive" refers to an instant recorded pictographically just before or just after a sequence of events begins; the "progressive" refers to the moment recorded in the midst of ritual proceedings; and the "completive" is the moment caught at the time of completion of the sequence of events. Snapshots in Maya iconography could be combined into a series of "stills," suggesting a progression in action, where the separate scenes depict different episodes of a ritual (as is apparent in the Bonampak Murals, see Miller 1986). An alternative was to record the ritual as a "simultaneous narrative," a narrative in which an image presents a number of events as a single moment, such as where a captive is portrayed at the instance of his capture already stripped and marked for sacrifice (Schele and Miller 1986: 38; see also Weitzmann 1970 and Reents-Budet 1989).

## 1.6. Text–image ratios and monumental art

Although they represent a relatively minor component of this investigation, the text–image ratios of sculpture associated with the Cross Group and the Temple of Inscriptions at Palenque were calculated. The objective of this procedure was twofold: (1) to test the proposition that biases exist in the relative proportion of test and imagery on monumental art, which varies depending on its spatial context; and (2) to validate this author's proposition that imagery was the primary method used by the Maya elite to communicate to the wider population at Palenque.

Text–image ratios were calculated by measuring the relative coverage of text over the surfaces of monuments and comparing it with the quantity of surface area occupied by the iconography. Various software packages were used to do this: Adobe Illustrator 9.0.1, Adobe Streamline Version 4.0, AutoCAD 2000, and Microsoft Excel. Scale drawings of monuments obtained from published sources were first scanned and converted to vector format before being adjusted and scaled to their specified sizes in Adobe Illustrator. The scanned monuments were then exported into AutoCAD where the surface area occupied by text and the surface area occupied by iconography could be traced and calculated; these data were then inserted in Microsoft Excel where they could be graphed. Together with access analysis, the power of this method is its ability to detect gradual shifts in the relative proportions of text and image on monuments and architecture with changes in relative accessibility of the art. As previously stated, it is generally agreed that literacy was one of the privileges of an educated minority (Houston and Stuart 1992: 592; Houston 1994: 27–50). Therefore, it is reasonable to argue that monumental art positioned in public spaces (areas of low elevation and ease of access) would contain a lower proportion of text than would those contained in private spatial contexts (see Chapter 3, Section 3.5, Text–Image Ratios and the Cross Group for more information).

2

# Investigative considerations and methodology

In this investigation, monumental sculpture assigned to private/elite spaces was compared and contrasted with sculpture positioned in public/civic spaces. Monuments and other iconographic media investigated included those that were/are positioned in areas classified as most-public and most-private as well as those that could be positioned along the "privacy gradient" in between (see Chermayeff and Alexander 1965: 121–126; Newman 1972: 9; Rapoport 1977: 290; Lang 1987: 155; Pearson and Richards 1994a: 40; Grahame 1997: 139–142). In my investigation, function and status were ascribed to Maya architecture by interpreting Maya monumental art and its spatial context. It is my proposal that public/civic and private/elite elements are present within sculpture and other media and therefore may be identified and interpreted. Furthermore, I argue that the content, behaviour, and symbolism as displayed in monumental art can be understood as behaviour restricted wholly or partially to these spaces and to certain individuals with status.

A premise underpinning my research was the view that "public space" and "private space" correspond to notions of "common space" and "elite space" in Maya society, where movement between these categories of space can be understood in terms of a gradient of accessibility. It is generally argued that "the demarcation of public/civic and private/elite precincts was . . . sharply and tangibly controlled by location and scale of features" (Jones et al. 1983: 165). The intention of the Maya elite, in the construction of specific building groups, was quite clear – the goal was to establish

> . . . a hierarchy of spaces . . . where the individual open spaces and buildings were arranged in a continuous succession of solids and voids which move in a progression from most accessible and most public, to least accessible and most private. The progression is controlled by distance and height; the higher we are allowed to go and the further away we move from the open public spaces the closer we get to the realm . . . where only the privileged may enter (Andrews 1975: 67).

In accordance with Andrews (1975), Schele and Mathews (1998: 29) propose that "the Maya controlled access and channeled movement by the use of stairways, constricted or blind entrances, causeways and other devices that were part of the spatial design of buildings." In addition, "progressive enclosure, additional doors, corridors, and rooms change the meaning of the space by making it more difficult to get to, so that openings and plazas go from unrestricted to claustral in their layout, perhaps with implications for the ever widening gulf between rulers and ruled" (Pendergast 1992: 62–63).

In line with the proposals just mentioned, it follows that fixed sculpture associated with distinct types of spaces (marked by controlled access, progressive enclosure, channelled movement, and changes in elevation) may also be sorted from most public to most private, whilst the scenes displayed on such media (whether portraying individuals, ritual activities, and/or cosmological imagery) may be characterised similarly. When attempting to discern intent in the configuration of Maya architecture, it is clear that "some vehicles of information inherent in Maya built forms (e.g., painted, carved, and inscribed murals; lintels and plaques inside temples and palaces) would have been accessible only to elites" (Webster 1998: 35).

The objectives of the present investigation were achieved by employing various analytical techniques. These were as follows: (1) the construction of a comprehensive image database of monumental art obtained from a range of spatial contexts; (2) the adaptation and implementation of techniques in "access analysis" (see Hillier and Hanson 1984; Foster 1989a, 1989b; Fairclough 1992; Moore 1992; Gilchrist 1994; Grahame 1995, 1997; Gould 1999; West 1999; Liendo Stuardo 2003); and (3) thematic–analytical (Clarkson 1979; Parmington 2003) and reductionist–comparative inquiry of sculpture (Proskouriakoff 1950; Parmington 2003).

Applying the investigative techniques just mentioned, the various proposals that I wanted to test were as follows: (1) Sculpture marks the exclusivity of associated activity areas by its positioning and iconographic content, signalling both common and elite accessibility; (2) function and exclusivity of activity areas are reflected in the compositions of spatially associated monuments; (3) public sculpture and its positioning mark common ritual space; and (4) private sculpture and its positioning mark elite ritual space.

## 2.1. Public versus private space, public versus private art

Defining what constitutes public and private space is somewhat problematic because of the diffuse nature of the concepts (Velecky 1978: 21; Grahame 1995: 28). The word "private" has its origins in the Latin *privatus*, meaning to "bereave, deprive" (Dev Sharma 1991: 33), terms that relate to states of being that are both select and withdrawn from society. In contrast, the word "public" comes from the Latin *populus*, a term which denotes inclusiveness; that which is "generally assessable, open to general observation, is manifest and not concealed, is [said to be] public" (Dev Sharma 1991: 35). Privacy relates to power, the power to choose whether or not to associate with others, to participate in shared activities or seek seclusion, to share information or withhold it, to communicate information to selected individuals without being observed or overheard (Young 1978: 4). Accordingly, privacy may also relate to "groups or institutions to determine for themselves when, how and to what extent information is communicated to others" (Velecky 1978: 21). Rapoport (1977) defines privacy as simply "the avoidance of unwanted interaction with other people" (Rapoport 1977: 289). To Westin (1967), privacy relates to "the voluntary and temporary withdrawal of a person from the general society through physical and psychological means, either in a state of solitude or small-group intimacy, or when among larger groups, in a condition of anonymity or reserve" (p. 7).

Privacy does not simply relate to the physical removal of the individual from the company of others, but has a psychological component as well. Privacy also concerns control over disclosure of the self and to what extent elements of people's personalities and innermost thoughts are communicated to others. Westin (1967: 31–32) describes four forms of privacy, which he terms (1) solitude, (2) intimacy, (3) anonymity, and (4) reserve. "Solitude" refers to a person's need to be removed from the observation of others, whereas "intimacy" is a state of being that a person has with another individual (or individuals) which is free from the observances of the outside world. "Anonymity" relates to a person's ability to be unknown even in the midst of others, whereas a person's "reserve" refers to the projection of psychological barriers to guard against intrusion. The objectives of attaining privacy depend on the individual's need for personal autonomy, release of emotions, self-evaluation, and restricting and controlling communication (see Westin 1967: 33–39).

A correlation may be drawn between the want for privacy, personal space, and territory, and the human need for security, affiliation, and esteem. The concept of personal space and territoriality are related in the sense that they are both mechanisms by which privacy is attained (Lang 1987: 145). The want and need for privacy varies depending on the social roles of individuals (Lang 1987: 155), whereas the type and degree of privacy attainable remain subject to cultural context, the personality and aspirations of individuals, and the socioeconomic status of persons (Lang 1987: 145). The effect that cultural factors have on the way that spatial attributes are conceptualised, abstracted, and expressed remains important (Hallowell 1977: 132) as "spatiality like temporality, coordinates patterns of behavior" (Hallowell 1977: 131).

Territoriality is defined here "as a set of behaviors an organism or group exhibits, based on the perceived ownership of physical space" (Bell and Fisher 1978: 169). Characteristics of territories include notions of ownership, personalisation, and rights to defend against intrusion (Lang 1987: 148). Through the personalisation of space, human beings mark their territory. This marking may occur as a means of aesthetic expression or control or as a way of altering an environment to accord with a pattern of activity. Privacy attained through territoriality fulfils several basic human needs; more specifically, it creates a sense of security in addition to functioning as a frame of reference from which the rest of the world is perceived. Human territories also provide organisational information relating to individual as well as group identity; they determine the placement of boundaries and/or limits of control, instilling a sense of order and continuity as well as contexts for host and visitor organization (Lang 1987: 147–149). Whereas boundaries function to delineate and secure territory (in addition to sheltering and containing), entrances mark thresholds or points of transition between opposing domains, some of which include inside/outside, sacred/profane, male/female, enemy/friend, initiated/uninitiated, public/private, and common/elite (Pearson and Richards 1994b: 24).

The term "personal space" refers to the (culturally variable) distance maintained between humans except in those times where intimate contact is either sought or required; in such cases a person or persons may be invited to encroach on an individual's personal space (Lang 1987: 147). Personal space has two principal uses: First, it functions as a defence against physical and psychological danger resulting from inadequate privacy, excessive intimacy, and threats of physical aggression. The second function of personal space relates to modes of communication and proximity, in the sense that distance between individuals and/or groups dictates whether more-intimate or less-intimate communicative methods are employed in a transaction. The distance that one chooses to maintain while interacting with others, in addition to the types of communication one uses, indicates varying degrees of concern for self-protection as well as signaling the nature of the relationship between individuals (Bell and Fisher 1978: 144).

The human requirement for public and private states of being have their origins in culture and human biology, manifesting as domains through which we as a species operate and interact: ". . . man's space requirements, his notions of territoriality, and his handling of persons and things around him all are rooted in his culture which in turn, and to a large extent, is rooted in . . . biology" (Dev Sharma 1991: 111). In general, proxemics is concerned with the ordering of space in terms of how the environment is structured and utilised on an interpersonal level (Watson 1970: 33).

In his seminal work entitled *The Hidden Dimension* (1966), Hall describes human proxemic behaviour as it relates to three levels of spatial organisation as influenced by culture and human physiology; these levels are referred to as "precultural," "infracultural," and "microcultural" (Hall 1966: 95–122). The precultural level relates to how space is interpreted on a sensory level, to which different cultures assign meaning (Hall 1966: 95). The human senses can be divided into "distance receptors," which include the eyes, ears, and nose (Hall 1966: 39–47), and the "immediate (tactile) receptors," which are the skin and muscles of the body (Hall 1966: 49–60). The infracultural level of spatial ordering is associated with territorial demarcation, whilst microcultural ordering concerns how culture effects its arrangement. There are three aspects to microcultural spatial ordering; they are referred to as "fixed-feature," "semi-fixed-feature," and "informal space" (Hall 1966: 94–104). Fixed-feature refers to the placement of immovable elements that articulate space, such as walls and barriers incorporated into the environment of a given culture. Semi-fixed-features are spatial arrangements of movable features such as furniture, screens, and partitions that may be used to modulate space (Hall 1966: 101–104). Informal space relates to shifts in the modes of communication subject to changes in the physical environment (Hall 1966: 105–120).

It is proposed that there are four distances or zones in which we as human beings interact and communicate depending on the type of relationship we share with others. These distances and corresponding behaviours are termed "intimate," "personal," "social," and "public" (Hall 1966: 110–120). Whilst these distances may vary culturally, "the specific distance chosen depends on the transaction, the relationship of the interacting individuals, how they feel, and what they are doing" (Hall 1966: 120). At intimate distances (0–0.45 metres), physicality becomes the primary mode of communicative interaction, which involves activities such as lovemaking or wrestling (Hall 1966: 110). At zero intimate distances, there is increased input through sensory receptors, such as smell and touch, and vocalisations are reduced in both frequency and volume. In addition, visual distortion occurs when individuals attempt to discern physical attributes of one another. At personal distances (0.45–1.20 metres), one may still hold or grasp another individual, but physical interaction tends to be reduced, and vocalisations become elevated to a moderate level. At personal distances, visual distortion between those persons interacting disappears and detailed physical attributes such as face and hair become apparent. In addition, facial expressions and eye movements become involved in communicative interactions. At social distances (1.20–3.60 metres), physical attributes are discerned in less detail and the voice becomes more elevated, representing the principal mode of communication. The overall form of the body (including clothing detail) becomes visible at this distance; no physical contact occurs at social distances. Persons who participate together in business dealings generally hold closer social distances when compared to those who do not. At public distances (3.60–7.50+ metres), evasive as well as defensive action is more readily possible in response to imminent threat. At this distance, the voice becomes louder and words are enunciated more carefully, visible detail in the general appearance of others is reduced, and individuals or groups are generally perceived as within a setting. When persons are communicating over greater distances, their gestures, which include shifts in stance, tend to be more exaggerated, although stance is characteristically static or "frozen" if persons

Table 2.1. *Distance, perception, and interaction*

| | Distance in Metres | | | |
|---|---|---|---|---|
| | 0 | 1 2 3 | 4 5 6 | 7 8 9 10 |
| Distance classes | Intimate/personal | Social | Public | |
| Communication Modes | Visual, touch, vocal, smell | Visual, vocal, less touch | Visual, vocal | Visual, vocal |
| Oral/Aural | Soft voice, whisper | Casual or consultative voice | Loud voice when talking to group | Full public speaking voice; frozen style |
| Detail Vision | Details of skin, teeth, face visible | Fine lines of face fade; wink visible | Eye colour not discernible; smile vs. scowl visible | Difficult to see eyes, subtle expressions |
| Scanning Vision | Whole face visible | Upper body; can not count fingers | Upper body and gestures | Whole body has space around it in visual field |
| Peripheral Vision | Head and shoulders | Whole body movement | Whole body visible | Other people become important in vision |
| Types of Interaction | Lovemaking, comforting, fighting | Contacts between close friends, family, and acquaintances | Impersonal and businesslike contacts and interactions | Formal contact between individuals and the public (e.g., public speaking) |

*Note:* Modified after Moore (1996: 791); see also Hall (1966: 117–118) and Bell and Fisher (1978: 147).

encountered are to remain as strangers (Hall 1966: 115) (see Table 2.1).

Drawing on Giddens's (1979, 1981, 1984) social theory, Grahame (1995, 1997) argues for the importance of encounters in ordering the environment, as it is through the dynamics of social encounters that space becomes shaped. In accordance with this view, the human body can be understood as a "generator" for ordering the environment, as it is by moving through space, and by the very act of perceiving it, that space is structured knowingly or not (Pearson and Richards 1994b: 10). Space is "the continuous physical domain upon which individuals and objects are located" (Grahame 1995: 11; see also Hillier and Hanson 1984: 26–51). Within this sphere, it is the architecture of the body, combined with the human emotion, that influences the way that we as a species perceive, articulate and control space (see Grahame 1995: 11–16). Some of the examples of physical differentiation or divisions we create relative to our physical proximity include front/back, top/bottom, left/right, and vertical/horizontal (Pearson and Richards 1994b: 10).

The objective of the individual is neither to be isolated nor to be forever exposed to the observances of others, but rather to find a balance between the need for communication (and disclosure of the self) and the want for individual privacy. As a result, there are gradations in public and private states of being to which individuals align themselves, satisfying the demands on the individual caused by society's surveillance (as a means of enforcing social norms) and to limit and/or control the availability of information obtained about oneself by others (Westin 1967: 7, see also Altman 1975: 11). "There is a scale from self display to extreme privacy with corresponding treatment, decorations and barriers admitting people depending on their relationships and status to various parts of the system" (Rapoport 1977: 293). Privacy gradients separate public and private domains, and operate to control the degree to which outsiders permeate a dwelling. To Rapoport (1977: 290), "privacy gradients clearly separate individuals and groups and determine where one interacts and with whom." Cross-culturally, the capacity to achieve and sustain private states of being remains dependent on the effectiveness of one or more of the following: (1) rules relating to manners, avoidance, and hierarchies; (2) behavioural cues; (3) undertaking activities at a particular time as a strategy of avoidance; (4) psychological means, such as mental withdrawal and depersonalisation; (5) physical devices, such as walls, doorways, curtains, and courtyards; and (6) spatial separation (Rapoport 1977: 289–290).

An idealised model of residential architecture has been proposed that views space as organised in what it terms a "defensible hierarchy" (see Figure 2.1), where spatial units change from "public circulation spaces at one end of the spectrum, to the private interior of the dwelling at the other" (Hanson and Hillier 1982: 6). Newman (1972: 3) defines a "defensible space" as

> . . . a surrogate term for a range of mechanisms (real and symbolic barriers, strongly defined areas of influence,

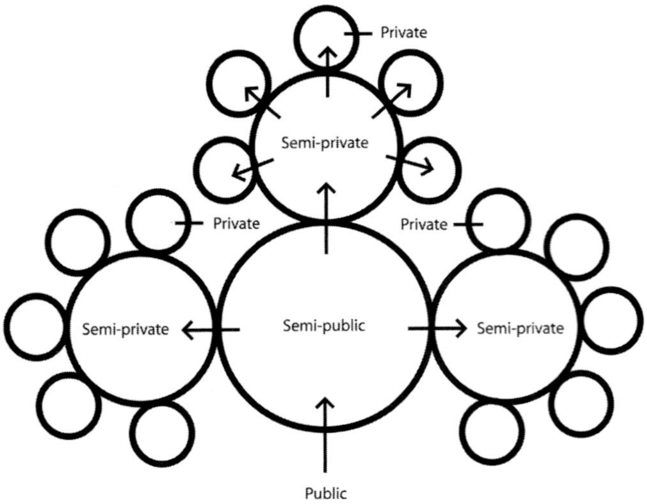


**Figure 2.1.** Model of Defensible Space Paradigm. After Newman (1972: 9), reprinted with permission.

> and improved opportunities of surveillance) that combine to bring an environment under the control of its residents . . . [it] is a living residential environment which can be employed by its inhabitants for the enhancement of their lives while providing security for their families, neighbors and friends.

Defensible environments display one or more of the following characteristics: (1) a hierarchical arrangement of spaces which moves progressively along a gradient from most public to most private; (2) openings (such as windows and doors) that provide a means by which residents can survey the more public (i.e., more accessible) areas of their environment; (3) an architectural form that permits individuals to differentiate between environments of vulnerability and seclusion; and (4) locations that are functionally sympathetic to occupants who are not likely be threatened (Newman 1972: 9).

One problem with the public–private axis, as represented by the Defensible Space Paradigm, is that depth on the public–private gradient is measured relative to a "principal entrance." Because most buildings display more than one entrance, defining what precisely constitutes a main entry point is somewhat problematic as the term may be used interchangeably to describe the grandest entrance or the entrance that is used most often (Grahame 1997: 140). Things that should be considered when attempting to discern formal from informal points of entry are the elaborateness of doorways and decorative façades, symmetry and alignment of structures, building orientation, and the presence or absence of stairways either signaling or controlling the direction of approach.

Privacy can be viewed as "operating in two dimensions: one that relates to the public world of 'strangers', to the private world of 'inhabitants', and a second which relate[s] . . . inhabitants with one another" (Grahame 1997: 143) (see Figure 2.2). This view of privacy "requires that the notion of 'public' [be] no longer confined to 'outside' but be brought . . . [inside], so that certain interior spaces can be seen to possess the characteristics of public space relative to the population of inhabitants" (Grahame 1997: 144). In line with Hillier and Hanson (1984: 146), an "inhabitant" is

> . . . if not the permanent occupant of a cell [space], at least an individual whose social existence is mapped into the category of space within that cell . . . [The] Inhabitant is thus a categoric concept, and therefore a transpatial entity, and in that sense the inhabitant is part of a global categoric reality as a result of being mapped in to the local bounded space of the cell, as well as being a member of a local spatial reality.

In contrast to the inhabitant, a stranger is described as "one who belongs to the public domain" (Dev Sharma 1991: 31), whereas visitors are defined as a subset of individuals drawn from a pool of strangers who exist externally to the domain of the inhabitant. Therefore, by extension, they are people "who may enter a building temporarily but may not control it" (Hillier and Hanson 1984: 146).

Chermayeff and Alexander (1965), in *Community and Privacy*, note that "in cultures both present and past . . . there is a clear physical expression of the need for varying degrees of privacy and the integrity of the domains corresponding to these" (p. 122). It is argued that our present understanding of public and private states of being have their foundation in the nineteenth-century development of the English middle class and the organisation of society into public and private spheres. During this period, public life was dominated by the political and industrious ventures of men, whereas private life encompassed activities undertaken in the home, which was principally the domain of children, women, and servants. The distinction made between public and private modes of being was enhanced with the growth of capitalism, which coincided with a shift in the arenas of manufacturing and production to outside the home. This shift effectively increased the capacity of the state to control the behaviour and conduct of civic populations. A consequence of this change was an increased concern for establishing acceptable limits on the state's capacity to supervise and regulate society's behaviour. It was at this time that "a dichotomy [was established] where 'public' has come to mean the official world of the state . . . whereas 'private' has come to mean precisely the opposite: removal from the sphere

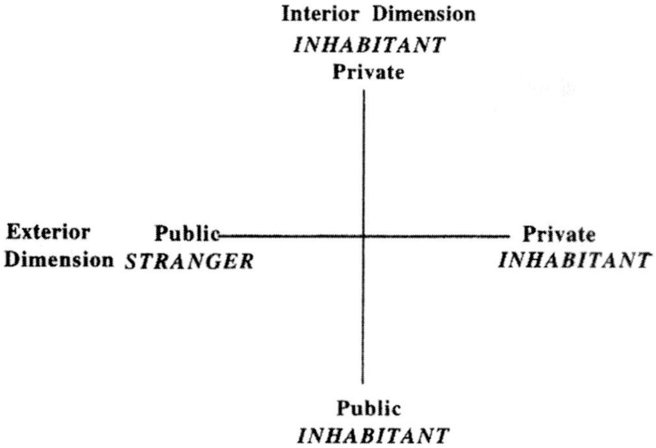


**Figure 2.2.** Diagrammatic Representation of the Dimensions of Privacy. Drawing by M. Grahame, in *Domestic Space in the Roman World* (Journal of Roman Archaeology Supplement 22, 1997, page 144). Reprinted with permission.

of public knowledge, and observation" (Grahame 1997: 138–139).

Cross-culturally, spatial patterning in architecture is influenced largely by the activities ascribed to different spaces. For example, in Western society it is common for different behaviours to be assigned to different areas of the home. More specifically, houses are partitioned into cooking, dining, sleeping, and living areas where "any intrusion of one activity into a space it is not designed for is immediately felt to be inappropriate" (Nanda 1994: 82). This system stands in contrast to the cultural tradition of the Japanese, where interior space is modulated to accord with different day, evening, and nightly activities. In Indian cities and homes, spaces are segregated, organised intentionally to separate high caste and low caste, male and female, sacred and secular activities. In addition, spaces are perceived as superior or inferior (polluted) based on the type of activities undertaken within their confines (Nanda 1994: 82).

Chermayeff and Alexander (1965) describe six domains in the modern Western urban environment which, when combined, form a gradient of public–private spaces. In order from most accessible to least accessible, these domains are termed "urban–public," "urban–semi-public," "group–public," "group–private," "family–private," and "individual–private." Urban public domains are described as spaces in public ownership and are therefore accessible to the general public; examples include civic parks, roads, and pathways. Urban–semi-public domains refer to spaces under government or institutional control, such as public schools and judicial courts. Group–public domains are situated between private property and spaces exhibiting public utilities necessitating cooperative public and private access. Group–private spaces are secondary domains managed on behalf of residents, such as community gardens and laundries. Domains categorised as family–private are those controlled by a family unit functioning as areas of shared and/or communal activity, which includes the receiving and entertaining of guests. Spatial types that fall within this category include lounge rooms, bathrooms, and dining areas. Finally, individual–private domains are those spaces to which individual family members may withdraw to retire or avoid public scrutiny, such as bedrooms.

Examples of ancient architecture that Chermayeff and Alexander (1965: 122–127) argue express hierarchical distributions of space in their configurations include the nuraghi citadels (e.g., the Fortezza d'Orrolli) in Sardinia (circa 700 B.C.), the Baphuon Temple in Cambodia (A.D. 1050–1066), a Japanese palace, and an Italian palace of the Renaissance period (see Figure 2.3). The Fortezza d'Orrolli in Sardinia is an example of a sequence in architecture designed to enhance security. Comprising three stages of development, the central tower was the first architectural feature to be constructed, followed by the inner then the outer wall of the building. Set on a hill, the configuration of this fortress would have been an effective deterrent against attack as the crossing of three defensive boundaries would have been required before access could be gained to the tower's interior. The Baphuon Temple in Cambodia is a building complex that displays beautiful symmetry and hierarchical order in its design. The building group has a single approach that necessitates the crossing of a moat. Overall, the temple comprises three main terraces that steadily increase in elevation and that terminate in a pyramidal structure at the centre of the complex. The Japanese palace is described as "a direct expression of a plan designed to serve a social hierarchy" (Chermayeff and Alexander 1965: 124). The entire structure is designed in such a way as to segregate the residential quarters from areas designed to serve ceremonial and courtly functions. Access to the palace is provided through an entrance bordered by courts, after which access is gained to residential pavilions and private gardens. The Italian palace comprises three principal subdivisions, each of which serves a different purpose. From the perspective of the palace entrance, access to each zone, which becomes progressively deeper, provides "a sequence of three functions: security, ceremony, and privacy" (Chermayeff and Alexander 1965: 126). Entry to the palace is provided via the guard quarters, and beyond this point is an area characterised by a series of reception rooms surrounding a court space, deeper still is the "ducal sanctuary" (Chermayeff and Alexander 1965: 122–127).

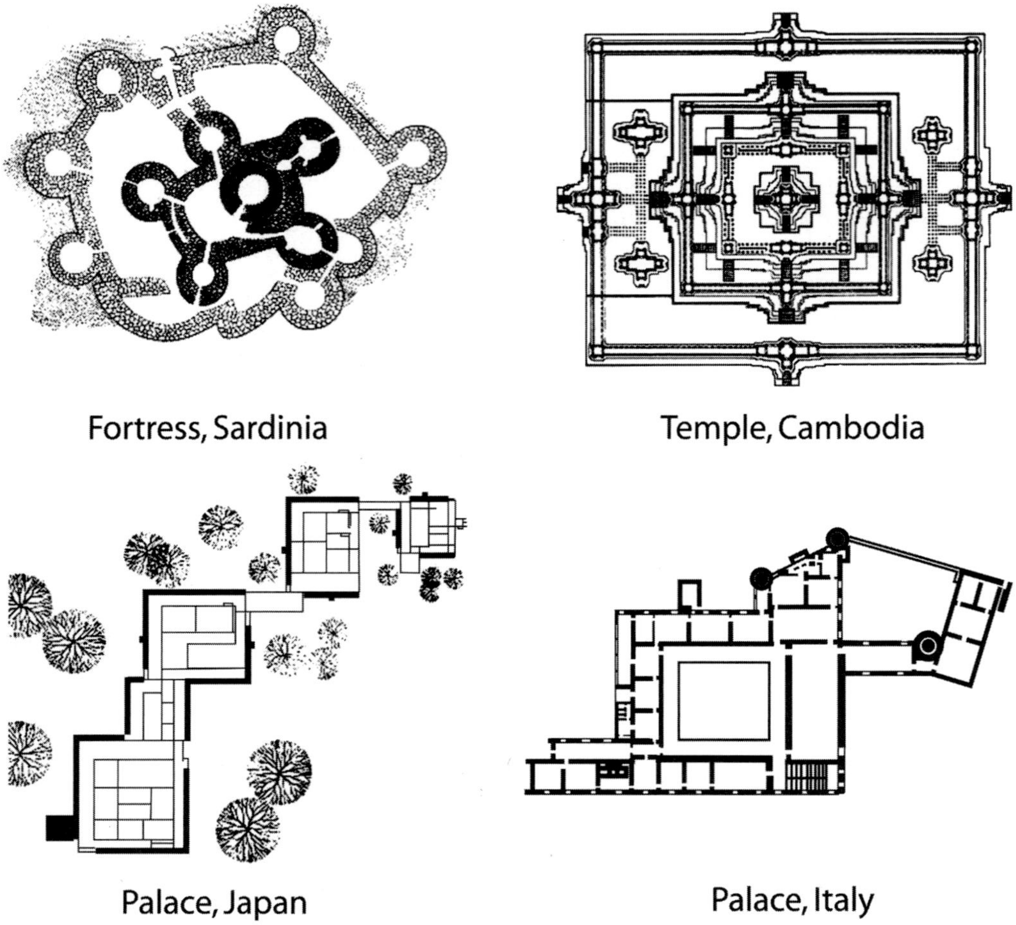


**Figure 2.3.** Examples of Ancient Architecture Displaying Hierarchical Arrangements of Space. Drawings from *Community and Privacy* by Serge Chermayeff and Christopher Alexander, © 1963 by Serge Chermayeff. Used with permission of Doubleday, a division of Random House, Inc.

In early European society, persons of standing did not necessarily go home to avoid public scrutiny, but rather to present themselves in the best way possible. In Girouard's (1978) study entitled *Life in the English Country House*, late-seventeenth- and early-eighteenth-century architecture is described from a social perspective, as a symbol of power and prestige reflecting (both architecturally and spatially) the cultural and political values of the powerful aristocracy of the time. Beyond functioning simply as a residence, the country house provided a decorative setting for political, business, and social ritual: Its purpose was to assert and reiterate societal superiority and control. Architectural elements, both real and symbolic, united architecturally to exemplify a system of hegemonic supremacy (Arnold 2002: 141–142). Formal houses in England, Russia, Sweden, and Sicily architecturally reflected the social order of the time. For example, the main rooms of English country houses were ordered sequentially in what Girouard describes as an "axis of honor," where gentlemen and lords could meet to negotiate business dealings whilst acknowledging the hierarchical status quo.

> The formal house was beautifully calculated as an instrument to both express etiquette and to back up negotiation. Since each room in a sequence of an apartment was more exclusive than the last, compliments to or from a visitor could be nicely gauged not only by how far he penetrated along the sequence, but also by how far the occupant of the apartment came along it – and then beyond it – to welcome him. The situation changed radically depending on whether the visitor was grander or less grand than the person he or she was visiting. The less grand visitor hoped to penetrate as far as possible along the line, but did not always succeed. The grander visitor was pressed to penetrate to the inner sanctuary, but could not always be tempted (Girouard 1978: 145–146).

In ancient Roman society, houses were also centres of public as well as private life, as the reputation of elites relied heavily on displays of opulence in and outside the home (Wallace-Hadrill 1988: 46). For Roman elites, space was partitioned emphasising gradations in privacy. For example, certain spaces were set aside for the reception of notable guests and associates, whereas others were reserved for dependents and for conducting business (Wallace-Hadrill 1988: 55, 58–59). In Roman society "low status individuals, dependents and clients, congregated in the atrium, whereas friends and closer associates were invited deeper into the more private area of the house" (Grahame 1997: 140).

There are various ways in which privacy can be attained and maintained architecturally, including the segregation and territorial demarcation of space with walls, doors, screens, as well as symbolic devices that may be used to differentiate one area from another (Lang 1987: 146, 149). Together with the privacy gradient (as described by Chermayeff and Alexander 1965: 121–126; Newman 1972: 9; Rapoport 1977: 290; and Grahame 1997: 139), Wallace-Hadrill (1988) offers another axis of differentiation that moves from "humble" to "grand" (Wallace-Hadrill 1988: 55). In reference to the architecture of the Roman house, Wallace-Hadrill (1988) notes that

> area[s] may be both public and grand (the magistrate's atrium) or private and grand (his triclinium or cubiculum). [They] may be private and humble (the slave's bedroom or farmer's storeroom) or even . . . public and humble (a shop, public lavatory, or service corridor) (Wallace-Hadrill 1988: 55).

Although the distribution of architectonic space may convey meaning in itself, it is through things such as signs, materials, decoration, and structural form that "different settings become indicators of social position and a way of asserting social identity to oneself and others" (Rapoport 1976: 19). The level and type of decoration in Roman architecture were fundamental to the controlling of interaction in a given area, providing a means of signaling the status of corresponding spaces; for example, the finer the architectural decoration, the less restricted the corresponding space (Wallace-Hadrill 1988: 58–59). A space that is simultaneously public and grand is so due to its function as a reception area, as "decoration is a symbolic medium and to be effective it requires an audience" (Grahame 1997: 141).

Privacy gradients may be correlated with variations in the types of images displayed in a sequence of spaces, functioning as a form of "impression manipulation" (Rapoport 1977: 290–292). For example, the art associated with Assyrian palaces (800–900 B.C.) was strategically positioned, displaying thematic variations subject to differences in spatial context. In *Ideology and Propaganda in Assyrian Art*, Julian Reade (1979) discusses the distribution of palace narrative art in terms of its accessibility; among the themes represented are Tribute, War, Feasting, and Courtly Activities (see Figure 2.4). Reade argues that the distribution of narrative art in Assyrian palaces signaled the function and status of activity areas (Reade 1979: 338), whilst operating as a "corpus of personal propaganda" (Reade 1979: 331). Reade notes that, whereas Assyrian palaces have their own unique character, they display similarities relating to the ruler's need to validate authority through artistic means. In addition to fulfilling magical as well as ornamental functions, Assyrian palace art also portrays rulers worshipping, receiving tributes, dispensing justice, and achieving victory in battle. As noted by Reade (1979: 335–336), in reference to provincial palaces such as Til-Barsip and Khorsabad, the narrative art was ordered sequentially, differing thematically with changes in the depth and accessibility of spaces to which they were assigned.

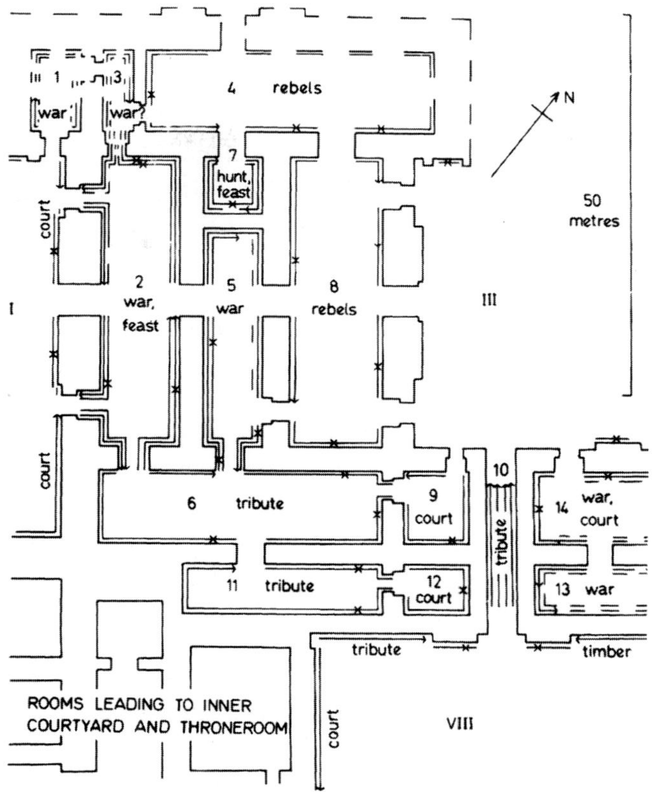


**Figure 2.4.** Distribution of Sculptural Themes in the Assyrian Palace at Khorsabad. After Reade (1979: 337, fig. 16), reprinted with permission.

> The exterior walls are largely plain but for magical decoration at the entrances and along the crenellations. This is all that the common people of Assyria

> normally saw . . . The sheer size of some of the magical figures, however, which weighed up to 16 tons, must have been intended to impress human as well as supernatural visitors . . . Those with business inside the palace entered through a gate-chamber, which was decorated internally . . . with narrative pictures of a military campaign. The palace gate-chamber opened into a huge outer courtyard surrounded by offices. Beyond lay another courtyard with more offices but also having, on one side, the magnificent façade of the royal throne room . . . showing the king at worship . . . in some relationship with the powers of earth and sky. On either side . . . leading the eye towards the throne room doors, were smaller slabs showing the king with a few attendants receiving groups of people (Reade 1979: 335–336).

In *Gender and Material Culture: The Archaeology of Religious Women*, Roberta Gilchrist (1994) argued for the potential of iconographic analysis and architecture in understanding gender relations within monasteries and nunneries. According to Gilchrist (1994), "space is the medium through which social relationships are negotiated" (Gilchrist 1994: 150), whilst artistic works contained within spaces function to signal boundaries and thresholds, assisting the navigation through spatial systems. Gilchrist proposes that monastic sculpture functioned, in part, as a mechanism of control by signaling the gender identities of delineated spaces. According to Gilchrist, the interior of a monastery was hierarchically ordered, and that "hierarchical spatial divisions were reinforced by the embellishment of archeological features" (Gilchrist 1994: 152). By assessing both the thematic content and distribution of various sculptural media (i.e., sculpture, glass, wall paintings, and ceramic tiles), Gilchrist sought to understand how boundaries informed the male and female inhabitants actually and symbolically. Buildings that consist of a number of spaces, and have corresponding sculpture, may be interpreted as a matrix of strategically positioned images. Within such a matrix, any understanding of the patterning of iconographic themes assists in establishing and cementing hierarchical relationships among its occupants (Gilchrist 1994: 151–152). Gilchrist also argues that a particular sculpture may have female associations. For example, the carved sacred heart from the Dartford Dominican nunnery is connected with female devotion. "To the well educated nuns of Dartford, the carved sacred heart represented a tradition of female worship linked to the Eucharist, which predated the image's wider popularity" (Gilchrist 1994: 153). At the Cistercian Sinningthwaite refectory, a carved serpent is positioned at its entrance; Gilchrist proposes that this sculpture was used to remind the nuns to be mindful of the demise of Eve through temptation (Gilchrist 1994: 155).

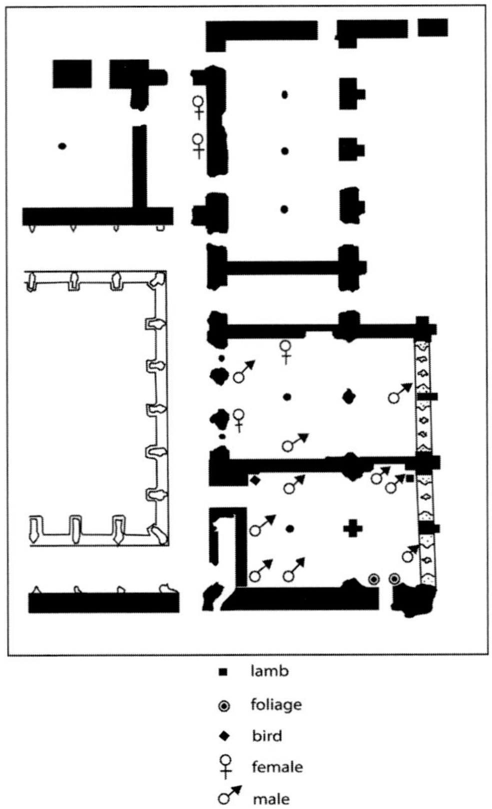


**Figure 2.5.** Spatial Matrix of Carved Corbels at Lacock. Image from *Gender and Material Culture: The Archaeology of Religious Women* by Roberta Gilchrist (page 159, fig. 64), © 1994 by Routledge, reprinted with permission of the publisher.

Carvings found within monasteries and nunneries did not always depict "scriptural or apocryphal iconography" (Gilchrist 1994: 159). Sculpture found on capitals and corbels often "depicted animals, foliage and the heads of secular men and women" (Gilchrist 1994: 159). Gilchrist argues that these images functioned to control movement through the structures affecting the meaning felt by the observer as one passes from one space to another (see Figure 2.5). Gilchrist notes that at Lacock, the sculptures of the male heads predominate in spaces frequented by the priests, as is the case with the sacristy. By using "access diagrams," Gilchrist compared the relative depths of men's and women's sacred and secular spaces, where it was discovered that sculptural embellishment was used to "reinforce the particular devotional practices of a house and confirmed its identity through reference to its dedication. More specifically, group membership and segregation were signaled by a spatial matrix which constructed meaning through the location and content of images" (Gilchrist 1994: 160). For example, a number of sculptures depicting male heads were found to be fewer in

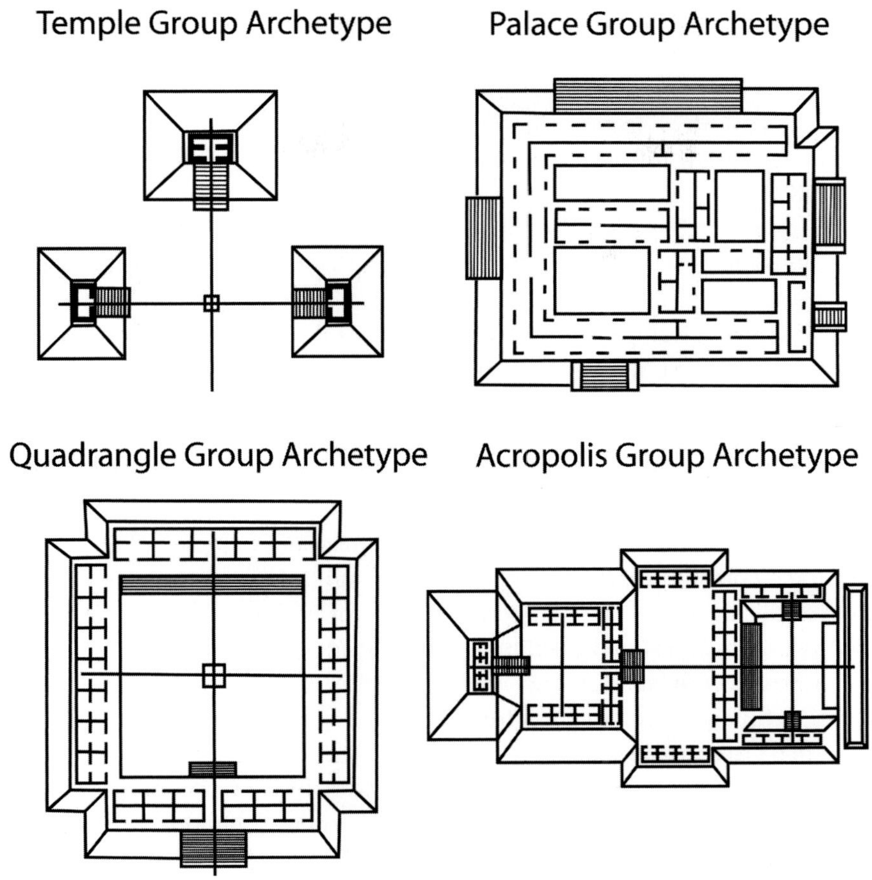


**Figure 2.6.** Maya Archetypal Building Groups. Images from *Maya Cities: Placemaking and Urbanization*, by George Andrews, © 1975 by the University of Oklahoma Press, Norman, reprinted with permission of the publisher. All rights reserved.

spaces frequented by men, such as the chapter houses. Furthermore, no secular male sculpture was found in the deeper areas of the nunnery, between the outer court spaces accessible by secular persons and the inner spaces reserved for the religious personnel and esoteric activities (Gilchrist 1994: 160).

## 2.2. Criteria for appropriate comparisons in Maya art and architecture

Generalising about patterns of behaviour associated with Maya art and architecture remains difficult as a result of variations in the chronology of built forms (Webster 1998: 24), dissimilar topography affecting site plans, differences in building materials, and disparities in artistic traditions. As Andrews (1975), however, points out, "it is possible to identify and characterize building groups, or complexes of structures, on both the basis of their form and organization." Andrews defines these basic building groups "as those archetypal forms of building complexes which tend to be repeated with minor variations throughout the Maya area, leading us to believe that they proved themselves to be exceptionally useful either on functional or symbolic grounds" (Andrews 1975: 54–55). In his analysis of twenty Maya sites, Andrews (1975) succeeded in identifying four basic building forms that occur most frequently. These structural archetypes were as follows: (1) Temple Groups; (2) Palace Groups; (3) Quadrangle Groups; and (3) Acropolis Groups (see Figure 2.6).

### 2.2.1. Temple Group

Although this type of building group may include additional buildings, "in its simplest form the Temple Group consists of a set of three [pyramid-type] temples situated on three sides of a rectangular plaza" (Andrews 1975: 56). The remaining open side of the central plaza establishes a unique set of relationships among the elements of the group. The "triadic architectural form" (see also Hanson 1998: 77–82) establishes a hierarchy, indicating an order of importance among the buildings represented. Typically, the main temple faces the open end of the plaza (i.e., the main entrance to the group), whereas the two peripheral temples face in toward the centre of the plaza. The triangular form of the Temple Group is believed to be representative of the three mythic throne stones (the jaguar, the snake, and the shark) that were laid down at the creation of the current world. The triangular shape appears in the constellation of Orion, which the Maya reproduced architecturally in the layout of their temples (Schele and Mathews 1998: 44).

### 2.2.2. Palace Group

The term "palace" is problematic given that the precise function of this structural form remains in question. The term *palacios* was used by the early Spaniards to describe a structural type that was large and displayed a relatively high degree of architectural decoration (Liendo Stuardo 2003: 194). Andrews (1975: 59) defines the Palace Group as an aggregate of structures spatially associated so as to form a boundary of buildings behind which additional structures were incorporated to create a series of inner courtyards, partitioned by intervening buildings. This building type was constructed on one or more raised platforms that served to distinguish the complex from the natural terrain and other surrounding structures (see also Chapter 1, Section 1.3, Common Forms in Maya Architecture).

### 2.2.3. Acropolis Group

An Acropolis Group is described as a number of spatially associated structures, similar to a palace or temple type that is organised hierarchically on a series of elevated platforms. Admission to the higher levels of the complex was via strategically positioned stairways that provided predetermined routes of approach; these stairways gave access to a number of structures and associated courtyards. "The sequence [of an Acropolis Group] culminates in the most important buildings within the complex, usually a temple that is farthest removed from the plaza below both in terms of height and distance" (Andrews 1975: 67).

The distinguishing elements of dynastic-elite architecture are elevation, hierarchical arrangement, and restrictiveness. In accordance with Andrews (1975: 67), Cohodas (1973: 215) argues that Maya buildings and platforms that collectively comprise acropolis-type configurations are intentionally ordered as a progressive accent that starts at the entrance of the group and terminates with a principal structure at its farthermost point. In this type of building arrangement, lines of sight remain open as one moves progressively deeper into the complex and approaches the most elevated structures. The objective of this building configuration was to allow rituals undertaken by elites in the upper reaches of the complex to remain visible to the lower classes occupying the more accessible plazas below. This configuration is in sharp contrast to palace-type building groups, which are characterised by a series of closely arranged, private courtyards where lines of sight are limited and buildings are less hierarchically organised (see Chapter 4, Architectural and Sculptural Programs of the Palenque Palace Group). In acknowledging the apparent differences in form between acropolis- and palace-type building groups, however, one can see that both share two characteristic features that define them as elite architectural forms – more specifically, their elevation and restrictiveness (Cohodas 1973: 216).

### 2.2.4. Quadrangle Group

A Quadrangle Group consists of a series of range-type structures ordered around a large, open courtyard. In this type of building configuration, it is not crucial that the four sides of the court space are closed off. In its simplest form, however, the Quadrangle Group consists of four buildings completely surrounding a central space. Entry into the courtyard is generally achieved through a door or archway located at the centre point(s) of one or more of the peripheral buildings. Like Palace Groups, Quadrangle Groups are also located on large, elevated platforms designed to distinguish them from the surrounding environment (Andrews 1975: 63).

Originally, my objective in this study was to analyse and interpret the distribution of sculptural themes associated with examples of all four of Andrews's (1975) building group archetypes (i.e., Temple Groups, Palace Groups, Quadrangle Groups, and Acropolis Groups); structures found to comply with Andrews's definitions included the Palenque Cross Group, the Palenque Palace, the Uxmal Nunnery Quadrangle, and the Copán Acropolis (see Figure 2.7). Time and space, however, permitted the inclusion of only two of the structural types just outlined. Subsequently, I have chosen to focus the analytical component of this study on the Palenque Cross Group and the Palenque Palace. Given the consistency at which these building forms are found at different Maya city centres, I saw value in their use for the purposes of my investigation, as both provided broad spatial complexes suitable for my study while permitting the formulation of generalised arguments concerning the behaviour associated with art and architecture among the Classic Maya.

## 2.3. Temporal controls

When undertaking an analysis such as this there needs to be awareness that temporally and spatially disparate

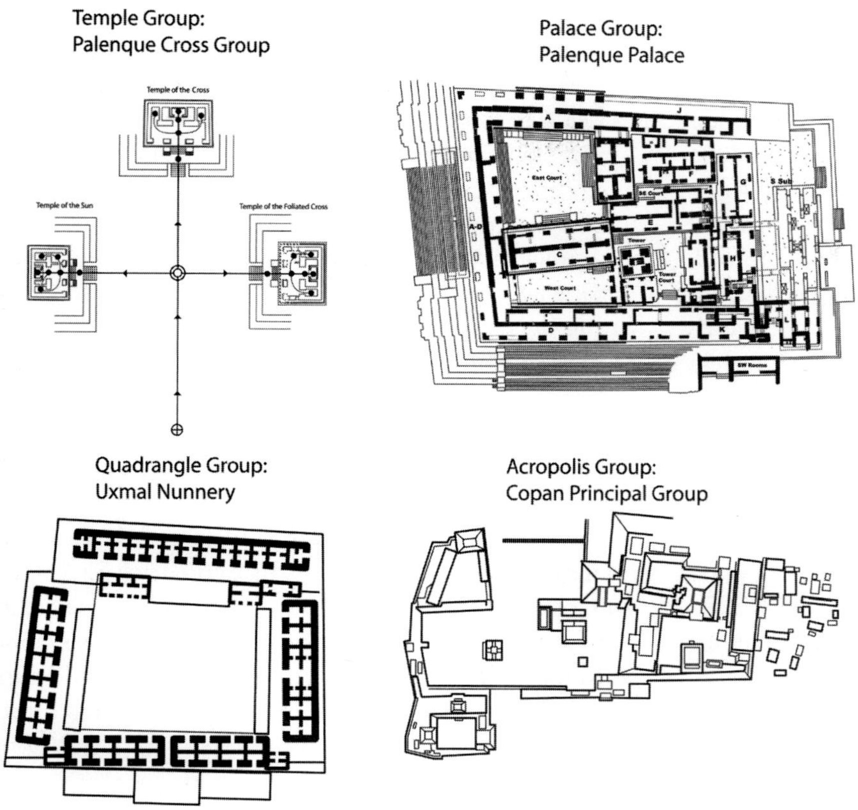


**Figure 2.7.** Examples of Maya Archetypal Building Groups in Accordance with Andrews (1975). Drawing of Palenque Palace by Merle Greene Robertson (1985b, fig. 1), reproduced with the permission of Merle Greene Robertson. Drawing of the Principal Group Copán modified after Fash (1991: fig. 8).

sculpture cannot be compared indiscriminately. The gradual accumulation of sculpture over time, including extension and renewal of building groups, affects the capacity of sculpture and associated architecture to provide information about relationships. It was for this reason that measures were taken to provide some temporal control over the investigation to assure that any correlation made among sculpture, architecture, and associated spaces was both appropriate and meaningful.

Structures and associated sculpture were sorted to reigns of kings to confine the periods investigated whilst providing a way of inferring the intention of different rulers in the placement of certain sculpture within specific spatial contexts. Graham Fairclough (1992), in a study of Edlingham Castle, Northumberland, broke a continuous sequence of architectural development into three main periods. Applying access analysis, the objective of Fairclough's study was to understand the effects of "warfare on Anglo Scottish marches and the development of the northern tower house and related traditions" (Fairclough 1992: 357). Similarly, the Palace at Palenque (as it appears today) was the result of a sequence of architectural development implemented over the reigns of consecutive rulers at Palenque. More specifically, the structures occupying the second terrace of the Palace alone represent approximately 180 years of architectural and sculptural development beginning during the reign of K'inich Janab' Pakal I (A.D. 615–683) and ending during that of Wak Kimi Janab' Pakal III (A.D. 799–?) (see Chapter 4, Architectural and Sculptural Programs of the Palenque Palace Group).

Arguments concerning intent in the architectural design of the Palace are based primarily on the interpretation of monumental art and the conscious choices made by rulers as to the context of its placement. As has been noted for Tikal, Uaxactun, and Copán, "many of the architectural patterns . . . documented at [Maya sites] . . . can shed light on the issue of the ideological and political strategies put to work by different dynasties and rulers in their respective dynastic building programs" (Fash 1998: 224–225) – a view that I hold for Palenque. General questions of interest are as follows: (1) What motivated the positioning of informative sculpture at various levels of common and elite accessibility? and (2) What are the differences between the information being transmitted within these different spatial contexts? With these objectives stated, it is acknowledged that it remains difficult, when analysing the iconographic content of monumental art from different periods, to discern elements that may represent changes in the artist and those that represent changes in the subject of the art. The advantage of an intrasite study of motifs, however, is that "variables of time are controlled, making more reliable the discovered regularities, continuities, and changes. Furthermore, comparisons of different forms manifested in the same contexts lead to the discoveries of equivalencies and oppositions which . . . may be used for the construction of a lexicon of images, whose value is provisionally restricted to the site under study" (Baudez 1986: 17–18).

## 2.4. Investigative procedures

For a long time archaeologists have used a range of analytical techniques to discern or decipher patterns in the distribution of artifactual material surviving in the landscape (e.g., Clark 1956; Hodder 1972; Washburn 1974; Hammond 1975; Hodder and Orton 1976; Blankholm 1991; Ebert 1992). The earliest attempts in spatial analyses relied on basic site maps and piece plots of artifacts within boundaries. Beginning in the 1960s, increasing focus was directed at designing more rigorous statistical techniques in spatial analyses to further assist the archaeologist. After the publication of Hodder and Orton's (1976) *Spatial Analysis in Archaeology*, quantitative spatial analyses steadily grew in popularity. More recently, however, approaches in spatial analysis have been directed at interpreting architecture in terms of the syntactic relationships of space (e.g., Hillier and Hanson 1984; Grahame 1997; Gould 1999; West 1999; Liendo Stuardo 2003). Studies in "space syntax" have led to more channeled attempts to assign social functions to architecture through the study of the distribution of "space" alone (Foster 1989a: 40; Fairclough 1992: 348). As noted by Fash (1998: 242), "a key to understanding the function of the buildings within their social context and the behaviors that took place in each one of them comes in the form of the architectural spaces themselves."

In this investigation, the analysis of internal and external spatial arrangements of structures and associated sculpture was undertaken in accordance with the interpretive methodology formally introduced by Hillier and Hanson (1984) and subsequently used by a range of other researchers (see Foster 1989a, Foster 1989b; Fairclough 1992; Moore 1992; Blanton 1994; Gilchrist 1994; Grahame 1995, 1997; Gould 1999; West 1999; Liendo Stuardo 2003; D. Mathews 2004). Of particular focus in this study is a technique originally referred to as "gamma analysis" (Hillier and Hanson 1984: 148), now more commonly described as access analysis. This approach investigates the layout of space within architecture as influenced by the distribution of boundaries and entrances as a means of discerning relationships between inhabitants and between inhabitants and "strangers" (Foster 1989a: 40). Hillier and Hanson's approach to the interpretation of space is supported by a number of premises: (1) "spatial organization in society is a function of differentiation" (Hillier and Hanson 1984: 142–143); (2) social structure controls the way in which space is arranged and allocated within architectural space; and (3) a correlation can be drawn between the organisation of a society and the degree of order imposed spatially and architecturally in that society (Hillier and Hanson 1984: 235). According to Hillier and Hanson (1984: 154)

> the essential proposition of gamma-analysis [access analysis] is the buildings transmit social information through their interior structures both through general variations in the basic syntactic parameters and also – perhaps primarily – through the variations in the syntactic parameters which appear when the complex is looked at from the points of view of its various constituent spaces.

Access analysis is viewed as an effective tool, providing a universal frame of reference (Liendo Stuardo 2003: 186). The power of this method lies in its "descriptive autonomy . . . societies which might vary in their type of physical configuration and degree to which the ordering of space appears as a conspicuous dimension of culture can all be compared on a similar basis" (Foster 1989a: 41), providing the means by which hypotheses can be derived and tested. One method of avoiding culturally predetermined histories of architecture is to focus on the space itself rather than the structures that articulate it (Arnold 2002: 134). In Foster's (1989a) work entitled *Analysis of Spatial Patterns as an Insight into Social Structure: Examples from the Scottish Atlantic Iron Age*, in addition to her follow-up article (Foster 1989b), she illustrated the power of Hillier and Hanson's methodology and how it can be used to interpret the configuration of entire settlements. Foster successfully demonstrated how access analysis could be applied to reveal how social and political relationships were ordered spatially. The social organisation of space in buildings was marked by the strategic placement of boundaries and entrances (such as doors, curbs, and ramparts) that restricted and/or controlled the movement of pedestrian traffic to varying degrees. Foster's study revealed a hierarchical use of space, demonstrating the degree to which strangers were segregated from higher-status building groups situated in the upper levels of the broch and how the outbuildings were effectively open to the outside world (Fairclough 1992: 349).

In Peter Harrison's (1970) work entitled *The Central Acropolis, Tikal, Guatemala: A Preliminary Study of Functions and Its Structural Components during the Late Classic Period*, Harrison attempted to discern the function and accessibility of architectural space in a Maya context, through the categorisation and analysis of structures and other architectural features. Although Harrison did not use access analysis to reveal what was a "deliberate attempt to preserve privacy" (Harrison 1970: 187), he argued that secondary controls were installed in the Acropolis at Tikal during the building group's expansion as a means of controlling the movement of pedestrian traffic. According to Harrison, "supplementary stairs" were constructed to provide access to secondary additions in buildings, "masonry screens" were installed to redirect traffic flow, "gateways" were designed to narrow passages and doorways, and "masonry blocks" were positioned to deter entry. In Harrison's study, buildings were sorted into four main categories "according to constants of room arrangement" (Harrison 1970: 96–98). These categories were described as: (1) Tandem present/Transverse present, (2) Tandem present/Transverse not present, and (3) Tandem not present/Transverse present. (For recent studies combining archaeology and access analysis, see Samson 1990; Banning 1996; Ferguson 1996.)

Maya elite architecture can be identified by its restrictiveness. The elevated nature of many Maya building groups, combined with their small interiors, suggests that access was reserved primarily for the elite in Maya society (Cohodas 1973: 215). Most Maya buildings were constructed on elevated platforms; these platforms supported architecture that ranged from simple houses made from adobe, poles, and thatch to palaces and temples made from quarried stone. The platforms themselves varied in size, height, and degree of decoration. The simplest earthen platforms usually supported the simplest of structures, whereas multiterraced and pyramidal platforms supported ritual, elite-residential, and administrative architecture often finished with a layer of masonry and decorative elements (Sharer 1994: 631).

According to Cohodas, one of the defining features of Maya public architecture is its "rigidity of form" (Cohodas 1973: 216). The orientation and arrangement of buildings and associated space were standardised and rarely subject to modification. Architectural assemblages that fall within this category include radial pyramids, twin temple complexes, and astronomical and building complexes such as E Groups. Characteristics of public architecture are its spaciousness, comparatively low elevation, and increased accessibility. For example, wide entrances promoted entry into plazas and subsequent participation in the activities undertaken within their confines (Cohodas 1973: 216).

One of the criticisms leveled at access analysis is that although it may be useful to decipher patterns in the arrangement of space, it fails to adequately deal with the meaning of those spaces (Gilchrist 1994: 51; Liendo Stuardo 2003: 199). Gould (1999) attempted to discern the hierarchy of space in industrial architecture by analysing routes of access, physical boundaries, and architectural decoration. Critical of the conventional approach of assessing nineteenth- and twentieth-century architecture based purely on methods of construction, phasing, and technological fittings, Gould argues that such interpretations "are effectively dead . . . no mention is made of the social actors who performed in these arenas of power" (Gould 1999: 144). Questions of social organisation, the movement of people, and hierarchies of control are often ignored as being beyond the scope of the archaeological data and therefore "the relationship between routes of access . . . and the subtle use of decoration and architecture to convey meaning are likely to be missed" (Gould 1999: 144).

The basic product of this technique is the "justified" access map, a type of flow chart that identifies properties of depth within structures. Access analysis treats space in terms of its formal and relational properties rather than dimensional attributes (Liendo Stuardo 2003: 188; see also Bavin 1996: 158), arranging space syntactically into patterns of permeability (Foster 1989a: 41). As described by Hillier (1985: 170), "depth [in architecture] exists to the extent that to go from one space to another it is necessary (i.e. not merely possible) to pass through intervening spaces." Access maps can also be used to construct various descriptive indices, more specifically, the Relative Asymmetry (RA) Index (a global measure of how deep or shallow, on average, a space is within a system of spaces) and the Control Values Index (a local measure of how many spaces to which a particular space controls access). For purposes of this study, methods in access analysis provide a means by which architectural space, and the artistic themes corresponding to those spaces, can be ranked hierarchically in terms of relative orders of accessibility (i.e., from most accessible to least accessible).

### 2.4.1. Access maps and access analysis

The basic requirement for the production of access maps is the well-preserved ground plan of the building or the architectural space being investigated where entrances into and within structures can be easily identified. The reason for this requirement is that clear partitioning between spatial units in architecture is required for an accurate application of access analysis. Internal and external units of space are defined by the presence of bounding walls or structures that inhibit or control access between spatial units; in access analysis, movement between spatial units is provided via doorways and/or gaps in barriers and partitions.

To produce an access map, the first stage is to represent each delineated space within a complex of spaces with a small circle ("node"), which can then be linked by lines ("edges") indicating routes between spaces where the architecture permits. In access analysis, the various spatial categories said to comprise architectural space are as follows: (1) transitional space, (2) open space, (3) closed space, and (4) carrier space (Grahame 1997: 147). The term "transitional space" refers to passages and hallways; "open space" refers to plazas and open courtyards;

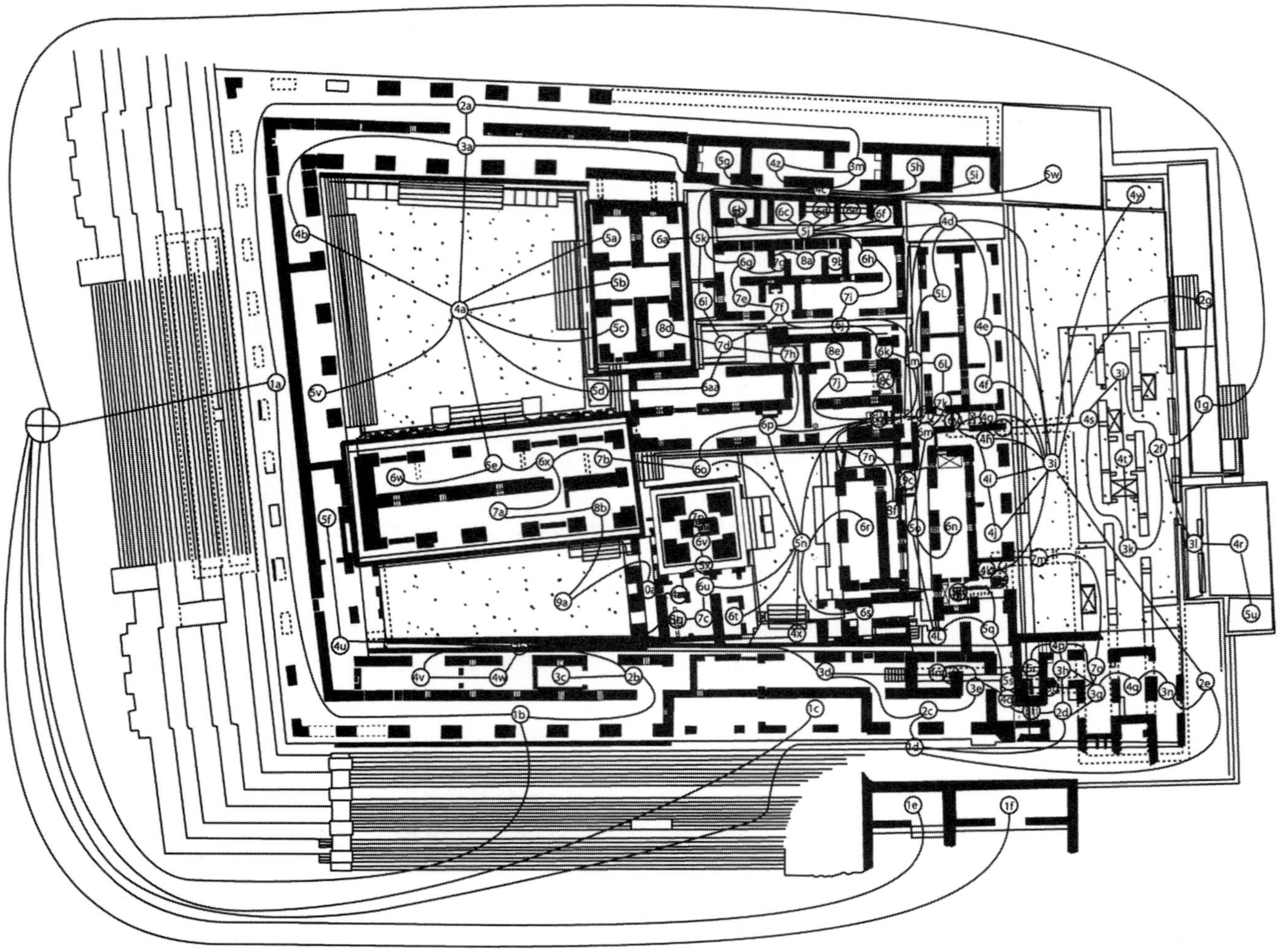

**Figure 2.8.** "Unjustified" Access Diagram of the Palace at Palenque. Drawing of Palenque Palace modified after Merle Greene Robertson (1985b: fig. 1), reproduced with the permission of Merle Greene Robertson.

"closed spaces" are the interior of rooms or single room structures. An access map can have only one "carrier space," as the "carrier" represents the area in which the entire spatial complex is contained. Therefore, although a building may have several entrances, it can only have one carrier space. By convention, the carrier space is represented by a circle with a cross, drawn in its centre. In the justified versions of the access maps presented in this investigation, a closed space will be represented by an open circle, a transitional space by a closed (coloured) circle, and an open space by a double circle. Following this method, access maps may be used to describe "the spatial configuration in any building" (Grahame 1997: 147).

In the diagram of the Palenque Palace Terminal Phase (see Figure 2.8), we see the end product of the procedure just described, an "unjustified" access map of a building where small circles are used to represent delineated spaces and lines represent routes of direct access between them where the architecture permits.

## 2.4.2. Principals of convexity

One problem with using access analysis to investigate the function and hierarchical placement of monumental art in Maya city centres is that a large proportion of the artwork found in Maya sites is positioned outdoors and is adhered to the exterior of buildings; examples of external sculptural media include stelae, altars, tablets, piers, roofcombs, and staircases. In addition, the layout of many Maya centres makes it difficult to discern units of external space, something that is precisely represented in the interior of buildings. Interior space is easier to investigate than exterior space as it is simpler to break it into analytical units due to the presence of well-defined boundaries (i.e., walls and doorways), which are not so readily found on the outside of buildings (Hillier and Hanson 1984: 16; see also Foster 1989a: 42). Subsequently, for purposes of this investigation, it was important to establish a method of discerning meaningful analytical units of external space.

The method used to address the problem was to divide, or break up, areas of external space by producing "convex maps" that can be converted into "access maps" (see Hillier and Hanson 1984: 97–98). A convex map designates units of external space by articulating boundaries at points where buildings or other structures encroach and recede on the spatial configuration, which would interrupt and/or influence the regularity of movement through a spatial system. Regularity of movement around building complexes may be affected in a number of ways; for example, structures may impinge on a space, either blocking movement or reducing the volume of traffic flow. Alternatively, structures may recede, or fall back, increasing the potential for larger numbers of people to occupy or pass through a given area. In an external spatial network, as represented by convex maps, buildings become boundaries analogous to the walls as seen in the interior of buildings. The gaps between the buildings and structures mark points of transition between units of space similar to the way doorways are associated with the interior of buildings.

By drawing the fewest number of "fattest" (most convex) shapes possible in all open areas on the plan-view maps, boundaries of external spaces and points of transition between those spaces are indicated where the circles or polygons intersect or touch. The size and shape of each convex space is dependent on the distribution of structures and natural features bounding the perimeters of the spaces. When a structure or topographical feature alters the space so that its convexity is interrupted, it marks the boundary of that space and the boundary of a new convex space. "Convexity" is defined by Hillier and Hanson (1984: 96) as "the maximum local extension of the system unified two-dimensionally" or alternatively "when straight lines can be drawn from any point in the space to any other point in the space without going outside the boundary of the space itself" (see Hillier and Hanson 1984: 97–98) (see Figure 2.9).

For further clarification, the procedure is best demonstrated by example. Figure 2.10 is a map of a town in the Var Region in France; the dark area represents the total area of external space associated with the settlement, revealing its overall configuration.

Figure 2.11 is a convex map of the same town in the Var Region in France where all the fattest (i.e., most convex) spaces comprising the exterior of the settlement are represented by intersecting squares, rectangles, and polygons.

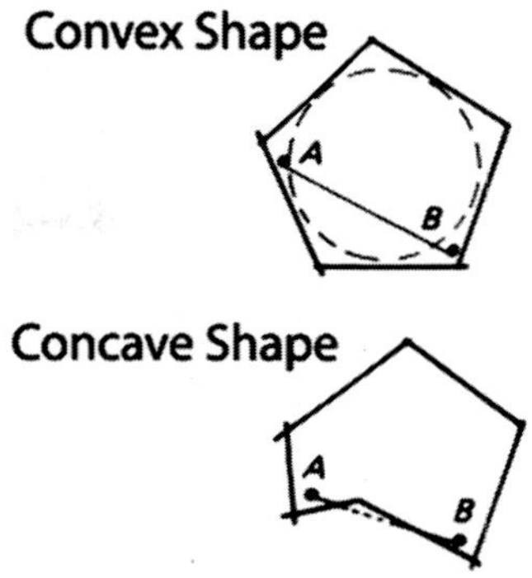


**Figure 2.9.** Convex versus Concave Shapes. Hillier and Hanson (1984: 97–98). Reprinted with the permission of Cambridge University Press.

After all convex spaces have been identified on a plan-view map, each is then represented by a small open circle in the same way that interior spaces are represented. Then, lines are drawn between each of the spaces where a relationship of direct adjacency is evident. In Figure 2.12, we see the result of this process where the convex map in Figure 2.11 has been converted into an "interface map," which is effectively an access map of the town, village, or building group being investigated (Hillier and Hanson 1984: 104–105). Applying this method, one can interpret external spaces in the same way that internal spaces can be interpreted in terms of accessibility and control, where each small circle represents a unit space that provides access to other spaces beyond it. Figure 2.13 is a combined convex and access map of Phase 1 of the Palace at Palenque (see Chapter 4, Architectural and Sculptural Programs of the Palenque Palace Group, for the complete construction sequence of the Palace at Palenque).

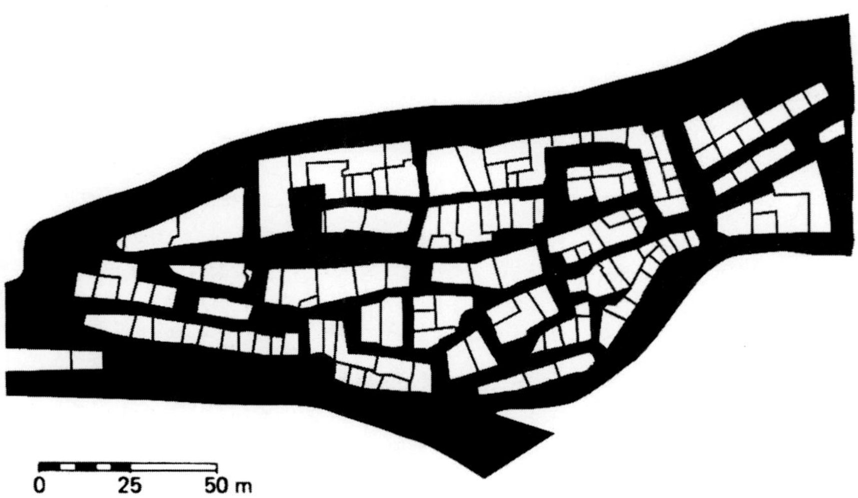


**Figure 2.10.** Space Structure of a Town in the Var Region of France. Hillier and Hanson (1984: 91). Reprinted with the permission of Cambridge University Press.

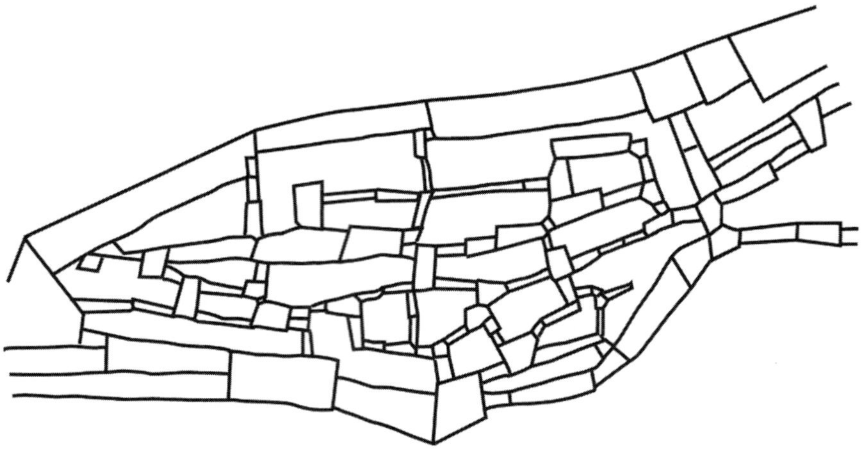

**Figure 2.11.** "Convex Map" of a Town in the Var Region of France. Hillier and Hanson (1984: 92). Reprinted with the permission of Cambridge University Press.

## 2.4.3. Justified access maps

After the access map has been produced, it can then be justified from the perspective of the carrier position (Foster 1989a: 41). Justified access maps provide a clear visual impression of the relational properties of each space in terms of their access to one another. In a justified access map, each horizontal layer represents different spaces found at common levels of permeation; space as it relates to increasing depth is represented vertically on consecutive levels (Hanson and Hillier 1982: 149). By justifying an access map, each space is assigned a depth value depending on its distance from the carrier position whilst preserving the links of permeability. The depth values of spatial units (nodes) are determined quantitatively by counting the number of boundaries (edges) that need to be crossed along a given route to reach the target space. This approach provides a useful way of assessing the relative accessibility to a given space from the perspective of a stranger and/or visitor entering the spatial system from the carrier position. In addition, this process allows one to determine those spaces that are found at common levels of permeation from the perspective of the carrier and those that are not. Figure 2.14 shows justified access maps of four building groups that are consistent with Andrews's (1975) definitions of Maya building group archetypes: the Cross Group and Palace at Palenque, the Nunnery at Uxmal, and the Principal Group at Copán. As mentioned earlier in text, each layer of dots in the diagrams represent units of space found at common depths from the carrier, and each consecutive layer of dots represents spaces found at increasing depths. The numbers and letters next to each dot allow for the easy identification of depth values from the carrier position as well as the location of each space on the original unjustified access maps.

## 2.4.4. Principals of distributedness–nondistributedness and symmetry–asymmetry

After formulating an access map, the spatial system represented may be described in terms of its

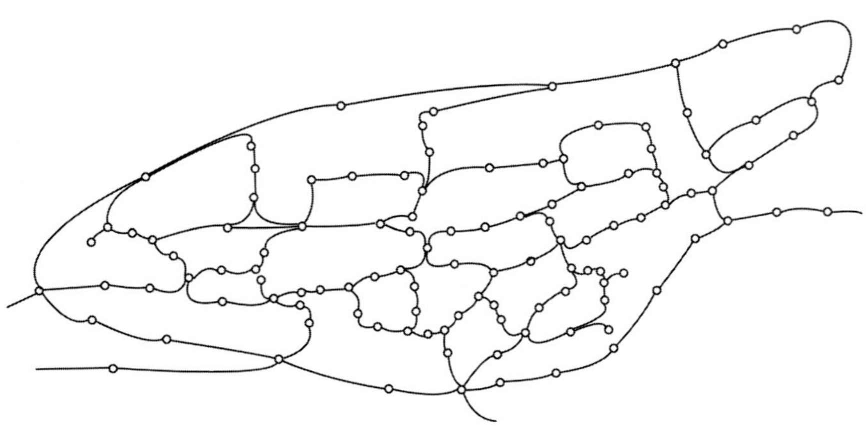

**Figure 2.12.** "Interface Map" of a Town in the Var Region of France. Hillier and Hanson (1984: 100). Reprinted with the permission of Cambridge University Press.

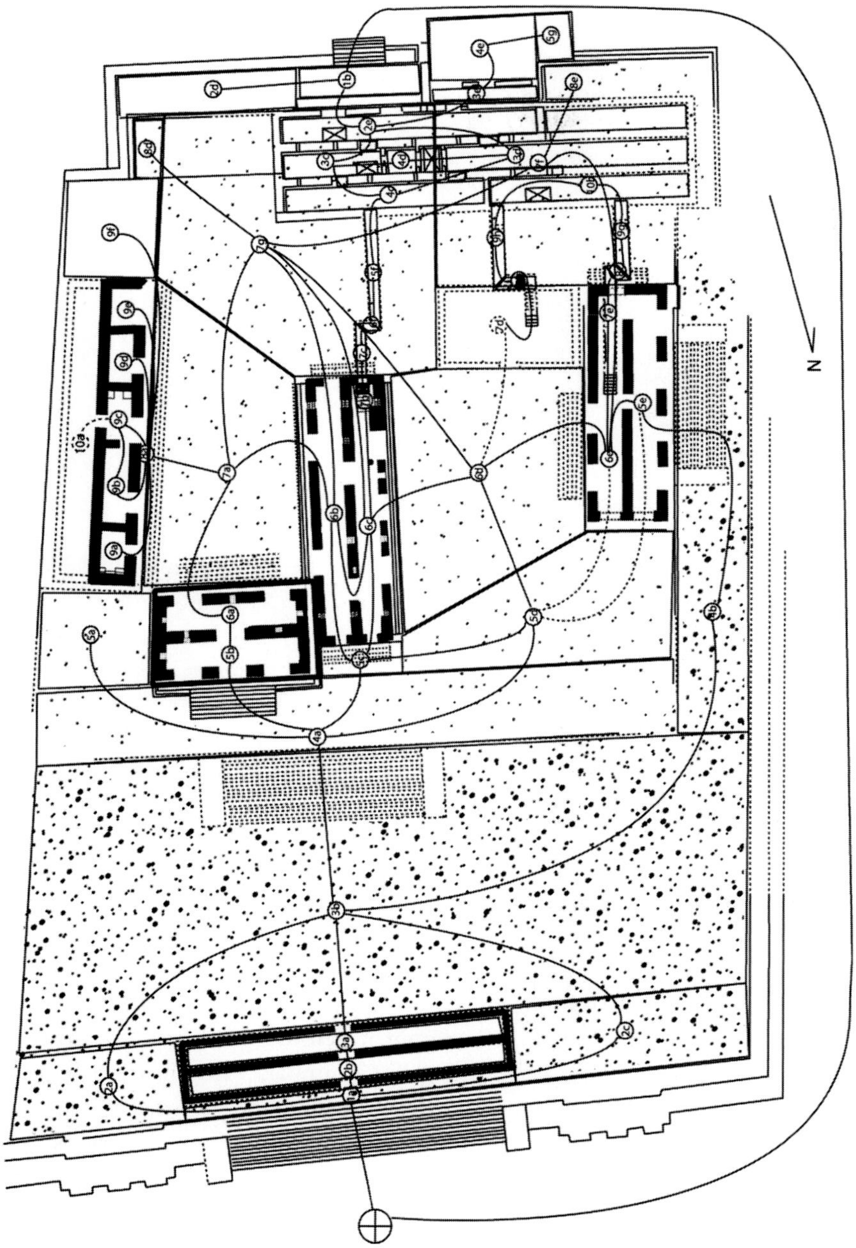


**Figure 2.13.** Combined Convex and Access Diagram of Palenque Palace Phase 1. Palenque Palace modified after Greene Robertson (1985b: fig. 1).

"syntactic relations," or more specifically in accordance with mathematical principals of "symmetry–asymmetry" and "distributedness–nondistributedness" (see Hillier and Hanson 1984: 93). A space or system of spaces is characterised as distributed if there is more than one route of access between spatial units that include passing through intervening space to get to the target space. Conversely, a space or system of spaces is categorised as nondistributed when there are fewer or no more than one route to a given space (Hillier and Hanson 1984: 108). Distributed and nondistributed relationships between spatial units function as indicators of how well a system of spaces is integrated (Blanton 1994: 32). Highly nondistributed access maps will often appear treelike ("dendritic") in shape, as they tend to comprise fewer connections (i.e., pathway options) between spatial units. In contrast, an access map comprising the same number of spaces, but with a greater number of connections, will often appear flatter in shape and comprise fewer permeation levels. A system that displays a high number of spaces but relatively few levels of depth would be described as a highly distributed. The flatter appearance of the diagram is a consequence of the increased number of route options increasing accessibility through the system. Justified access maps totaling a relatively high number of permeation levels would be described as deep and suggestive of a more restricted spatial system. In a deep system, persons must pass through a greater number of spaces to gain access to the target space. In the most restricted of systems, only one space controls access to the next. Distributed systems will usually form diamonds or rings on access maps (Hillier and Hanson 1984: 94, 148; Bavin 1996: 159).

A unit of space within the system of spaces is defined as symmetrical if it has a large number of similar relationships with other spaces in the system, an indication that, on average, fewer boundaries have to be traversed to access that space (Grahame 1997: 148–149). Conversely, a unit of space within a system of spaces is defined as highly asymmetrical if it has a low number of similar relationships with other spaces within the system. In line with Grahame (1995: 74–75), this concept is best described by example. In Figure 2.15, two spatial configurations are represented, each comprising the same number of spaces (nodes) and boundaries (edges). In Example 1, the spatial arrangement would be described as highly symmetrical, as the relations of a, b, c, and d are equivalent with respect to their level of accessibility from other spaces in the system. More specifically, in Example 1, one would have to cross no more than two boundaries, starting from any point in

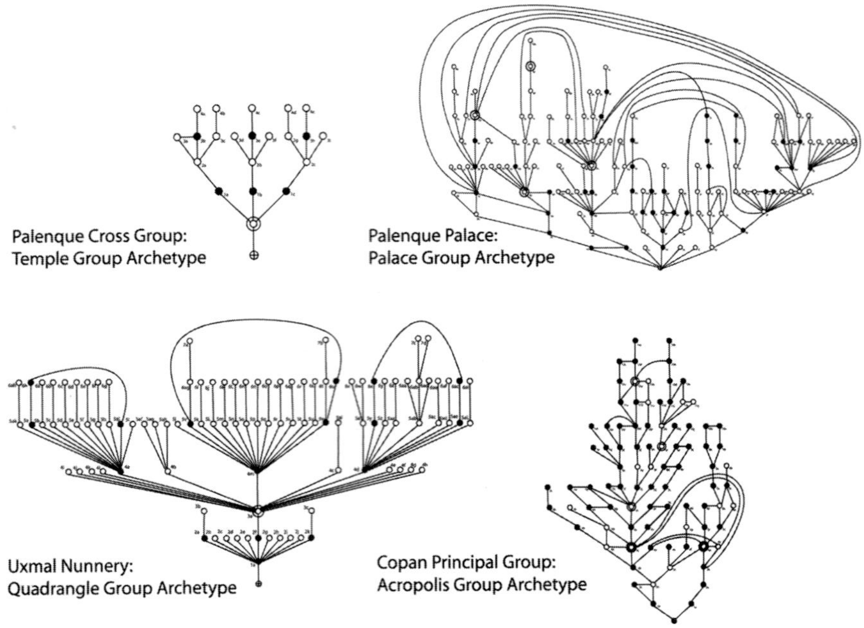


**Figure 2.14.** Access Diagrams of Examples of Andrews's (1975) Building Group Archetypes.

the system, to access any other space within the spatial assembly. If we randomly take position a as our destination, the number of boundaries that need to be crossed to reach this space, starting from all other positions, is as follows: from b to a, two boundaries; from c to a, two boundaries; from d to a, two boundaries; from e to a, one boundary. In contrast to Example 1, the spatial configuration of Example 2 would be best described as asymmetrical, as the spatial relationships among a, b, c, d, and e vary distinctly. For example, the relationship between b and a is different than the relationship between e and a. Whereas access from b to a would be achieved by crossing one boundary, three additional boundaries would have to be traversed to access a from e, including passing through b. Consequently, this difference defines the spatial configuration as asymmetric. Spaces seen to have asymmetric relations "always involve some notion of depth, since we must pass through . . . [intervening spaces] to go

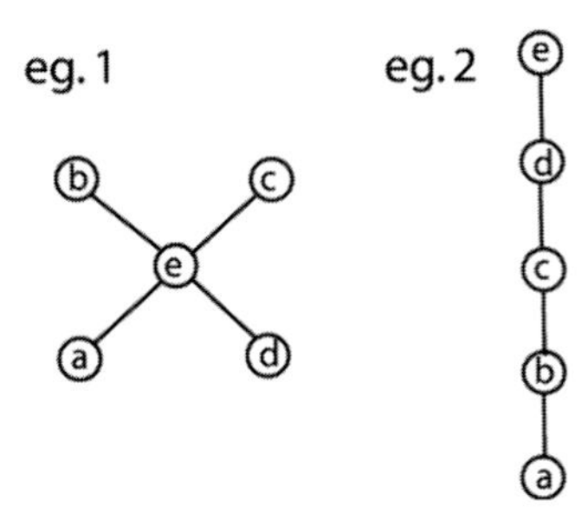


**Figure 2.15.** Example of "Symmetric" and "Asymmetric" Relationships between "Nodes."

from one space to another" (Hillier and Hanson 1984: 94).

In Figure 2.16 are some simple examples of justified access maps as offered by Hillier and Hanson (1984) indicating depth relationships between nodes as well as patterns of symmetry, asymmetry, distributedness, and nondistributedness (Hillier and Hanson 1984: 148–149). In these examples, symmetric relationships between spaces will be apparent if neither space a nor b controls access to the other, for example, where a is to b as b is to a with respect to c. Alternatively, spaces a and b will be considered to have asymmetric relationships if a is not to b as b is to a, or, more specifically, if one of these spaces (a or b) controls access to the other from a third space, designated c (Hillier and Hanson 1984: 148–149; see also Foster 1989a: 43 and Bavin 1996: 160).

(1) a and b are in an asymmetric and nondistributed relationship with respect to c
(2) a and b are in a symmetric and nondistributed relationship with respect to c
(3) a and b are in a symmetric and distributed relationship with respect to c
(4) a and b are in a symmetric relationship with respect to c, but d is asymmetric to both with respect to c
(5) d is in a nondistributed and asymmetric relationship to a and b, which still remain symmetric to each other with respect to d (or to c).

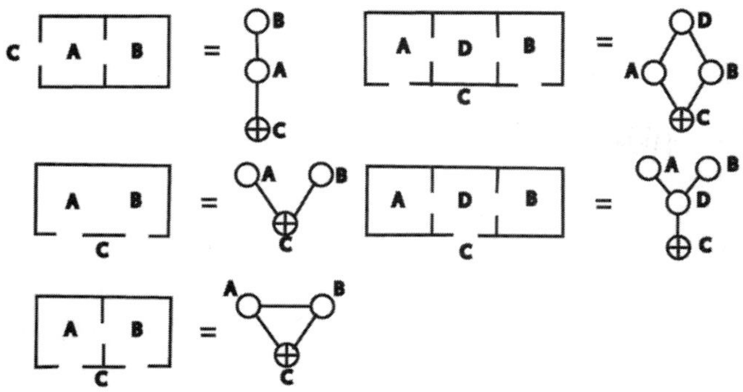

Figure 2.16. "Justified" Access Maps Showing Symmetric–Asymmetric and Distributed–Nondistributed Relationships. Hillier and Hanson (1984: 94). Reprinted with the permission of Cambridge University Press.

## 2.4.5. Measuring control values and RA analysis

For the purposes of this investigation, justified access maps provided a useful platform for the calculation of various indices as offered by Hillier and Hanson (1984) that have assisted in the interpretation and description of space and associated sculpture. Two of these indices are referred to here as the RA Index and the Control Values Index. To fully comprehend these indices, we should first understand that every space within a network of spaces has both local relationships with its neighbouring spaces and global relationships with all other spaces in the system (Grahame 1997: 147). The method of discerning the local relationships that a space has with its neighbours is achieved by calculating its control value (CV), which is a relative measure of the amount of control that a space possesses over access to other spaces in the system in localised proximity; the more spaces a unit of space controls access to, the higher the level of control that space is said to have.

The method for discerning the global relationships of a space within a spatial system is through RA analysis, an approach that measures how deep a spatial configuration is, on average, from each unit of space within that configuration. In this approach, mean depths (MDs) are calculated then standardised to discern which spaces are deepest from the perspective of the spatial system (i.e., the domain of the inhabitant) and which are shallowest. Differences in the relative depths of spatial units are argued to indicate which spaces are least and most easily accessible from the perspective of the system to which they belong. RA analysis functions effectively as a measure of integration: The higher the resulting value derived from a calculation, the more segregated that space is from the spatial configuration; the lower the resulting value for a given space, the more integrated the space (Hillier and Hanson 1984: 109).

Table 2.2. *Path matrix for building floor plan (Figure 2.17)*

| | C | a | b | c | d | e | f | g | h | i |
|---|---|---|---|---|---|---|---|---|---|---|
| C | x | 1 | 2 | 3 | 3 | 3 | 3 | 3 | 4 | 4 |
| a | 1 | x | 1 | 1 | 1 | 2 | 2 | 2 | 3 | 3 |
| b | 2 | 1 | x | 2 | 2 | 3 | 3 | 3 | 4 | 4 |
| c | 2 | 1 | 2 | x | 2 | 3 | 3 | 3 | 4 | 4 |
| d | 2 | 1 | 2 | 2 | x | 1 | 1 | 1 | 2 | 2 |
| e | 3 | 2 | 3 | 3 | 1 | x | 2 | 2 | 3 | 3 |
| f | 3 | 2 | 3 | 3 | 1 | 2 | x | 2 | 3 | 3 |
| g | 3 | 2 | 3 | 3 | 1 | 2 | 2 | x | 1 | 1 |
| h | 4 | 3 | 4 | 4 | 2 | 3 | 3 | 1 | x | 2 |
| i | 4 | 3 | 4 | 4 | 2 | 3 | 3 | 1 | 2 | x |

To calculate the RA values of a spatial configuration first requires determining the MD of the system from the perspective of all spaces within that system. This determination is made by first assigning a score of 1 to every space within the system that necessitates the crossing of one boundary to reach a target space, a score of 2 for those spaces that require the crossing of two boundaries to reach a target space, a score of 3 for those spaces that require the crossing of three boundaries, and so on (see Hillier and Hanson 1984: 108 and Grahame 1995: 148–149). A simple, although time-consuming way of compiling these distance values is to construct a path matrix (Blanton 1994: 34–35) for the spatial system being investigated. In a path matrix, each depth value represents the minimum number of boundaries that have to be crossed to gain access from one space to another (see Table 2.2); it is from each of the columns in the path matrix that the RA values are derived (see Table 2.2 derived from Figure 2.17).

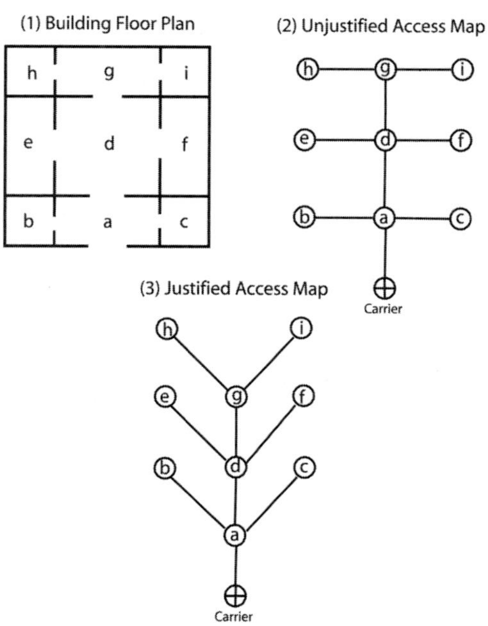


Figure 2.17. Example of a Building Floor Plan.

Table 2.3. *Ranked MD and RA values for building floor plan (Figure 2.17)*

| Rank | Node | MD | RA | Sum |
|---|---|---|---|---|
| x | Ca. | 2.6667 | 0.4167 | 24 |
| 1 | d | 1.6667 | 0.1667 | 15 |
| 2 | a | 1.7778 | 0.1944 | 16 |
| 3 | g | 2.0000 | 0.2500 | 18 |
| 4 | e | 2.4444 | 0.3611 | 22 |
| 4 | f | 2.4444 | 0.3611 | 22 |
| 5 | b | 2.6667 | 0.4167 | 24 |
| 6 | c | 2.7778 | 0.4444 | 25 |
| 7 | h | 2.8889 | 0.4722 | 26 |
| 7 | i | 2.8889 | 0.4722 | 26 |

For example, if we wish to calculate the RA value for column h (space h) in Table 2.2, we first add together all spaces found at common levels of depth and then multiply these totals by their corresponding depth values. For column h, we count 1 space at a depth of 1 (space g); 2 spaces at a depth of 2 (spaces d and i); 3 spaces at a depth of 3 (spaces a, e, and f); and 3 spaces at a depth of 4 (spaces b, c, and the carrier): $1 \times 1 = 1$, $2 \times 2 = 4$, $3 \times 3 = 9$, and $3 \times 4 = 12$. The totals are then added together (in this example equating to 26) and divided by the total number of spaces in the system, less the carrier space. The process has been summarised by the following formula: $MD = \frac{\sum d_k}{k-1}$, where $\sum d_k$ refers to the sum of the depth values as represented by $d$, for each of the spaces represented by $k$ (Grahame 1997: 149). As h is part of a spatial system totaling 10 spaces (including the carrier space), the MD of the system, from the viewpoint of position h, will be as follows: $\frac{26}{9} = 2.8889$ Boundaries (MD). After the MD of the spatial system from the perspective of space h has been ascertained, its RA value can be calculated by using the following equation: $RA = \frac{2(MD-1)}{k-2}$. Therefore, the RA value for space h will work out to be $\frac{2(2.8889-1)}{10-2} = 0.4722(RA)$. The resulting RA value will always be a number that falls between zero and one. The lower the resulting value, the more integrated (or shallower) the space is said to be from the rest of the system, and the higher the resulting value, the less integrated (or deeper) the space is from the rest of the system (Hillier and Hanson 1984: 108–109).

After all the RA values for the building floor plan (Figure 2.17) have been calculated, they can then be ranked accordingly (see Table 2.3), providing a gradient of integration for all spatial units represented.

Table 2.3 provides an example of how MD values (measured in numbers of boundaries crossed) as well as RA values (standardised MDs) can be ranked together in order from most integrated to least integrated. In Table 2.3, space d works out to be the most integrated space of the system, with an MD value of 1.6667 and a corresponding RA value of 0.1667. In this example, h and i work out to be the most segregated of spaces in the configuration, each with an MD of 2.8889 and corresponding RA values of 0.4722. It should be noted here that, in Table 2.3, the carrier space is not included in the overall ranking of values. As you will recall, the carrier space is external to the system representing the space in which the entire spatial complex is contained.

**Calculating CVs**

The CV of any space within the complex of spaces is calculated by first assigning a value of 1 to each space; subsequently, "the total amount of control, which is invested in any [spatial] configuration, is equal to the number of spaces comprising that configuration" (Grahame 1997: 147). After a score of 1 has been allocated to each space, the procedure is then to divide each space neighbouring the unit space in question by its immediate neighbours and then add them together. For example, taking unit or node a in Figure 2.17 as our working example, we see that this space has four immediate neighbours: nodes b, c, d, as well as the carrier space. Node b has only one neighbour node: a (the space in question). Therefore, the calculation for node b would be $1 \div 1 = 1$, a result shared by node c as well as the carrier space; as they too have only space a as their immediate neighbour. The exception is node d, which has three immediate neighbours in addition to node a, necessitating the calculation $1 \div 4 = 0.25$. After all calculations have been made for all neighbouring spaces, the results are added together to give the CV for the space in question: $1 + 1 + 1 + 0.25 = 3.25$ (CV). Applying this procedure, node a of Figure 2.17 is found to have a CV of 3.25 (see Table 2.4 for other CVs for the building floor plan). In this type of calculation,

Table 2.4. *Ranked CVs for building floor plan (Figure 2.17)*

| Rank | Node | CV |
|---|---|---|
| x | Ca. | 0.2500 |
| 1 | a | 3.2500 |
| 2 | d | 2.5833 |
| 3 | g | 2.2500 |
| 4 | h | 0.3333 |
| 4 | i | 0.3333 |
| 5 | b | 0.2500 |
| 5 | c | 0.2500 |
| 5 | e | 0.2500 |
| 5 | f | 0.2500 |

Table 2.5. *Ranked RA values and corresponding CVs for building floor plan (Figure 2.17)*

| Rank | Node | RA | CV |
|---|---|---|---|
| x | Ca. | 2.6667 | 0.2500 |
| 1 | d | 1.6667 | 2.5833 |
| 2 | a | 1.7778 | 3.2500 |
| 3 | g | 2.0000 | 2.2500 |
| 4 | e | 2.4444 | 0.2500 |
| 4 | f | 2.4444 | 0.2500 |
| 5 | b | 2.6667 | 0.2500 |
| 6 | c | 2.7778 | 0.2500 |
| 7 | h | 2.8889 | 0.3333 |
| 7 | i | 2.8889 | 0.3333 |

any space that has a CV of 1 or greater is considered to be a controlling space, any space with a CV less than one is considered to be a controlled space (Hillier and Hanson 1984: 109; Grahame 1997: 147). After all the CVs have been calculated for the entire spatial system under study, like the RA values, they can also be ranked hierarchically (see Tables 2.4 and 2.5).

After the RAs and CVs have been calculated for all spaces in a system, any artistic media assigned to those spaces can, like the spaces themselves, be interpreted or understood in terms of the ordering of these values. For example, by ranking spatially disparate artwork according to RA calculations, we can determine which are more or less accessible from the perspective of all spaces in the system that collectively represent the domain of the inhabitant. A similar process can be undertaken for the CVs, in which we can understand the distribution of sculpture and other artistic media in terms of its placement in either controlling or controlled spatial contexts.

A problem with the values derived from access analysis is that the calculations do not distinguish between spaces of different sizes and elevations. Unfortunately, two spaces that display the same depth, integration, and control, but are at different heights and are of different sizes, cannot be assumed to have the same value. To address this problem, differences in the dimension of interior and exterior spaces were considered as well as the relative heights of both indoor and outdoor spaces.

### 2.4.6. Reductionist–comparative and thematic–analytical analysis of Maya art

Problematically, "graph analysis [access analysis] does not yield any information about the meaning and the use of specific spaces" (Liendo Stuardo 2003: 199). By combining additional sources of information, however, access analysis can be an effective way of developing methods of pattern recognition and understanding how space was used in Maya city centres. More specifically, this type of approach may be further enhanced with the consideration of additional information: room dimensions, the presence of decoration, and a consideration of building materials (Liendo Stuardo 2003: 200). As stated at the beginning of this inquiry, the objective in this investigation was to determine whether access analysis of archetypical building groups, combined with the thematic–analytical investigation of fixed sculptural media (positioned within specific spatial contexts) would provide a way of assessing the function and hierarchical division of bounded space.

By combining a reductionist–comparative approach (Proskouriakoff 1950) with thematic–analytical methods (Clarkson 1979), steps were taken to classify and sort sculpture by theme as suggested by accompanying texts and iconography (see also Parmington 2000, 2003). In this investigation, function and status were ascribed to Maya buildings and associated spaces by interpreting trends in the iconographic content of various artistic media whilst considering their spatial contexts. In this study it is my supposition that public/civic and private/elite elements are present within Maya iconography and therefore may be identified and interpreted. In addition, the content and behaviour as displayed on Maya monumental art may be viewed as reflective of behaviour restricted wholly or partially to certain spaces and to certain status individuals. For example, "costume and associated paraphernalia [carved on monuments] frequently indicate the activity that is taking place" (Clarkson 1979: 1).

Proskouriakoff (1950), in her work entitled *A Study of Classic Maya Sculpture,* detected shifts in the stylistic traits of figural art over time. By considering compositional elements in monumental art such as figural position, line, perspective, general composition, and dress motifs, Proskouriakoff succeeded in correlating variations in the figural art of monuments with the periods of their fabrication. For the purposes of my own investigation, Proskouriakoff's analytical approach was viewed as a useful way of detecting trends in the way that individuals are depicted in monumental art subject to variations in spatial context. Iconographic and compositional elements considered in my own study included the position of individuals portrayed, perspective, furniture, general composition, as well as costume.

In *Classic Maya Attire as Indicators of Status, Role and Function* (1979), Clarkson sought to discern how costume functioned to signal status and ritual among the Classic Maya. Drawing on iconographic depictions of costume and associated accoutrements appearing on ceramics, stone, wood, bone, and shell, her analysis of artworks identified seven main themes: Human Sacrifice, Purification, Accession, Ballgame, Deer Hunting, Warfare, and Confrontation. Additional themes derived from my own study of stone sculpture undertaken in 2000 (see Parmington 2000, 2003) included Auto Sacrifice, Captive Display, Dressed for War, Ritual Dance, Royal Audience, and Solo Portraiture. Clarkson's study concluded that the status of individuals portrayed could be determined only in relative terms by assessing differences in the degree of costume ornamentation and its quality. My more recent investigation reaffirmed conclusions arrived at by Clarkson – that Maya dress was used to signal gender, relative status, and the ritual role of the wearer. Points to consider when attempting to establish relative status of individuals depicted together are: (1) the degree of embellishment in costumes; (2) the degree of embellishment in associated paraphernalia; (3) the number of costume elements; (4) furniture (e.g., the presence of thrones, benches, etc.); (5) figure orientation (whether individuals are depicted frontally or in profile); (6) elevation (e.g., whether individuals are portrayed crouching, seated, standing, or elevated); and (7) posture (whether persons portrayed are depicted in assertive or submissive pose). Within the context of this investigation, assessment of costumes following these criterion is seen as a valuable method to determine the status and behaviour of individuals portrayed within specific spatial contexts.

Before proceeding, it should be noted that arguments presented follow the view that elements portrayed in Maya art are "naturalistic images of real objects" (Tate 1992: 50) and that all terminology presented in the discussion of iconography, costume, sculpture, and architecture is in accordance with prior research on the subjects (Proskouriakoff 1950; Pollock 1965; Cordry and Cordry 1968; Andrews 1975; Clarkson 1979; Anawalt 1981; Taylor 1983; Schele 1984; Schele and Miller 1986; Tate 1992; Schele and Mathews 1998).

## 2.4.7. Some categories for thematic analytical investigation

Maya sculpture can be ranked according to its location on the access diagrams and interpreted in terms of its public and private accessibility. Discerning patterns in the distribution of artistic themes, however, requires dealing with the variability present in Maya artwork. Twenty-two themes were defined which broadly characterise the iconographic content of palace and temple sculpture at Palenque (see Chapters 4 and 5). Expanding on the sculptural themes outlined by Clarkson (1979) and myself (Parmington 2000, 2003), these themes (which appear singularly and in various combinations) are as follows: (1) Accession; (2) Ancestor Portraiture Solo; (3) Ancestor Portraiture with Attendants; (4) Captive Display; (5) Text, Captive Display; (6) Cosmological Underworld; (7) Cosmological Terrestrial; (8) Cosmological Celestial; (9) God Portraiture Solo; (10) God Portraiture with Attendants; (11) Human Sacrifice; (12) Auto Sacrifice; (13) Mat Motif; (14) Portraiture Solo; (15) Sacrificial Dance; (16) Text, Cosmological; (17) Text, Historical; (18) Text, Mythological; (19) Text, Dedication; (20) Royal Audience; (21) Ritual Dance; and (22) Royal Portraiture Solo. The approach used to thematically sort sculpture is clarified in the following sections in the descriptions of five of the principal themes: For more detailed examples of the remaining themes, see descriptions of sculpture and other artworks in Chapter 4, Architectural and Sculptural Programs of the Palenque Palace Group. (Note: Three of the sculptural themes outlined in the following sections [i.e., Royal Audience, Ritual Dance, and the sacrificial themes are consistent with descriptions I presented in previous research [see Parmington 2000, 2003]).

### Captive display

Capture iconography in monumental art generally falls into two main categories, each one relating to different stages in the ritualised processing of prisoners. One way that captives are portrayed in Maya art is at the moment of their capture in battle; the other is as victims in subsequent ceremonies (Schele 1984: 17). The glyph for capture, which may or may not accompany such scenes, reads *chukaj* ("is captured"; Miller and Houston 1987: 49–50). Generally, the act of capture is implied in Maya iconography rather than represented directly (Schele 1984: 9), where captives are portrayed after the fact either seated or kneeling at the feet of their captors. The clothing of captives usually comprised simply a plain loin- or hip-cloth; in addition, a cloth may have been draped over the forearm of the captive (see Figure 2.18). There are examples in which captives are portrayed wearing war costume, although most often they are depicted stripped of all symbols of authority. Carvings found on stelae suggest that captives were used as seats or pedestals, as a form of public humiliation (Schele 1984: 16). Often captives are portrayed bound by rope or cloth; in addition, paper or cloth is sometimes portrayed pulled through the victim's ear perforations marking them for sacrifice (P. Mathews 2000: personal communication).

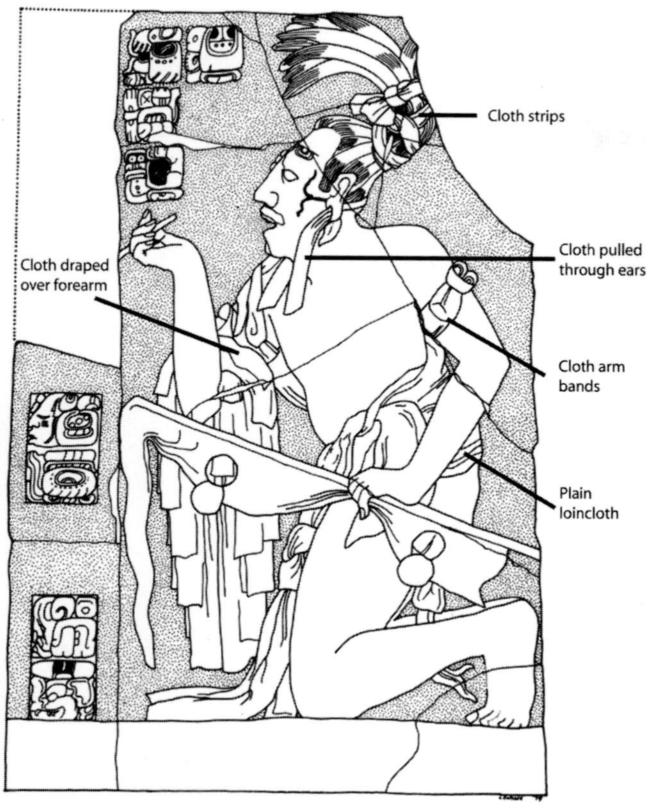


**Figure 2.18.** Tablet of the Scribe (GR 75). Sculptural Theme Represented: Captive Display. Drawing by Linda Schele (No. 147), © David Schele, courtesy of the Foundation for the Advancement of Mesoamerican Studies, Inc., www.famsi.org.

**Human sacrifice and auto sacrifice**

Two types of sacrifice are portrayed in Classic Maya art: Auto Sacrifice (bloodletting) as well as ritualised execution. By convention, sacrificial themes appearing on monuments and pottery portray the higher status individual undertaking the activity, where scenes depict the moment just before or just after the act has taken place. Some of the clearest examples of both fatal and nonfatal forms of sacrifice appear on the Bonampak Murals (see Schele 1984: 11).

In Maya art it is costume and associated accoutrements rather than the act itself that signal the act of Auto Sacrifice. Elements of dress and other paraphernalia that mark the activity of bloodletting generally include one or more of the following: knotted cloth (i.e., blood knots); the God C fret apron; bundles; bowls; deified perforators; thorned rope; bloodstained cloth or paper, in addition to godly manifestations, such as "vision serpents" (see Clarkson 1979: 51, 80; Tate 1992: 56; Miller and Taube 1993: 47; Parmington 2000: 67–69). As noted by Diego de Landa, the act of bloodletting has been documented historically among descendants of the Classic Maya:

> They offered sacrifices of their own blood, sometimes cutting themselves around in pieces and they left them in this way as a sign. Other times they pierced their cheeks, and other their lower lips. Sometimes they scarify certain parts of their bodies, at other times they pierced their tongues in a slanting direction from side to side and passed bits of straw through the holes with horrible suffering; others slit the superfluous parts of their virile member leaving it as they did their ears, on account of which the general historian of the Indies was deceived saying that they practiced circumcision (de Landa, in Tozzer 1941: 113).

As with Auto Sacrifice, Classic Period monuments rarely portray the act of human sacrifice; rather, it is pre- and postsacrificial activity that was generally portrayed (see Figure 2.18 for an example sacrificial representation in Maya art). Diego de Landa (in Tozzer 1941) described human sacrifice as a highly ritualised activity during which specific costumes were worn and specific forms of torture were inflicted on the victim. Among the postcontact Maya, various methods of sacrifice included death by spear, disembowelment, heart incision, axe, or arrow. According to Diego de Landa:

> And then the day arrived, they came together in the court of the temple, and if the victim was to be sacrificed with arrows, they stripped him naked, and anointed his body with a blue colour, and put a coroza on his head . . . And making a certain sign to the dancers they began one after the other to shoot . . . And in this way they make his whole chest one point like a hedgehog of arrows (de Landa, in Tozzer 1941: 117–118).

**Ritual dance**

Scenes in monumental art, categorised as ritual dance, were determined through the identification of the verb "to dance" (as observed in accompanying hieroglyphic texts) or through the interpretation of iconographic elements in scenes implying this activity (Parmington 2000: 54–55). For example, the tilted posture and raised heels of individuals as depicted imply dance in Maya iconography (Grube 1992: 201). Artists portrayed dancing on pottery as well as on stone monuments and mural art. For the Maya, dance was a means of transformation and communication; by the very act of dancing, kings and nobles believed that they transformed into gods and supernatural beings. It was through the act of dance that participants transformed themselves into their "soul companions" (*wayob*), transforming into creatures such as birds and the jaguar as represented in costume (Freidel and Schele 1993: 260–270).

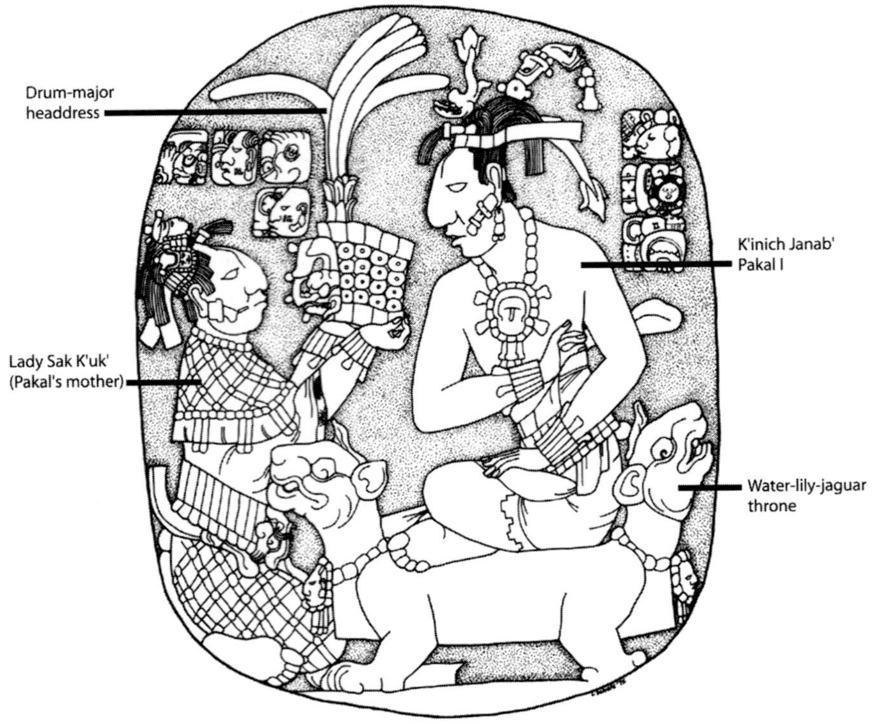


**Figure 2.19.** House E Oval Tablet (GR 5). Sculptural Theme Represented: Royal Audience (Accession). Drawing by Linda Schele (No. 143), © David Schele, courtesy of the Foundation for the Advancement of Mesoamerican Studies, Inc., www.famsi.org.

**Royal audience**

Scenes on monuments categorised as royal audience "generally depict palace scenes and formal gatherings. The context has been defined so as to include scenes depicting two or more people and where the status of individuals portrayed can be identified as subordinate in the company of a ruler as indicated by title glyphs" (Parmington 2000: 52). Scenes often depict individuals paying homage or making offerings to a ruler protagonist. Scenes may also depict the offering of tribute. Typically, the ruler (who is most often frontally oriented) is the focus of the composition, whereas attendants are depicted standing or seated in submissive poses undergoing various subordinate activities (Parmington 2000: 52–53) (see Figure 2.19).

## 2.5. Description of primary data sources

All arguments presented in the following chapters are drawn from the analysis of residual sculpture preserved in situ – more specifically, from imagery adhered to architecture, minimising the likelihood that an artwork has been repositioned. It is acknowledged that the absence of looted monuments and those that may have been repositioned in antiquity may affect some of the arguments presented. Although interpretive problems may exist, however, structures like the Palace Group at Palenque display a "wealth of iconographic information" (Miller 1988: 154). As Fash (1991: 26) proposed for Copán, "elaborate and distinctive sculpture adornments [appearing] on . . . structures indicate that they all had discrete meanings, and possibly different uses, in ancient times." Overall, this inquiry has examined indoor and outdoor sculpture, the sculpture of entrances, sculpture and its depth within structures, and has considered the sculpture of stairways and elevated architecture. Data for the investigation were obtained primarily from photographs and drawings. The most important iconographic sources were friezes, murals, piers, jambs, panels, roofcombs, benches, stairways, lintels, altars, and thrones. The next section provides general definitions and descriptions of the various iconographic media investigated.

### 2.5.1. Architectural sculpture

**Façades**

The term façade, applying the broadest possible definition, is the front of a building. A number of elements comprise building façades in Maya architecture, all of which may carry sculpture in one form or another; some

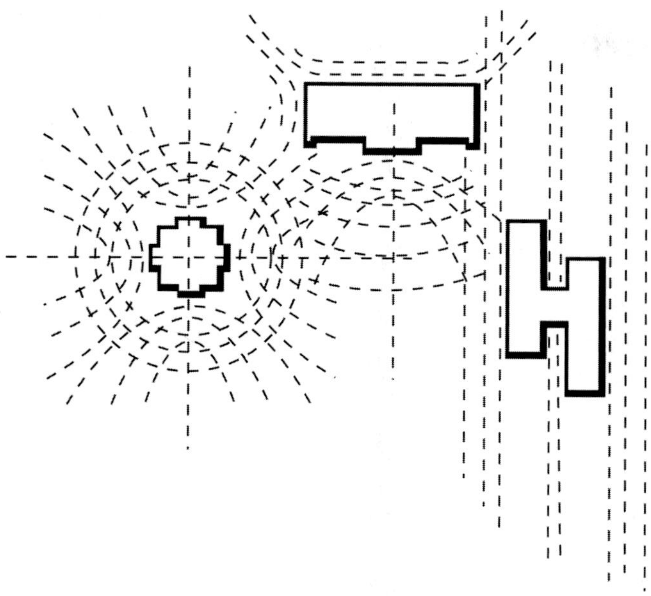

**Figure 2.20.** Radiance Fields and Architecture. Pierre von Meiss, *De la forme au lieu*, 1986, Presses polytechniques et universitaires romandes (page 108, fig. 155). Reproduced with publisher's authorization. All rights reserved.

of the features include substructure terraces, vertical bearing walls, piers, entablatures, and roofcombs. In Classic Maya architecture, as in most cultures both past and present, the façade was usually the most elaborately decorated part of a building. The reason for this is that the façade represents the "face," or main front of the building marked by the presence of a formal entrance (Schele 1998: 479). There is an "underlying anthropomorphic significance [to the façade]: the façade, the face, the human face – the face of the building" (von Meiss 1990: 94). For Baroque architects, the façade functioned well beyond the reality of the enclosed space that it preceded. In an attempt to control the space in front of the building, architectural elements were added to structures to enhance the effect of the structures. For example, church façades were extended beyond the height of the main structure to more effectively radiate their governance over the surrounding space. Similarly, the Maya added roofcombs to buildings and pyramidal structures to enhance the visual power of the structures. It remains difficult to define the boundaries or the extent of the radiance projected by a façade. For this reason, a space may be bounded by other structures to delineate the limits of its radiance (see Figure 2.20). The effect is to establish a restriction underpinning the authority of the composition (von Meiss 1990: 94–96).

One of the roles of building façades among the Maya was the public display of political and religious information (Schele 1998: 479). The most common form of architectural sculpture found in the Maya lowland is the "mask façade." The imagery displayed on the mask façade often comprises several portraits with a central or main mask depicting either a ruler or a zoomorphic or anthropomorphic supernatural being. Whether it was a ruler or a supernatural being, by convention, the main portrait was adorned with symbolic accoutrements; among these were flowers and maize foliation, knots and mat signs, mirrors, headbands, and earflares. Earflares and headbands often adorned the main individual in the portrait, identifying the character or person portrayed (Schele 1998: 481).

The information transmitted by a building façade may vary depending on the proximity of the observer. For example, larger sculptural or iconographic elements may be designed to transmit information over a greater distance, whilst smaller sculptural features may become visible only on approaching the structure or entering the building. Often the intention of the builder in the construction of certain building components was to draw the visitor's attention to the function of the building. For Andean cultures, façades (combined with other forms of architectural adornment) may have been used to communicate prosaic information, perhaps drawing the viewer's attention to the purpose of the building and providing clues on how to navigate through structures (Moore 1992: 96).

#### Friezes

A frieze is a broad strip of decoration running along the wall on the outside of a building between the roof and the ceiling. An alternative definition would be a wide horizontal band of sculpture filling in the portion of entablature running between the architrave and the cornice. Kubler (1985) defines a frieze as "a band of decoration composed of repeating elements" (Kubler 1985: 258), which functioned to draw the viewer's attention to the structure and to reiterate its message through the repetition of its design. In Maya architecture, generally the most decorative part of a frieze is the area above the main entrances to buildings. Thematically, the content of this type of carving or sculpture is most often cosmological in nature, displaying gods, supernatural beings, as well as rulers.

#### Roofcombs

One of the functions of the roofcomb in Maya architecture was to provide added height to structures as a means of increasing their overall grandeur. "The roofcomb gives height to what otherwise would be very squat monuments" (Willey 1980: 415). Roofcombs were sometimes positioned along the midline of structures as seen at Palenque, or positioned to stand vertically at the rear of buildings as at Tikal. They were constructed either of solid masonry, or in a lattice configuration; the latter case was a means of minimising their weight. Called "flying façades," some examples were built to line up with the front of the building (Sharer 1994: 637). The lines of the roofcomb had the effect of drawing the viewer's gaze up. Because of their size and elevation, some of the more massive roofcombs are visible from great distances with their large front and rear surfaces making the structure stand out amidst other architecture as well as the surrounding environment. The best extant examples of Maya roofcombs include those found crowning the Temple of Inscriptions and the Temple of the Sun at Palenque; the K-5 Pyramid at Piedras Negras; and Temples I, II, IV, and VI at Tikal. Maya roofcombs were often heavily decorated with stucco or mosaic sculpture. Displaying text and imagery, the roofcomb was a characteristic part of Classic Maya architecture throughout the Maya lowlands; however, there is a distinct bias in the surface areas occupied by iconography.

**Murals**

Murals are decoratively painted walls that typically appear in the interior of buildings, such as the Bonampak Murals (see Miller 1986). Since their discovery, the murals in Structure 1 at Bonampak have been an important source for the reconstruction of social structure and societal behaviour among the Classic Maya. The Bonampak Murals represent the most complete pictorial record of the progression of rituals associated with bloodletting and sacrifice of individuals (Schele 1984: 11). Other activities portrayed in the mural scenes include ritual dance, warfare and capture. On Wall 4 of Room 2, two captives obtained in a raid appear to be involved in a judgement ritual where torture and decapitation are represented as part of the ceremony. The murals provide evidence of activities once associated with Structure 1, indicating that the staircases were used as backdrops for ritual performances. Room 3 appears to depict the climax of festivities, where a large dance is taking place against the façade of the building (see Miller 1986).

**Piers and Doorjambs**

Piers are the supporting pillars of buildings and are most often located between windows and bounding the entrances to structures. In Classic Maya architecture, piers were often decorated with iconography and passages of text. Some of the best examples of pier sculpture appear on the Palace houses at Palenque, where rulers are portrayed in formal garb in the company of subordinate individuals (see Chapter 4, Section 4.3.3, The Buildings of the Palace Phase 3, Houses A-D: Description and Sculpture). Maya pier sculpture was positioned internationally to be in public view, to be interpreted by the passersby or those attempting to enter corresponding structures. In contrast, text and imagery appearing on doorjambs is less visible because one has to pass through the entrance to view the sculpture.

**Lintels**

Maya lintels are supporting slabs of stone located in the undersides of doorways; often carved, they represent one of the more restricted forms of monumental sculpture associated with Maya architecture. Embellished lintels are most often associated with palaces and temples, where their more restricted positioning limits exposure to large audiences. The purpose of Classic Maya lintels was to record events and to document lineage, titles, and conquests of individuals (Marcus 1992a: 81–84). In addition, these more private forms of event-recording also depict appeals to deceased royal ancestors in the form of bloodletting rituals, meetings among elites, and the taking of captives and sacrificial victims (Parmington 2000: 3–4). Some of the finest examples of carved lintels are found associated with structures at Yaxchilan (see Tate 1992).

**Panels and tablets**

Panels and tablets are flat stone carvings attached to walls and pillars located in the interior of buildings. Most often positioned above the surrounding floor space, they appear carved in either high or low relief or sculptured from stucco. Maya panels and tablets display imagery as well as long passages of text documenting the achievements of rulers during their respective reigns. Some of the best examples of Maya wall panels appear in the Temple of Inscriptions and the Temples of the Cross Group at Palenque (see Greene Robertson 1984, 1991).

**Staircases**

Compared to wall panels, hieroglyphic staircases are located in more accessible areas; subsequently they were "clearly intended for public (vertical) propaganda and wall panels [were] intended for less public (horizontal) propaganda" (Marcus 1992a: 84). The finest example of a hieroglyphic staircase in the Maya region is associated with Temple 26 at the site of Copán, in Honduras (see Fash 1991). As Houston (1998) remarks, "monumental stairways . . . do many things beyond allowing people to move up and down a building or platform. Iconographic and textual data show that stairways . . . display tribute, booty, and captives. By sitting at various levels, the Maya establish or confirm relative social ranking" (Houston 1998: 522).

### 2.5.2. Furniture

**Thrones and benches**

Thrones and benches functioned as seats of power that were used by rulers throughout Mesoamerica as early as 1200 B.C. until the Spanish conquest. Late Classic thrones appearing at sites such as Palenque, Yaxchilan, and Piedras Negras were intricately carved and had characteristically downward tapering, wedge-shaped legs. Depictions of thrones found in Maya sculptures and paintings reveal that they were often painted and draped with animal pelts and cushions. In Maya art, both gods and rulers are portrayed using thrones. Rulers are often seen reclining on cushions made from jaguar skins (Miller and Taube 1993: 165–166). It is likely that not all benches functioned as thrones or were used as the focal points for political gatherings. It remains probable that some benches were used as beds, workbenches, seats, and altars as well as shelves for storage (Christie 2003a: 329). When attempting to define the function of a bench, it is important to consider the spatial context in addition to the type and degree of embellishment present over its surfaces. For example, the presence of a bench in a confined space may indicate a more private function than that of a bench positioned adjacent to an open courtyard. A bench positioned adjacent to a courtyard or plaza signals a more public use, most likely functioning as a seat of power used by rulers for the receiving of guests or for the overseeing of ritual proceedings.

3

# Access analysis of the Palenque Cross Group and its sculpture

Applying the methods discussed in Chapter 2, my goal in Chapter 3 is to demonstrate, using a simple example, how "access analysis" of architecture and associated sculpture can be used to distinguish societal boundaries in Classic Maya city centres, before proceeding to more complicated architecture and methodological procedures in later chapters. The building group focused on in this chapter is the Cross Group at Palenque. Following a brief introduction to the site of Palenque, a general description of the buildings and sculpture that comprise the Cross Group will be presented in accordance with Schele and Freidel (1990: 237–261), Greene Robertson (1991: 9–54), and Stuart (2006: 86–190). Following the introductory section, the results of a spatial analysis of the Cross Group temples and associated sculpture will be presented.

Situated in the southern Mexican state of Chiapas, the site of Palenque is positioned in the foothills of the northern slope of the Sierra de las Naranjas, on the southwest periphery of the Maya region (Maudslay 1889–1902b: 8; Sharer 1994: 276). Placed strategically on a high natural platform overlooking low fertile plains (Andrews 1975: 168), Palenque's core measures approximately 2.2 $km^2$. The site's settlement density is estimated to have once been the second highest in the Maya area, averaging 673 structures per square kilometre (Copán is purported to have had the highest density at 1,449 structures in a 0.6 $km^2$ area) (Barnhart 2001: 73). At its peak, population estimates have placed the occupation of Palenque's core during the Late Classic Period at between 4,147 and 6,220 people (Barnhart 2001: 76; see also Rice and Culbert 1990).

The undulating topography of the core of the site severely hampered attempts to impose architectural order on the overall plan of the city. Consequently, the layout of Palenque is described as highly organic, suggesting that the siting of many of the structures was determined opportunistically rather than through the imposition of "abstract or formal notions of order" (Andrews 1975: 168). It appears that, although the landscape at Palenque was altered to accommodate buildings, it was never drastically modified to conform to more regimented architectural configurations as expressed in northern Maya lowlands sites where cities were constructed on flatter terrain.

Without a doubt, the most famous structures at Palenque are the Palace, the Temple of Inscriptions, and the Temples of the Cross Group, structures that collectively take pride of place at the core of the site. Buildings such as these inspired many of the early explorations at Palenque, and later prompted more formal investigations of the site by scholars. Explorer Ramon de Ordoñez y Aguilar was the first to report on Palenque; his investigations led to further surveying expeditions during the late eighteenth and early nineteenth centuries (Maudslay 1889–1902b: 7). In 1832, Waldeck undertook an extensive, two-year survey of the ruins; his survey resulted in many architectural plans and drawings of the indigenous people of the region. After Waldeck's visit, in 1840, Stephens and Catherwood arrived to further investigate and document the Palenque ruins, of which some limited discussion was included in the famous book entitled *Incidents of Travel in Central America, Chiapas, and Yucatan*. Today, Alfred Maudslay's late-nineteenth-century drawings and photographs remain important sources for modern archaeologists and epigraphers (see Maudslay 1889–1902a, 1889–1902b, and 1889–1902c). Eminent twentieth-century scholars who have worked at Palenque include Eduard Seler, Sylvanus Morley, and Frans Blom; more recent scholars include Floyd Lounsbury, David Kelley, Peter Mathews, and Linda Schele (Sharer 1994: 278).

Figure 3.1 portrays the core area of Palenque, which functioned as the centre of ritual and administrative

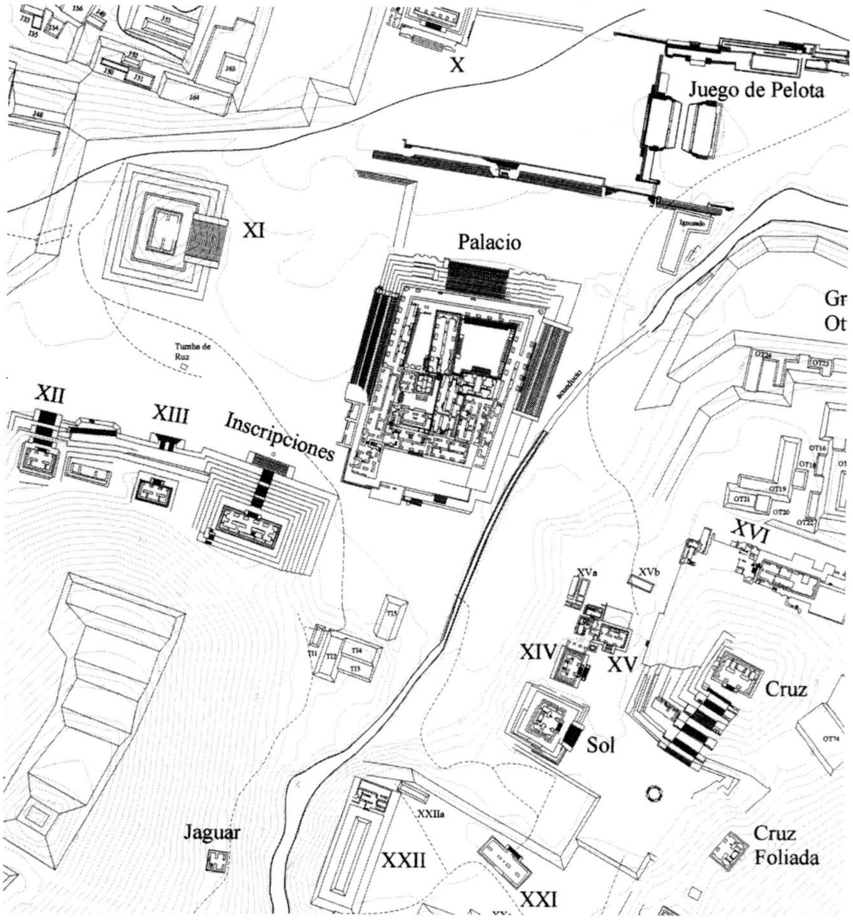


**Figure 3.1.** Map of the Central Precinct at Palenque. Image from Urbanism at Palenque (2002), by Edwin Barnhart, in *The PARI Journal* 4(1):10–16, map detail reprinted with the permission of Edwin Barnhart.

activity. Positioned centrally on the map detail is the Palace Group, an architectural feature that "occupies a dominant position in the Main Plaza" (Andrews 1975: 168). To the northeast of the Palace is the ballcourt, to the southwest is the Temple of Inscriptions (the great funerary monument K'inich Janab' Pakal I), and to the southeast are the Cross Group temples constructed by K'inich Kan B'alam II, son and successor of K'inich Janab' Pakal I.

## 3.1. The Cross Group general description

Positioned on a high platform, east of the Otolum River, the Temple of Inscriptions, and the Palace, is the Palenque Cross Group. The form and layout of these three structures accords with Andrews's (1975: 56–59) definition of a Temple Group archetype (see Figures 2.6 and 2.7), where three temples are configured around three sides of a square or rectangular plaza (see Figure 3.2). The Temple of the Cross is positioned on the north side of the plaza, the Temple of the Sun on the southwest side, and the Temple of the Foliated Cross on the southeast side (Sharer 1994: 284). All three of the temple entrances face inward towards an altar situated in the plaza at the centre of the group to which access is provided from the south (Schele and Freidel 1990: 237).

The configuration of the Cross Group is hierarchical, the Temple of the Cross representing the focal point of the architectural composition, whereas the Temples of the Sun and Foliated Cross hold peripheral positions (Cohodas 1973: 218). Resembling each other closely in their general layout and configuration, the surviving entablature of the Temple of the Cross portrays sculptured *witz* monsters, identifying the temples that form this triadic group as "sacred mountains" (Schele and Freidel 1990: 239; Sharer 1994: 284). In all three buildings, inscriptions appearing on the inner jambs give the names of the temples: the Temple of the Cross is referred to as *wak ? chan*, the Temple of the Sun as *mah k'ina kab*, and the Temple of the Foliated Cross as *na te k'an*. As suggested by their names, the Temple of the Cross had celestial associations (Stuart 2006: 110), the Temple of the Sun was connected with death and the setting sun, and the Temple of the Foliated Cross is thought to be associated with the earthly realm (Sharer 1994: 287; see also Schele and Freidel 1990: 237–252).

Once flanking each of the temple's three doorways were piers displaying hieroglyphic texts, with the two inner piers on either side of the temple's central entrances once lavishly decorated with modeled stucco. The function of the sculpture was "to carry messages to the court space in front of it" (Schele and Mathews 1998: 31). Standing independently within each of the Cross Group temples are sanctuaries that functioned as "symbolic sweatbaths, places of purification" (Martin and Grube 2000: 169). Referred to as "the *pib na* (the 'underworld house' within the 'sacred mountain')" (Sharer 1994: 284), each of the Cross Group sanctuaries houses a finely sculptured tablet adhered to the rear wall of the spaces, "each composition representing one of the three paths to *Xibalba*" (Schele and Freidel 1990: 239). The exterior of the sanctuaries also once displayed varying proportions

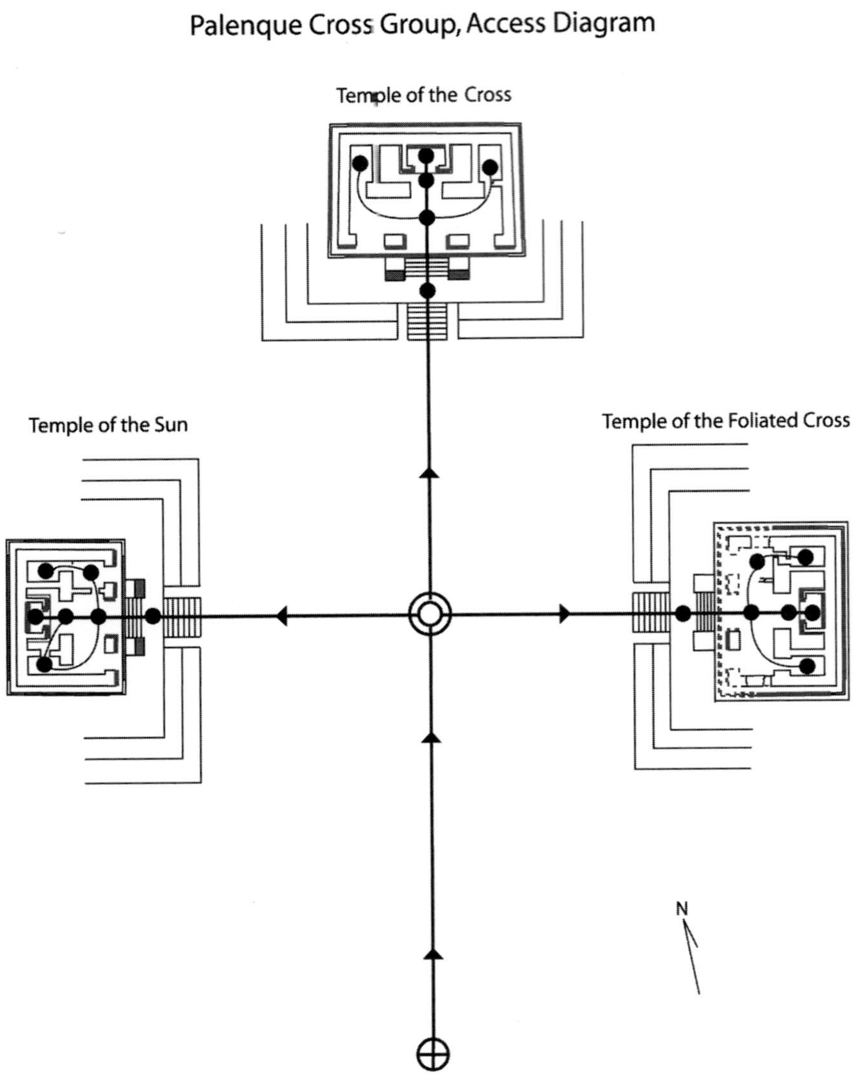


**Figure 3.2.** Palenque Cross Group Plan and Access Diagram.

of text and sculpture. The restricted nature of the sanctuary tablets suggests that they were to be read only by elite individuals. "Although the temples themselves were built for the populace as a whole to view, the inner sanctuaries were private art meant to be viewed by only a privileged few, the Palenque elite" (Greene Robertson 1991: 10).

Designed to be read like the pages of a book, the Cross Group tablets record historical and mythological events in Palenque's past, using both text and image to do so (see Greene Robertson 1991: 12–13 and Stuart 2006 for alternate reading orders). For example, the text on the left side of the Tablet of the Cross records mythological events, such as the births of the "First Mother" (3121 B.C.) and "First Father" (3122 B.C.), which occurred before the current world age. In addition, the text refers to the birth of one of the three gods who comprise the Palenque Triad (GI, GII, GIII): The Temple of the Cross records the birth of GI (God 1) in 2360 B.C., a god also referred to as *Hunahpu* or the Venus Deity; the birth of GIII (God 3), also known as the Jaguar Sun God or *Ajaw K'in*, is recorded in the Temple of the Sun as occurring four days later; and the birth of GII (or *Unen K'awiil* [Stuart 2006: 88]), also referred to as *Bolon Tza'cab* or God K, is recorded in the Temple of the Foliated Cross sixteen days later (Sharer 1994: 288). GI and GIII represent the Hero Twins in Maya origin mythology as described in the *Popol Wuj;* they were born to the First Mother Goddess early in the current world age. The Hero Twins' antics established the sky and earth realm for human occupation, magically defeating the lords of the Underworld and giving rise to the human race through the planting of maize. According to Maya mythology, GI manifested as Venus and/or the sun whereas GIII was represented by the moon (see Greene Robertson 1991: 12–19).

Rulers dressed to impersonate GI, GII, and GIII, as the Palenque Triad were believed by the ancient Maya to be the patron gods of Palenque. Today it is thought that GII (God K), as well as being the patron deity of the Temple of the Foliated Cross, also represented the patron deity of the royal lineage. During the Classic Period, the Maya claimed to be direct descendants of these gods, and this blood tie was used to justify their authoritative status (Tate 1992: 50). Costume elements identifying GI impersonation include the shell earplug, the cross-banded diadem, and the quadripartite headdress. The main costume element indicating GIII impersonation is the water lily–jaguar headdress. Busts of GIII can be identified by the presence of a jaguar ear often allied with sky-bands (Tate 1992: 52–55). GI was a celestial god of royal ancestry, exhibiting aquatic attributes and those symbolic of the rising sun. GII was the embodiment of royal power and agricultural fertility as symbolised in new growth, maize, and fresh water. GIII was a solar deity and was connected with warfare and the underworld (Stuart 2006: 88).

K'inich Kan B'alam II's objective in constructing the Cross Group was "to assert the ancient and pristine quality of his legitimacy" (Schele and Freidel 1990: 238). In addition to mythological events, the tablets of the

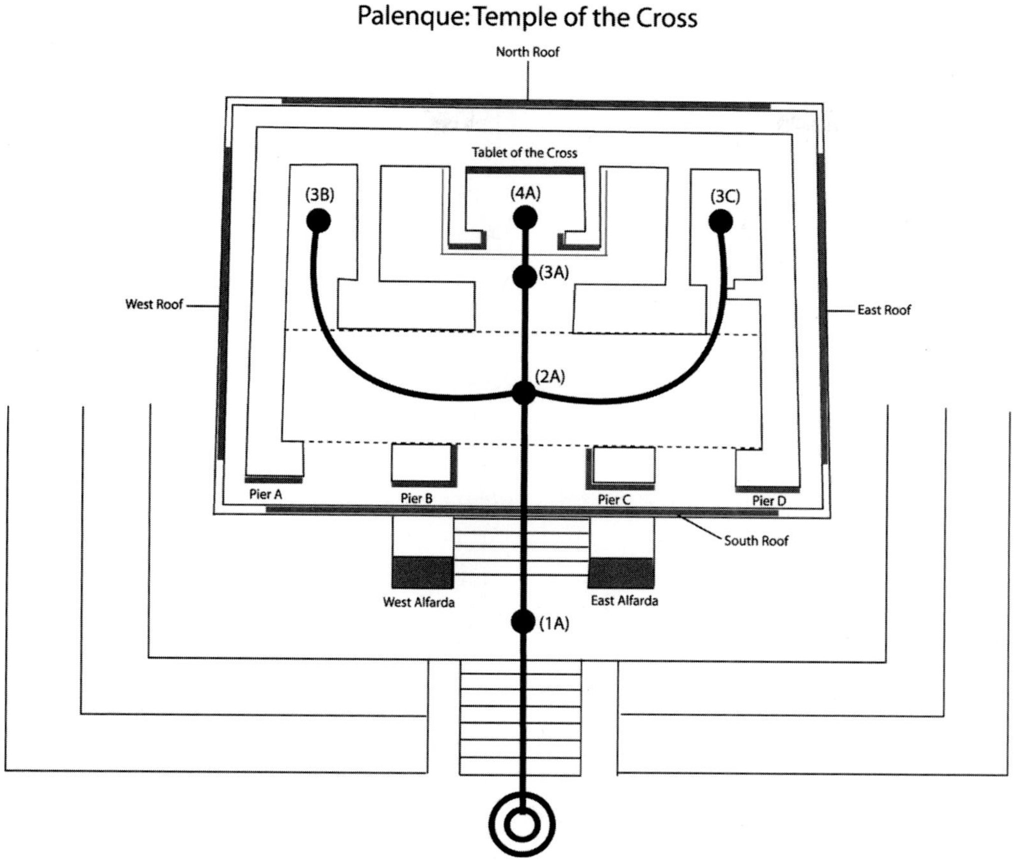


**Figure 3.3.** Plan and Access Diagram, Temple of the Cross.

Cross Group record historical episodes in the reign of K'inich Kan B'alam II, including the transfer of power from K'inich Janab' Pakal I on the date 9.12.11.12.10 8 Ok K'ayab'(A.D. 683). The Cross Group tablets also record rites associated with the dedication of the temples, beginning in A.D. 690 (9.12.18.5.1) and ending in A.D. 692 (9.12.19.14.12) (Schele and Freidel 1990: 242; Sharer 1994: 288). The dedicatory date of the Cross Group has been identified as 9.13.0.0.0 (A.D. 692), a date that falls nine years after K'inich Janab' Pakal I's death (see Appendix 1, The Palenque Cross Group Dates).

The Cross Group was built by K'inich Kan B'alam II, who reigned from 7 January 684 A.D. (9.12.11.12.10 8 Ok 3 K'ayab') until his death on 16 February 702 A.D. (9.13.10.1.5 6 Chikchan 3 Pop) and is believed to have taken no longer than nine years to build. This proposal is supported by the view that K'inich Kan B'alam II would not have commenced his own memorial before his father's death on 28 August 683 A.D. (9.12.11.5.18 6 Etz'nab' 11 Yax). Evidence of the authoritative capacity of K'inich Kan B'alam II is implied by the speedy construction of the Cross Group. Needless to say, for these structures to have been completed within nine years, the project would have required a vast, well-coordinated workforce (Greene Robertson 1991: 9–10).

### 3.1.1. Temple of the Cross

On entering the Cross Group plaza one is first confronted by the façade and massive stairway of the Temple of the Cross. The Temple of the Cross is the largest of the three pyramidal structures that collectively comprise the Cross Group, displaying the remains of what was probably the tallest roofcomb in Palenque (Cohodas 1973: 218). The superstructure measures 15.3 metres across, and 9.45 metres from the front to the back. The temple's interior is broken up into a series of small rooms. The front half of the structure comprises a single space fronted by piers that were once elaborately decorated, whereas the rear half of the structure is divided into three rooms, one of which houses a freestanding sanctuary containing a base-relief tablet referred to as the Tablet of the Cross. The two smaller peripheral rooms measure 200 cm wide, and the central space that contains the inner sanctuary measures 810 cm × 320 cm. The external walls of the temple are 90 cm thick, and the width of the interior medial wall measures 127 cm wide; keyhole openings are present in the central wall of the structure (Greene Robertson 1991: 22–24) (see Figure 3.3).

The Temple of the Cross is the most elevated of all the Cross Group temples (Schele and Freidel 1990: 237);

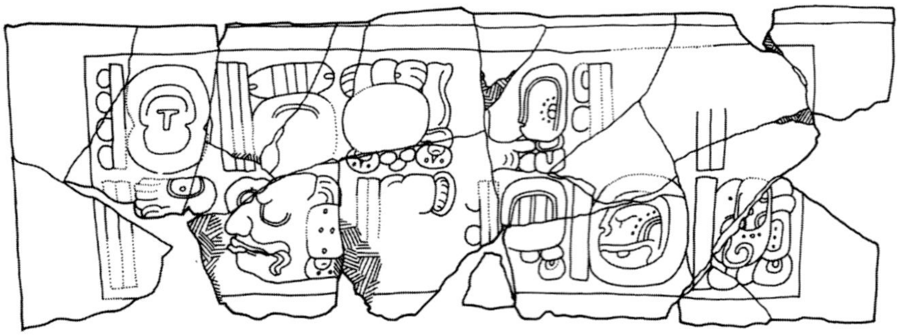

**Figure 3.4.** The Temple of the Cross, West *Alfarda*. Drawing by Linda Schele (No. 197), © David Schele, courtesy of the Foundation for the Advancement of Mesoamerican Studies, Inc., www.famsi.org.

entering the Temple of the Cross necessitates climbing seventy-two stairs fronting a large pyramidal substructure and stepping onto a narrow platform that surrounds the summit. It is from here that access is provided to the interior of the structure through piered entrances. Narrow doorways in the temple's medial wall provide access to the northeast and northwest rooms. On the interior wall of the northeast room are two elements resembling cord holders positioned 60 cm and 194 cm above the floor. In addition, several *Ik'* (T-shaped) openings positioned 257 cm and 255 cm are present in the walls of the structure; their elevation above the floor suggests that they probably provided ventilation to the interior of the building. Corbelled vaults span the inner and outer rooms; holes visible in these vaults indicate that once there were two parallel beams stretched across the upper portion of the temple's interior. Three large post holes, measuring 29 cm and 34 cm deep, were found on the inner north side of Pier A, and six post holes of the same dimensions were found associated with Pier B. Curtain holes have also been identified, indicating that the interior spaces were once cordoned off. These holes suggest that all the interior walls of the temple were once furnished with curtains (Greene Robertson 1991: 22–25).

### Temple of the Cross sculpture

Elaborate sculpture and bas-relief carvings once decorated the Temple of the Cross; today, little decorative stucco remains on the temple's exterior. Surviving sculpture associated with this structure include the Tablet of the Cross, elements found on the exterior roof of the temple's inner sanctuary, the east and west sanctuary jambs, as well as the *alfardas* flanking the upper temple stair. There is little or no sculpture surviving on the external piers of the temple.

All that survives of the roofcomb on the Temple of the Cross is the frame that once supported a mass of stucco sculpture. Remnants of sculpture suggest that an Underworld theme once adorned the roofcomb. Among the surviving sculptural elements are the remnants of an Underworld monster, a serpent scale border, a large cross-hatched fish, and the curved section of a serpent (Greene Robertson 1991: 37–38).

Flanking both sides of the upper stairs are the temple *alfardas*, both of which are entirely glyphic (see Figure 3.4). The *alfardas* record the birth of GI (i.e., "touching the earth") on 1.18.5.3.2 and K'inich Kan B'alam II's participation in the "house event" on 7.14.14.11.10, a day on which blood was let. Both of K'inich Kan B'alam II's parents are also named on the monument: K'inich Janab' Pakal I and Lady Tz'akb'u Ajaw (Schele and Mathews 1979: figs. 271 and 272; Schele and Freidel 1990: 246–247; Greene Robertson 1991: 38; Stuart 2006: 107).

The south sanctuary roof of the Temple of the Cross once displayed an Underworld sun monster as indicated by the remains of the creature's feathered wings. Sculptured on the upper border of the west sanctuary roof are the remnants of a cosmic monster, comprising a sky-band body terminating at each end with the head of a serpent. The cosmic monster combines both celestial and Underworld symbolism; whereas the forward head of the creature represents Venus in the heavens, the trailing head of the creature symbolises the Underworld aspect of the sun (Miller and Taube 1993: 45). Just below the sky-band is a quatrefoil cartouche resembling symbolic representations of cave portals in Maya iconography (see Miller and Taube 1993: 57). Inside the cartouche are the remains of a human portrait, perhaps representing an ancestor. The heads of snaggle-tooth serpents are visible emerging from either side of the cartouche framing the figure. Today, all that remains discernible of this sculptured feature are the chest and shoulders of the individual, in addition to hair ribbons and the remains of a beaded necklace adorning the figure's neck. The east sanctuary roof once mirrored the west sanctuary roof; today only hair ribbons remain visible on what is left of the figure sculptured on the east side of the sanctuary (Greene Robertson 1991: 35–36).

On the temple sanctuary's west jamb is a portrait of ruler K'inich Kan B'alam II impersonating ancestor U Kokan (?) Chan (Stuart 2006: 112); he is displaying symbols of rulership, auto-sacrifice, and the Underworld. Among the many costume elements worn by K'inich Kan B'alam II are a multitiered feathered headdress consisting of, among other things, a jawless "Long-lipped" God and stingray spine; the Long-lipped God displays a

Figure 3.5. Southwest Jamb, Temple of the Cross. Drawing by Linda Schele (No. 175), © David Schele, courtesy of the Foundation for the Advancement of Mesoamerican Studies, Inc., www.famsi.org.

netted brow, cross-bands, and jaguar spots. K'inich Kan B'alam II also wears a beaded necklace, shoulder piece, bat pectoral, spotted leg straps, wristlets, and earplugs. Hanging from K'inich Kan B'alam II's belt at the rear of his body is a representation of the Jaguar God of the Underworld. One arm of the Jaguar God is threaded through a *cauac* cartouche, symbolising the Underworld (see Greene Robertson 1991: 32–34). Portrayed in a profile stance, K'inich Kan B'alam II faces inward towards the sanctuary entrance holding an inverted Quadripartite God in his right hand. Comprising a shell, *k'in* sign, cross-bands, and a stingray spine, water or blood can also be seen falling from the brow of the god (Figure 3.5).

God L, one of the principal gods of the Underworld, is depicted in profile on the east sanctuary jamb of the Temple of the Cross (see Figure 3.6). The god is portrayed smoking a pipe and, like K'inich Kan B'alam II, faces the sanctuary entrance. Resembling a toothless old man, this deity reinforces the view that the sanctuary entrance functioned as a portal to the Underworld (see Schele and Freidel 1990: 239, 241). In this representation, God L wears a hipcloth that has a pleated and feathered edge as well as *cauac* symbol. Over the hipcloth, God L wears a cross-banded belt similar to that which is worn by K'inich Kan B'alam II, but with a *quincunx* as its fastener. God L also wears a scarf displaying a shell-and-mat symbol, in addition to an Underworld *muwan* bird headdress; the bird headdress has scroll eyes, a curved beak, and a *kimi* death sign on its cheek. There are Long-lipped Gods depicted in front, and behind God L they have bell assemblages dangling from their jaws and ribbons with mirrors attached to them (Greene Robertson 1991: 34–35). God L also has also been "associated with . . . tribute and scenes of presentation" (Stuart 2006: 112).

Figure 3.6. Southeast Jamb, Temple of the Cross. Drawing by Linda Schele (No. 176), © David Schele, courtesy of the Foundation for the Advancement of Mesoamerican Studies, Inc., www.famsi.org.

**Figure 3.7.** Centre Panel from the Tablet of the Cross. Drawing by Linda Schele (No. 170), © David Schele, courtesy of the Foundation for the Advancement of Mesoamerican Studies, Inc., www.famsi.org.

Compositionally, the tablets of the Cross Group can be divided into three main parts: mythological and historical texts and the imagery that separates them (Sharer 1994: 288). The text documents the creation of the house, the rebirth of GI, and the nine generations of rulers that proceeded "Palenque's religious foundation" (Stuart 2006: 116). (For complete transcriptions of the Tablet of the Cross text, see Schele and Freidel 1990: 246–247 and Stuart 2006: 117–131.) The theme of the imagery present at the centre of the Tablet of the Cross is principally cosmological (see Figure 3.7), where the central image combines symbolic elements of all three Maya realms (Heaven, Earth, and the Underworld) into an integrated whole. The main pictorial element on the Tablet of the Cross is a stylised tree, which is an iconic representation of the ceiba or "World Tree" (Schele and Freidel 1990: 242) still sacred among the Maya today (Sharer 1994: 33). Emerging from deep within the earth, the ceiba in Maya mythology separates the heavens and earth whilst simultaneously connecting them (Schele and Freidel 1990: 66–67). Perched on a three-petal floral disk at the top of the tree is the celestial bird, symbolising the heavens. Emerging from the branches of the World Tree are the heads of two jeweled serpents, which together resemble a "double-headed serpent bar" purported to be one of the primary symbols of kingship (Sharer 1994: 282). The jeweled serpents are portrayed fleshed, an indication that they dwell in the earthly realm. At the base of the image is a Quadripartite God, from which the ceiba tree emerges. The creature represents all three Maya realms, "he is of the Underworld and the Middle World and is the vehicle for connections to the Upper World, or the heavens" (Greene Robertson 1991: 26). The monster also represents the sun in its transition from the heavens to the Underworld (*Xibalba*). The god displays a stingray spine, a shell, cross-bands, and stylised serpents that have bone (or *ajaw*) elements attached to them, in addition to a skeletal lower jaw. The monster also has bone and maize foliation emerging from its earplugs. The sky-band at the base of the scene comprises symbols of the heavens, which include the moon, Venus and the sun (*k'in*).

On each side of the stylised tree, at the centre of the tablet, stands a figure. The shorter of the two figures (on the viewer's left) holds a "Scepter-God" staff (Perforator God); he stands opposite a larger K'inich Kan B'alam II whose arms extend in a gesture of offering. Perhaps representing a manifestation of K'inich Janab' Pakal I in *Xibalba* (Benson 1978: 45; Schele and Freidel 1990: 242), the shorter individual wears a twisted muffler, an apron, a chest-hugging garment, a necklace with streamers and rattlesnake segments bound with a mat symbol, earplugs, and a sevenfold-embroidered hat with tassels. The figure also wears cloth anklets and bracelets tied with cord like those worn by gods. He is depicted barefoot, standing on a skeletal God M (Nine Death), a deity of the Underworld. Overall, the costume worn by this individual is not common to the Maya region; it resembles more the style of Central Mexico (Greene Robertson 1991: 27–28).

The individual on the right of the composition has been identified as K'inich Kan B'alam II. He is portrayed scantily dressed in a hipcloth and a knee-length triangular apron. Other elements adorning K'inich Kan B'alam II include a large beaded necklace; anklets and bracelets made from long, tubular, and bone-shaped beads; as well as ornate pendulous earplugs. On his head he wears a cloth turban or wrap, from which feathers can be seen emerging at the back. A mat symbol also decorates his

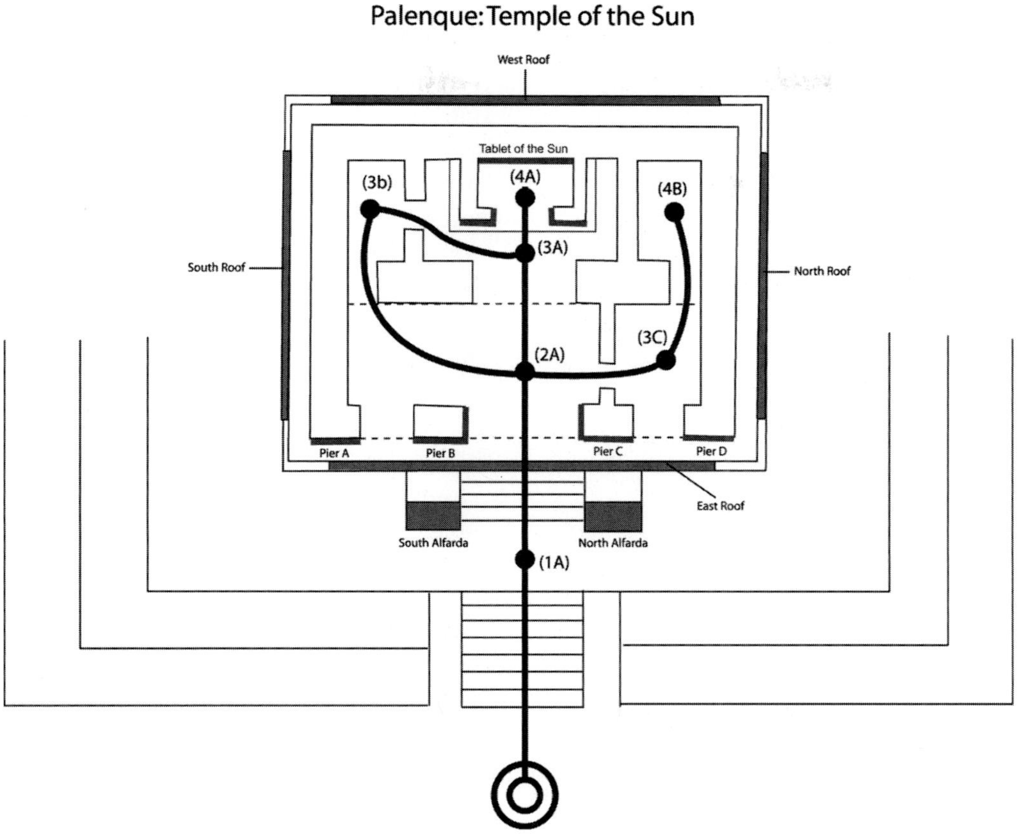


**Figure 3.8.** Plan and Access Diagram, Temple of the Sun.

hair, which is shaved at the back just below his head-dress, from which a pigtail pulled through a bead is visible emerging from just above the shaved area. In his hands, he holds a Jestor God seated on a cushion (Greene Robertson 1991: 29).

## 3.1.2. Temple of the Sun

The Temple of the Sun is the least elevated and best preserved of the three Cross Group temples. Supporting an intact roofcomb, the position of this building marks the western boundary of the Temple Group, blocking access to the Cross Group plaza from the direction of the Temple of the Inscriptions and the Palace complex (Greene Robertson 1991: 39). The floor plan of the Temple of the Sun resembles that of the Temple of the Cross and that of the Foliated Cross, where the interior of the building is divided into one large front space and three smaller rear spaces (Figure 3.8). The more-open front half of this temple was at one time divided into two rooms. Directly in front of the Temple's piered entrance is a narrow platform that measures 64 cm wide. Today, accessing the interior of the Temple of the Sun requires climbing twenty-nine stairs. Like the other Cross Group temples, the rear central room contains an inner sanctuary flanked by two smaller rooms. The small room at the northwest corner of the temple has a vaulted entrance; in addition, there is also a shelf in the southeast corner of the room. Cord holders are present under the eaves suggesting that banners may have once adorned the exterior of the structure (eighteen on the south side, twenty-one on the west side, more than sixteen on the north side, and twenty-two to twenty-six on the east side). In addition, cord holders present over entrances suggest that they were probably once cordoned off with curtains. As with the Temple of the Cross and the Foliated Cross, to the rear of the sanctuary is a bas-relief carving called the "Tablet of the Sun" (see Greene Robertson 1991: 39–40).

### Temple of the Sun sculpture

Due to relatively good preservation, some of the sculpture that once decorated the interior and exterior of the Temple of the Sun remains visible today, parts of which include elements found on the roofcomb, fragments of the *alfardas*, and some pier sculpture, in addition to remnants of sculpture on the sanctuary roof and jambs. A reconstruction of the Temple of the Sun (as offered by Greene Robertson 1991: fig. 146) indicates that the imagery once present on the entablature of the temple included two bicephalic serpents. At the centre of the

**Figure 3.9.** Fragments of the Temple of the Sun *Alfardas*. Drawing by Linda Schele (No. 151), © David Schele, courtesy of the Foundation for the Advancement of Mesoamerican Studies, Inc., www.famsi.org.

composition, beneath one of the bicephalic serpents, is K'inich Kan B'alam II in the guise of God K seated on a bench or throne (Greene Robertson 1991: 46). The ruler is depicted here wearing a cross-banded belt, a loincloth, beaded collar, bracelets, anklets, and a feather headdress. The God of Number 7 and the God of Number 9 comprise the heads of the serpent running behind K'inich Kan B'alam II. In addition to these sculptural elements, the entablature also portrays two kneeling human figures depicted in profile presenting offerings to K'inich Kan B'alam II.

A sculpture of a *Cauac* Monster (a zoomorphic representation of a mountain [Miller and Taube 1993: 120]) once occupied the lower-central position on the Temple of the Sun roofcomb. Seated above this creature was a figure that may or may not have displayed human facial features. Flanking the *Cauac* Monster and the upper-seated figure, four *bacabs* (two above and two below) are portrayed supporting a sky-band that runs around the outer edge of the roofcomb and horizontally through its centre (see Greene Robertson 1991: 46–48, for more information).

Surviving fragments of the Temple of the Sun *alfardas* (see Figure 3.9) record the birth of GIII (God 3) as October 25, 2360 B.C. as well as the dedication of the structure on March 18, 692 B.C. following its completion by K'inich Kan B'alam II (for transcription of *alfardas* fragments, see Schele and Mathews 1979: figs. 333 and 334; also see Schele and Freidel 1990: 250–251; Stuart 2006: 154–155).

Today, little sculpture survives on external piers of the Temple of the Sun; however, remaining stucco does indicate that the two centre piers of the temple were primarily iconographic whereas the two outermost piers were principally textual. Pier A once portrayed a text framed within two large quatrefoil cartouches. The top cartouche once contained three glyphs, whereas the lower cartouche comprised six glyphs; combined, they once gave an initial series date. Although too little of the text survives to discern precisely the date once represented on the pier, part of the date has been reconstructed and is in accordance with what is recorded on the jambs and *alfardas* of the temple: 5 Eb' 5 K'ayab' (Greene Robertson 1991: 45).

All that survives on Pier B are the ankles and the lower calves of an individual. Remnants of stucco on the outer border of the pier suggest that a segmented serpent similar to what appears on Pier A once decorated it. On the north side of the same pier, the upper portion of an individual is discernible; the remains of a beaded collar are visible, including what appears to be a backrack of some kind and remnants of a feathered headdress. On Pier C of the Temple of the Sun are the remains of a frontally oriented individual in formal costume. The figure wears a huge feathered headdress surmounted by a Jester God; no other feathers of the type found in this headdress appear at Palenque. In addition to the elements just mentioned, attached to the individual's left wrist is a square shield bordered by short feathers (see Greene Robertson 1991: 45–46).

Residual stucco on the south sanctuary roof of the Temple of the Sun suggests that a sculpture of an earth monster once decorated the exterior of the inner sanctum. Presently, the bottom border of the south sanctuary roof displays a series of inverted zoomorphic heads sculptured close to one another, whereas the top border displays remnants of a sky-band. In addition, a representation of a Quadripartite Monster is also visible on the vertical panel at the west end of the south sanctuary roof (Greene Robertson 1991: 40).

The jambs of the Temple of the Sun portray K'inich Kan B'alam II (Stuart 2006: 156) (Figures 3.10 and 3.11). On the southeast jamb, K'inich Kan B'alam II wears cross-banded leggings, jaguar sandals, and a jaguar pelt around his waist, as suggested by a hanging tail. A feathered backrack with a god is portrayed spewing water or blood from its mouth. Other costume elements adorning K'inich Kan B'alam II include a voluminous feathered headdress and a headband with a tuft of hair pulled through a bead at his forehead. The border framing the composition portrays a maize plant emerging from the head of an inverted Long-lipped God. In addition, a *Cauac* Monster is portrayed beneath the maize foliation with a spear in its mouth.

**Figure 3.10.** Southeast Jamb, Temple of the Sun. Drawing by Linda Schele (No. 238), © David Schele, courtesy of the Foundation for the Advancement of Mesoamerican Studies, Inc., www.famsi.org.

The northeast jamb of the Temple of the Sun, recently reconstructed by Peter Mathews, portrays K'inich Kan B'alam II formally dressed and in profile. K'inich Kan B'alam II is seen here wearing jaguar skin sandals, a beaded necklace and shoulder piece, a short hipcloth and loincloth, as well as a jaguar pectoral. In addition, he carries an incense bag on his left wrist and a spear that he grips with his left hand. To his right, a *Cauac* Monster is portrayed towards the edge of the jamb, in addition to a maize plant that can be seen emerging from the headdress of an inverted human individual; a row of gods' heads comprises the left border of the composition. The surviving text associated with the jamb documents K'inich Kan B'alam II's parentage (Greene Robertson 1991: 43–44).

The text flanking both sides of the central image of the Tablet of the Sun is concerned primarily with lineage and divinity, recording rites involved in procurement of the seat of power at Palenque. One of the events recorded on the Tablet of the Sun concerns K'inich Kan B'alam II's designation as heir to the throne of Palenque, which occurred six years after his birth on 9.10.8.9.3 (14 June, 641) (Schele and Freidel 1990: 251). The text carved over the shield at the centre of the tablet records K'inich Kan B'alam II's accession, which occurred on 9.12.11.12.10 (7 January, 684). On that day, K'inich Kan B'alam II was given the title "B'alam-Ajaw" and named as the son of his father and mother. The text also records the birth of GIII on 1.18.5.3.6 13 Kimi 19 Keh (see Greene Robertson 1991: 43) (for more detailed transcription of the text, see Schele and Freidel 1990: 250–251; Stuart 2006: 162–171).

The central imagery portrayed on the Tablet of the Sun is symbolic of warfare (Stuart 2006: 161). Representing the Jaguar God of the Underworld, the shield at the centre of the tablet (see Figure 3.12) is the element of principal importance in the composition (Schele and Freidel 1990: 243). The Jaguar-God shield displays scroll eyes, ear spools, and a "cruller" on the bridge of its nose; fangs can also be seen protruding from the mouth, in addition to water flowers and feathers which stick out from the corners of the shield. Behind the shield element

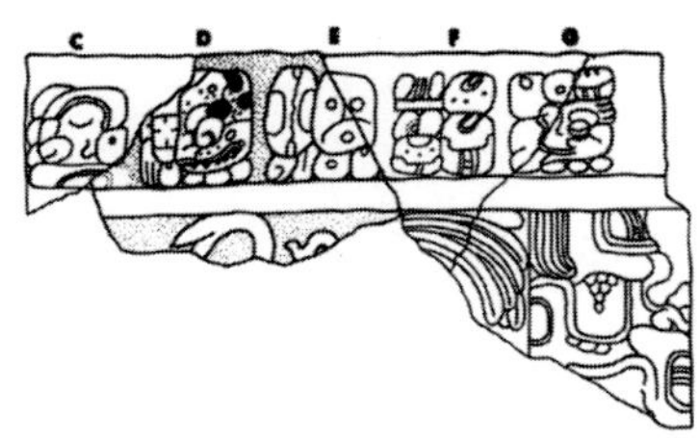


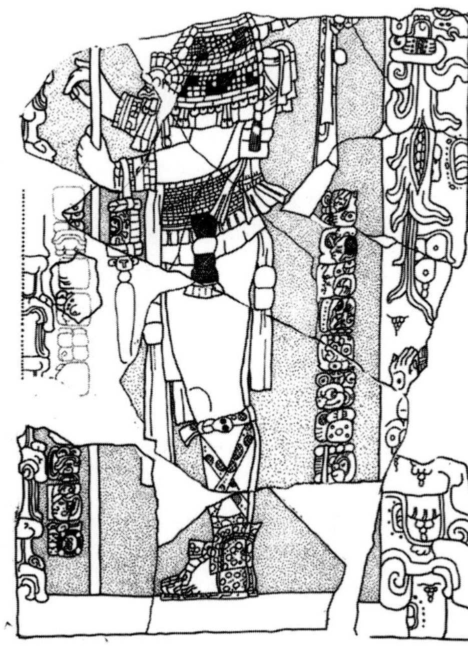

**Figure 3.11.** Northeast Jamb, Temple of the Sun. Drawing by Linda Schele (No. 415), © David Schele, courtesy of the Foundation for the Advancement of Mesoamerican Studies, Inc., www.famsi.org.

**Figure 3.12.** The Tablet of the Sun. Drawing by Linda Schele (No. 171), © David Schele, courtesy of the Foundation for the Advancement of Mesoamerican Studies, Inc., www.famsi.org.

are two crossed spears that terminate with the heads of snaggle-tooth dragons; flint points can be seen protruding from the dragons. A serpent bar is depicted supporting the shield and the dragon spears. God L and God M, two gods connected with death and warfare, are portrayed holding the serpent bar. God L is portrayed wearing shell earplugs, a jaguar pelt as a cape, a skirt, and a cross-banded belt. This god also wears a shell that hangs from a mat-sign pectoral and an elaborate *muwan* bird headdress. God M is also portrayed wearing a mat pectoral, in addition to a patterned cape, shell earplugs, and a bird headdress with quadripartite badge included among the motifs (Greene Robertson 1991: 41–42).

On both sides of the central shield motif is a figure that stands in profile. The shorter individual is argued to be either K'inich Janab' Pakal I or perhaps a younger portrait of K'inich Kan B'alam II (see Benson 1978: 45; Schele and Freidel 1990: 242; Greene Robertson 1991: 40). The costume adorning the shorter of the two figures is generally consistent with depictions on the other Cross Group tablets (Figures 3.7 and 3.17) with the exception of the Flint-Shield God held by one of the individuals. The taller figure depicted on the Tablet of the Sun is K'inich Kan B'alam II, who is portrayed here minimally dressed, holding a seated God K in his outstretched arms. He wears a fringed loincloth, an apron, feathered sandals, a large beaded necklace, pendulous ear spools, bracelets, and a plain turban with a mat symbol and "leaf of ancestry" positioned at the top (Greene Robertson 1991: 41).

### 3.1.3. Temple of the Foliated Cross

The Temple of the Foliated Cross differs somewhat in appearance from the other Cross Group temples as it was built into the side of a hill rather than on a pyramidal platform (Greene Robertson 1991: 48). A staircase comprising seventy rises has to be ascended before the temple's interior can be accessed. In terms of its elevation, the Temple of the Foliated Cross would be considered an intermediate structure, between the higher-set Temple of the Cross and the lower-set Temple of the Sun. As in the other Cross Group temples, the floor plan of this structure is divided in half by a thick medial wall. The front half of the temple originally consisted of a single room that was later divided into two separate rooms with the addition of a curtain wall. The rear of the temple is divided into three rooms, one that is central and two that are peripheral. Like the Temple of the Cross and the Temple of the Sun, the central room houses a freestanding sanctuary that was once lavishly decorated.

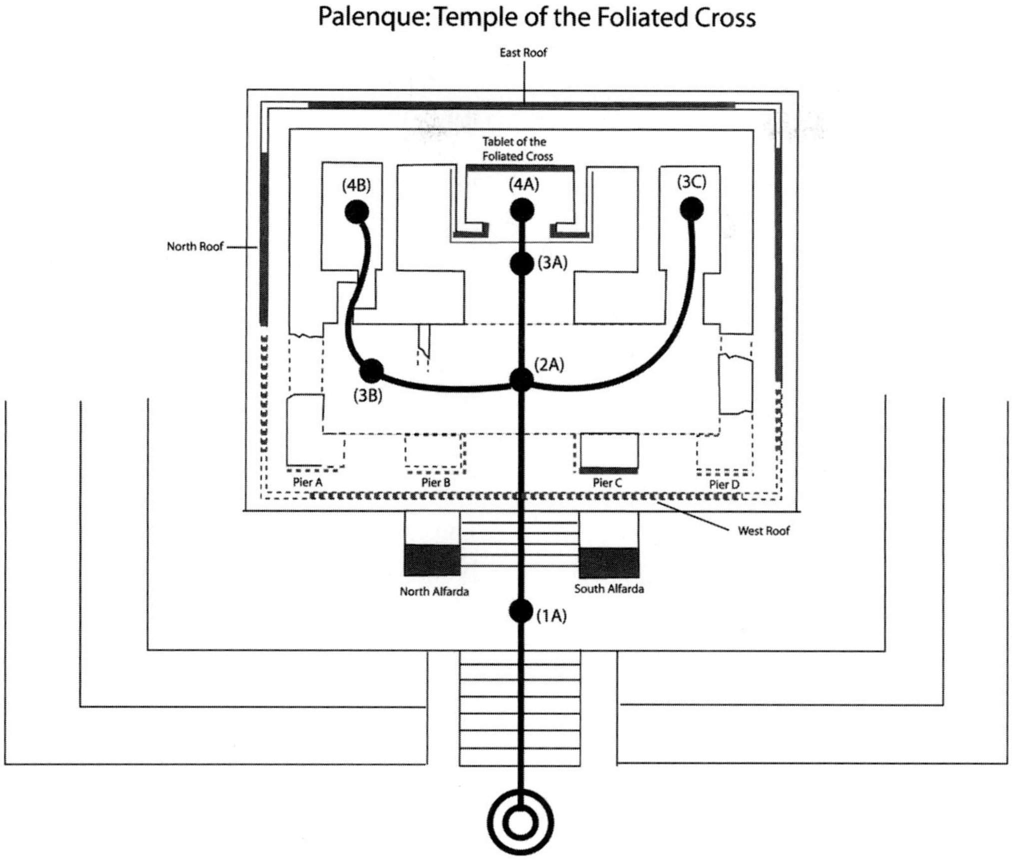


**Figure 3.13.** Plan and Access Diagram, Temple of the Foliated Cross.

Within this sanctuary is the delicately carved Tablet of the Foliated Cross (Figure 3.17).

The small northeast room of the Temple of the Foliated Cross was at some stage narrowed after its initial construction. The addition of two staggered notches, one on the forward left and the other on the rear right of the entrance, would have blocked the view of activities once undertaken within its confines. Inside this room on the south wall are two niches positioned 210 cm off the floor; there is also an *Ik'* (T-shaped) opening in the east wall of the room. The small southeast room at the rear of the structure resembles the small northeast room; the only difference between the two is that the entrance to the south room is marginally wider. Today, the Temple of the Foliated Cross is in a poor state of preservation; unfortunately the entire roofcomb and west façade of the structure have collapsed (Greene Robertson 1991: 48–49).

**Temple of the Foliated Cross sculpture**

Among the few surviving sculptural elements remaining on the Temple of the Foliated Cross (Figure 3.14) are the *alfardas*, parts of the northwest sanctuary jamb and

**Figure 3.14.** The Temple of the Foliated Cross *Alfardas*. Drawing by Linda Schele (No. 203), © David Schele, courtesy of the Foundation for the Advancement of Mesoamerican Studies, Inc., www.famsi.org.

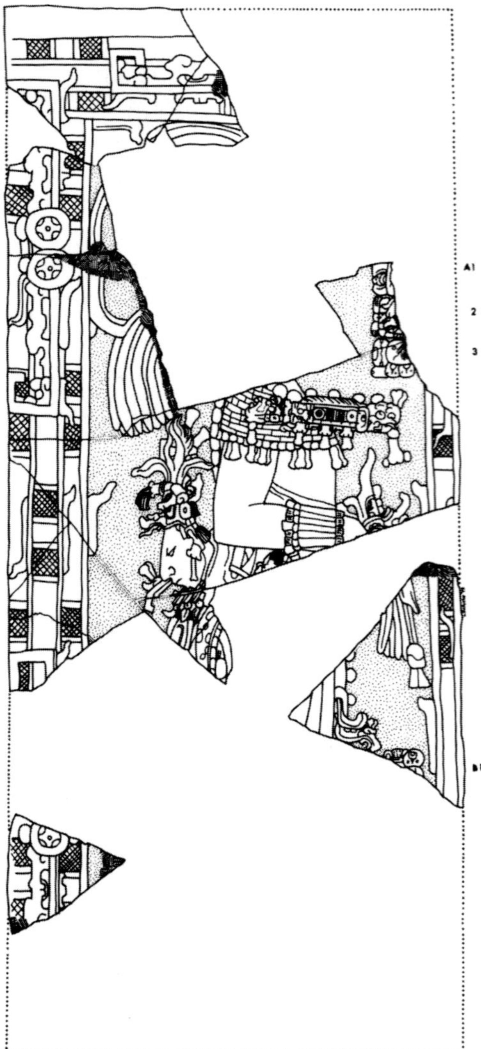

**Figure 3.15.** North Sanctuary Jamb, Temple of the Foliated Cross. Drawing by Linda Schele (No. 182), © David Schele, courtesy of the Foundation for the Advancement of Mesoamerican Studies, Inc., www.famsi.org.

roof, as well as the sanctuary panel referred to as the Tablet of the Foliated Cross. The *alfardas*, positioned on either side of the stairs fronting the Temple of the Foliated Cross (Figure 3.14), record the birth date of God K (GII), which occurred on November 8, 2360 B.C. The monuments also name K'inich Kan B'alam II as the son of K'inich Janab' Pakal I and Lady Tz'akb'u Ajaw, as well as record the name of the temple as '*na te kan*' ("Foliated Cross") (Schele and Freidel 1990: 248–249). The angle and position at which the *alfardas* were set on either side of the upper temple stairs suggest that they were designed to be read as one was entering the structure (for complete transcription, see Stuart 2006: 134).

The north and south sanctuary jambs found in the Temple of the Foliated Cross record rites connected with the transfer of power from K'inich Janab' Pakal I to his son, K'inich Kan B'alam II. An inverted perforator god held by K'inich Kan B'alam II on the south sanctuary jamb indicates that auto-sacrifice is the activity represented (Figure 3.16). On the now fragmented north jamb, the individual (probably K'inich Janab' Pakal I) wears a bird pectoral over a beaded shoulder piece, bracelets, and a feathered headdress. On his back he wears a backrack comprising a human head adorned with a zoomorphic headdress (Figure 3.15). On the south sanctuary jamb, K'inich Kan B'alam II wears sandals with long leg straps, a cross-banded belt with a jaguar element, a god in a quatrefoil cartouche hanging from a string of beads, and a pleated jaguar skin hipcloth. The patterned borders on both the north and south sanctuary jambs represent stylised serpents (Greene Robertson 1991: 53–54; see also Schele and Mathews 1979: 301, 302). Both figures display deified perforators symbolising bloodletting. The surviving inscription on the right panel "records the parentage of K'inich Kan B'alam II . . . [and]

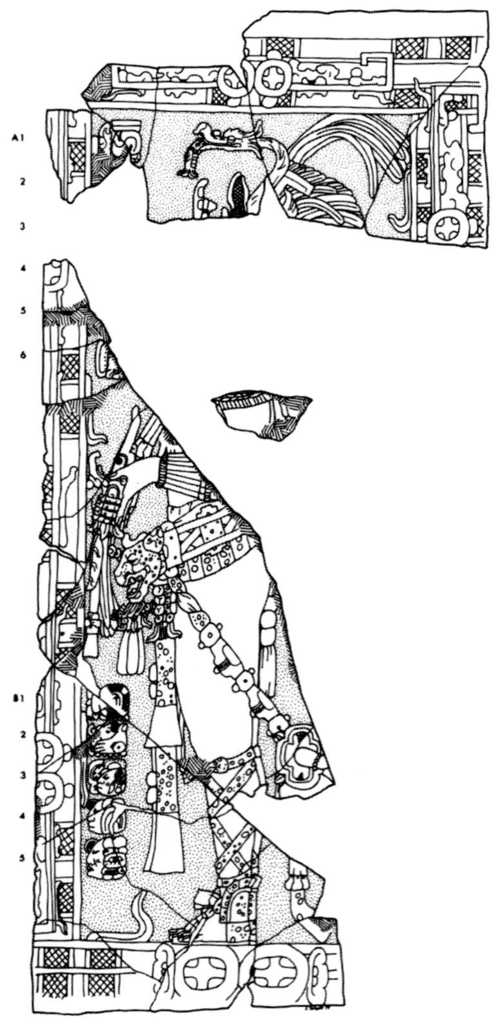

**Figure 3.16.** South Sanctuary Jamb, Temple of the Foliated Cross. Drawing by Linda Schele (No. 180), © David Schele, courtesy of the Foundation for the Advancement of Mesoamerican Studies, Inc., www.famsi.org.

**Figure 3.17.** The Tablet of the Foliated Cross. Drawing by Linda Schele (No. 172), © David Schele, courtesy of the Foundation for the Advancement of Mesoamerican Studies, Inc., www.famsi.org.

Lady Tz'akb'u Ajaw' title 'The Matwiil Noble women'" (Stuart 2006: 137).

The north and south sanctuary roof of the Temple of the Foliated Cross portrays zigzagging foliated bands; above and below these bands are *K'an* crosses symbolising fertility and growth. Similar bands frame the standing figures portrayed on the sanctuary panels. Combined, the sculptural elements communicate the glyphic name of the structure: '*na te k'an*' (see Stuart 2006: 135).

"The Tablet of the Foliated Cross focuses on the ideology of kingship as it relates to maize agriculture and fertility" (Stuart 2006: 140). The text of the Tablet of the Foliated Cross (see Figure 3.17) records the birth of GII (1.18.5.4.0 1 Ajaw 13 Mak), a deity born to "Lady Beastie" through ritual bloodletting. The tablet text also records the dedication of the temple by K'inich Kan B'alam II, the "divine lord of Palenque." The dedication occurred one day after a celestial conjunction between GII and GIII on July 23, 690. As on the Tablet of the Sun, on the Tablet of the Foliated Cross is also documented the accession of K'inich Kan B'alam II (Schele and Freidel 1990: 248) (for detailed transcription of tablet text, see Schele and Freidel 1990: 248–249 and Stuart 2006: 141–152).

The central motif on the tablet is a foliated cross emerging from an earth monster (*Ah-Uuc-ti-Cab*). The earth monster displays Sun God elements that include *tau* (teeth), fangs, square eyes, and a *k'an* (water) cross headdress with maize foliation emerging from it. Signifying rebirth, human heads are visible growing from the branchlike arms of the foliated cross. Heads are portrayed in profile facing upward; each wears ear and nose ornaments and has long hair hanging from the scalp. Beneath the Principal Bird Deity perched at the top of the tree is a representation of the Sun God; the trunk of the foliated cross forms the torso of the god. The Sun God wears a beaded necklace with a pectoral of a human head displaying a stepped haircut and nose plug. A mirror element set into a bellflower is visible hanging from beneath the pectoral. Just below the pectoral is another smaller Sun God that has bones set into its nose. The Principal Bird Deity, portrayed at the top centre of the tablet, is similar to the one represented on the Tablet of the Cross – with some minor differences. The creature in this example wears a Long-lipped God mask and a beaded necklace with shell pectoral, and has plant foliation emerging from its mouth (Greene Robertson 1991: 51–52).

On the Tablet of the Foliated Cross, two figures stand in profile flanking the central imagery. As with the Cross Group tablets previously discussed, the taller individual is argued to be K'inich Kan B'alam II, this time portrayed on the left of the composition. The figure portrayed on the right of the scene is purported to be either the deceased K'inich Janab' Pakal I (Benson 1978: 45; Schele and Freidel 1990: 242) or a more youthful K'inich Kan B'alam II (Stuart 2006: 140). The shorter of the two figures is depicted wearing a costume similar to that worn by the same figure on the Tablets of the Cross and Sun (Figures 3.7 and 3.12). In his hands, the individual holds an inverted perforator god that comprises three blood knots and a cluster of feathers attached to its head. The shorter figure also stands on a shell that has maize foliation growing from it. A human head, perhaps that of an ancestor, is visible growing out of the maize leaves, and a representation of God K emerges from the shell beneath.

On the Tablet of the Foliated Cross, the taller K'inich Kan B'alam II is portrayed wearing a turban or cloth wrap around his head, bracelets, a beaded necklace, pendulous earflares, and a plain loincloth, all of which he wears on the other Cross Group tablets. Elements of costume unique to his representation on the Tablet of the Foliated Cross include a beaded hipcloth or apron and a belt consisting of what appear to be tubular beads finished at the lower edge with round beads and a *xoc* fish emblem. Hanging from the *xoc* fish is a Long-lipped God and a bellflower combination. In K'inich Kan B'alam II's hands is a Jestor God (GI) seated with his arms folded across his chest. Both K'inich Kan B'alam II and the Jestor God face the foliated cross at the centre of the composition. K'inich Kan B'alam II stands on a *Cauac* Monster, symbolising the watery Underworld. The monster has glyphs for eyes and maize foliation growing from its head (Greene Robertson 1991: 50–51).

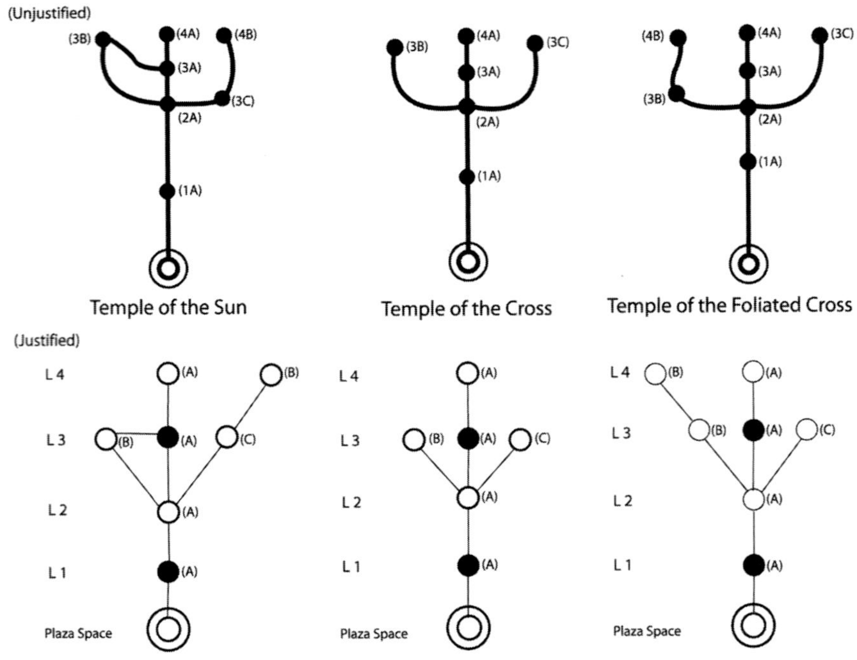

**Figure 3.18.** Access Diagrams of Individual Cross Group Structures, "Justified" from the Plaza Position.

## 3.2. Access analysis and the Cross Group temples

In my investigation, access maps were constructed for all three temples of the Cross Group at Palenque (Figures 3.3, 3.8, and 3.14). Discernment of the floor plans of these structures, in addition to the positioning of associated sculpture, was achieved by combining information from a number of sources (Blom 1923; Marquina 1964; Schele and Freidel 1990; Greene Robertson 1991; Stuart 2006). As outlined in Chapter 2, the first stage in drawing an access diagram is to represent each delineated space within the complex of spaces with a small circle, which can then be connected by lines indicating where direct access between spatial units is possible. Spatial units (or "nodes") are articulated by the presence of walls or other permanent architectural features that bound spaces. Access between the spatial units is provided via doorways and/or gaps in barriers and partitions, permitting movement between adjacent spaces. Relative accessibility between spatial units is determined by assessing the minimum number of boundaries that needed to be crossed to access the space in question. In my investigation, "justified" access maps were useful in articulating "permeation (depth) gradients" from the perspective of the individual entering the spatial system from the "carrier" position. In justified access maps, each horizontal layer represents different spaces at common degrees of permeation; space as it relates to increasing depth is represented vertically on consecutive levels.

Applying the procedure just mentioned, all three temples of the Cross Group were found to comprise a total of five levels of depth from the carrier position (i.e., from the perspective of an individual standing externally to the building group) and a total of four levels of depth from the Cross Group plaza. Given that access to each of the Cross Group temples is provided exclusively from the Cross Group plaza, in the discussion that follows, the plaza space (as represented by the double circle) has been taken as the starting point of each temple's access diagram. (Note: In my analysis, the presence of cord holders suggesting the use of impermanent partitioning was not considered, as the precise number and configuration of such additions remain in question.)

The access diagrams of the Cross Group buildings indicate that the spatial configuration within each structure is highly restricted; this is revealed by the treelike configuration of the access maps (Figure 3.18). In all three examples, the spatial system within each of the structures was found to be "nondistributed," as there is no more than one route to a given space (with the exception of the small southwest room in the Temple of the Sun that has an additional point of entry from an alternative space). The Temples of the Cross, the Sun, and the Foliated Cross were all found to have one bounded space at the first level of permeation (L1A). The first level of depth for each of the Cross Group temples is the platform directly adjacent to the building's entrances, access to which is provided via staircases fronting the structures. Level two in each of the buildings is the space directly behind the piered entrances (L2A). The deepest space in each of the temples is the inner sanctuary contained within the rear-central space of each structure (L4A). In the Temple of the Sun and the Temple of the Foliated Cross, however, there is an additional space positioned at a depth of level four (L4B) (see Figures 3.8 and 3.14).

The front interior space of the Temple of the Sun and the Temple of the Foliated Cross were at one time partitioned by an additional wall that did not appear in the front interior space of the Temple of the Cross. In the Temple of the Sun, remnants of this wall are visible on the north side of the front room running in an east to west direction; in the Temple of the Foliated Cross, the remains of a similar wall are also found running east to west at the northern end of the room. The effect of these additional walls on the spatial configuration of the two temples is to further divide the front interior spaces. These walls provided an additional boundary and unit of space that had to be permeated before access could be gained to the small northwest room of the Temple of the Sun and the small northeast room of the Temple of the Foliated Cross. On acknowledging this similarity, it appears that the Temple of the Sun and the Temple of the Foliated Cross share the same spatial configuration. Flora Clancy (in Greene Robertson 1991: 13) proposed that the reversed positioning of the carved figures on the Tablet of the Sun and Foliated Cross suggests that the monuments may in fact be mirror images of each other (see also Clancy 1986). My analysis of the Temple of the Sun and the Temple of the Foliated Cross supports Clancy's observation, although it takes it a step further: As indicated by the justified access maps, it also appears that the configurations of the Temple of the Sun and the Temple of the Foliated Cross reflect one another spatially.

## 3.3. Public and private sculpture of the Cross Group

Due to the stratified nature of Classic Maya society and the subsequent segregation of the populous, it can be reasonably argued that monumental art and architecture contained within common and elite categories of space would have been viewed by corresponding common and elite audiences. More specifically, whereas easily accessible sculpture was for the wider population to see, less accessible sculpture would have been visible only to elites. In accordance with this view, Greene Robertson argues that, "although the temples themselves were built for the populous as a whole to view, the inner sanctuaries were private art meant to be viewed by only the privileged few, the Palenque elite" (Greene Robertson 1991: 10). As noted by Greene Robertson (1991: 47):

> The roofs of the Cross Group temples [at Palenque] were elaborately sculpture[d] . . . like a series of billboards . . . This was definitely public art . . . People could have their secret ceremonies in . . . the subterranean chambers of the Palace and the inner sanctuaries . . . where only the privileged few could enter or witness solemn events. Tablets, murals and sculpture in these hidden places can be considered private art . . . But Palenque did have great public art . . . [among these were the] sculptured piers . . . towering roofcombs embellished with painted stucco sculpture.

Differences have been identified in how the rulers chose to represent themselves on monuments positioned in different spatial contexts at Palenque. These differences appear to be dependent on whether representations were positioned in public (common) view or confined to more private (elite) display. By assigning monumental art of the Cross Group to "depth values" as prescribed by their position on access diagrams, it was possible to detect variations in "more-common" and "more-elite" categories of art. Employing this methodology, it is also possible to detect shifts in the relative proportion of text and images found on monuments as one moves progressively deeper into the Cross Group (see Section 3.5, Text–Image Ratios and the Cross Group).

The use of Maya text and iconography was purposeful in the sense that it was a means by which a ruler could instill obedience and loyalty among his subjects, functioning in part as a petition to the gods, promoting a ruler's successes and validating his authoritative position (Marcus 1992a: 14). Results of my analysis indicate that the information transmitted by Maya iconography and sculpture at Palenque varied depending on its intended audience. Public sculpture, which is placed intentionally in spatial contexts where access to its viewing is uncontrolled and/or promoted, tends to portray rulers authoritatively. In contrast, private sculpture, access to which is more restricted (controlled by progressive enclosure, channeled movement, and changes in elevation) tends to portray elite individuals as subordinates.

Table 3.1. *Architectural features of the Palenque Cross Group displaying surviving sculpture ordered to corresponding depth values starting from the plaza position*

| | Temple of the Cross sculpture | Temple of the Sun sculpture | Temple of the Foliated Cross sculpture |
|---|---|---|---|
| Level 1 | Roof (entablature)<br>Roofcomb (no sculpture present)<br>*Alfardas* Stela 1? | Roof (entablature)<br>Roofcomb *Alfardas* Piers A, B, and C | *Alfardas* |
| Level 2 | [No sculpture present] | [No sculpture present] | [No sculpture present] |
| Level 3 | Sanctuary roof east and west sanctuary jambs | Sanctuary roof southeast and northeast jambs | Sanctuary roof north jamb and northwest inner surface |
| Level 4 | Tablet of the Cross | Tablet of the Sun | Tablet of the Foliated Cross |

Standing in the plaza at the base of all three Cross Group temples one is first confronted by the public sculpture of the temple façades (L1 sculpture) (see Table 3.1). The message transmitted by the exterior of these structures is one of power. In Greene Robertson's (1991: figs. 74, 134, and 146) reconstruction of the Temple of the Sun, a centrally positioned, frontally orientated K'inich Kan B'alam II is portrayed seated on a *Cauac* Monster. The image is sculptured high on the roofcomb of the temple. Flanking K'inich Kan B'alam II are two *bacabs* ("sky bearers"), which support sky-bands whilst simultaneously grasping a meandering serpent. Depicted in profile and proportionately smaller than K'inich Kan B'alam II, the *bacabs* are represented as subordinates. The fact that K'inich Kan B'alam II is depicted centrally and larger and is in frontal orientation implies that he is of principal importance in the composition. In this representation, K'inich Kan B'alam II wears an elaborate zoomorphic feathered headdress and a cross-banded belt, and he holds a serpent bar against his chest. On the entablature of the same structure, K'inich Kan B'alam II is again depicted, this time publicly impersonating God K (Greene Robertson 1991: 46). To the left and right of K'inich Kan B'alam II are two human figures kneeling on benches and presenting offerings to the king. As on the temple roofcomb, the peripheral figures are portrayed smaller and more humbly than K'inich Kan B'alam II and wearing simply loincloths and turban-style headdresses.

If one had the authority to climb the stairs to the platform and stand before the entrances to the Cross Group temples, one would be confronted by the *alfardas* positioned on either side of the upper stairs. The text on the *alfardas* of the Cross Group temples gives the birth dates of the Palenque Triad: GI, GII, and GIII (the Temple of the Cross records the birth date of GI; the Temple of the Foliated Cross records the birth date of GII; and the Temple of the Sun records the birth date of GIII). The placement of the *alfardas* is strategic in the sense that they signal that one is about to enter the sacred birthplaces of patron gods of Palenque.

On passing into the interior of the temples, the sanctuary jambs (L3 sculpture) are the next sculptural features to become visible, all of which portray elaborately dressed individuals, in profile, facing in towards the entrance to the sanctuaries. The sanctuary jambs of the Temple of the Cross portray K'inich Kan B'alam II in full royal regalia; the elaborateness of the costume is comparable to what was once worn by the figures on the two central piers of the temple. The most notable difference between the pier sculpture and the sanctuary jambs is the change in the orientation of the figures represented in the art from a frontal orientation (as seen on the external piers of the building) to a profile stance (as displayed on the sanctuary jambs). This change in the direction that the figures are facing suggests a form of homage to the inner sanctuary and/or the threshold through which one had to pass to enter the sacred space beyond.

Maya rulers used monumental art to instill and/or reinforce a perception of the world and themselves among the broader population, a perception influenced by the agenda of the king and the intended audience of the art. For example, the public representation of K'inich Kan B'alam II as God K on the entablature of the Temple of the Sun differs from how he is represented on the more private Cross Group tablets, where he is portrayed in profile, more humbly dressed, and in human form. Residual sculpture indicates that elaborately dressed individuals once adorned the public piers at the entranceways to the Cross Group temples, whereas less elaborately dressed figures are portrayed in the temples' inner sanctuaries. Comparative analysis of iconography has revealed a distinction between public and private dress as displayed in public and private art. More specifically, monumental stelae and other public sculpture typically portray rulers in elaborate costumes, as they would have appeared before large crowds or for important public occasions, whereas simpler dress is typically found on private artistic media where access to their viewing would have been restricted. One explanation for this observed disparity is that although elaborate "dress . . . [was] important for public occasions . . . it [was] less necessary when members of the same class or family meet in private, but among mixed groups, lavish dress projects social status, activities and sexual gender" (Taylor 1983: 13–15). Given the proximity of the piers, and the elaborateness of the costumes represented, it is likely that the pier sculpture portrayed the ruler K'inich Kan B'alam II as he would have appeared publicly. This representation differs from that of K'inich Kan B'alam II on the Cross Group tablets, where he is portrayed in simpler costume undertaking more private activities. As has been previously noted, "on each of the inner panels . . . [K'inich Kan B'alam II] is dressed simply in the Maya equivalent of underwear" (Schele and Freidel 1990: 242). In contrast, remaining stucco on Pier C portrays him in elaborate attire.

The external sculptures of the Cross Group temples portray K'inich Kan B'alam II authoritatively among subordinate individuals. Conversely, all three of the Cross Group tablets portray the ruler of Palenque subordinate to the gods and the cosmological imagery represented on the monuments. The private and restricted nature of the Cross Group sanctuaries is suggested by the stance and understated nature of the costume worn by K'inich Kan B'alam II in all Cross Group tablets (L4 sculpture) (for general descriptions of tablets see Sections 3.1.1, Temple of the Cross, 3.1.2, Temple of the Sun, and 3.1.3, Temple of the Foliated Cross). All three tablet images portray the human characters in profile to the left and right of the stylised gods and cosmological imagery placed at the centre of the composition. In all examples, K'inich Kan B'alam II is making an offering to the central effigies, behaviour which parallels royal courtly activities where courtiers and visiting dignitaries are portrayed presenting tribute to rulers in a similar manner.

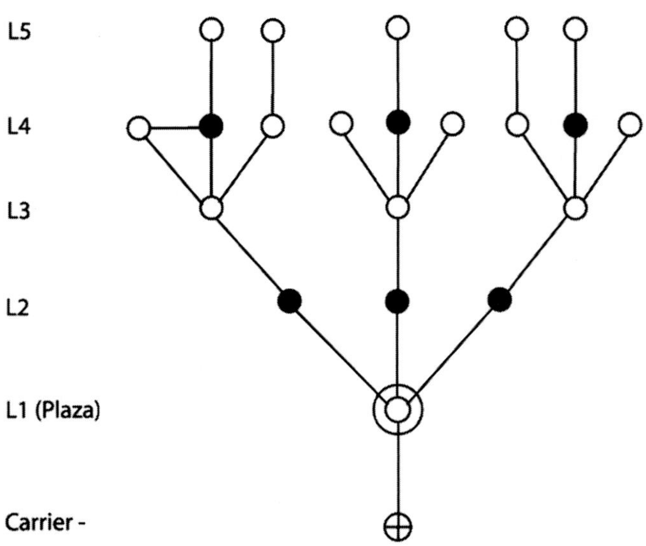


**Figure 3.19.** "Justified" Access Diagram of the Palenque Cross Group.

Table 3.2. *Palenque Cross Group: Spatial units ranked according to their RA values (ranked from least integrated to most integrated)*

| Spatial unit (node) | RA value |
|---|---|
| Carrier | x |
| 5c | 0.4571 |
| 5d | 0.4381 |
| 5e | 0.4381 |
| 5b | 0.4143 |
| 5a | 0.4095 |
| 4d | 0.3714 |
| 4f | 0.3714 |
| 4e | 0.3619 |
| 4i | 0.3524 |
| 4a | 0.3429 |
| 4c | 0.3429 |
| 4g | 0.3429 |
| 4h | 0.3429 |
| 4b | 0.3381 |
| 3b | 0.2762 |
| 3a | 0.2571 |
| 2b | 0.2190 |
| 2a | 0.2095 |
| 2c | 0.2095 |
| 3c | 0.2095 |
| 1a | 0.1714 |

## 3.4. Control values and relative asymmetry analysis of the Cross Group

On ranking the relative asymmetry (RA) values derived from the complete access diagram of the Cross Group (Figure 3.19), there appears to be a correlation with the building groups "permeation gradient" and shifts in the values calculated with increasing depth (see Table 3.2). More specifically, as one moves progressively deeper into the Cross Group, spatial units become less integrated from the rest of the spatial system. As outlined earlier in this investigation, RA analysis is a standardised measure of how deep, on average, a space is from every other unit of space in that configuration. The higher the resulting value derived from a calculation, the more segregated (i.e., private) that space is from the configuration; the lower the resulting value for a given space, the more integrated (i.e., public) that space is said to be. Differences in the relative depths of spatial units are argued to indicate which units of space, on average, are more or less accessible from the perspective of the entire spatial system under study (Hillier and Hanson 1984: 109). In this investigation, a correlation is drawn between the increasing segregation of spaces and increases in relative privacy of those spaces and their corresponding sculpture.

When considering data derived from the complete access diagram for the Cross Group, the least integrated spaces are the deepest spaces, nodes 5c, 5d, 5e, 5b, and 5a, respectively. The inner sanctuary of the Temple of the Cross (node 5c) is the least integrated and, subsequently, the most private space of the Cross Group, followed by the northeast room (node 5d) and the sanctuary of the Temple of the Foliated Cross (node 5e). The northwest room of the Temple of the Sun (node 5b) and the sanctuary (node 5a) of the same temple are the next most segregated spaces of the Temple group. The most integrated space is the Cross Group plaza (node 1a), as access to all of the Cross Group temples is provided exclusively from this position.

The most "controlling" spaces of the Cross Group are the rooms fronting all three of the temples (nodes 3b, 3c, and 3a) because access to the temples' inner sanctuaries (nodes 5a, 5c, and 5e), as well as the rooms to either side of them (nodes 5b and 5e), are given via these spaces (see Table 3.3). In addition to the front rooms of all three structures, the Cross Group plaza (node 1a) was also calculated as being a controlling space, as access to all three of the temple interiors necessitates passing through this space. As outlined in Chapter 2, a "control value" is a relative measure of the amount of control a space possesses over access to other spaces in the system in localised proximity. Put simply, a controlling space is any unit of space that has to be penetrated before access can be attained to those spaces beyond it; the more spaces a particular space controls access to, the higher the level of control that space is said to have. In the calculation of control values, any space that has a value of one or greater is considered to be a controlling space, and any space with a control value less than one is considered to be a controlled space (see Section 2.4.5, Measuring Control Values and RA Analysis).

In summary, the privacy of the inner sanctuaries, which house the texts and imagery of the Tablets of the Sun, Cross, and Foliated Cross (nodes 5a, 5c, and 5e), is marked by the relatively high depth of the monuments in relation to the carrier position as well as the humility displayed by the royal actors in the artwork. Conversely, the shallower locations of the temple façades, with their displays of opulence and power, indicate the public nature of the corresponding sculptures. On the Cross Group's access maps, the exterior sculptures of the temple sanctuaries were found to hold intermediate positions. As well as occupying controlled spaces, the sculptures of the sanctuary interiors are the least integrated (i.e., most segregated) of the Temple Group, a further indication of the highly private character of the sculpture.

Table 3.3. *Palenque Cross Group: Spatial units ranked according to their control values*

| Spatial units (node) | Control values |
|---|---|
| Carrier | x |
| 3b | 3.0000 |
| 3c | 2.5000 |
| 1a | 2.5000 |
| 3a | 1.8333 |
| 4b | 1.7500 |
| 4e | 1.2500 |
| 4c | 1.2500 |
| 4g | 1.2500 |
| 4h | 1.2500 |
| 4a | 0.5333 |
| 5c | 0.5000 |
| 5d | 0.5000 |
| 5e | 0.5000 |
| 5b | 0.5000 |
| 2b | 0.5000 |
| 2a | 0.5000 |
| 2c | 0.5000 |
| 5a | 0.3333 |
| 4d | 0.2500 |
| 4f | 0.2500 |
| 4i | 0.2500 |

*Note:* Values below 1.0000 indicate controlled spaces, whereas a value of 1.0000 or above indicates controlling spaces.

## 3.5. Text–image ratios and the Cross Group

It can be reasonably argued that, at sites where literacy is low among the general population, public/civic areas (i.e. areas characterised by their low elevation and ease of access) would contain a lower proportion of glyphic texts than would more restricted spatial contexts. Given that the greater proportion of Maya society was unable to read (Brown 1991: 494; Houston and Stuart 1992: 592; Houston 1994: 27–50), one would expect a higher proportion of communicative imagery to be positioned in spatial contexts that would be more accessible and subsequently more public. The benefit of iconography over text, as a vehicle of communication, is that imagery conveying information about political and religious doctrines may be more readily understood by the larger proportion of the population. This proposal seems apparent for

Table 3.4. *Cross Group text–image ratios*

| | Text % | Image % |
|---|---|---|
| **Temple of the Cross** | | |
| Temple External Elevation | 1.30 | 98.70 |
| External Sanctuary | 15.80 | 84.20 |
| Internal Sanctuary | 47.00 | 53.00 |
| **Temple of the Sun** | | |
| Temple External Elevation | 3.40 | 96.60 |
| External Sanctuary | 17.00 | 83.00 |
| Internal Sanctuary | 34.10 | 65.90 |
| **Temple of the Foliated Cross** | | |
| Temple External Elevation | 2.10 | 97.90 |
| External Sanctuary | 16.50 | 83.50 |
| Internal Sanctuary | 35.40 | 64.60 |

Table 3.5. *Temple of Inscriptions text–image ratios*

| | Text % | Image % |
|---|---|---|
| **Temple of Inscriptions** | | |
| Piers (*N*%) | 1.80 | 98.20 |
| Tablets | 100.00 | 0.00 |

Terminal Classic Chichen Itza, where the spatial positioning of text and image suggests that "glyphic monuments had a narrow audience whilst pictorial/iconographic monuments had a much broader appeal and aim than what was typical of similar vehicles of official propaganda elsewhere in the Maya Lowlands" (Lincoln 1986: 155).

Although levels of literacy may have varied from site to site, parallels with this observation are found when comparing the ratios of image and text on monuments appearing on architecture at Palenque, where a correlation was found between variations in the relative proportions of text and image on sculpture and the accessibility of the corresponding spaces. More specifically, primary buildings, such as those that comprise the Cross Group, all contain higher proportions of text on interior private sculpture than on exterior public sculpture. This result was also found for the Temple of Inscriptions (see Tables 3.4 and 3.5).

In determining the text–image ratios of sculpture at varying levels of depth into the Cross Group, it was necessary to assign the various sculptural media to their corresponding depth values. This process required grouping a range of sculptural features to a single permeation level in some instances. As the sculptural elements that collectively comprise the façades of the Cross Group can be viewed simultaneously, each was treated as a single composition. In all three Cross Group temples, a total of three levels of depth were identified as displaying relative quantities of sculpture and text: Levels 1, 3, and 4. Level 1 encompasses the temple piers, the roof, the *alfardas*, and the roofcomb; Level 3 includes the external sanctuary sculpture, which comprises the sanctuary jambs, panels, and roof; and Level 4 sculpture is represented by the sanctuary tablet.

Given the poor condition of the Cross Group temples, some of the information used for this analysis has relied to a certain extent on published reconstructions (Blom 1923; Marquina 1964; Greene Robertson 1991) combined with some assumptions on how the Cross Group would have appeared when architecturally and sculpturally intact. More specifically, it was assumed that where elements of sculpture and architecture were found to be missing, the appearance of these missing elements could be reasonably surmised based on arguments of symmetry and continuity of structural form. Consequently, what I offer here cannot be viewed as a precise assessment of the relative percentages of text and image on sculptured surfaces of the Cross Group, but rather an approach designed simply to illustrate a trend.

4

# Architectural and sculptural programs of the Palenque Palace Group

## 4.1. Palenque Palace: General description

In accordance with Andrews (1975: 59–63), the Palace at Palenque has been defined as a Palace Group archetype. In this chapter, I will first present a general description of the Palace Group followed by a phase-by-phase explanation of each building and associated sculpture. As outlined earlier in this investigation, temporally and spatially disparate sculpture and architecture cannot be compared indiscriminately: The gradual accumulation of monuments over time and the extension, modification, and renewal of buildings affect the capacity of architecture and associated sculpture to provide telling relationships. To address this problem, architectural features and sculptures of the Palace were assigned to phases and the reigns of kings. These assignments provided a way of confining the periods investigated, as well as placing any correlation made between sculpture, architecture, and space in context. With the help of "access analysis," the objectives in articulating the various Palace phases were threefold: (1) to more effectively reveal the progression of sculptural and architectural development of the Palace Group; (2) to provide a platform for inferring the possible intent of Maya rulers in their commissioning and placement of particular sculptural media within specific spatial/architectural contexts; and (3) to establish a method of comparing the different rulers' sculptural and architectural programs.

The first stage in the Palace Group investigation required establishing the relative chronology of the Palace houses followed by the breaking up of the proposed sequence into phases that correspond to the reigns of the Palenque kings. Following this, plan-view maps of each building phase were produced, and the various corresponding sculptural media were tabulated. The objective in doing this was to discern how the Palace appeared sculpturally and architecturally during the reigns of those persons who presided over each phase. In this study, Palace phase maps were used as the basis for the construction of access diagrams that were implemented to order and interpret the architecture and sculpture of each Palace phase.

Six construction phases for the buildings of the Palace at Palenque are proposed in this study. In the following sections, Palace Phases 1 through 6 are discussed sequentially and combined with descriptions of associated sculpture. Appearing at the beginning of each section, the maps provided reveal the likely appearance of the Palace over the reigns of different Palenque kings. The descriptions of the individual Palace houses and other structures are in accordance with Andrews's (1974) unpublished survey of the Palenque Palace and Greene Robertson's *The Sculpture of Palenque: The Early Buildings of the Palace and the Wall Paintings* (1985a) and *The Sculpture of Palenque: The Late Buildings of the Palace* (1985b) As Greene Robertson's (1985a, 1985b) studies remain the most detailed and thorough catalogue of the Palace sculpture yet published, all general descriptions of the sculpture of the Palace will be drawn from these publications unless otherwise stated. On the Palace phase maps, the heavy black lines represent exposed architectural features, whilst the buildings composed of transparent lines represent buried architecture. Dotted lines featured on the plan-view maps represent probable architectural features that are yet to be verified archaeologically.

## 4.2. Evolutionary sequence of the Palace

The Palace as it appears today was not planned and constructed as a single architectural concept but is instead the result of a gradual accumulation of buildings over time

(Maudslay 1889–1902b: 11; Holmes 1897: 169; Martin and Grube 2000: 164). Today, the Palace at Palenque represents one of the highest architectural achievements of the Classic Maya, in which advancements in engineering and design permitted larger, more spacious rooms to be constructed (Miller 1999: 38). Comprising a series of "range" structures, in its totality the Palace at Palenque is large, measuring ninety-seven metres north to south and seventy-three metres east to west. In addition, the building group measures approximately ten metres to the top of its basal platform (Martin and Grube 2000: 164). Some of the more notable characteristics of Classic Period Palenque architecture are multiple piered entrances providing access into buildings, sloping upper façades, and open-work roofcombs (Sharer 1994: 278). Most of the Palace buildings are configured as a series of vaulted galleries that define the boundaries of inner courtyards, many of which display a high level of interior and exterior decoration. Among other things, the modeled stucco and low relief sculpture of the Palace displays texts, portraits of gods and ancestors, lords, subordinates, and captives (Coe 2000: 131), in addition to an array of cosmological imagery. Some building interiors also portray the remains of painted narrative scenes (e.g., Palace House B), as well as various motifs that decorate their exterior surfaces (e.g., Palace House E).

The platform that supports many of the buildings of the Palace was constructed in several phases that were built up gradually to coincide with the vertical and northward expansion of the building group (see Palace cross-section diagram, Figure 4.1). The northern Palace platform, on which House C was built, was constructed from huge slabs braced with mortar and earth, which is different from the earth and rubble that was later used to support House A-D (Schele 1986: 119). The Palace buildings, designated Houses A, A-D, B, C, D, E, F, G, H, I, J, K, L, and the Tower (Maudslay 1889–1902b: 1–38), all occupy the level of the Second Palace Terrace and its associated extensions, which collectively overlie the remains of earlier structures (see Ruz 1958: 69–116). In the Palace phase sequence proposed in this investigation (see Section 4.3, The Palace at Palenque, Phases 1–6, this chapter), Houses B, E, F, G, H, I, J, K, L, in addition to the Tower and Toilets all occupy the Second Palace Terrace, whereas Houses A, C, and D occupy Extension A of the Second Palace Terrace, and House A-D occupies Extension B of the Second Palace Terrace. The north and south subterranean buildings of the Palace occupy the level of the First Palace Terrace (see Table 4.4 and Palenque cross-section diagram [Figure 4.1]).

Over the years, several construction sequences have been offered for the Palace houses (see Table 4.1). The sequence that is proposed here generally accords with those of Andrews (1974), Greene Robertson (1985b: fig. 5 and table IV), Schele (1986: 118–119), Miller (1999: 35–38), and Martin and Grube (2000: 164). By considering aspects of each of these scenarios, in addition to all data available to this author, an attempt has been made to refine the sequence to more accurately represent the order in which the Palace at Palenque was formulated (see Table 4.1).

The bulk of the Palace's expansion and renovation occurred over an approximately 180-year period, between the reigns of K'inich Janab' Pakal I (A.D. 615–683) and K'inich K'uk' B'alam II (A.D. 764–799?) (Martin and Grube 2000: 164). In line with P. Mathews (2004: 154–157), these rulers are listed in chronological order in Table 4.2 (Palenque rulers' birth, accession, and death dates).

It has been the combination of epigraphic information, stratigraphic and ceramic data, as well as stylistic variations in the art architecture that have been used to support arguments for the order of construction of the Palace buildings (see Section 4.3, The Palace at Palenque, Phases 1–6, this chapter). Given the minimal amount of relevant stratigraphic information and ceramic data available, however, some of the strongest evidence for determining the construction sequence of the Palace Group buildings comes in the form of epigraphic information (i.e., texts found on architecture and associated monuments that display rulers' names and dedicatory dates that fall within particular reign periods). Table 4.3 and the far right column of Table 4.1 list the Palace houses in the order in which I believe they were constructed. Table 4.4 shows the corresponding ceramic phases (for a detailed discussion on the ceramics of Palenque, see Rands and Rands 1957, 1959; Rands 1974, 1985). Table 4.3 also provides the Long Count and Calendar Round dates that appear on monuments adhered to architecture. As will be outlined in the following discussion of the Palace phase sequence, these dates were used to assign the various Palace buildings to the reigns of different Palenque kings (see Appendix 2, The Palenque Palace Dates).

When comparing the more detailed of the Palace sequences proposed (Andrews 1974; Greene Robertson 1985b: fig. 5, table IV; Schele 1986: 118–119), there is general agreement concerning the building order of most of the houses, more specifically, that the construction of House E occurred before House B, House B preceded that of House C, which was later followed by the construction of Houses A, D, and A-D; the Tower; and Houses I, F, and L, respectively. The main differences between each of the construction scenarios proposed concern the addition of Palace Houses J, K, and L.

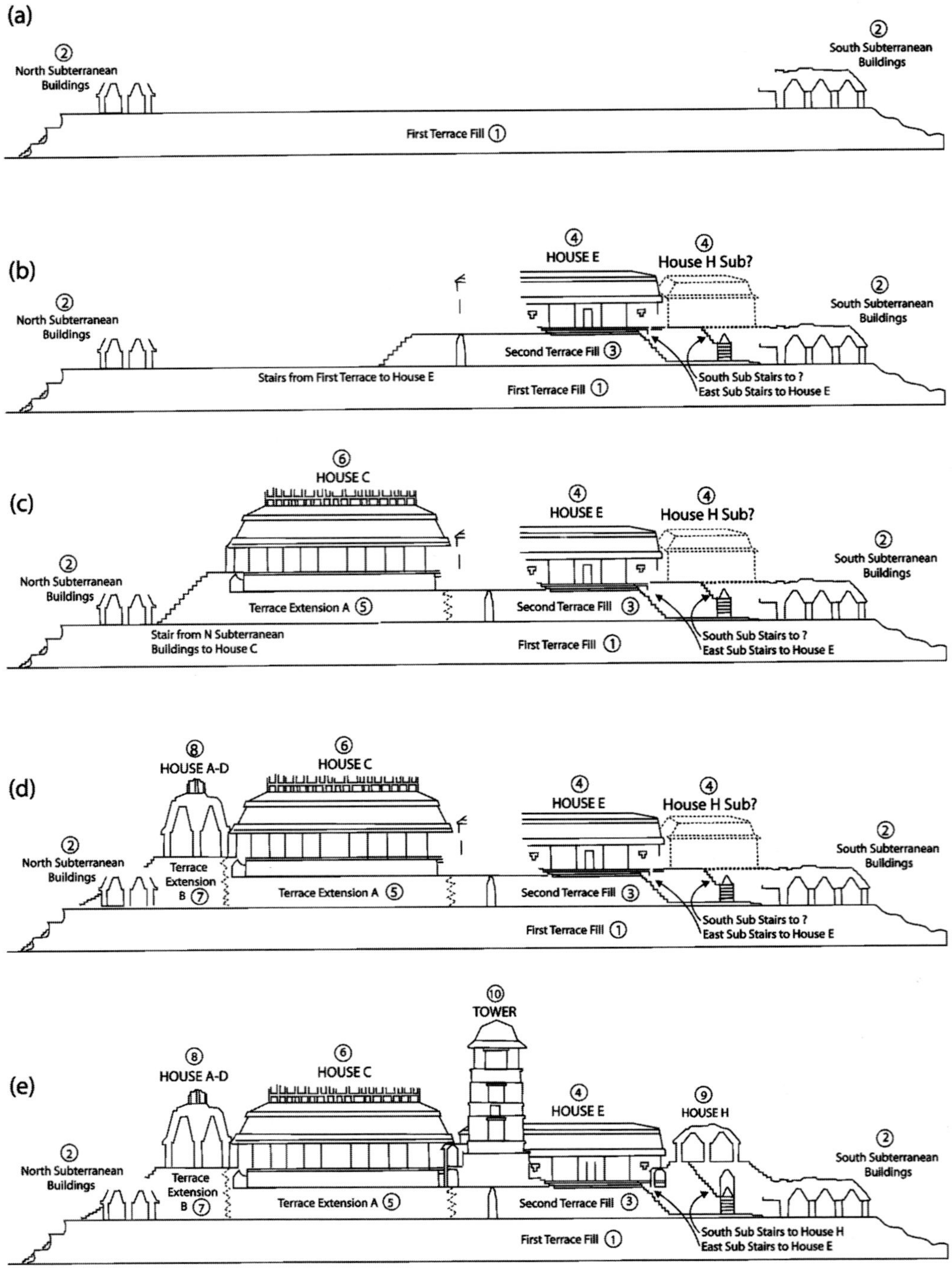


**Figure 4.1.** Palenque Palace Cross Section Looking East, Showing Construction Sequence of Platforms and Buildings. Modified after Greene Robertson (1985b: fig. 7), reproduced with the permission of Merle Greene Robertson.

For example, in Andrews's (1974) sequence, the construction of House K is broken into two separate phases: a southern portion of House K and its northern addition. The northern extension of House K was added after the construction of House D, which occurred at a later stage of the Palace's development; the effect was to combine Houses D and K into a single structure (see Section 4.3.1, The Buildings of the Palace Phase 1, House K: Description, this chapter). In addition, Andrews places the construction of House K's southern portion between that of Houses E and B and the construction of the northern extension after the erection of the Tower. Greene Robertson (1985b: fig. 5, table IV) offers two construction sequences for the Palace, placing House K differently in each scenario. The positioning of House K in Greene Robertson's (1985b: fig. 5) first sequence agrees with

Table 4.1. *Various proposals for the relative chronology of Palace structures at Palenque*

| Andrews (1974) | Greene Robertson (1) (1985b:Vol. III, fig. 5) | Greene Robertson (2) (1985b: table IV) | Linda Schele (1986: 118–119) | Alex Parmington, (this volume) |
|---|---|---|---|---|
| Subterranean Galleries | Subterranean Galleries | Subterranean Galleries | Subterranean Galleries | Subterranean Galleries |
| | | | | **Phase [1]** |
| House E | House E | House E | House E | House E, |
| House K (southern portion) | Houses J and K | | | House K, and H-Sub? |
| House B | House B | House B | House B | House B |
| | | House J | | House J |
| | | House K | | |
| | | | | **Phase [2]** |
| House C | House C | House C | House C | House C |
| | | | House K | |
| | | House L? | House L | |
| House A | House A | House A | House A | House A |
| | | | | **Phase [3]** |
| Houses D and A-D | House D | House D | Houses D and A-D | Houses D and A-D |
| | House A-D | House A-D | | |
| House J (including East Gallery) | | | | |
| | | | | **Phase [4]** |
| Tower | Tower | Tower | | Tower |
| House K (north addition) | | | | |
| | Toilets | | | Toilets |
| House G | Houses H and G | Houses G and H | Houses G and H | Houses G and H |
| House H | | | | |
| | | | | **Phase [5]** |
| House I | House I | House F | | House F |
| House F | House F | House I | | House I |
| | | | | **Phase [6]** |
| House L | House L | SW Rooms (plaza level) | | House L |
| 2 SW Rooms (plaza level) | 2 SW Rooms (plaza level) | | | SW Rooms (plaza level) |

Andrews's (1974) placement of the southern portion of House K (i.e., predating that of House B and postdating that of House E); however, Greene Robertson coincides its fabrication with that of House J. In Greene Robertson's (1985b: table IV) second scenario, both Houses J and K are constructed after House B, and House K postdates House J. Schele (1986: 118–119) argues that Houses B, C, and E (including the Subterraneos), were built between 9.10.15.0.0 (A.D. 647) and 9.11.15.0.0 (A.D. 667), whereas House K was built after House C and before House L. Andrews places House L after Houses F and I, which accords with Greene Robertson's first scenario. Greene Robertson's second sequence places House L much earlier, predating the construction of House A.

As just demonstrated, when compared, the Palace construction scenarios show some variation, which is a consequence of the limited data available to categorically verify the sequences proposed. Before proceeding, it is conceded that the Palace sequence offered in this investigation, in addition to the analysis drawn from it, can only be as accurate as the evidence that supports it. Like the earlier construction sequences proposed (Andrews 1974; Greene Robertson 1985b: fig. 5, table IV; Schele 1986: 118–119; Miller 1999: 35–38; Martin and Grube 2000: 164), the one that is presented here is also based on limited archaeological, epigraphic, and stylistic evidence, in addition to some probabilistic arguments. For this reason, the analysis that is offered here should be viewed as a practical demonstration of an analytical technique rather than a presentation of definitive results. The primary objective of this chapter is to illustrate by example how a broad sequence in Maya architecture, once identified, may be combined with methods in access analysis to divulge shifts in the distribution of sculpture and architectural space over different periods of time. The power of this procedure lies in its capacity to be reapplied given the availability of new information. For example, in the advent of any further archaeological investigation(s) being undertaken, to more precisely determine the evolution of the Palenque Palace, some refinement of the architectural sequence offered here may be undertaken, after which amended access maps may be overlaid and

Table 4.2. *Palenque rulers' birth, accession, and death dates*

| | | Birth | | Accession | | Death | |
|---|---|---|---|---|---|---|---|
| Ruler | Reign | Long Count | Julian Date | Long Count | Julian Date | Long Count | Julian Date |
| K'uk' B'alam I | 431 – ~435 | 8.18. 0.13.6 | 30 Mar 397 | 8.19.15. 3. 4 | 10 Mar 431 | 8.19. ?. ?. ? | ~435 |
| "Casper II" | 435 – ~487 | 8.19. 6. 8.8 | 8 Aug 422 | 8.19.19.11.17 | 9 Aug 435 | 9. 2. ?. ?. ? | ~487 |
| B'utz'aj Sak Chik | 487 – ~501 | 9. 1. 4. 5.0 | 9 Aug 459 | 9. 2.12. 6.18 | 28 Jul 487 | 9. 3. ?. ?. ? | ~501 |
| Akul Mo'Nab' I | 501 – 525 | 9. 1.10. 0.0 | 5 Jul 465 | 9. 3. 6. 7.17 | 3 Jun 501 | 9. 4.10. 4.17 | 29 Nov 524 |
| K'an Joy Chitam I | 529 – 665 | 9. 2.15. 3.8 | 3 May 490 | 9. 2.15. 3. 8 | 23 Feb 529 | 9. 6.11. 0.16 | 6 Feb 565 |
| Akul Mo'Nab' II | 565 – 570 | 9. 4. 9. 0.4 | 3 Sep 523 | 9. 6.11. 5. 1 | 2 May 565 | 9. 6.16.10. 7 | 21 Jul 570 |
| Kan B'alam I | 572 – 583 | 9. 4.10. 1.5 | 18 Sep 524 | 9. 6.18. 5.12 | 6 Apr 572 | 9. 7. 9. 5. 5 | 1 Feb 583 |
| Ix Yol Ik'nal | 583 – 604 | 9. 4. ?. ?. ? | ~530 | 9. 7.10. 3. 8 | 21 Dec 583 | 9. 8.11. 6.12 | 4 Nov 604 |
| Ajen Yol Mat | 605 – 612 | 9. 6. ?. ?. ? | ~555 | 9. 8.11. 9.10 | 1 Jan 605 | 9. 8.19. 4. 6 | 8 Aug 612 |
| Ix Sak K'uk' | 612 – 615 | 9. 7. ?. ?. ? | ~580 | 9. 8.19. 7.18 | 19 Oct 612 | ? | ? |
| K'inich Janab' Pakal I | 615 – 683 | 9. 8. 9.13.0 | 23 Mar 603 | 9. 9. 2. 4. 8 | 26 Jul 615 | 9.12.11. 5.18 | 28 Aug 683 |
| K'inich Kan B'alam II | 684 – 702 | 9.10. 2. 6.6 | 20 May 635 | 9.12.11.12.10 | 7 Jan 684 | 9.13.10. 1. 5 | 16 Feb 702 |
| K'inich K'an Joy Chitam II | 702 – 721? | 9.10.11.17.0 | 2 Nov 644 | 9.13.10. 6. 8 | 30 May 702 | 9.14. ?. ?. ? | 721? |
| K'inich Akul Mo'Nab' III | 721 – ~740? | 9.12. 6. 5.8 | 13 Sep 678 | 9.14.10. 4. 2 | 30 Dec 721 | 9.15. ?. ?. ? | ~740 |
| U Pakal K'inich Janab' Pakal II | r.742 – ~750 | 9.13. ?. ?. ? | ~700? | 9.15. ?. ?. ? | r.742 | 9.15. ?. ?. ? | ~750 |
| K'inich Kan B'alam III | r.751 – ~763? | 9.13. ?. ?. ? | ~710? | 9.16. ?. ?. ? | r.751 | 9.16. ?. ?. ? | ~763? |
| K'inich K'uk' B'alam II | 764 – ~799? | 9.14. ?. ?. ? | ~715? | 9.16.13. 0. 7 | 4 Mar 764 | 9.18. ?. ?. ? | ~799? |
| Wak Kimi Janab' Pakal III | 799 – ? | ? | ? | 9.18. 9. 4. 4 | 13 Nov 799 | ? | ? |

*Note:* Correlation constant: 584285. r = reconstructed date

Table 4.3. *Palace construction sequence and date correlations*

| Structures | Julian Correlation | Long Count and Calender Round Dates | | | Monument and event |
|---|---|---|---|---|---|
| Subterranean | 11 Oct 652 | 9.11. 0. 0. 0 | 12 Ajaw | 8 Kej | Tableritos 1, Dedication |
| Passages to Houses E, | 9 Jun 654 | 9.11. 1.12. 6 | 7 Kimi | 4 Xul | Tableritos 2, Dedication |
| K, and H-Sub | 11 Jun 654 | 9.11. 1.12. 8 | 9 Lamat | 6 Xul | Tableritos 3, Dedication |
| House E | 1 Nov 654 | 9.11. 2. 1.11 | 9 Chuwen | 9 Mak | Dedication, Tablet 96 Glyphs |
| House B | | | – | | – |
| House J | | | – | | – |
| House C Foundation | 7 Aug 659 | 9.11. 6.16.11 | 7 Chuwen | 4 Ch'en | Dedication? Hieroglyphic Stairs |
| House C | 22 Dec 661 | 9.11. 9. 5.19 | 4 Kawak | 2 Pax | Dedication, Eaves |
| House A Foundation | 25 Mar 662 | 9.11. 9.10.12 | 6 Eb' | 10 Woh | Auto Sacrifice, *Alfarda* 1 |
| | 26 Mar 662 | 9.11. 9.10.13 | 7 B'en | 11 Woh | Auto Sacrifice, *Alfarda* 2 |
| House A | 12 Mar 668 | 9.11.15.11.11 | 1 Chuwen | 19 Pop | Event? Pier A |
| | 7 Jan 692 | | or | | |
| | | 9.12.19.14.12 | 5 Eb' | 5 K'ayab' | Event? House A, Lintel, and Pier Text |
| Houses D and A-D | 10 Aug 720 | 9.14. 8.14.15 | 9 Men | 3 Yax | Dedication, Palace Tablet |
| Tower | 2 Jan 725 | 9.14.13. 5. 0 | 2 Ajaw | 8 K'ayab' | Sacrifice, Tablet Scribe |
| | | 9.14.13. 5. 0 | 2 Ajaw | 8 K'ayab' | Sacrifice, Tablet Orator |
| Houses G and H | | | – | | – |
| House F | | | – | | – |
| House I | | | – | | – |
| House L | | | – | | – |
| SW Room Plaza Level | | | – | | – |
| Intensive Partitioning | | | – | | – |

*Note:* Correlation constant 584285.

Table 4.4. *Relative chronology of Palace buildings and corresponding rulers*

| Ruler | Reign AD | Building/Structure | Ceramic Phase | AD |
|---|---|---|---|---|
| Akul Mo'Nab' II | 565–570 | ? | Motiepa | 400 |
| Kan B'alam I | 572–583 | ? | | |
| Ix Yol Ik'nal | 583–604 | North Substructure Galleries?<br>South Substructure Galleries? | | |
| Ajen Yol Mat | 605–612 | ? | Otolum | 600 |
| Ix Sak K'uk' | 612–615 | ? | | |
| K'inich Janab' Pakal I | 615–683 | (South, East, West Subterranean Passage Construction)<br>House E<br>House K, H-Sub?<br>(Post 654 Construction)<br>House B<br>(Northwest Court First Platform Occupation)<br>House J<br>(Second Terrace Construction)<br>House C<br>(Post 659 Construction)<br>House A? | | |
| K'inich Kan B'alam II | 684–702 | House A? | Murcielagos | 700 |
| K'inich K'an Joy Chitam II | 702–721? | House D & A-D<br>(East Court Completed)<br>North Substructure Stairs and Tiers | | |
| K'inich Akul Mo'Nab' III | 721–740? | House G & H<br>Toilets<br>Tower<br>Tower Court | | |
| U Pakal K'inich Janab' Pakal II | 742–750? | House F<br>House I<br>Sweat Baths (bracketed for the three rulers 742–799?) | | |
| K'inich Kan B'alam III | 751–763? | | | |
| K'inich K'uk' B'alam II | 764–799? | | Balunte | 770 |
| Wak Kimi Janab' Pakal III | 799– ? | (No Known Building Assigned)<br>Remodeling and partitioning of existing Palace structures<br>House L??<br>Rooms on Southwest corner at Plaza level, west side?? | | |

Modified after Greene Robertson (1985b: table IV).

reanalysed, and subsequent conclusions drawn. Although problems remain in determining the precise order in which the Palace buildings were constructed, the six phases presented in the following sections collectively represent a thorough attempt to accurately identify and assign both the sculptural and architectural programs of the Palenque Palace to their corresponding rulers.

## 4.3. The Palace at Palenque, Phases 1–6

### 4.3.1. The buildings of the Palace Phase 1

Phase 1 of the Palenque Palace (see Figure 4.2), offered in this investigation, comprises Houses B, E, J, K, and "H-Sub" (?), which were constructed during the first half of K'inich Janab' Pakal I's reign (circa A.D. 615–650). Earlier structures used, although not constructed during this phase, include the then-exposed northern subterranean structures (referred to here as the Northern Galleries) and the Subterraneos. As seen in the Palace cross-section diagram (Figure 4.1), this phase comprises two principal stratigraphic levels, where the southern upper half of the Second Palace Terrace fill supports Houses B, E, J, K, and H-Sub, which overlie the Southern Subterranean Structures. Combined, the "upper" Palace Houses form an S-shaped configuration, which defines the boundaries of two small courtyards, one opening to

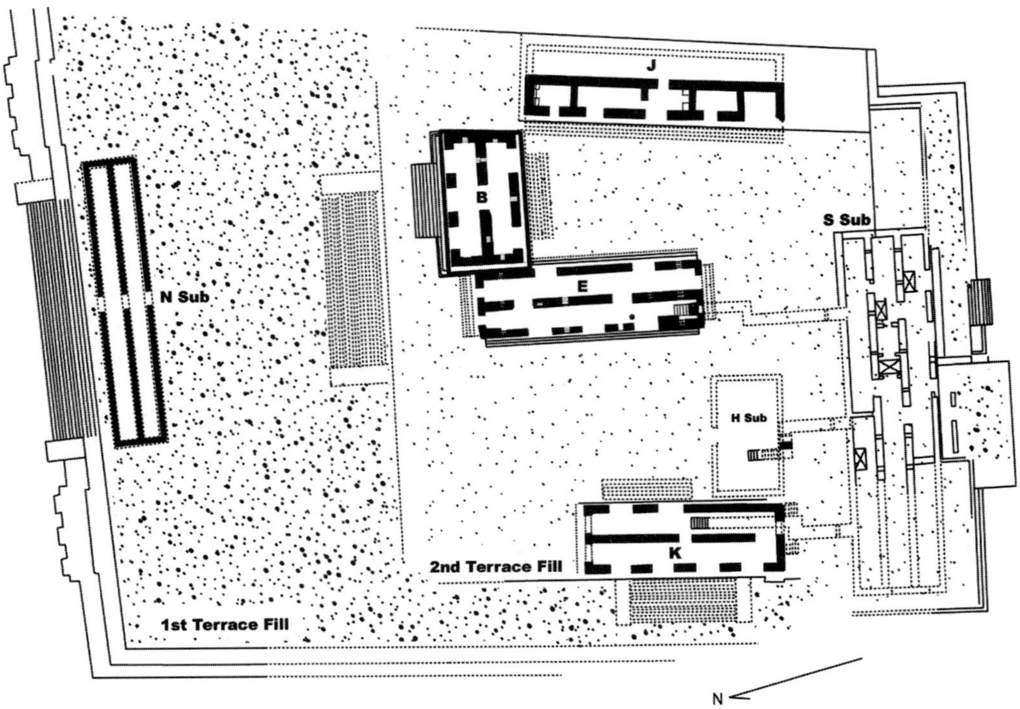


**Figure 4.2.** Map of the Palenque Palace Phase 1. Palenque Palace modified after Greene Robertson (1985b: fig. 1).

the north and one opening to the south. Houses B, E, and J define the perimeter of what later becomes the Southeast Court, and Houses E, H-Sub, and K form the boundary of what later becomes the Tower Court.

At this phase of the Palace's construction, the northern half of the First Palace Terrace was not yet covered by the Second Palace Terrace extensions A and B, which left the Northern Galleries exposed. The effect of combining these terrace platforms was to provide a progressive change in elevation from the perspective of someone entering the building group from the plaza fronting the northern Palace stairs.

#### The northern and southern subterranean galleries and passages

The northern and southern substructures (see Figure 4.3) represent the remains of what was probably an earlier phase of the Palace (Miller 1999: 37). It has been suggested that many other buildings once occupying the level of the First Palace Terrace were destroyed and/or buried to raise the height of the Palace to its current ten-metre level (Martin and Grube 2000: 164). At a later architectural phase, the Northern Galleries were buried beneath a massive stairway commissioned by K'inich K'an Joy Chitam II, which occurred just prior to the construction of House A-D (see Section 4.3.3, The Buildings of the Palace Phase 3, this chapter). Ruz (1958: 118–123, 1962: 46–53) has noted the presence of a razed building, now under today's northern Palace stair; more specifically, the remnants of a staircase and a medial wall have been identified, including a pier that occupies the Palace's lower platform. At this early stage of the evolution of the Palace, the northern stair once provided access to the level of the First Palace Terrace and to the Northern Galleries interior. Given the close proximity of this building to the northern stairs, it is likely that the structure functioned in part to control access to those structures and/or plaza spaces positioned behind the building. It is reasonable to argue that at least one doorway would have existed in the northern, medial, and southern walls of this northern building, providing access through the structure and on into the Palace interior. It is also possible, however, that the northern and southern façades of the Northern Galleries once comprised multipiered entrances configured in a similar way to the upper Palace houses (e.g., Houses E, B, and K).

The earliest structure at Palenque is argued to be the Subterraneos (the Southern Subterranean Structures; Figure 4.4), positioned at the southern end of the Palace; they predate the Olvidado Initial Series, which reads 9.10.14.5.10 3 Ok 3 Pop (2 March 647; Schele 1986: 117). The Northern Galleries are also argued to be early, correlating stratigraphically with the Subterraneos (see Figure 4.3). As noted by Andrews (1974), presently the Southern Subterranean Structures comprise three parallel galleries, or corridors, which measured approximately

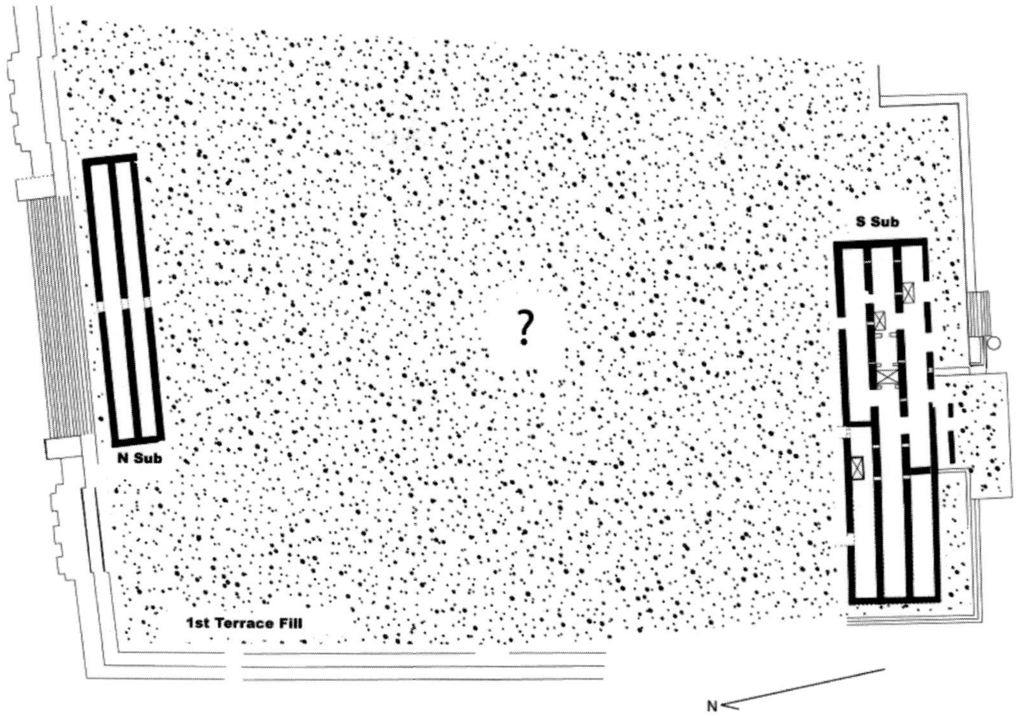


Figure 4.3. North and South Subterranean Buildings of Palenque Palace, modified after Greene Robertson (1985b: fig. 1).

forty-four metres in length before their later partitioning (see Table 4.5). At some later stage, probably just after the burial of the building, additional curtain walls were installed segmenting the corridors into a number of smaller rooms. Among other things, a transverse wall was added immediately to the east of the South Subterranean Passage entrance, breaking the northern galleries of the Subterraneos into two narrow rooms (Rooms 1 and 2). In addition, according to Greene Robertson's (1985b: fig. 1) plan of the Palace, the doorway just southeast of

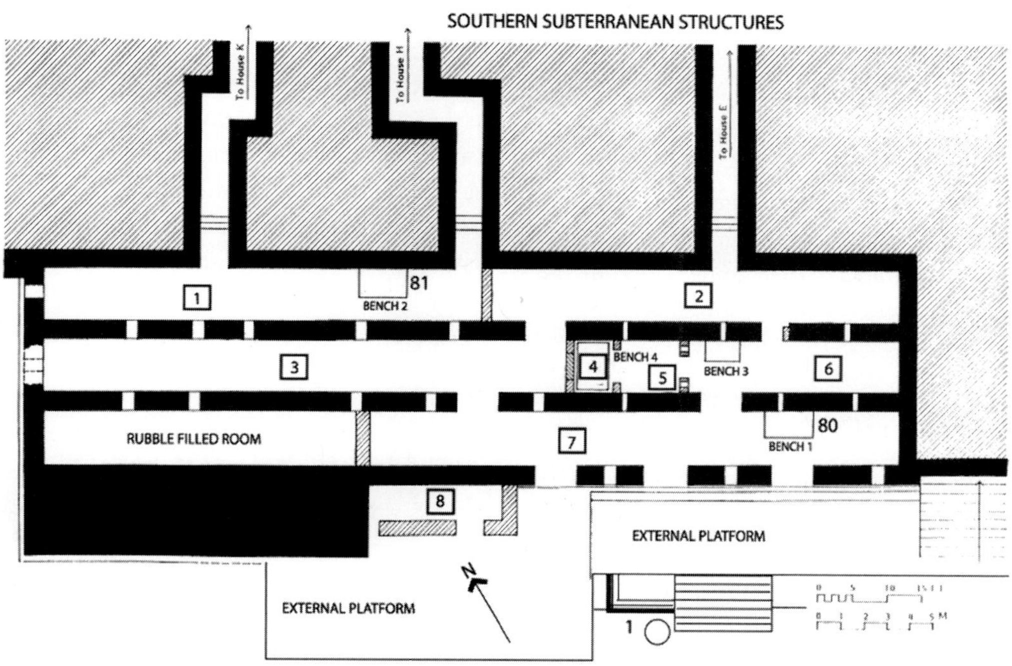


Figure 4.4. South Subterraneos Plan, Palenque Palace. Modified after Andrews (1974). George F. and Geraldine D. Andrews papers. The Alexander Architectural Archive, the University of Texas Libraries, the University of Texas at Austin.

Table 4.5. *South Subterraneos room dimensions*

| Room | Length | Width | Area |
|---|---|---|---|
| 1 | 19.46 m | 2.31 m | 44.95 $m^2$ |
| 2 | 18.09 m | 2.31 m | 41.78 $m^2$ |
| 3 | 23.40 m | 2.32 m | 54.28 $m^2$ |
| 4 | 1.95 m | 2.31 m | 4.50 $m^2$ |
| 5 | 3.03 m | 2.31 m | 6.99 $m^2$ |
| 6 | 9.58 m | 2.32 m | 22.22 $m^2$ |
| 7 | 23.47 m | 2.31 m | 54.21 $m^2$ |

*Note:* Dimensions in accordance with Andrews (1974).

**Figure 4.5.** Two of the Six Subterranean Passage Tableritos. Drawing by Linda Schele (No. 4060), © David Schele, courtesy of the Foundation for the Advancement of Mesoamerican Studies, Inc., www.famsi.org.

Bench 2 (GR 81; for GR definition see Chapter 5, Section 5.1, Access Analysis of the Palace Phases 1–6) in Room 1 was also blocked. It appears that, combined, the purpose of these two secondary additions may have been to close off access to the west and south subterranean passages via the Subterraneos. As well as the additions just mentioned, for reasons unknown, the southern end of the south corridor of the Subterraneos was intentionally filled with rubble and walled up (see Chapter 5, Section 5.1.6, Analysis of Palace Phase 6).

Attached to the south of the Subterraneos is a terrace that is accessible from the southeast room of the substructure (Room 7) in addition to Room 8, which is positioned on the terrace itself. Leading from the northeast and northwest rooms (Rooms 1 and 2) of the Subterraneos are three passages, one providing access to House E (i.e., the East Subterranean Passage), one leading to House H (i.e., the South Subterranean Passage), and one leading to House K (i.e., the West Subterranean Passage).

**THE PALACE SUBTERRANEAN PASSAGE TABLERITOS 1, 2, AND 3: DESCRIPTION.** Three dates have been recovered from six carved stone tablets once located in three subterranean passages of the Palace at Palenque (see Figure 4.5 and Table 4.6): two in the East Subterranean Passage, two in the South Subterranean Passage, and two in the West Subterranean Passage (Schele and Mathews 1979: fig. 36). Referred to as the "Tableritos," the principal theme of these textual monuments is "Text, Dedication" (TDed). In each of these corridors, two limestone panels once faced each other from opposite sides of the passage in close proximity to the stairs (P. Mathews 2004: personal communication). Although today it is unclear precisely which pair of Tableritos corresponds to which passage, several points concerning these monuments seem sure: (1) all three Tableritos dates fall within two years of each other; (2) the dedication of House E postdates the latest Tableritos date by less than five months; and (3) all three dates, including the House E dedication (identified in the Tablet of 96 Glyphs), fall within the reign of K'inich Janab' Pakal I (see Tables 4.2 and 4.6).

Due to the early nature of the Tableritos dates, in addition to Andrews's (1974) and Greene Robertson's (1985b) positioning of House H after that of the Tower in their Palace construction scenarios, it appears that there must have been an earlier structure underlying House H (i.e., H-Sub). At this stage of the Palace's evolution, the "closing events" recorded on the Tableritos panels have been taken to represent the dedicatory date of the subterranean passages, signaling the approximate period of their completion in addition to those of the buildings to which they lead (Houses E, K, and H-Sub). In my view, it is exceedingly unlikely that the South Subterranean Passage leading from the Subterraneos emerged onto open ground (which would be the implication if the construction of House H postdates that of the Tower, and also if there was no structure preceding House H). Consistent with the logic of this proposal, Seler (1976: 53) noted that House E "must be considered one

Table 4.6. *East, West, and South Subterranean Passage Tableritos dates*

| Structure | Long Count and Calendar Round dates | | | Julian date |
|---|---|---|---|---|
| Subterranean Passages Leading to Houses E, K, and H-Sub | 9.11.0. 0. 0 | 12 Ajaw | 8 Kej | 11 Oct 652 |
| | 9.11.1.12. 6 | 7 Kimi | 4 Xul | 9 Jun 654 |
| | 9.11.1.12. 8 | 9 Lamat | 6 Xul | 11 Jun 654 |
| House E Dedication (Tab 96 Glyphs) | 9.11.2. 1.11 | 9 Chuwen | 9 Mak | 1 Nov 654 |

*Notes:* From Schele and Mathews (1979). Correlation constant: 584285.

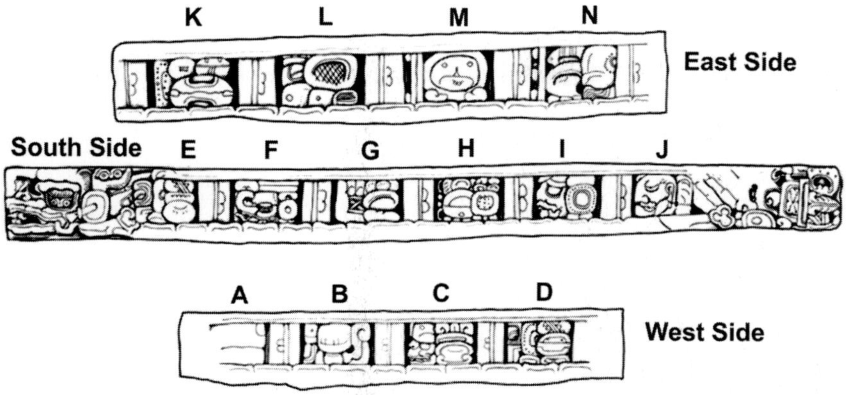


**Figure 4.6.** Southern Subterraneos Bench 1 (GR 80). Drawing by Linda Schele (No. 123), © David Schele, courtesy of the Foundation for the Advancement of Mesoamerican Studies, Inc., www.famsi.org.

of the oldest [buildings] in the Palace complex, not only on the basis of its location . . . [and the] distinctive characteristics of its form [but] because of one of the underground passages that leads from the south side of the plateau ends in it." According to Schele (1986: 119), "the inscriptions and the iconography associated with the lower galleries were designed to define their function in terms of the upper building. Each of the three passages that lead from the subterranean galleries to the upper level is marked with plaster sculpture identifying it with an entry to the Underworld."

**THE SOUTHERN SUBTERRANEOS BENCH 1: DESCRIPTION (GR 80).** Four benches were once located in the Subterraneos (i.e., Benches 1, 2, 3, and 4). Bench 1, now residing in the Palenque Museum, was found in the southeast room (Room 7); Bench 2 is in the northeast room (Room 1 [precise position unknown]); Bench 3 is in the east-central room (Room 6); and Bench 4 is in the small alcove towards the centre of the structure (Room 4). Bench 1 (GR 80) displays an inscription that is carved around the periphery of the throne (Figure 4.6). The sculptural theme of the throne text has been defined here as "Text, Cosmological Celestial Underworld" (TCosCU). The inscription, together with other iconographic elements, represents a "Cosmic Monster." A diagnostic element of the Cosmic Monster (also known as the Bicephalic or Celestial Monster) is the presence of two heads, often found at each end of a reptilian or sky-band body. By convention, the front head of the creature is depicted fleshed whereas the rear head is portrayed inverted and skeletal. In all examples, the skeletal head displays quadripartite symbolism; more specifically, a spondylus seashell, cross-bands, and a stingray spine are depicted associated with a bowl. In addition, the creature often displays either a cross-band or Venus symbol as one of its eyes. The creature combines aspects of the heavens and Underworld: Its forward head represents Venus and the rear head the Underworld aspect of the sun (Miller and Taube: 1993: 45).

The Cosmic Monster carved on Bench 1 displays a zoomorphic head with Venus (*Ek*) eye and scaled underbelly; at the tail end of the creature is another head surmounted by a combination of celestial and Underworld symbols: a *k'in* (sun) sign; a sea shell; stingray spine; a bowl. The inscription is early in style and comprises fourteen glyphs carved in a band that forms the creature's body. Stuart's (2003: 3) transcription of the throne text reads:

A: [?-?}]-**ba**
B: **a-je-ne** (*ajen*)
C: **?-SAK-IK'** ("white wind")
D: **nu-mu li-TA-CHAN-na** ("passing in the sky")
E: **nu-mu-li-TA-ka-ba** ("passing in the earth")
F: **?-NAL**
G: **yo-to-ti** ("the house of")
H: **2-WINIKHAAB?-pi-tzi-la**
I: **K'INICH-JANAB**
J: **pa-ka-la**
K: **K'UHUL-BAAK-AJAW**
L: **tu-u-pa-ti** ("on back of . . . ")
M: **12-AJAW**
N: **8-CHAK-SIHOM-CHUM-TUUN-ni** (*chumtuun* "stone seating")

The Celestial Monster's association with heaven and the earth is noted in the text (see Glyphs D and E), where *numil ta chan, numil ta kab* ("passing in the sky, passing in the earth") refers to the Milky Way as it progresses across the night sky and into the Underworld (Stuart 2003: 2). K'inich Janab' Pakal I is named on the throne at positions I and J and is followed by a Calendar Round date located towards the end of the text (Glyphs M and N). Reading 12 Ajaw 8 Kej, and correlating to a Long Count date of 9.11.0.0.0 (A.D. 652), the date marks the throne as one of Palenque's earliest monuments. It remains unclear whether the southeastern gallery of the Subterraneos was the original location of Bench 1. It remains possible that Bench 1 might have originally been positioned in the northwestern room of House E, below K'inich Janab' Pakal I's coronation scene (i.e., the Oval Tablet; P. Mathews [2004: personal communication]). Functioning as K'inich Janab' Pakal I's principal throne, it was later moved to the Subterraneos by K'inich K'an

Figure 4.7. East Subterranean Passage Vaults 1 and 2 (GR 7a and 7b, respectively). Drawing by Linda Schele (No. 130), © David Schele, courtesy of the Foundation for the Advancement of Mesoamerican Studies, Inc., www.famsi.org.

Joy Chitam II and replaced with the Del Río Throne (GR 6) to coincide with the later king's inauguration in 9.13.10.6.8 (A.D. 702).

**THE SOUTH, EAST, AND WEST SUBTERRANEAN PASSAGE VAULTS: DESCRIPTION (GR 7, 78, AND 79).** In addition to the Tableritos, located near the stairs in the South, East, and West Subterranean Passages, are elaborately decorated stucco vaults (GR 7, 78, and 79). The iconographic evidence found in both House E and the Subterraneos passage leading to and from the building supports the view that their function should be understood in terms of their association with one another. It has been proposed that "the Subterranean Galleries were a symbolic replication of the Underworld, and the corridors that led from the upper levels in the Palace to them are sculpturally defined as points of transition and transformation" (Schele 1986: 122). The Underworld iconography of House E sculpture has been found to correlate with the Underworld symbolism found in the three passages leading to the Subterraneos. The East, West, and South Subterranean Passage vaults all display iconography, signaling each passage as an entrance to the Underworld.

**THE EAST SUBTERRANEAN PASSAGE VAULT 1: DESCRIPTION (GR 7A).** Positioned in the East Subterranean Passage leading to the western room of House E, the sculptural theme of Vault 1 is defined as "Cosmological Celestial Underworld" (CosCU). The vault displays some iconographic similarities with the Bicephalic Room of House E (Schele 1986: 121–122). The bicephalic creature sculptured at the northern end of the eastern corridor of the house is similar to the creature represented on the outer vault of the East Subterranean Passage. Each monster displays shells and water symbols, as well as serpent scales. In addition to these elements, there is a sun (*k'in*) sign carved at the centre of the east subterranean vault. The monster's head, located on the viewer's left of the arch, has a large nose with an infixed mirror, and a flint blade or stingray spine, which is visible emerging from its mouth; the body of the monster makes up the arch of the vault. The presence of cord holders, in close proximity to the decorated vault, suggests that the vault functioned to signal a threshold of some kind (Greene Robertson 1985a: 32–33).

**THE EAST SUBTERRANEAN PASSAGE VAULT 2: DESCRIPTION (GR 7B).** The sculptural theme of Vault 2 (Figure 4.7), also located in the East Subterranean Passage, is defined as "Cosmological Underworld" (CosU). The carving displays a two-headed reptilian creature spanning the arch of the vault. On the viewer's left of the image, a deer can be seen lying on the back of the serpent facing a crouching, Underworld aspect of the Sun God, which is present on the viewer's right of the scene. Growing between the deer and the Sun God is plant foliation of some kind, perhaps maize (Greene Robertson 1985a: 33–34).

**THE WEST SUBTERRANEAN PASSAGE VAULT 3: DESCRIPTION (GR 78).** Like that of the vaults appearing in the East Subterranean Passage, the sculptural theme of the west subterranean vault (Figure 4.8) is also "Cosmological Underworld" (CosU). The scene comprises the upper torso of the Maize God either tending or emerging from a

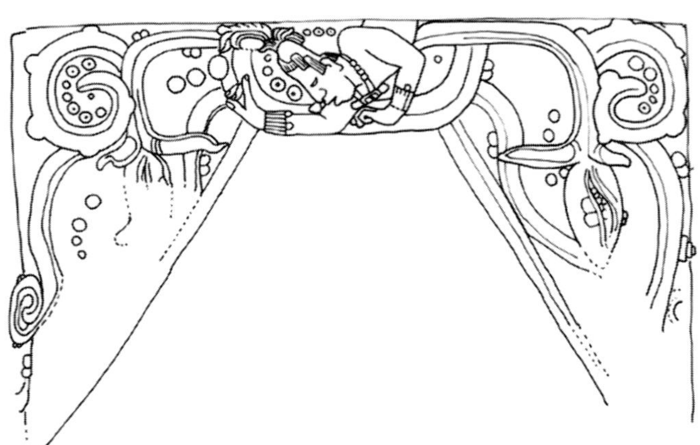

Figure 4.8. West Subterranean Passage Vault 3 (GR 78). Drawing by Linda Schele (No. 137), © David Schele, courtesy of the Foundation for the Advancement of Mesoamerican Studies, Inc., www.famsi.org.

maize plant. The sides of the vault depict budding maize foliation and scrolls of flowing water. The Maize God wears tubular bead bracelets with an *ajaw* bone attached to one edge, pendulous earrings, and a beaded necklace with a pectoral ornament. Originally the image was painted blue, which was a sacred colour among the Maya. The presence of water symbolism in the scene suggests that this "birth" of the Maize God is occurring in the Underworld (see Greene Robertson 1985a: 34).

**THE SOUTHERN SUBTERRANEAN PASSAGE VAULT: DESCRIPTION (GR 79).** Very little sculpture survives on the South Subterranean Passage vault; however, enough remains to tentatively define the sculptural theme as consistent with the vaults found in the East and west subterranean Passage. More specifically, the sculptural theme is CosU. For example, the inner vault displays the remains of maize foliation. In addition to a cross-band, the southern side of the vault also shows the remains of a square-nosed serpent (Greene Robertson 1985a: 35; Schele 1986: 122).

### House E: Description and sculpture

Set on a lower platform than the other Palace houses, House E is the first building to be constructed on the upper platform of the Palace (Maudslay 1889–1902b: 23; Andrews 1974; Seler 1976: 53). The construction of House E was most likely commissioned by K'inich Janab' Pakal I (Schele 1986: 118): The dedicatory date

Table 4.7. *House E, East and West Gallery dimensions before partitioning*

| Gallery | Length | Width | Area |
|---|---|---|---|
| East | 22.19 m | 2.37 m | 52.6 $m^2$ |
| West | 22.19 m | 2.33 m | 51.7 $m^2$ |

*Note:* Dimensions in accordance with Andrews (1974).

recorded for House E on the Tablet of the 96 Glyphs reads 9.11.2.1.11 9 Chuwen 9 Mak (A.D. 654). Originally painted white, House E was named *sak nuk naah* ("white skin house"); the other Palace buildings were painted predominantly red (Greene Robertson 1985a: 15; Martin and Grube 2000: 163). As noted by Andrews (1974), House E originally comprised two parallel galleries separated by a medial wall (see Table 4.7 for the gallery dimensions). Part of House E's original plan included narrow doorways seen at each end of both galleries; these small doorways are unusual for Palenque, ranging from 50 to 70 cm in width. House E also displays wider entrances (located in its east and west walls) than are displayed in all the other Palace houses (Figure 4.9).

Unlike Houses B and C, there are no piers associated with the entrances of House E (Greene Robertson 1985a: 10); in this structure, doorways are separated by walls. The northern doorway to the northeast room of House E was once used as a main entrance providing access to and from the East Court (see Section 4.3.2. The Buildings of the Palace Phase 2, Section 4.3.3. The

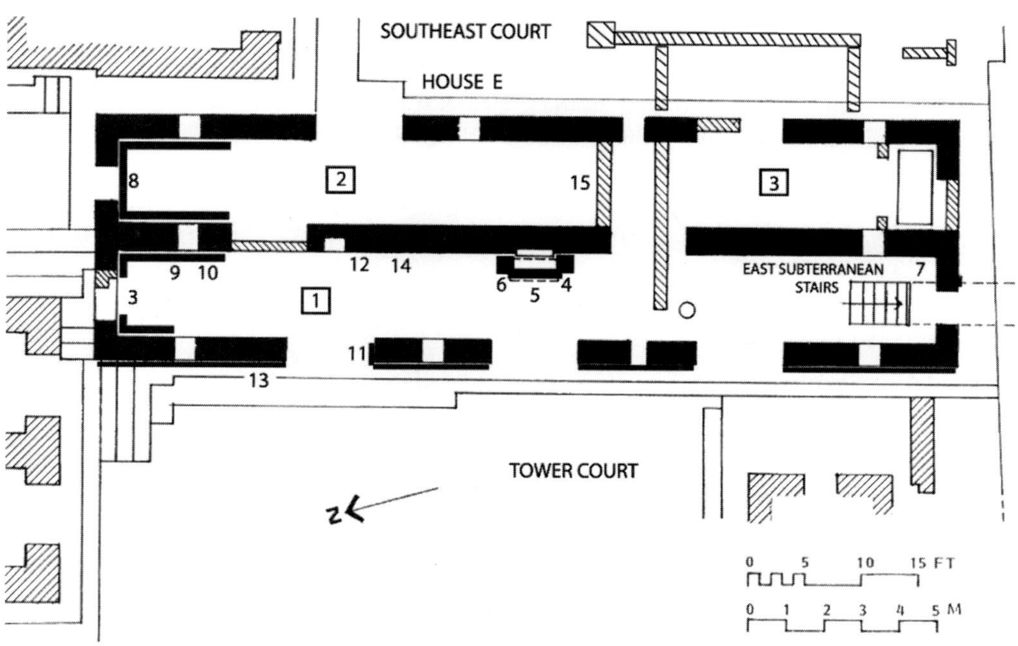


**Figure 4.9.** House E Plan, Palenque Palace. Modified after Andrews (1974). George F. and Geraldine D. Andrews papers. The Alexander Architectural Archive, the University of Texas Libraries, the University of Texas at Austin.

Buildings of the Palace Phase 3, Section 4.3.4. the Buildings of the Palace Phase 4, Section 4.3.5. The Buildings of the Palace Phase 5, and Section 4.3.6. The Buildings of the Palace Phase 6, this chapter). With the later addition of transverse walls (as well as the extension assigned to the southeast of the building), the number of rooms in House E was increased to five. The later modifications suggest that House E was never intended to be used by the general population (Greene Robertson 1985a: 9). Although it remains unclear how the temporary partitioning was configured in House E, the high number of curtain ties (numbering 237) suggests a high level of concern towards concealing the activities once undertaken within this structure (see Greene Robertson 1985a: 36–40).

Architecturally, House E differs from Houses B and C, representing what was probably an earlier style in the structural evolution of the Palace. For example, House E never supported a roofcomb (Maudslay 1889–1902b: 23), an architectural feature common to the later structures of the Palace. This difference may have been the end result of an architectural experiment undertaken during the construction of House E and the Olvidado, which led to the later developments in roof style at Palenque (Greene Robertson 1985a: 7).

Eleven *Ik'* openings are present in the exterior and interior walls of House E; eleven is the highest number associated with any single structure at Palenque. The openings range in size (between 40 cm and 58.4 cm in width and between 41 cm and 51 cm in height), and there is still much speculation as to their use; some of their proposed functions include ventilators, lineage symbols, as well as sighting devices (Greene Robertson 1985a: 10–11). Many of the *Ik* openings are positioned at approximately eye level, an indication that they may have been used for sighting between buildings and courtyards. Interestingly, several of the Palace *Ik'* openings were sealed with mortar and masonry whereas others were made to never completely penetrate the walls (Peterson 1983: 20–22).

With the construction of House C, the northwest entrance to House E became blocked, after which a bench was placed in the old southeast entrance to the building (see Section 4.3.6, The Buildings of the Palace Phase 6, this chapter). Functioning symbolically as a portal to the Underworld, access from the southern end of the western corridor of House E to the southern subterranean building (eastern end) was/is given via a stair and underground vaulted passageway (Schele 1986: 121) (see Section 4.3.1, The Buildings of the Palace Phase 1, The Northern and Southern Subterranean Galleries and Passages, this chapter). Equivalent to the number of levels comprising the Maya Underworld, "nine steps lead down from the opening of the eastern-most underground passage near the southern end of House E's west hall" (Seler 1976: 65).

**HOUSE E FLORAL MOTIFS: DESCRIPTION (GR 13).** On the western face of House E are the remains of six rows of painted motifs; originally there were 132 painted on the exterior of House E, although only remnants of 46 remained at the time of Greene Robertson's (1985a) publication. Even though the precise meaning of these motifs remains unclear, thematically they have tentatively been defined as "Cosmological Terrestrial Underworld" (CosTU), due to the abstract representation of what is probably dry land flora combined with Underworld symbolism. The motifs run horizontally and in vertical columns, with Rows 2, 3, and 4 resembling quatrefoil flowers. Seler (1976: 77) described these repeated motifs as "an open flower-part blossoms with paired protrusions on the four sides and an eye in the center." The central eye elements vary slightly in that some are portrayed lidded, lashed, open, as well as closed. Other examples of the flower motif have an additional four-petal element in the interior of the flower. It has been suggested that the paired protuberant elements extending from the petals of the flowers may represent, in part, the stamen and stigma of the flower (Greene Robertson 1985a: 14–15). In my view, the apparent similarity between the quatrefoil flowers painted on the exterior walls of House E and the way in which cave entrances were depicted in Mesoamerican art suggest ideological associations (Figure 4.10). More specifically, the forms of the quatrefoil flowers resemble depictions of cave portals that were considered by the ancient Olmec to be "open to the Underworld," a view that was also held by the Classic Maya (Miller and Taube 1993: 57).

In addition to the quatrefoil flower motif is the hanging flower motif also painted on the exterior of House E. The depiction of this flower is similar to the morning

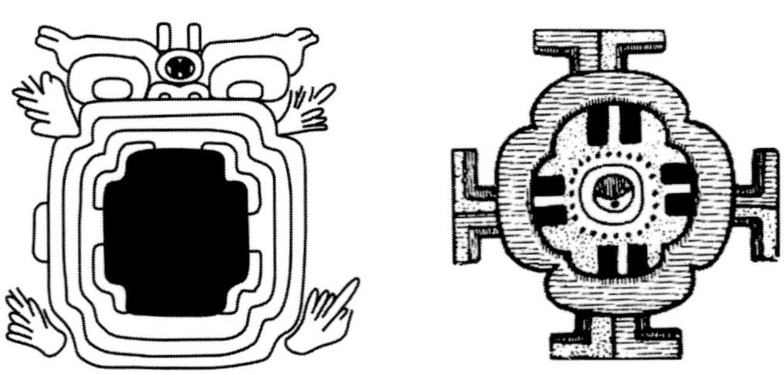

**Figure 4.10.** Left: Olmec Quatrefoil Cave Portal from Chalcatzingo. Right: Morelos and House E West Wall Quatrefoil Flower (GR 13). Left drawing by Karl Taube, from Miller and Taube (1993: 57), reprinted with permission. Right drawing after Seler (1915: fig. 132).

**Figure 4.11.** House E Hanging Flower Motifs with Creatures Emerging (GR 13). After Seler (1915: figs. 136 and 137).

glory, a plant found naturally in the region. The seeds of the morning glory flower have psychotropic properties when ingested; the Maya may have used these seeds for ritual purposes (Greene Robertson 1985a: 17). Insects and animals appear to be emerging from the centre of some of these flowers (Figure 4.11). For example, a half-insect and half-human with a large proboscis appears to be emerging from a flower on Wall 4 of House E, in addition to a deer and a quetzal from which speech scrolls seem to be emerging. Another bird, most likely a vulture, is also portrayed projecting from the top of a down-turned flower (see Seler 1976: fig. 134; Greene Robertson 1985a: 16–17; Schele 1986: 121). A deer is also depicted on the second vault of the East Subterranean Passage that runs beneath the southwest room of House E.

On the exterior roof of House E are the remains of a thatch design (Greene Robertson 1985b: 7–8), which symbolically identify the building as a house, as opposed to a temple (Schele 1986: 121). In addition to this thatch pattern, just below the eaves is a decorative band that runs the length of the building (Figure 4.12). Comprising a single repeated motif, measuring 23 cm to 27 cm wide, this pattern resembles a row of eyeballs consisting of a conjunctiva, pupil, and eyelid as well as something that has been interpreted as an eyebrow (Seler 1976: 75). Representing "death eyes" (Beyer 1937: 151-2; Thompson 1950: 45), each eye section is framed in a white painted diamond framed in blue. There are half-circle sections at the top and bottom coloured red or orange with a black centre. This "under-eave motif" is repeated on the medial

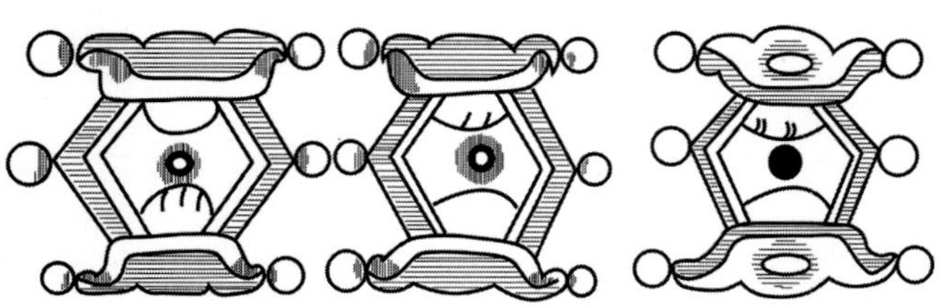

**Figure 4.12.** Painted Border on the Upper Exterior Wall of House E (GR 13). After Seler (1915: fig. 128).

**Figure 4.13.** House E Oval Tablet (GR 5). Drawing by Linda Schele (No. 143), © David Schele, courtesy of the Foundation for the Advancement of Mesoamerican Studies, Inc., www.famsi.org.

wall vault apron on the western side of the structure (see Greene Robertson 1985a: 12–14).

**OVAL PALACE TABLET: DESCRIPTION (GR 5).** Measuring 117 cm high by 95 cm wide, the sculptural theme of House E's Oval Tablet (Figure 4.13) has been defined as "Royal Audience Accession" (RAudAcc). Once positioned above the Del Río Throne (GR 6), against the western surface of the medial wall in House E, the presence of this monument suggests that one of the functions of the western gallery of House E was to inaugurate kings (Martin and Grube 2000: 163). Identified by his name caption, the Oval Tablet portrays the accession of K'inich Janab' Pakal I on 9.9.2.4.8 (A.D. 615) at the age of twelve. K'inich Janab' Pakal I is portrayed on the viewer's right of the scene in the company of his mother, Lady Sak K'uk'. Whereas both individuals are seated cross-legged and in profile, K'inich Janab' Pakal I's torso is turned in a frontal orientation. K'inich Janab' Pakal I is also depicted seated on a two-headed water lily–jaguar throne in royal pose. K'inich Janab' Pakal I wears minimal attire comprising simply a fretted hipcloth, pendulous earflares, *ajaw* bone wristlets, and a belt made from beads. Other costume elements include an *ajaw* bone pectoral with an *Ik'* sign at its centre, as well as a Jestor God headband with a length of cloth projecting from the back. A *le* leaf emerges from the rear of the headband, a symbol believed to be associated with lineage and accession (Schele 1976:14); an additional leaf emerges from the forehead of the Jestor

God. On the top of K'inich Janab' Pakal I's head balances a hand and mirror element supporting beads and an *ajaw* bone in a drop configuration.

The central focus of the scene is the crown of rulership, which is being presented to K'inich Janab' Pakal I by his mother, Lady Sak K'uk'. Attached to the front of this mosaic crown, referred to as the "drum-major headdress," is a Jestor God; a spray of feathers can also be seen rising vertically from the top of the headdress. In the scene, Lady Sak K'uk' is depicted in profile, in a subordinate pose, and on a lower register. One of the heads of the jaguar throne overlaps Lady Sak K'uk', an indication that K'inich Janab' Pakal I is the more important subject of the image. Lady Sak K'uk' wears a netted, tubular-beaded cape and skirt over an embroidered garment. Her belt and bracelets are similar to the ones that K'inich Janab' Pakal I wears, but with the addition of a *xoc* fish positioned at the front of the belt. Lady Sak K'uk' also wears a beaded headband supporting three anthropomorphic Jestor Gods, one at the front and one on each side of her head (Greene Robertson 1985a: 29–30).

Once positioned beneath the Oval Tablet, it would appear that Del Río Throne was commissioned by K'inich K'an Joy Chitam II (see Section 4.3.3, The Buildings of the Palace Phase 3, this chapter) to replace a preexisting throne (i.e., Bench 1), which was moved from House E to the southeast room of the Subterraneos (see Section 4.3.1, The Buildings of the Palace Phase 1, The Northern and Southern Subterranean Galleries and Passages, this chapter).

**HOUSE E SERPENT DOOR: DESCRIPTION (GR 3).** The theme of the House E Serpent Door is "Cosmological Underworld" (CosU). Painted around the interior surface of this entrance are the remains of two serpent heads, the lower parts of which continue onto both the eastern and western walls of the northwest room of House E. Painted many times, the image represents the gaping mouth of the serpent (Schele 1986: 121), and the right side of the doorway precisely mirrors the left side. The effect of this composition is similar to elements appearing on the sarcophagus lid found in the tomb of the Temple of Inscriptions, where K'inich Janab' Pakal I is portrayed emerging from the maw of the Underworld. For the Maya, the imagery of the Serpent Door probably marked the entrance it surrounds as a point of transition, where an individual crossing this threshold was perceived as symbolically entering, or emerging from, the domain of the Underworld (Schele 1986: 121).

With the construction of House C at Phase 2 of the Palace, the northeast door to House E was blocked (see Section 4.3.2, The Buildings of the Palace Phase 2, this chapter). Following this renovation, a bench was placed at the base of the old entrance. Today the bench is missing, but marks are present in the floor indicating where the bench once resided. The presence of a royal bench together with the serpent iconography may have been intentional; the serpent may have been designed to frame persons who once seated themselves on the throne (Greene Robertson 1985a: 24).

**BICEPHALIC ROOM: DESCRIPTION (GR 8).** The sculptural theme of the north wall of the Bicephalic Room (GR 8) is "Cosmological Underworld" (CosCU). Just below the course spring over the north entrance is a stucco representation of the Cosmic Monster (Schele 1986: 121) (see Figure 4.14), elements of which resemble those found associated with the bicephalic creature represented on Vault 1 in the East Subterranean Passage (Figure 4.7). Extending onto the east and west walls of the room, the monster comprises a sky-band body terminating at both ends with a zoomorphic head (one fleshed and the other skeletal), elements that are consistent with the Miller and Taube (1993) description of the creature. Combining celestial and Underworld symbolism, the forward head of the creature represents the heavens, "the front head being identified with Venus, pulling behind it the fleshless head of the sun in the Underworld" (Miller

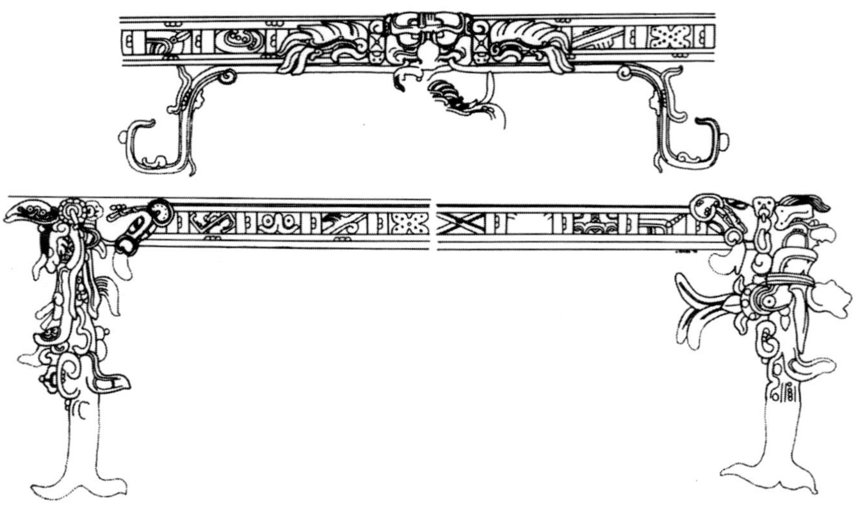

**Figure 4.14.** Top: Serpent-Bird Depicted on the North Wall of the Bicephalic Room of House E. Bottom: Venus Monster Depicted on the West Wall of the Bicephalic Room of House E and the Quadripartite Monster, East Wall of Bicephalic Room House E (GR 8). Drawing by Linda Schele (No. 129), © David Schele, courtesy of the Foundation for the Advancement of Mesoamerican Studies, Inc., www.famsi.org.

and Taube 1993: 45). The sky-band is comparable to what frames K'inich Janab' Pakal I's sarcophagus lid as well as the piers that front House A and the Temple of Inscriptions (Greene Robertson 1985a: 25). Portrayed on the western wall of the room is the forward head of the Venus Monster, whilst the Underworld sun (displaying a shell, cross-bands, and a stingray spine in its forehead) is sculptured on the eastern wall. Identified by a Venus symbol in its left ear, the Venus Monster has a long, curled muzzle; in addition, it wears a netted cap and round, beaded earplugs from which an *ajaw* bone hangs at the back. Water or blood can also be seen falling from the head of the creature. The left arm of the monster is scaled and has a water scroll shell for an elbow; in addition to the shell are cross-bands and a stingray spine. The Quadripartite God at the rear of the serpent displays a *k'in* (sun) sign on his forehead. The fleshless lower jaw of the god symbolises death and the Underworld. Water also appears to be falling from the headdress of the creature.

Centred over the doorway at the midpoint of the sky-band is the head of the Serpent Bird Deity. Comprising large, curling serpent fangs, the monster is adorned with square earplugs with an inverted *ajaw* underneath; from each earplug, inverted serpent wings can be seen emerging. Positioned centrally, but this time just below the sky-band, are the painted remains of the "Principal Bird Deity," similar in its appearance to those carved at the top of K'inich Janab' Pakal I's sarcophagus lid and on K'inich Kan B'alam II's Tablets of the Cross and Foliated Cross. Elements of the creature include a feathered body, supraorbital plate, *tau* tooth, and curled nose (see Greene Robertson 1985a: 25–28).

**House B: Description and sculpture**

Although there is no date for the dedication of House B (Figure 4.15), it can be persuasively argued that it was built sometime during the middle of K'inich Janab' Pakal I's reign. Epigraphic data combined with stratigraphic evidence provide a sequence of construction for Houses B, C, and E, indicating that House B was built after the completion of House E and before that of House C. As recorded on the Tablet of the 96 Glyphs (GR 73), the dedication of House E occurred on 9.11.2.1.11 9 Chuwen 9 Mak (1 November 654). The date for the completion of the House C foundation is argued to correlate closely with the latest date recorded on House C's Hieroglyphic Stair (H.S.), which is 9.11.6.16.11 (A.D. 659) (P. Mathews 2004: personal communication), while the completion of House C probably occurred just before the 9.11.9.5.19 date recorded on the eaves at the northeastern corner of the house (see Table 4.8). In addition to the House C dates, it is important to note that House C was constructed only after the completion of Second Palace Terrace extension A on which the building stands together with House A (Schele 1985: 119; see Figure 4.1, Palenque Palace Cross Section). Other indicators that the construction of House E predates that of House B are that the foundation of the western extremity of House B abuts the northeast corner of House E (Schele 1985: 119) and that the eastern external wall of House E (the wall that sides against the western extremity of House B) was painted repeatedly. Furthermore, an *Ik'* window in the northeast corner of House E, now blocked, once opened onto the area that is now occupied by House B (Greene Robertson 1985a: 41).

The internal partitioning of House B postdates the closing of the north end of the Palace with House A-D

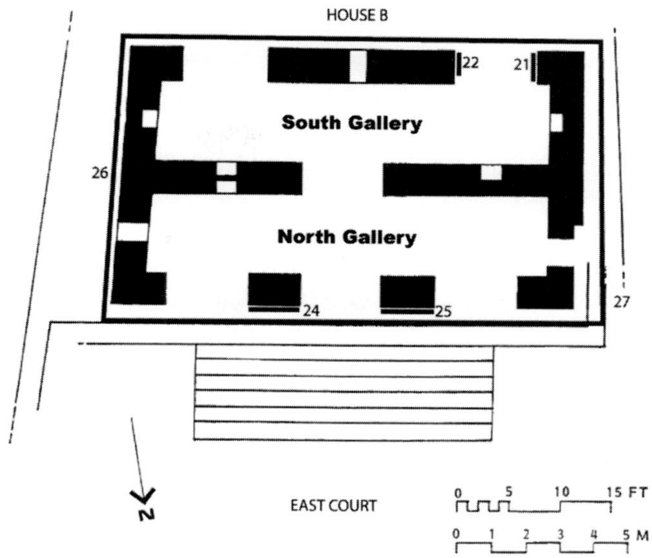


**Figure 4.15.** House B Phase 1, Palenque Palace. Modified after Andrews (1974). George F. and Geraldine D. Andrews papers. The Alexander Architectural Archive, the University of Texas Libraries, the University of Texas at Austin.

Table 4.8. *Construction sequence for Palace Houses E, B, and C*

| | | | |
|---|---|---|---|
| House E | 9.11.2. 1.11 | 1 Nov 654 | Dedication, Tablet 96 Glyphs |
| House B | – | – | – |
| House C Foundation | 9.11.6.16.11 | 7 Aug 659 | Dedication? Hieroglyphic Stairs |
| House C | 9.11.9. 5.19 | 22 Dec 661 | Dedication? Eaves |

*Notes:* Sequence in accordance with Andrews (1974). Correlation constant: 584285.

Table 4.9. *House B interior gallery dimensions*

| Gallery | Length | Width | Area |
|---|---|---|---|
| North | 11.30 m | 2.50 m | 28.25 $m^2$ |
| South | 11.50 m | 2.40 m | 27.60 $m^2$ |

*Note:* Dimensions in accordance with Andrews (1974).

(Schele 1986:119; see also Miller 1998: 199). The purpose of the renovation was to prevent access from the East Court through House B into the southern courts of the Palace (Greene Robertson 1985b: 50). This proposal places the refurbishment of House B, at its earliest, late in the reign of K'inich K'an Joy Chitam II (A.D. 702–721?), the second son to K'inich Janab' Pakal I. K'inich K'an Joy Chitam II oversaw the completion of House A-D on or around 9.14.8.14.15 9 Men 3 Yax (A.D. 720) (see Section 4.3.3, The Buildings of the Palace Phase 3, House A-D: Description and Sculpture, this chapter).

Oriented in an east–west direction, House B was the first building of the Palace to be constructed with a roofcomb (Greene Robertson 1985a: 48). House B is smaller than House E, but it is set marginally higher than House E. The eastern limit of House B faces the junction between Houses A and J. House B originally comprised two galleries separated by a medial wall running in an east–west direction (see Table 4.9 for the gallery dimensions). A doorway in the central wall provided access between the northern and southern galleries of the building. Additional walls were later constructed within the corridors between the southern surfaces of Piers B and C and the northern side of the medial wall flanking either side of the central doorway (Andrews 1974). The effect was to break up the northern and southern galleries of House B into five separate rooms, three of which became accessible only from the East Court and two only from the Southeast Court (see Section 4.3.3, The Buildings of the Palace Phase 3, this chapter). On the north side of House B there are three doorways separated by four piers; on the south side of the structure there are two entrances.

It is clear that both the interior and exterior of House B were once elaborately decorated with stucco sculpture and paintings, remnants of which can still be seen today. Because of later modifications undertaken in the House B interior, some of the artworks correspond to different phases of construction. As noted earlier in this chapter, there are at least two building stages associated with House B – the original double gallery plan as envisaged by K'inich Janab' Pakal I and the subdivided version of the building put into operation during the reign of K'inich K'an Joy Chitam II. In the following sections, I will first list all the House B sculpture and paintings, after which a more detailed discussion of the House B Phase 1 artworks will be presented.

All the surviving interior artworks associated with House B are found in Rooms 4 and 5; there are no paintings or sculptures present in Rooms 1, 2, and 3 in the northern half of the structure. Among the surviving decorative elements appearing in Room 4 are the incised and sculptured figures (GR 16), the sculptured *Ik'* figures (GR 17), the hieroglyphic inscription (GR 18), the *cauac* throne (GR 19), and the wall painting (GR 23). On the north wall of Room 5 is a step fret, sculptured in stucco (GR 23). The exterior of House B also displays sculptural residues that are located on the piers and roof of the building. On the northern exterior of House B, two human stucco figures remain discernible on the surfaces of the central Piers B (GR 24) and C (GR 25), flanking the entrance to Room 2. The east and west jambs of Room 4 (GR 22 and GR 21) display the lower portions of standing figures sculptured in stucco. In addition, the west roof of House B displays a central mask with two serpents emerging from Venus symbols to the left and right of the mask (GR 27). Appearing on the roof on the east side of House B are the remains of a serpent-bird (GR 26) sculptured in stucco (for more detailed descriptions of the House B embellishments, see the next section).

**HOUSE B, PIER B: DESCRIPTION (GR 24).** It is highly probable that the entire north façade of House B would have been modeled immediately after the building was completed, placing its construction during the reign of K'inich Janab' Pakal I between A.D. 654 and 659. Currently, little stucco sculpture survives on the north side of Pier B; however, the outline of a sculptured male individual remains discernible. Probably representing a lone portrait of K'inich Janab' Pakal I, the sculptural theme of House B, Pier B, has been defined as "Royal Portraiture Solo" (RPSolo). The figure measures approximately 230 cm high and stands in frontal orientation with his head turned to the viewer's right towards the central entrance of the house. Both arms of the figure are bent at the elbow and turned in across the abdomen. A stingray spine can be seen attached to the upturned snout of a serpent in the individual's headdress. The stingray resembles the one worn by K'inich Kan B'alam II on the west jamb of the Temple of the Cross. Folded feathers can also be seen coming out of the rear of the headdress. In addition, the outlines of wristlets are visible on the figure's arms as well as a Jestor God diadem that is attached to his forehead. The stance of the figure on House B mirrors that of the figure represented on Pier C. More specifically, the figure stands with his head turned towards the entrance of Room 2 (see House B, Pier C discussion in the next

Table 4.10. *Dates for Palace Houses A and C*

| | | | | | |
|---|---|---|---|---|---|
| House C Foundation | 9.11. 6.16.11 | 7 Chuwen | 4 Ch'en | 7 Aug 659 | Dedication? Hieroglyphic Stairs |
| House C | 9.11. 9. 5.19 | 4 Kawak | 2 Pax | 22 Dec 661 | Dedication, Eaves |
| House A Foundation | 9.11. 9.10.12 | 6 Eb' | 10 Woh | 25 Mar 662 | Auto Sacrifice, House A *Alfarda* 1 |
| | 9.11. 9.10.13 | 7 B'en | 11 Woh | 26 Mar 662 | Auto Sacrifice, House A *Alfarda* 2 |
| House A | 9.11.15.11.11 | 1 Chuwen | 19 Pop | 12 Mar 668 | Event? Pier A |
| | | or | | | |
| | 9.12.19.14.12 | 5 Eb' | 5 K'ayab' | | Event? House A, Lintel, and Pier Text |

section). The head and/or the entire body of the figures sculptured on piers at Palenque are generally turned in towards the central entrance of the building on which they appear (e.g., see Section 4.3.2, The Buildings of the Palace Phase 2, House A: Description and Sculpture and House C: Description and Sculpture, this chapter) (Greene Robertson 1985a: 42–43).

**HOUSE B, PIER C: DESCRIPTION (GR 25).** The sculptural theme of Pier C, which is located on the northern façade of House B, is "Royal Portraiture Solo" (RPSolo). Like Pier B, the remaining sculpture on Pier C probably depicts K'inich Janab' Pakal I in the same pose. One of the few differences is that the head of the figure is turned towards the viewer's left rather than the right of the composition. Both Piers B and C flank the central entrance in the northern façade of House B. Discernible costume elements adorning the figure on Pier C include a feathered headdress and a "nose-bridge-enhancer." The head of the figure appears conical in shape, showing a rather severe taper that starts wide at the base of the skull and narrows towards the top. At the bottom right of the image is a sculptured shell together with a plant element comprising three parts. To the right of the plant element the remains of a serpent rise up; the serpent's body can be seen curling under the left foot of the figure (Greene Robertson 1985a: 43).

Piers B and C of House B were designed as a single composition. This type of figural sculpture is most commonly assigned to the external surfaces of piers fronting gallery-type architecture, where the principal figures face in towards the central entrance of the structures. As would be the case for Pier B, Pier C probably dates to the period of House B's original construction, commissioned by K'inich Janab' Pakal I probably between A.D. 654 and 659.

**HOUSE B, SOUTHWESTERN JAMBS: DESCRIPTION (GR 21, 22).** The sculptural themes of House B's southwestern jambs have been defined as "Portraiture Solo" (PSolo) as well as "Ritual Dance" (RitDan). On the southwestern jambs of House B are the remains of two frontally oriented individuals sculptured in stucco. The only surviving sculpture on both of these jambs is the lower legs of the individuals (Greene Robertson 1985a: 46). Southwestern Jamb A of House B displays the remains of a figure standing in a static pose with his feet turned out. Conversely, Southwestern Jamb B displays the lower legs of a frontally oriented figure standing with one foot flat on the ground and one uplifted heel. The uplifted heel creates an angle in the calf indicating that the knee of the same leg would have been bent. This posture signals the act of dancing in Maya iconography (Grube 1992: 201). A similar pose appears on Pier D of House D (see Section 4.3.3, The Buildings of the Palace Phase 3, House D: Description and Sculpture, this chapter), where the figure portrayed performs a serpent dance ritual. As is the case with Piers B and C, the southwestern jambs of House B were probably commissioned during the reign of K'inich Janab' Pakal I, placing the addition of these sculptures at approximately the same time that the exterior of House B was completed.

**HOUSE B, WESTERN ROOF: DESCRIPTION (GR 27).** Due to the combination of heaven and earth elements, the sculptural theme of the western roof of House B has been defined as "Cosmological Terrestrial Celestial" (CosTC). The presence, however, of the mat motif at the top of the structure suggests that the building may have also functioned as a *popol nah* ("mat house") (Miller 1998: 199, 1999: 38), fulfilling both administrative as well as private ritual purposes. Studies associated with Structure 10L-22 at Copán (see Fash 1992; Fash et al. 1992) suggest that buildings displaying mat motifs may have functioned as "council houses" where the "council of lords congregated to undertake various political and/or ritual activities" (Schele and Mathews 1998: 44; Sharer 1994: 329). In addition to the mat weave pattern, the iconography of the western roof of House B comprises a central mask measuring 17 cm deep by 48 cm wide, and the remnants of two serpents emerging from Venus symbols to the left and right of the mask. Outlines present where the stucco is now missing suggest that the serpents' mouths were probably open. The central-basal position and the

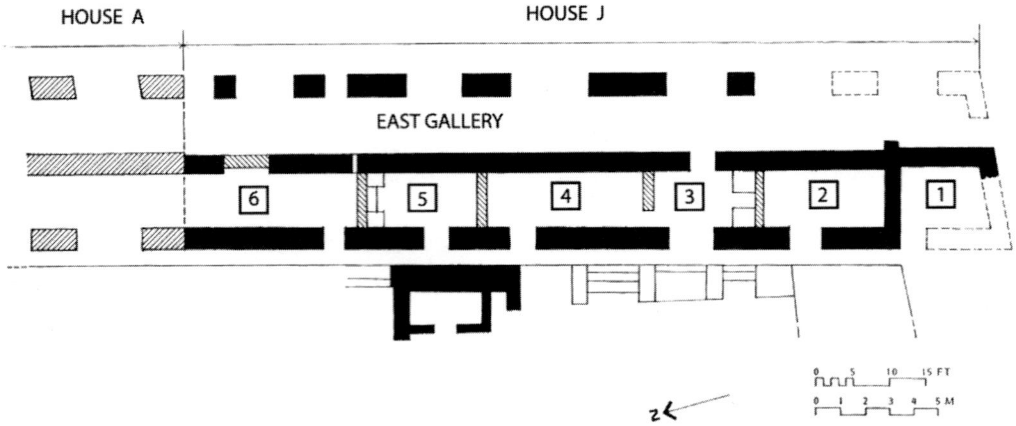


**Figure 4.16.** House J Plan, Palenque Palace. Modified after Andrews (1974). George F. and Geraldine D. Andrews papers. The Alexander Architectural Archive, the University of Texas Libraries, the University of Texas at Austin.

remaining detail present indicate that the mask represents an earth-bound *Cauac* Monster (Greene Robertson 1985a: 49) that Miller and Taube (1993: 120) defines as a zoomorphic representation of a mountain. The space directly above the mountain monster probably once displayed additional imagery, perhaps an additional mask or a seated figure.

**HOUSE B, EASTERN ROOF: DESCRIPTION (GR 26).** The theme of the sculpture assigned to the east roof of House B is "Cosmological Underworld" (CosU). Like the western roof, the remains of a mat motif are present on the upper roof of the building, as well as the remnants of a feathered serpent and a death eye border. As noted earlier in text, it is generally argued that this type of pattern had important cultural significance among the Maya, marking the structure as a *popol nah* (Kowalski 1998: 405), an elite structure reserved for Maya nobility and ritual administrative activities. The serpent-bird represented below the mat pattern comprises elements not seen anywhere else at Palenque. The shape of the feathers suggests that the creature depicted was a screech owl, a bird that is native to the region. In Maya cosmology, the screech owl is a creature of the Underworld. The head of the bird in this example comprises elements symbolising death; among these elements are a fleshless lower jaw and bulging death eyes (Greene Robertson 1985a: 49–50).

### House J: Description

Oriented in a north–south direction and located in the southeast quadrant of the Palenque Palace, House J comprises six rooms on the west side of the structure, two of which contain benches (see Rooms 3 and 5, Figure 4.16). In addition, a long gallery once ran along the east side of the building. The east and west halves of the building are separated by a medial wall where access between them was/is provided via a single doorway adjacent to Room 3. At some stage, a more northern doorway (appearing in the medial wall of the building) was constructed and subsequently blocked. House J once stood as an independent structure as evidenced by a construction join and the change in vault heights where Houses A and J intersect. The construction join runs perpendicular to the north–south orientation of Houses A and J, lining up with the southeastern-most pier of House A (Andrews 1974).

House J was built under the auspices of K'inich Janab' Pakal I at approximately the middle of his reign (A.D. 654–657). More specifically, there appears to be little doubt that the construction of House J postdates the construction of House E (A.D. 654). It is also probable that House J was constructed before the completion of the northern Palace platform, which supports Houses A and C (A.D. 657). The discovery of large quantities of Otolum ceramics in the southern wall of the house further supports the argument for the early fabrication of this building (Greene Robertson 1985b: 4).

House J is an unusual building in the sense that it lacks the bilateral symmetry apparent in most other buildings of the Palace. The structure comprises piers and/or sections of wall of varying widths, which are separated doorways of equally dissimilar widths. The lack of balance expressed in the House J configuration (combined with the proximity to and the relative lowness of its vault compared to that of House A), suggests that this building predates a tradition of piered architecture that emphatically expressed the importance of symmetrical planning and construction (see Houses A, C, and A-D). In addition, the irregularity of the House J doorways indicates that the interior partitioning of the house was considered

Table 4.11. *House J, approximate interior room dimensions*

| | Length | Width | Area |
|---|---|---|---|
| Eastern Gallery | 31.8 m | 2.4 m | 76.3 $m^2$ |
| Room 1 | 3.0 m | 2.2 m | 6.6 $m^2$ |
| Room 2 | 4.9 m | 2.2 m | 10.7 $m^2$ |
| Room 3 | 4.2 m | 2.2 m | 9.2 $m^2$ |
| Room 4 | 4.5 m | 2.2 m | 9.9 $m^2$ |
| Room 5 | 3.6 m | 2.2 m | 7.9 $m^2$ |
| Room 6 | 7.0 m | 2.2 m | 15.8 $m^2$ |

*Note:* Dimensions in accordance with Andrews (1974).

as a part of the structure's design. There is no evidence to suggest that the exterior of House J was ever embellished with sculpture (see Table 4.11 for the room dimensions).

**House K: Description**

House K comprised two structural phases referred to as the "southern portion" of House K and its "northern extension" (Andrews 1974). Measuring a total of 21.18 metres, the southern portion was built by K'inich Janab' Pakal I (Figure 4.17). The first phase of House K comprised an eastern and western gallery and five piers separated by four entrances that made up the western façade of the building (see Table 4.12 for the gallery dimensions). In addition, two doorways were built into both the northern and medial walls of the structure with one located at the southern end. At some later stage, several partitions were added to the building's interior, dividing the western gallery into two rooms and constricting access through the eastern gallery. The first phase of House K represents a relatively early architectural addition to the Palace corresponding temporally to that of House E (Andrews 1974), which was completed by K'inich Janab' Pakal I on 9.11.2.1.11 9 Chuwen 9 Mak or 1 November

Table 4.12. *House K, interior gallery dimensions*

| Gallery | Length | Width | Area |
|---|---|---|---|
| West | 20.0 m | 02.3 m | 46.0 $m^2$ |
| East | 20.0 m | 02.7 m | 54.0 $m^2$ |

*Note:* Dimensions in accordance with Andrews (1974).

654 (see Section 4.3.1, The Buildings of the Palace Phase 1, House E: Description and Sculpture, this chapter). Structural evidence in support of this proposal includes: (1) comparatively similar vault heights between Houses E and K; (2) the presence of *Ik'* windows in the medial walls of both buildings; (3) the absence of sculptured piers; (4) similar wall thickness (i.e., House K medial wall 0.71 m, House E medial wall 0.74 m, House K exterior wall 0.61 m, House E exterior wall 0.60 m); and (5) the presence of an underground stairway leading from both structures to the Subterraneos (see Section 4.3.1, The Buildings of the Palace Phase 1, The Northern and Southern Subterranean Galleries and Passages, this chapter). The addition of House K's northern extension occurred only after the construction of House D, and possibly House A-D. As illustrated by Palace Phase 3 diagram, the absence of the House K extension created a seven-metre gap between House D and House K, which has been interpreted here as an early entrance to the Palace interior (see Section 4.3.3, The Buildings of the Palace Phase 3, this chapter). On the western side of this entrance there are two narrow piers separated by three openings located at the top of the Palace's eastern monumental staircase.

## 4.3.2. The buildings of the Palace Phase 2

Palace Phase 2 (Figure 4.18) was the next largest construction effort undertaken in the Palace Group at

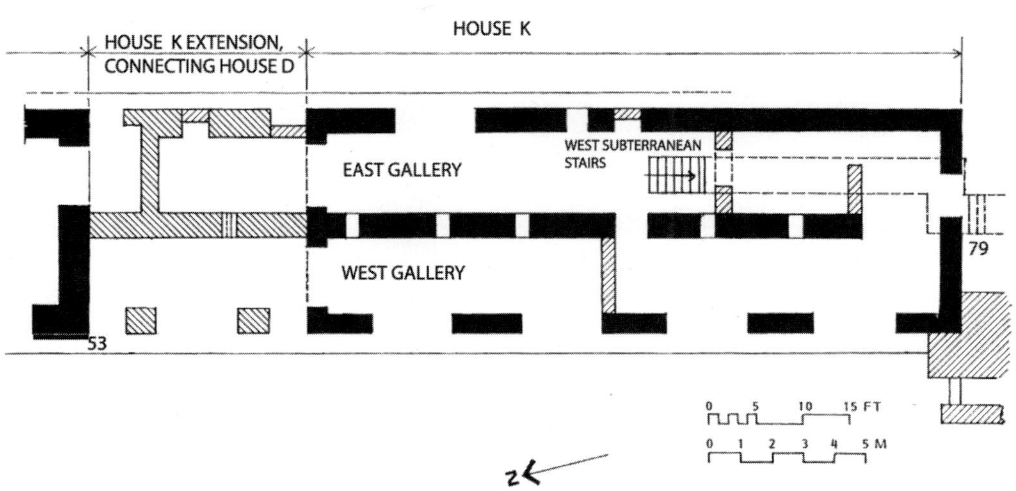


**Figure 4.17.** House K Plan, Palenque Palace. Drawing by George Andrews (1974). George F. and Geraldine D. Andrews papers. The Alexander Architectural Archive, the University of Texas Libraries, the University of Texas at Austin.

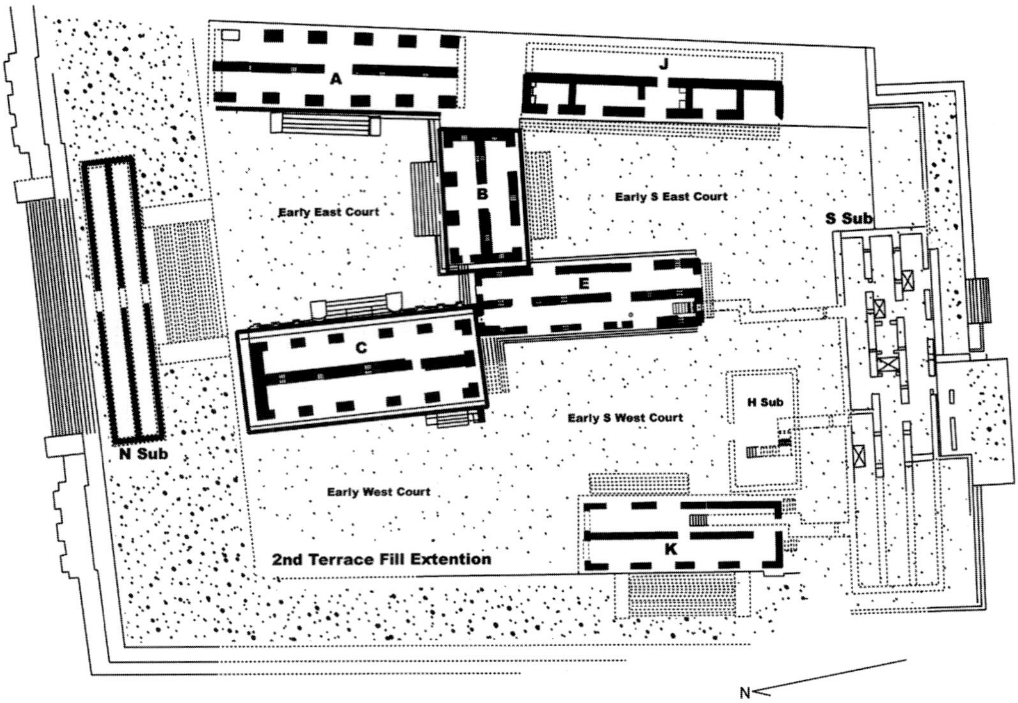


**Figure 4.18.** Map of the Palenque Palace Phase 2. Palenque Palace modified after Greene Robertson (1985b: fig. 1).

Palenque, which included the addition of House C and House A. The building of these two structures required that the southern platform that supported Houses B, E, J, K, and H-Sub be extended. As noted earlier in this chapter (see Section 4.1, Palenque Palace: General Description, this chapter), the upper Palace platform comprises three sections. As previously pointed out, "the construction of House C required that the Palace platform in the north be raised to the level of the southern area . . . the new platform was built of large slabs carefully laid with a minimum of soil and mortar" (Schele 1986: 119). This fill differs markedly from the northern platform extension that supported House A-D, which is described as consisting of an earth and rubble mix (see Figure 4.1). With the addition of House A and House C, an early version of the East Court was formulated with House A defining its eastern boundary, House B its southern boundary, and House C its western boundary. At this early stage of the Palace's evolution, House A-D did not yet define the northern periphery of the East Court (see Section 4.3.3, The Buildings of the Palace Phase 3, this chapter), effectively leaving it open. At Palace Phase 2, access to the new East Court and into the group's interior was provided via the Northern Galleries. At this time the Northern Galleries remained exposed and in use until the addition of House A-D and the construction of the northern monumental stair, which later buried the structure.

As the correlated dates for Houses C and A suggest, K'inich Janab' Pakal I is the most likely ruler to have presided over the northward expansion of the Palace's upper platform, which occurred towards the end of his reign (circa A.D. 650–683). The latest date appearing on the Palace's H.S. is being interpreted here as the approximate time that the foundation of House C was completed (A.D. 659), whereas the eave date (GR 62) found at the northeast corner of the building is viewed as correlating closely to the completion of the House C superstructure (A.D. 661). In reference to House A, the dates occurring on the foundation *alfardas* (GR 58) are being interpreted as the approximate time of the completion of the House A foundation (A.D. 662), whereas the date recovered from Pier A of House A is being construed as corresponding to the completion of the building (Table 4.12).

Although the above-mentioned dates do suggest that K'inich Janab' Pakal I built House C and House A, it remains possible that his son and successor, K'inich Kan B'alam II, constructed or at least contributed to the construction of one of these structures. It seems unlikely that K'inich Kan B'alam II, even with the momentous task of the Cross Group, never undertook any construction and/or renovation in the Palace over an approximate eighteen-year reign. To date it remains unclear whether House A was built by K'inich Janab' Pakal I or by K'inich Kan B'alam II.

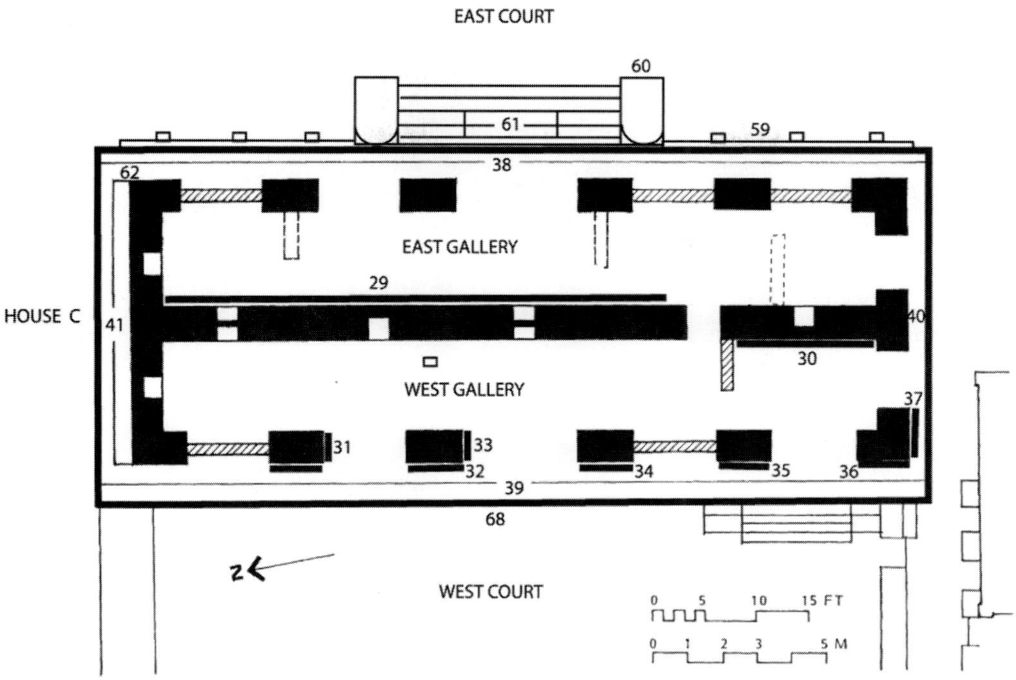


**Figure 4.19.** House C Plan, Palenque Palace. Modified after Andrews (1974). George F. and Geraldine D. Andrews papers. The Alexander Architectural Archive, the University of Texas Libraries, the University of Texas at Austin.

**House C: Description and sculpture**

House C, or the *yet nah* ("companion house") as it is referred to on the Palace's H.S. (Martin 2003: 94–95), was erected during the reign of K'inich Janab' Pakal I. The completion of the house probably corresponds to the dedicatory date located on the eaves at the northeast corner of the building: 9.11.9.5.19 4 Kawak 2 Pax (22 December 661). The interior of House C comprises an eastern and western gallery partitioned by a medial wall where access between these rooms is provided via a single doorway (Figure 4.19; see Table 4.13 for gallery dimensions). House C has a relatively open plan when compared to House B and House E, measuring approximately twenty-five metres in length by eleven metres in width. The structure runs north–south with one side adjacent to the East Court and the other adjacent to the West Court. The position of House C in relation to the main entrance of House A suggests that this building was a semi-public structure, perhaps functioning to receive important guests. House C is larger and more grandiose than House B and House E; the highly decorated and spacious nature of House C suggests that it was designed to receive and impress larger numbers of people, perhaps visiting nobility from neighbouring sites. The original design of the building was an open plan as suggested by the absence of curtain holders. Sometime later, however, additional walls were constructed between the piers at the northern and southern ends of both eastern and western galleries of the building. It appears that the objective of constructing the walls was to block the view through the central wall doorway of the building into neighbouring courts, as well as to create several smaller rooms. Access to House C's western gallery was/is provided by stairs found at the southwestern corner of the structure. The H.S. (GR 61) provides access between the eastern gallery of House C and the East Court. An earlier southern stair was also used to access House C and an earlier version of the West Court.

House C was once lavishly decorated with sculpture covering most of the building's exterior as well as parts of its interior. There are twelve piers associated with House C, many of which were highly decorated; six of the piers are on the east side of the building and six are on the west side. The eastern piers to House C once displayed sculptures of standing figures, where several of the western piers still portray the remains of seated individuals. In the interior of House C's eastern gallery, a row of portraits depicting gods of some kind are also portrayed

Table 4.13. *House C, interior gallery dimensions*

| Gallery | Length | Width | Area |
|---|---|---|---|
| West | 20.97 m | 02.68 m | 56.19 m$^2$ |
| South | 20.97 m | 02.82 m | 59.13 m$^2$ |

*Note:* Dimensions in accordance with Andrews (1974).

in addition to some sculptures appearing on the exterior roof of the structure. Surviving decorative elements also include an array of cosmological, ancestral, and captive imagery that appears on the roof, external piers, and the medial wall of the building. In addition, there is more imagery as well as historical texts located on the west basement wall and eastern staircase of House C. Some of the most impressive sculpture survives on the western façade of House C, on Piers C, D, E, and F; today Pier A (northwest corner of House C) and Pier B (next to Pier A) have no surviving sculpture. The eastern piers of House C once supported sculpture of some kind, probably standing royal portraits, as this side of the structure represents the formal entrance to the building. The sculptures on the four surviving piers all portray individuals seated on thrones, in frontal orientation, and in semi-elaborate attire. It has been suggested that all the House C piers portray K'inich Janab' Pakal I in various guises (Greene Robertson 1985a: 54). It is more likely, however, that the House C piers portray K'inich Janab' Pakal I in the company of his royal predecessors, a view that I also hold regarding the sculpture appearing on the external surfaces of the east piers of House A. Evidence in support of this proposal includes variations in costumes and throne designs, in addition to the representation of a woman on Pier E adorned with a *le* motif described as "a lineage symbol of ancestors" (Greene Robertson 1985a: 58).

**Figure 4.20.** House C, Pier C, Palenque Palace (GR 32). Maudslay (1889–1902a: Volume IV, plate 28).

**HOUSE C, PIER C: DESCRIPTION (GR 32).** The sculptural theme of Pier C, House C (Figure 4.20), has been defined as "Ancestor Portraiture Solo" (APSolo). The pier portrays a youthful individual seated cross-legged in frontal orientation on a throne, his head turned to the viewer's right towards the central entrance of the building. The figure's left arm is raised slightly, and his right arm hangs away from his body. The headdress adorning the seated individual is elaborate, comprising a beaded or mosaic helmet with a large spray of feathers emerging from the top and back. An element comprising two stacked flowers is visible at the top of the helmet. To the front of the headdress is an additional floral element with a shell emerging from its centre. Just above this element is a bird's head with an *ajaw* bone sculptured on the curl of its beak; a string of beads also hangs from the beak. To the rear of the helmet is a small head of a god crowned with a symbol consisting of a large outer circle and a smaller central inner circle that is surrounded by four smaller outer circles that are evenly spaced around it: A stack of five beads also rises from the long nose of the god. Below the front flower–shell element on the headdress is a mask that resembles a cormorant. In its beak, the cormorant holds a string of beads that has a bone attached at each of its ends. In contrast to the headdress, the costume worn by the seated person is relatively understated, comprising simply a beaded shoulder piece, bracelets and anklets, leg bands, a loincloth, and a cross-banded belt displaying maskets and celts.

There are two legible glyphs on Pier C, one of which survives on the front of the throne reading *tun*, and the other in the title caption at the top right corner of the pier which reads 11 Ajaw (Greene Robertson 1985a: 54–55).

**HOUSE C, PIER C, SOUTH SIDE: DESCRIPTION (GR 33).** The sculptural theme of Pier C's south side is "Portraiture

**Figure 4.21.** House C, Pier D, Palenque Palace (GR 34). Maudslay (1889–1902a: Volume IV, plate 28).

Solo" (PSolo); the figure represented on the pier is believed to represent a hunchback dwarf. The male individual is portrayed in a profile stance facing west and is simply dressed, wearing only a loincloth. Displaying little hair, the physical deformation of this figure is signaled by the projection of the individual's chest and the hump on his back. The importance of this sculpture is suggested by the fact that it was painted as many as eight times (Greene Robertson 1985a: 59). As evidenced by scenes appearing on polychrome vessels, Maya dwarves were often portrayed as members of the royal court and/or participants in courtly rituals (Schele and Freidel 1990: 252).

**HOUSE C, PIER D: DESCRIPTION (GR 34).** The sculptural theme of Pier D, House C (Figure 4.21), is "Ancestor Portraiture Solo" (APSolo). Pier D is positioned on the southern side of the central doorway of the western façade of House D overlooking the Palace's West Court. The general composition of Pier D mirrors that of Pier C. Like that on Pier C, the figure on Pier D is seated on a throne in frontal orientation with his head turned towards the central entrance of the building. The individual is wearing an elaborate headdress combined with what may be described as relatively simple attire. Elements appearing in the headdress include a *Tlaloc*, which is a symbol of self-sacrifice or bloodletting, and a mat motif positioned just behind the *Tlaloc* symbol. Above the *Tlaloc* element are two stacked Jestor Gods that are attached to a jaguar tail wrapped in a lattice pattern. A voluminous spray of feathers, some of which appear to be trimmed short, can be seen emerging from the rear of the headdress together with *ajaw* bones that are attached to the ends of rods. The rods appear to be emerging from a serpent-wing motif. Other costume elements adorning the figure on Pier D include a loincloth, leg bands, and anklets.

Although the upper torso, head, and arms of the portrait are gone, part of a mask can still be made out. A serpent-like protrusion extends down from the mask almost to waist level, first curling out and then back in towards the wearer. *Ajaw* bones, together with cylinder beads, emerge from a bell-shaped feature attached to the upper curve of the serpent-like feature. Although there was once an inscription on the pier (viewer's top left), none of it is legible today; above the pier is a glyph thought to mean "completion" (Greene Robertson 1985a: 57).

**HOUSE C, PIER E: DESCRIPTION (GR 35).** The sculptural theme of House C, Pier E (Figure 4.22), is "Ancestor Portraiture Solo" (APSolo). Pier E portrays an individual seated cross-legged and in a frontal orientation on a throne. The remaining two hieroglyphs still present on the pier name K'inich Janab' Pakal I as the individual represented (Mathews 2005: personal communication). In this image, K'inich Janab' Pakal I wears a voluminous feathered headdress with a jaguar tail emerging from the top. A Jestor God is attached to the base of the tail where cropped feathers can also be seen emerging. The angle of the headdress indicates that K'inich Janab' Pakal I's head (now missing) was turned to the viewer's left of the composition, so that it faced the central piered entrance of the structure. Other remaining costume elements still adorning K'inich Janab' Pakal I in this portrait include wristlets, a loincloth, and two maskets attached to the waist of the

**Figure 4.22.** House C, Pier E, Palenque Palace (GR 35). Maudslay (1889–1902a: Volume IV, plate 28).

individual, indicating that the figure was wearing a belt of some kind.

The throne portrayed on Pier E is different from the thrones appearing on the other piers of House C. More specifically, the inner leg of the throne on the viewer's right is curved whereas the throne legs portrayed on the other piers are straight. On the left of the throne is the head of a *xoc* fish with a miniature *Ajaw* God positioned behind its curled nose; foliation also appears to be emerging from the head of the *Ajaw* God. In addition, on the right arm of the enthroned individual is the head of the Principal Bird Deity (Greene Robertson 1985a: 57).

**HOUSE C, PIER F: DESCRIPTION (GR 36).** The sculptural theme of House C, Pier F, is "Ancestor Portraiture Solo" (APSolo). The general composition of Pier F appears to be consistent with that of the other House C piers. Surviving sculpture portrays the upper torso and arms of an individual; its height, size, and orientation suggest that the image was of a person seated cross-legged on a throne. Although the head of the person represented is now missing, the direction of the figure's glance can be surmised from remnants of sculptured hair. The hair on Pier F is projecting towards the viewer's right of the image, so the individual head was probably turned towards the viewer's left. The individual portrayed has long hair, which is bound in a ponytail at the rear of the head, as well as falling loosely behind the body and over the figure's left shoulder. Long hair in Classic Maya art is indicative of female personages, as seen on the periphery of K'inich Janab' Pakal I's sarcophagus in the tomb of the Temple of Inscriptions (Schele and Mathews 1998: 119–121). This is an important observation given that it has been proposed that all the piers of House C are portraits of K'inich Janab' Pakal I (Greene Robertson 1985a: 54). Under the left arm of the individual on Pier F is the lineage symbol. "The *le* motif, a leaf-like element with three circles at the end . . . is the lineage symbol of ancestors, and designates that the person wearing it . . . has a legitimate ancestral relationship with a royal person" (Greene Robertson (1985a: 58).

**HOUSE C, PIER F, SOUTH SIDE: DESCRIPTION (GR 37).** The sculptural theme of the south side of Pier F is "Portraiture Solo" (PSolo). Surviving sculpture on this pier portrays the waist, torso, and upper arms of what was probably a male individual. The general posture of the torso is similar to that of the sculpture found on the front of Piers C, D, E, and F, although the raised position of the torso indicates that this figure is standing rather than sitting. Surviving sculpture also reveals that the figure was portrayed in a frontal orientation, with his right arm partially raised whereas his left falls by his side. No costume elements are discernible on this pier, with the exception of a hip- or loincloth and part of what may have been a beaded shoulder piece. Red paint is present on the pier background and on the right arm of the individual (Greene Robertson 1985a: 59–60).

**HOUSE C, EASTERN GALLERY MASKS: DESCRIPTION (GR 29).** Nine grotesque stucco masks once decorated the eastern gallery of House C, each of which was painted a combination of colours: red, yellow, and blue. Sculptures

**Figure 4.23.** House C, East Gallery, Mask 3 (GR 29). Maudslay (1889–1902a: Volume IV, plate 24).

Mask 3 (Figure 4.23) appears to be a portrait of a god that has plant foliation with embedded Venus symbols emerging from the top of its head. Locks of hair are visible beneath the foliation that appears to be growing out of the circular elements positioned above large ear ornaments. The cartouche present in the forehead of the mask portrays leaf elements. Each of the leaves displays three central dots down its central axis. Remaining stucco indicates that the sculptured mask measured 88 cm high by 90 cm wide.

Mask 4 portrays the head of a god with an element resembling a mirror embedded in its forehead. The flattened head of the god appears to have maize foliation growing out of it. The central element of the plant comprises small beads of stucco that resemble kernels of maize. Measuring approximately 40 cm high by 60 cm wide, Mask 4 still shows residues of paint. Whereas the cartouche was painted yellow, the perimeter and forehead of the god were painted blue.

Mask 5 (Figure 4.24) is the most intact of the seven masks still appearing in House C; sculptured elements present on Mask 5 include fleshless cheekbones, a skeletal jaw,

**Figure 4.24.** House C, East Gallery, Mask 5 (GR 29). Maudslay (1889–1902a: Volume IV, plate 24).

that were painted blue signaled their divine and sacred nature (Greene Robertson 1979: 166, see also Schele 1985: 36). Set horizontally in a line on the medial wall of the building, only seven of the masks survive today. Unfortunately, the masks that held positions 2 and 9 are now missing. Collectively representing what were probably members of the Maya pantheon at Palenque, the sculptural theme of each mask has been subsequently defined "God Portraiture Solo" (GPSolo). In the following section, brief descriptions of the House C gallery masks will be presented in accordance with Greene Robertson (1985a: 51–53). (For a broader discussion of gods in Maya iconography, see Taube 1992.)

Mask 1 portrays what is probably the head of a god with a mirror embedded in its forehead. The god has large eyes with central spirals in addition to a gaping mouth with large, outward curling fangs. Foliation of some kind appears to be emerging from the top of its head. Remaining stucco indicates that the mask originally measured 85 cm in height and 85 cm in diameter.

Figure 4.25. House C, East Gallery, Mask 7 (GR 29). Maudslay (1889–1902a: Volume IV, plate 24).

spiral or hook-centred eyes, and large ear spools with spiraling elements positioned above them. In the forehead of the god is an *ajaw* sign on which rests a cap comprising a flat comb feature. Unfortunately, the nose and central mouth area of the creature have eroded away; however, remaining stucco indicates that, fully intact, Mask 5 probably measured 85 cm high by 90 cm wide as well as being painted red, yellow, and blue.

Measuring 90 cm high by 65 cm wide, the most prominent feature on Mask 7 (Figure 4.25) is the cross-band motif embedded in the forehead of the god. Mask 7 also displays a fleshless lower jaw and a flattened head that once would have supported additional iconographic elements. In addition, at the centre of the god's lidded eyes are three dots; the creature would have also once displayed ear ornaments similar to those adorning Masks 3 and 5.

In the forehead of Mask 8 (Figure 4.26) is a pattern that comprises a combination of small circles and wavy cross-hatched lines symbolising either a water lily or the shell of a turtle; mirror elements are also visible in the cheek areas of the mask. As on Mask 7, the ear ornaments on Mask 8 are missing, as are the nose and the central jaw area of the god. Furthermore, striations appearing around the chin area indicate that the god was probably bearded. The dimensions of Mask 7 are generally consistent with those of the masks previously discussed, measuring approximately 90 cm high by 65 cm wide.

**HOUSE C NARRATIVE SCENE: DESCRIPTION (GR 30).** Remnants of a painting narrative scene cover portions of the medial wall in the western corridor of House C, the limits of which are marked by a transverse wall positioned just south of the medial doorway providing access between the eastern and western galleries of House C. The theme of this mural has been tentatively defined as "Royal Audience" (RAud). Surviving elements of the mural portray two seated individuals making offerings to an elevated figure seated centrally on what appears to be a throne. The two figures seated peripherally each offer a bowl with unknown contents in their outstretched arms to the central figure. The central individual is undoubtedly the most authoritative of the three, as signaled by the elaborateness of his costume and his elevated positioning

Figure 4.26. House C, East Gallery, Mask 8 (GR 29). Maudslay (1889–1902a: Volume IV, plate 24).

between the two subordinate characters. The central figure wears a feathered headdress, a cape, and wristlets and holds a staff in his right hand. The remnant of a painted inscription (now illegible) is present just above the head of the right subordinate figure.

To the left of the scene are the remains of a standing figure sculptured in stucco that overlies the earlier mural. The presence of this figure indicates that the southwest wall of House C was probably decorated several times. There was once an inscription sculptured to the right of the standing individual that is now obliterated with the exception of the raised border that framed it. The profile stance and position (left of centre) of the figure suggests that he was subordinate to whatever decorated the area of the wall in front of him. This figure is compositionally consistent with an attendant standing in the company of something or someone of greater authority. Today, all that survives in the area of the wall in front of the standing figure is an *Ik'* window that has a raised stucco border, the top half of which was at some stage filled in.

The standing figure is portrayed in profile wearing an elaborate backrack-style headdress, a costume element more commonly associated with the eastern Petén region of Guatemala (Greene Robertson 1985a: 62). The unusual nature of the headdress suggests that the identity of the individual may be that of a visiting noble from the Petén region. Inset in the rear of the headdress is a person standing holding a parasol over a seated figure; a ceremonial bar with mat signs also can be seen emerging from between the standing individual's legs. In addition to the above-mentioned elements, a painted sky-band (representing one of the earliest painted layers) once ran along the base of the scene; this band was later replaced with a stucco version that ran beneath the feet of the standing figure. The presence of this repeated motif suggests that the later refurbishment of House C's narrative scene was thematically consistent with the mural that underlies it (see Greene Robertson 1985a: 61–63).

**HOUSE C, EAST ROOF: DESCRIPTION (GR 38).** The sculptural theme of the roof of House C has been defined as "Cosmological Terrestrial" (CosT). On the eastern and western roof of House C are fourteen *cauac* masks flanked by human figures (Greene Robertson 1985a: 63). Miller and Taube (1993: 120) define these *Cauac* Monsters as zoomorphic representations of a mountain. Evenly spaced, each representation of the creature has large eyes and curved fangs, in addition to large ear ornaments with inverted *ajaw* bones hanging beneath. Flanking the *cauac* masks are the remnants of life-sized human figures portrayed in profile holding various subordinate poses. Several of the figures are depicted

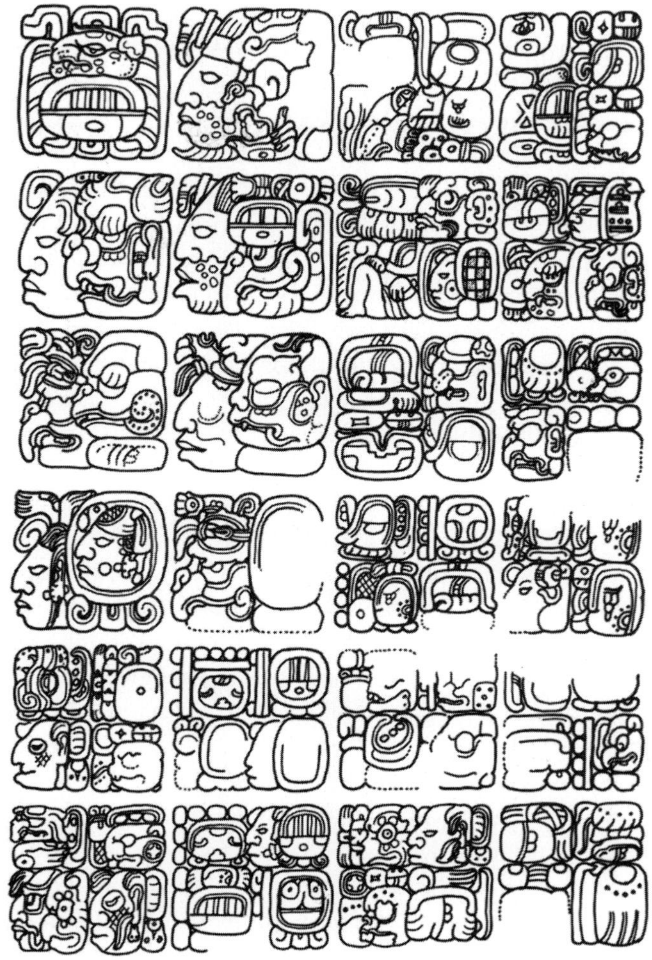

**Figure 4.27.** House C, Hieroglyphic Stair Text (GR 61). Drawing by Linda Schele (No. 138), © David Schele, courtesy of the Foundation for the Advancement of Mesoamerican Studies, Inc., www.famsi.org.

kneeling and paying homage to the *cauac* masks; they are plainly dressed wearing simply loincloths and turbans. A cross-band border is also visible running along the top of the roof of the building (Greene Robertson 1985a: 63).

**HOUSE C, H.S.: DESCRIPTION (GR 61).** The theme of the hieroglyphic stairs (H.S.) of House C (Figure 4.27) is tentatively defined as "Text, Historical" (THist). Due to some damage and the unusual nature of the syntax, the precise order of the reading of the H.S. has remained the subject of some debate among epigraphers. A recent transcription, offered by Simon Martin (2003), begins with the Long Count date 9.8.9.13.0 8 Ajaw 13 Pop (23 March 603) followed by a reference to the birth of K'inich Janab' Pakal I. The inscription then records a distance number that correlates temporally with K'inich Janab' Pakal I's accession, which occurred on 9.9.2.4.8 5 Lamat 1 Mol (26 July 615). After this date there is another date reconstructed by Peter Mathews that reads 9.11.1.16.3 6 Ak'b'al 1 Yax (25 August 654) (Mathews in Martin 2003: 87). The verb *ch'ak*, in the inscription, may be correlated with this date recording an "axing" event at Lakam Ha', which was "supervised" (*u-kab'jiiy*) by a Calakmul Lord (Kan Ajaw) named "Sky Witness" (Martin 2003: 88). Consistent with this proposal, the H.S. inscription also documents a "throwing down" (*yalej*) event connected with the patron gods of Palenque, implying the fall of the Palenque Triad (Gods I, II, and III) at this time; the position of this event in the inscription suggests that it occurred after the attack by Calakmul (Martin 2003: 89). Shield Jaguar of Yaxchilan (Itzamnah B'alam) is also recorded on the monument, together with the Split-Sky sign of Yaxchilan. The inscription names Nuun Ujol Chaak, a predecessor of Ruler A from Tikal as well. The verb *chukaj,* meaning "to seize" or "to capture," is followed by the date 7 Chuwen 4 Ch'en on the monument, which correlates with a Long Count date of 9.11.6.16.11 (7 August 659). The *yetej* glyph is also found in the inscription followed by the name K'inich Janab' Pakal I.

According to Martin, a "*yet naah*" reference on the H.S. of House C may mark the structure as a "companion house." Accordingly, this reference is followed by a glyph reading *u-k'ab'a* ("it is the name of"). An eroded glyph at the end of the inscription is thought to read *patlaj,* a term used to denote "the building of the house" (see Martin 2003: 90–95).

**HOUSE C FOUNDATION EAST SIDE FIGURES AND GLYPH BLOCKS: DESCRIPTION (GR 59).** The sculptural theme of House C's east basement figures and corresponding glyph blocks are defined as "Captive Display" (CapDis). Flanking each side of the H.S. along the basement wall of House C are carvings of six captive vassals, three of whom are represented on each side of the stairs. Each figure is depicted next to an inscription recording his or her name and place of origin (see Figure 4.28 for detail of Captive Figures 1–5, ordered sequentially from top to bottom). According to Martin (2003: 97), all the individuals represented were captured on the same day, one of whom is identified as coming from Pomona (Captive Figure 4). Interestingly, all of the captives' headdresses contain elements that correspond to their name captions. All figures are portrayed kneeling in frontal orientation and holding subordinate poses with their heads turned towards the H.S. that are positioned centrally against the eastern foundation of House C.

Captive Figure 1, also appearing on the eastern foundation of House C, portrays a male individual depicted in frontal orientation and with his head turned towards the viewer's right of the composition. The arms of this kneeling individual are crossed at the wrists and in front of his body. Behind the hands of Captive Figure 1 is a bar pectoral with *ajaw* bones that project from one of its ends. In addition to the bar pectoral, Captive Figure 1 wears *ajaw* bone wristlets and has a piece of material pulled through each of the perforations in his ears. Around his waist, Captive Figure 1 wears a plain hipcloth and a bell ornament that is attached at the front panel of the figure's loincloth. At the centre of the bell is an embedded mirror; the beads at the base of the bell ornament are tubular in shape and have *ajaw* bones attached at their ends. The headdress of Captive Figure 1 represents the head of a creature that contains a *ma* superfix, which corresponds to the first glyph (A1) of the figure's neighbouring caption. A2 of the text reads *aj* ("he of") followed by a title of some kind. Overall, the posture, pose, orientation, and general costuming of Captive Figure 1, with the exception of the headdress, are similar to those of Captive Figure 6 on the opposite side of House C's H.S. (Greene Robertson 1985b: 68).

Like Captive Figure 1, Captive Figure 2 also kneels with his head turned towards the viewer's right. In this image, the left hand of the captive is depicted flat against a bar pectoral that hangs about his neck, whereas the captive's right arm hangs by his side. In addition to the bar pectoral, Captive Figure 2 wears a hipcloth displaying a bell ornament identical to those worn by Captive Figures 1, 3, 4, 5, and 6. Captive Figure 2 also wears tubular beaded wristlets and a headdress representing a skull of some kind. As with the other captive figures represented, a piece of cloth is visible pulled through the ear perforations of the individual. The B1 glyph of the figure's neighbouring caption is a "shell-fist" title; the B2 glyph is an *ajaw* title, and the A2 glyph reads *aj* ("he of") "Bird-head Place" (Martin 2003: 97).

Captive Figure 3 wears a hipcloth, wristlets, and a loincloth that has a bell ornament attached at the front. In addition, the remains of a zoomorphic headdress are also discernible ornamenting the figure. Captive Figure 3's neighbouring caption records the individual title as *a* ("he of") the "split earth" (Martin 2003: 97). As with the other captives represented, Captive Figure 3 also appears kneeling in frontal orientation with his head turned towards the H.S. of House C. Although the left arm of Captive Figure 3 is obliterated, it is clear that the pose of this individual would have closely matched that of Captive Figure 2 (see Captive Figure 2 description).

Captive Figure 4, appearing on the north side of the H.S. of House C, mirrors the orientation and pose of Captive Figures 2 and 3 located on the south side of the stairs (see Captive Figure 2 and 3 descriptions). More

**Figure 4.28.** House C East Foundation Captives and Glyph Blocks, Details of Figures 1, 2, 3, 4, and 5, Respectively (GR 59). Drawing by Linda Schele (No. 134), © David Schele, courtesy of the Foundation for the Advancement of Mesoamerican Studies, Inc., www.famsi.org.

specifically, although the body of Captive Figure 4 is portrayed in frontal orientation, the head of the kneeling individual is turned to the viewer's left rather than the viewer's right. Furthermore, it is the captive's right arm, rather than his left, that is portrayed crossing the body. In addition to the Celestial Monster in the figure's headdress (Greene Robertson 1985b: 68), Captive Figure 4's costume comprises a bar pectoral, a plain hipcloth, wristlets with *ajaw* bones attached, and a bell feature (with an embedded mirror) that is attached to the front panel of his loincloth. At the top and sides of the bell feature are round beads of various sizes; below the bell feature are tubular beads with *ajaw* bones attached at the ends. A strip of cloth can be seen pulled through each ear perforation that once would have held ear ornaments of some kind. Captive Figure 4's neighbouring inscription identifies the individual as a *Pip-á Ajaw* from Pomona named *Ayiin Chan Ahk* (Martin 2003: 94–97).

As was the case with Captive Figure 4, the pose and orientation of Captive Figure 5 also mirrors that of Captive Figures 2 and 3 that appear on the south side of the H.S. Again, most of the costume elements worn by Captive Figure 5 are consistent with those worn by all the other House C foundation captives. Costume elements worn by Captive Figure 5 include a hipcloth, a loincloth with a bell ornament, a bar pectoral, wristlets, as well as a cloth pulled through each ear perforation. The only unique element in Captive Figure 5's costume inventory is a headdress comprising a Jaguar God (Greene Robertson 1985b: 69). The figure's neighbouring caption identifies the individual as he of the "Earth Place" (Martin 2003: 97).

The pose and orientation of Captive Figure 6 resemble those of Captive Figure 1 found at the southern side of the H.S. of House C. Captive Figure 6 is portrayed kneeling and in a frontal orientation with both arms raised, crossing in front of the torso (see Captive Figure 1 description). The head of the individual is depicted in profile, facing towards the viewer's left of the composition. The costume inventory of Captive Figure 6 includes elements worn by all other captive figures appearing on the foundation of House C: These elements include the same bar pectoral, wristlets, hipcloth, loincloth with bell ornament, and ear cloths. On his head, Captive Figure 6 wears a zoomorphic headdress with a *k'in* sign attached (Greene Robertson 1985b: 69). The corresponding text refers to the individual as an *Ajaw* named *Aj K'in* from "Turtle Place" (Martin 2003: 94–97).

**HOUSE C, EAST FOUNDATION *ALFARDAS* (GR 60).** The theme of the House C foundation *alfardas* are defined as "Captive Display" (CapDis). The *alfardas,* adhered to the balustrades on the left and right of the H.S. of House C (GR 60), each portray a frontally oriented captive whose head faces in towards the staircase. The north *alfarda* is severely eroded, displaying little detail. The remaining sculpture suggests, however, that the figure was dressed similarly to the figure carved on the south *alfarda*. Both individuals wore simple cloth headdresses, rectangular pectorals, and loincloths that displayed stacks of knotted bars (see Greene Robertson 1985b: 67).

**Figure 4.29.** House C West Foundation Glyph Blocks (GR 68). Maudslay (1889–1902a: Volume IV, plate 29).

**HOUSE C WEST FOUNDATION TEXTS: DESCRIPTION (GR 68).** The theme of the House C west foundation texts (Figure 4.29) are defined as "Text, Captive Sacrifice" (TCapSac). The seven texts carved on the western foundation of House C refer to six captives from Santa Elena and one from Pomona. The first glyph in each of the seven inscriptions records the name of the captive, and the remaining glyphs in each of the columns record their place of origin and their titles, respectively. Following the name glyph, Panel 1 displays the *Wa*-Bird emblem of Santa Elena followed by a *Ch'ok* glyph identifying this individual as a prince; Panels 2, 3, 5, and 6 all follow this same pattern. Instead of a *Wa*-Bird emblem, Panel 4 records a *Wa* that, together with a *Ch'ok* title, overlaps an *ajaw,* signaling the higher status of this individual relative to the others recorded on the wall. The seventh panel records 13 Manik', End of Yaxk'in, a date that has been assigned to a Long Count currently argued to be 9.11.10.16. 7 (13 July 663) (Martin 2003: 114–116). The remainder of the text on Panel 7 records the death of a *Pip-á* ("lord") from Pomona, the verb *och b'ih* meaning "he entered the road," which has been described as "a metaphor for death" (Martin 2003: 114).

Martin (2003) points out that the House C west foundation texts are unusual as they do not have accompanying portraits of the captives to which they refer – a trend that is seen on the eastern foundation of House C (see Section 4.3.2, The Buildings of the Palace Phase 2, House C: Description and Sculpture, this chapter). Even with the absence of these sculptures, Martin poses the possibility that the west foundation texts once had accompanying portraits that were at some stage moved from the vicinity of the inscriptions. If this proposal is accurate, the question remains: Where might these portraits be positioned today?

The seven captives referred to on the western foundation of House C may be those represented on the captive panels found resting against the western foundation of House A. A total of nine individuals are carved in bas-relief on limestone panels just below House A on the east side of the East Court: Four appear on the northern side of the stair, and five on the southern side of the stair. It is interesting to note that if one discounts the "*alfarda*-style" Panels 4 and 5 positioned immediately to the left and right of House A's western foundation stairway (see Section 4.3.3, The Buildings of Palace Phase 3, this chapter), the number of the remaining captive portraits corresponds to the number of texts carved into the west basement wall of House C (Figure 4.29). As noted in the House A discussion (see next section), House A was originally designed to architecturally reflect House C, and the addition of the *alfardas* (immediately to the left and right of the House A staircase) was undertaken to reinforce the symmetry expressed by the two structures. Importantly, the two *alfardas* are the only captive panels that appear to be compositionally connected (see Section 4.3.2, The Buildings of the Palace Phase 2, House A West Foundation *Alfardas*: Description, this chapter). Irregularities between the remaining seven panels (more specifically, variations in style, scale, and the relative proportions of figures represented) indicate that they were probably carved independently of one another and by more than one artist. Today, questions remain as to the origin of these sculptures; stylistically the sculptures are unique, having no parallel at Palenque.

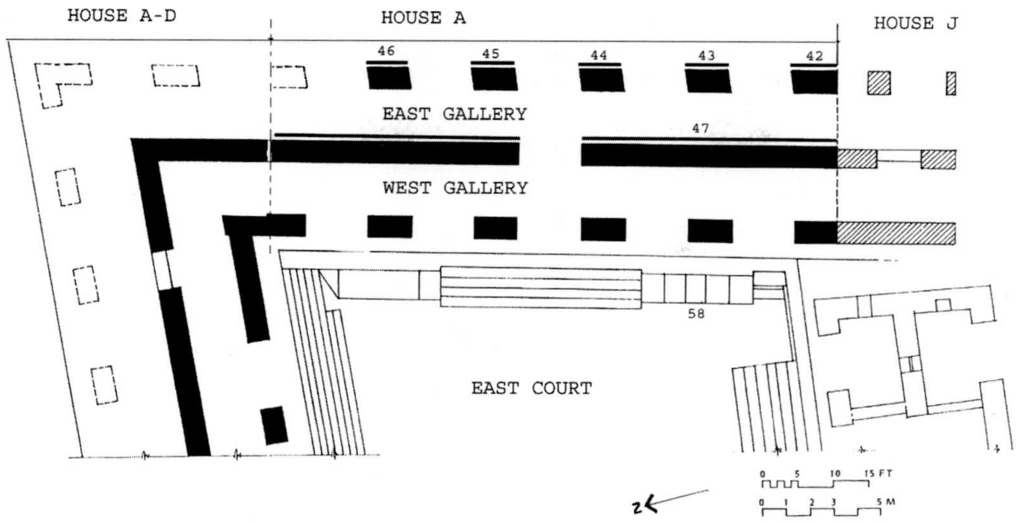


**Figure 4.30.** House A Plan, Palenque Palace. Drawing by George Andrews (1974). George F. and Geraldine D. Andrews papers. The Alexander Architectural Archive, the University of Texas Libraries, the University of Texas at Austin.

One possibility is that they may be portraits of conquered rulers that were carved by the members of the vanquished population and given as tribute as a form of public humiliation (P. Mathews 2004: personal communication). It seems clear, by the way that the nine panels fit together against the foundation of House A, that their current locations do not represent the sculptures' primary contexts. Each of the nine figures was carved separately on slabs of limestone of varying grades. In addition, the cropping of the tops and sides of the sculptures suggests that they were cut down to fit their current locations (Greene Robertson 1985b: 61–63). It was only after the addition of the *alfardas* that I believe the seven remaining captive panels were added to the foundation wall of House A. The cramming-in of Panels 1, 2, and 3 between House A's western foundation stair and the southeast corner of House A-D's foundation platform suggests that they were placed in their current position only after the addition of House A-D (see Section 4.3.4, The Buildings of the Palace Phase 4, this chapter). There are certain implications if this is the case, notably that the placement of these sculptures was undertaken under the auspices of K'inich K'an Joy Chitam II rather than his father K'inich Janab' Pakal I.

**House A: Description and sculpture**

House A is the most northeast of the Palace structures; it is oriented in a north–south direction with its southern extremity connecting with House J and its northern end with the eastern limits of House A-D (see Figure 4.30). The building was originally designed to stand independently of other Palace structures (Maudslay 1889–1902b: 13), only to be later connected to Houses A-D and J into a long, continuous building defining the northern and eastern Palace periphery. Among the lines of evidence that support this proposal are the presence of construction joins in the medial and western external walls of House A and differences in the vault heights between Houses A and J.

In general, House A comprises two vaulted galleries separated by a medial wall; the northern extremity of House A is marked by a fracture in the medial wall 10.96 metres north of the central doorway (see Andrews 1974; see also Table 4.14 for gallery dimensions). This vertically oriented split lines up with a construction join at the northern end of the western wall of House A, which in turn lines up with the northern side of Pier 6 on the eastern side of the structure (Maudslay 1889–1902b: 13). Unfortunately, the sixth pier of House A does not survive today, although the "footprint" of the pier is still visible, providing a means of verifying this alignment. Marking the southern extremity of House A are two additional construction joins that line up with the two southernmost piers of the building. The joining of Houses A and J was achieved by demolishing the ends of both buildings and then adding an extra section of medial wall to connect the

Table 4.14. *House A, interior gallery dimensions*

| | Length | Width | Area |
|---|---|---|---|
| West Gallery | 36.40 m | 02.28 m | 83.00 $m^2$ |
| East Gallery | 30.15 m | 02.28 m | 68.74 $m^2$ |

*Note:* Dimensions in accordance with Andrews (1974).

structures. The height of the House J vault is lower than the vault of House A, indicating that both structures were probably associated with different construction phases. Today, nothing remains of the eastern half of House J; however, there must have been an eastern portion to the building that lined up with the eastern passage of House A (see Section 4.3.1, The Buildings of the Palace Phase 1, House J: Description, this chapter) (Andrews 1974). As with the southern extremity of House A, the northern end of the building was at one stage opened to connect it to House A-D. Presently, only four piers define the western edge of House A: Each pier faces in on the East Court of the Palace. On the eastern side of House A, five of what was a total of eight piers survive (Piers 6, 7, and most of 8 are missing); as a consequence, the northeast corner of the building has completely collapsed.

House A reflects House C in its general layout and configuration (Figure 4.31). There are several ways in which the two structures resemble each other: (1) both Houses A and C are oriented in a north–south direction; (2) both buildings have stairs providing access to and from the East Court flanked by *alfarda*-style sculptures; and (3) both structures had/have six piers on their eastern and western sides separated by a medial wall. The even number of piers made it possible for the placement of a central doorway between Piers C (GR 44) and D (GR 45) of House A. The structural similarities between House A and House C suggest that both buildings may have been designed to complement the other architecturally, whilst simultaneously demarcating the eastern and western limits of the East Court.

At the centre of the medial wall of House A is an "ogee-trefoil arch" (Greene Robertson 1985b: 7), which provides access between the eastern and western galleries of the building. Greene Robertson (1985b: fig. 285) has proposed that the House A medial doorway once functioned as the principal entrance to the Palace. Late in the history of the Palace (see Section 4.3.4, The Buildings of the Palace Phase 4, this chapter), the northern monumental Palace stairway provided the formal access into the Palace interior, given that the stairway on the eastern side of the Palace complex never reached the structures above (see Maudslay 1889–1902b: 12). To access the House A interior, one entered the northern gallery of House A-D and followed it east before turning south into the eastern gallery of House A (Greene Robertson 1985b: 51). The House A medial entrance is directly adjacent to the East Court, the largest and most accessible of the Palace courtyards. In ancient times, standing at the threshold to the East Court, one would have had a view of a large, elaborately decorated, semi-public space, bounded on the west by the lavishly decorated façade of House C. The size and elaborateness of the entrance suggest that its purpose was to impress persons who passed through it, providing less restricted access to the space beyond. The shape of the upper portion of the ogee-trefoil arch resembles the niches that are positioned to the left and the right of the entrance in the eastern corridor vault (Greene Robertson 1985b: 7); currently the function of these niches remains in question. On the eastern side of the eastern corridor vault the niches are deeper, perhaps once functioning to hold censers or lamps to illuminate the corridor. Below the niches are eleven stucco medallions (GR 47) that are displayed on the eastern side of the medial wall of House A. Three large "hinge sockets" also flank both sides of the arch. The heavy-duty nature of these sockets indicates that they probably supported large, heavy, wooden doors (Greene Robertson 1985b: 25).

Although no precise date for the construction of House A has yet been determined, it is likely that it postdates the construction of House J and predates the construction of House A-D. On Pier A (GR 42) of House A are the remains of a date reconstructed by Berlin (1965: 340–341) and concurred by Peter Mathews (Mathews in Greene Robertson 1985b: 13); this date reads 9.11.15.11.11 1 Chuwen 19 Pop (20 March 668). If accurate, this date may be viewed as circumstantial evidence that the construction of House A may well have taken place during the reign of K'inich Janab' Pakal I. As the reconstructed date is located, however, in the middle of the Pier A text, there is a chance that the date may have functioned retrospectively and that the building was erected some time after this date. Furthermore, the Pier A inscription may have continued at the southeast end of House A, on Pier 6, 7, or 8, which may have displayed a later date during the reign of K'inich Kan B'alam II. It was common practice for sculptors at Palenque to bracket the inner piers displaying figurative art with piers exhibiting larger proportions of text. This trend is apparent when viewing the façades of many of the structures at the site (e.g., the Temples of the Cross Group and the Temple of Inscriptions). For a fully balanced effect, one would expect Pier 6 rather than Pier 7 or 8 of House A to have displayed a text similar to Pier 1 (Pier A), as this would have provided an equivalent number of sculptured piers on both sides of the medial doorway of the building.

Evidence in support of the view that K'inich Kan B'alam II may have constructed House A is found in a long hieroglyphic inscription that once ran the length of House A just below the lower eastern cornice of the building. This inscription ran horizontally along the top of the sculptured piers and across the wooden lintels that were once positioned above the doorways between the House A piers. Once totaling sixty-eight glyph blocks,

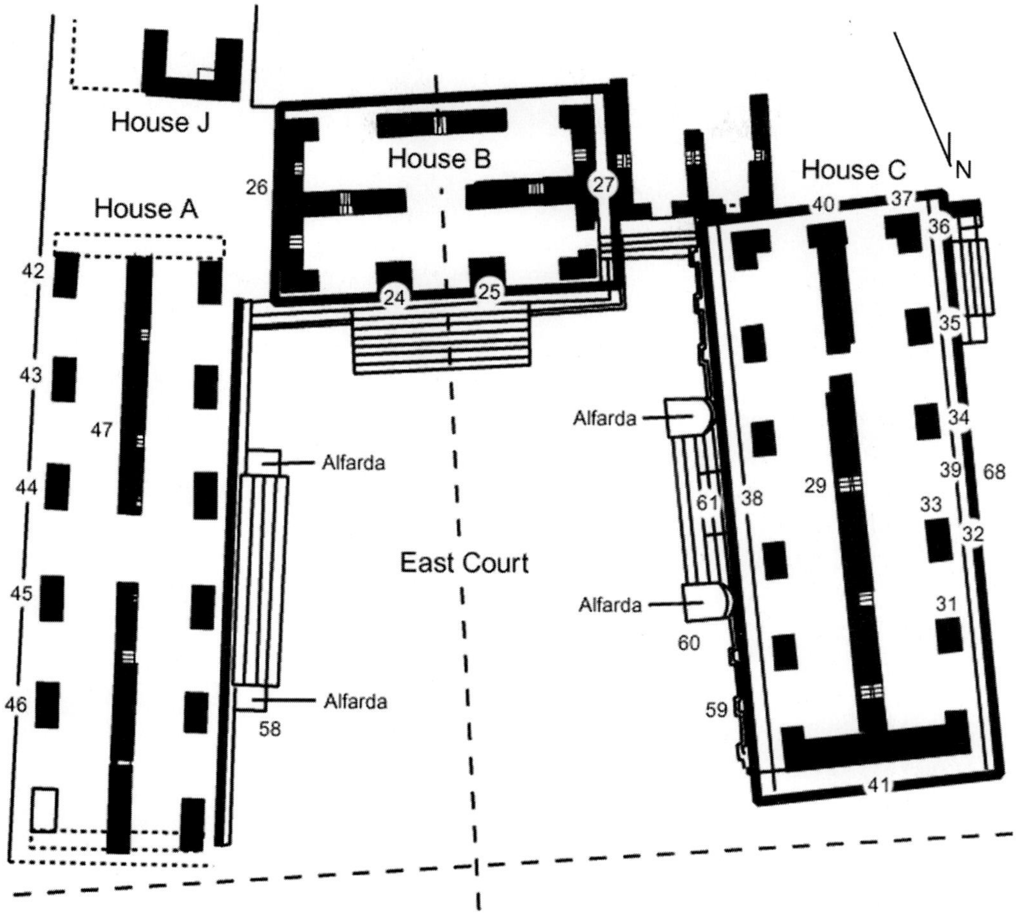


**Figure 4.31.** Early Plan of Palace Houses A and C and the East Court.

among the few surviving glyphs is an eleven- or twelve-*k'atun* glyph holding position 3 of the inscription, in addition to a "house" glyph holding position 27 in the text. If the *k'atun* coefficient of the glyph holding position 3 is twelve, then it remains possible that the Initial Series (I.S.) date to which this number belonged may have once read 9.12.19.14.12 5 Eb' 5 K'ayab' (A.D. 692), a date which falls well within K'inich Kan B'alam II's reign. In addition, if an 819-day count station were included in the inscription, the house glyph would hold the approximate verb position of the I.S. date. Combined, the date and verb may refer to a house event that occurred nine years after K'inich Kan B'alam II's accession (Mathews in Greene Robertson 1985b: 5). If K'inich Kan B'alam II did oversee the construction of House A, the ogee-trefoil or "keyhole" niches present in the eastern gallery vault of House A may represent an architectural development introduced by K'inich Kan B'alam II after the death of his father K'inich Janab' Pakal I, as the feature is also present in the vault above the medial wall of the Temple of the Cross (see Holmes 1897: 200). In addition, according to Greene Robertson, the roof of House A once displayed individuals seated on thrones, a theme that was repeated on the roof of the Temple of the Sun (Greene Robertson 1985b: 8).

**HOUSE A PIERS: GENERAL DESCRIPTION.** Originally, eight piers comprised the lower portion of the eastern façade of House A, of which only five survive today. As indicated by the remaining stucco sculpture, the southernmost pier of the house was primarily textual, indicating that the sixth, seventh, or eighth pier on the same side of the structure was probably of a similar nature (Green Robertson 1985b: 14). Piers B, C, D, and E of House A are predominantly pictographic, displaying the remains of elaborate figural scenes (see Figures 4.32, 4.33, 4.34, and 4.35). Compositionally, all four piers are similar: Each displays three individuals, one standing and two seated peripherally at the feet of the central figure. All three individuals in each of the pier scenes are portrayed in profile and in static pose. The seated figures are depicted facing one another whereas all the standing figures face the gap between the two central piers of House A (the central figures on Piers B and C face towards the viewer's right, whereas the central figures represented on Piers D and E face towards the viewer's left). Although there are

**Figure 4.32.** House A, Pier B, Palenque Palace (GR 43). Maudslay (1889–1902a: Volume IV, plate 8).

**Figure 4.33.** House A, Pier C, Palenque Palace (GR 44). Maudslay (1889–1902a: Volume IV, plate 8).

**Figure 4.34.** House A, Pier D, Palenque Palace (GR 45). Maudslay (1889–1902a: Volume IV, plate 9).

**Figure 4.35.** House A, Pier E, Palenque Palace (GR 46). Maudslay (1889–1902a: Volume IV, plate 9).

parallels in the way that the figures are represented on House A, it appears that each pier portrays a different protagonist, suggesting that they may in fact be ancestral portraits of different rulers undertaking the same activity. The general compositions of all four piers were standardised to portray a single static activity, in contrast to the House D piers that display various activities in the process of being enacted. Comparatively speaking, the costumes of the peripheral characters are relatively understated (comprising fewer, less elaborate elements) when compared to those of the central protagonists. In addition, the stances of the central individuals are formal and authoritative, whilst the seated figures hold subordinate poses (see Section 4.3.3, The Buildings of the Palace Phase 3: House D: Description and Sculpture, this chapter).

**HOUSE A, PIER A: DESCRIPTION (GR 42).** Thirty-two glyphs once covered the surface of Pier A, of which only a small proportion survive today. Due to the incomplete nature of this text, Pier A has been tentatively defined "Text, Historical?" (THist?). The size of the glyphs (approximately 25 cm in height by 30 cm in diameter) suggests that they were designed to be legible from a distance, probably from ground level just below House A. The remains of a sky-band and serpent scale border can be seen framing what is left of the inscription. Part of a date recovered from Pier A has been reconstructed, reading 9.11.15.11.11 1 Chuwen 19 Pop, which potentially places this pier, and by association the structure, within the reign of K'inich Janab' Pakal I. Several layers of paint have been identified where the glyph blocks were once attached to the pier, suggesting that the text was subject to repair and/or modification over time (Greene Robertson 1985b: 12–13).

**HOUSE A, PIER B: DESCRIPTION (GR 43).** The principal theme of House A, Pier B (Figure 4.32), positioned just to the left of the southernmost pier of the House (i.e., Pier A), has been defined as "Ancestor Portraiture with Attendants" (APWA). Pier B portrays three individuals framed by a sky-band; one is centrally positioned whereas the other two are seated cross-legged on either side. The central figure is depicted standing, in profile, and in an authoritative pose facing towards the viewer's right, in the direction of the medial wall entrance of House A. The male individual is portrayed wearing a large feathered headdress that has two back-to-back Jestor Gods positioned at the top. Short feathers are visible emerging from the heads of the gods. The remains of a serpent wing are also visible at the top rear of the headdress, beneath which a long tail of hair can be seen hanging. Around the central figure's waist is a jaguar or ocelot skin hip garment worn over a loincloth. The front panel of the loincloth can be seen hanging in front of the body; behind this panel is additional material represented by cross-hatching. The central protagonist also wears a mask on his back comprising a human face. Mat motifs as well as celts with embedded mirrors are attached just below the mask. The figure also wears sandals; on the figure's calves are the remains of leg straps in a cross-band configuration. In his left hand, the individual holds an incense bag comprising an inverted human head and draping feathers. In his right hand, the protagonist holds a God K staff with an axe head. A smouldering cigar or mirror is visible emerging from the head of the god, which appears at the top of the staff; there is also a mirror in the forehead of the god.

To the left and right of the central figure is a seated individual portrayed in profile. Both figures are minimally dressed wearing only loincloths and what are probably wide tubular beaded belts. The figure on the viewer's right wears a beaded necklace, and his gaze is fixed on the midpoint of the central figure. The figure on the viewer's left has foliage in his hair that projects upwards. Just above the seated individuals are the remains of what was probably an identifying name caption now obliterated (see Greene Robertson 1985b: 13–16).

**HOUSE A, PIER C: DESCRIPTION (GR 44).** Like Pier B, the sculptural theme of House A, Pier C (Figure 4.33), is "Ancestor Portraiture with Attendants" (APWA). Positioned just to the left of Pier B, Pier C is the most intact of the House A pier sculptures. Pier C portrays three individuals framed by a sky-band; one is centrally positioned, and the remaining two are seated cross-legged to the left and right of the central figure. As on Pier B, the central figure is depicted standing in profile, in an authoritative pose facing the medial entrance of House A. This male individual is portrayed wearing a feathered mosaic helmet with two Jestor Gods positioned at the top. As on Pier B, short feathers are visible emerging from the heads of the gods. At the front of his headdress can be seen emerging a bent sky-band, which terminates with the head of the Sun God. Bound with ribbon, a long ponytail can also be seen falling from the rear of the central figure's head, above which is a serpent wing and a fish nibbling water lilies. Around the central figure's waist is a jaguar or ocelot skin hipcloth worn over a loincloth. Attached to the front panel of the loincloth a zoomorphic god can be seen. The protagonist also wears a human mask on his back; below the mask is a mat motif that has celts attached. Wrapped around the central figure's calves are leg straps in a cross-band configuration. In the left hand of the figure is a human head incense bag that has feathers hanging from it. The figure's right hand holds a God K staff with an axe head. A smoking cigar or mirror is also visable emerging from the forehead of the god.

The two peripheral individuals on Pier C are portrayed in profile and are seated facing in towards the central figure. The male on the viewer's right is minimally dressed, wearing simply a loincloth, wristlets, and a wide tubular beaded belt. As suggested by the upward tilt of his head, the individual's gaze is fixed at the midpoint of the central figure; he is seen gripping the bottom of the God K staff held by the central figure. The figure on the viewer's left of the scene is female, as signaled by her costume. Her identity is purported to be Lady Sak K'uk', mother of K'inich Janab' Pakal I (Greene Robertson 1985b: 17). She wears a Quadripartite God headdress, beaded wristlets, and a necklace supporting an *Ik'* pectoral. Her headdress comprises a Long-lipped God missing a lower jaw and with a sun (*k'in*) sign visible in its forehead. At the top of the headdress, a cross-banded element projects forward with foliation at its end. Over her shoulders Lady Sak K'uk' wears a tubular beaded shoulder piece with a feathered hem that matches her skirt. She also wears a belt similar to the one worn by the figure seated to the left. As on Pier B, Pier C also shows evidence of once displaying name captions (see Greene Robertson 1985b: 17–20).

**HOUSE A, PIER D: DESCRIPTION (GR 45).** The sculptural theme of House A, Pier D (Figure 4.34), is "Ancestor Portraiture with Attendants" (APWA). Pier D is compositionally similar to Piers B and C with some minor differences. The central figure on Pier D, like that on neighbouring Pier E, is facing the opposite direction to the central figures portrayed on Piers B and C (i.e., towards the viewer's left). The central figure has been argued to be K'inich Kan B'alam II, the eldest son of K'inich Janab' Pakal I (Greene Robertson 1985b: 21). Much of the clothing worn by the central figure is now missing, with the exception of the individual's enormous headdress, jaguar or ocelot skin hip garment, and sandals. In addition to these elements, the central figure also wears a mask-back-ornament from which hangs a mat motif and mirrored celts. Objects held by the central figure are a God K staff and an inverted human head incense bag with draping feathers. The mosaic helmet worn by the central figure is identical to the one worn by the central figure on Pier C with a minor difference – the head at the end of the crooked sky-band (at the front of the headdress) is of an aged Sun God with a *k'in* sign in his forehead – not a youthful god as is sculptured on Pier C. Like Piers B and C, the God K staff held by the central figure on Pier D comprises a gaping zoomorphic head that has curled eyes. In addition, the zoomorphic head displays a long turned-up nose, a flint axe, smoking cigar, or mirror element emerging from its forehead. A small *Ajaw* God can also be seen crowning the God K at the top of the staff (Greene Robertson 1985b: 20–22).

As noted by Greene Robertson (1985b: 22), little costume is discernible on the two-seated figures on Pier D. Early photographs taken by Maudslay (1889–1902a), however, reveal that the seated figure to the right of the central individual wore a tubular beaded skirt and a short matching cape similar to what is worn by the female on Pier C.

**HOUSE A, PIER E: DESCRIPTION (GR 46).** House A, Pier E (Figure 4.35), has also been defined as "Ancestor Portraiture with Attendants" (APWA). As on Pier D, the central figure on Pier E faces the viewer's left towards the medial entrance to House A. The individual is portrayed wearing the same mosaic helmet headdress as worn by the central figures on Piers B, C, and D, which comprises two back-to-back Jestor Gods, a serpent wing, a fish nibbling water lilies, and a large spray of feathers emerging from the back. The central figure also holds the same style of God K staff and feather incense bag as those brandished by the central characters on the other House A piers. Other similar costume elements include the jaguar or ocelot skin hip garment and sandals, as well as a loincloth and belt that has a masket attached at its front (although only the outline of the masket is visible today). Some of the other costume elements that are discernible on the central figure of Pier E include a short, cross-hatched patterned cape and pectoral supported by a beaded necklace (see Greene Robertson 1985b: 22–25).

Due to extensive erosion, little costume is discernible on the figures seated to the left and right of the central individual. Some of the few costume elements still visible on the peripheral figures are their loincloths and remnants of their belts. The poses held by the peripheral figures on Pier E are similar to those of subordinates depicted in other artwork at Palenque and other sites (e.g., Palenque Temple XIX bench, Piedras Negras Lintel 3, Piedras Negras Stela 12, and La Mar Stela 1), where subordinates are represented with one or both arms folded across their chests.

When comparing Pier E with Piers B, C, and D, it is found that the general composition of all the piers is the same. The most noticeable differences are in the size of the central figure and the poses of the peripherally seated characters. When compared to the other piers of House A, the central figure of Pier D appears to be of noticeably shorter stature and of a more compact build.

Figure 4.36. House A Medallions 3 (GR 47). Maudslay (1889–1902a: Volume IV, plate 6).

**HOUSE A MEDALLIONS: GENERAL DESCRIPTION (GR 47).** The sculptural theme of the House A Medallions (see Figure 4.36) is "Ancestor Portraiture Solo" (APSolo). Representing what were probably portraits of deceased kings, they appear in the eastern corridor of House A adhered to the medial wall of the building. Originally numbering thirteen, only eleven medallions partially survive today; more specifically, these are Medallions 1, 2, 3, 5, 6, 7, 8, 10, 11, 12, and 13. Evenly spaced in a horizontal row, seven medallions are positioned to the south of the central doorway, whereas the remaining six were placed north of the doorway. Iconographic elements common to each of the medallions were a central portrait of a frontally oriented human individual sculptured within a cartouche from which four skeletal snaggle-tooth serpents' heads emerge at 45°, 135°, 225°, and 315°. Three cross-hatched triangular elements are present on the inner edge of each cartouche pointing in towards the central portraits at 0°, 90°, and 270°. Greene Robertson (1985b: 26) argues that, combined, these triangular elements form the edges of *k'in* signs which framed the central portraits. Facing east in the direction of the rising sun, their proximity behind the eastern piers of House A suggest that they would not have been easily seen if one were attempting to view them from the plaza below. Because each measures approximately 75 cm in diameter, one would have had to be directly in front of them (in the eastern gallery of House A) to identify any detail of the individuals once represented.

The few sculptured elements that distinguish each of the House A Medallions from one another are those located between the snaggle-tooth serpents projecting from the outer edge of the cartouches. More specifically, on Medallion 3 (Figure 4.36), just below the serpent positioned on the viewer's left, is a serpent-bird perched on a Venus sign similar in appearance to the creatures appearing on the Venus signs on the rear wall of Room 5 House B (see Section 4.3.3, The Buildings of the Palace Phase 3, House B: Further Description and Sculptural Additions, this chapter). On Medallion 5, elements resembling bundles are present around the periphery of the cartouche located between the serpents. On the surfaces of the bundles are water signs (small bars configured in tapered stacks) symbolic of the Underworld. The bundle positioned at the base of the cartouche also has a *kimi* ("death") sign sculptured next to it (Greene Robertson 1985b: 28). Located between the two surviving snaggle-tooth serpents on Medallion 6 is a "turkey glyph," which also has a *kimi* sign sculptured immediately to its left. Medallion 7 has the remains of a cross-hatched and oval-shaped glyph present between the surviving serpent heads. On Medallion 8 there are three glyphs sculptured at the top of the cartouche. Whereas the two peripheral glyphs are illegible, the central glyph reads *imix*; water signs symbolising the Underworld also appear on top of two of these glyphs (Greene Robertson 1985b: 29). Between the serpent heads on Medallion 10 are *yax* signs from which foliation can be seen emerging (Greene Robertson 1985b: 29). Medallion 11 has two inward spiraling scrolls between the remains of the snaggle-tooth serpents, and Medallion 12 has stacked, U-shaped elements – one of which has a Venus sign attached – projecting from the left and right of the cartouche. At the top and to the right of the cartouche on Medallion 13 are *k'in* signs framed in inverted U-shaped borders with *ajaw* bones projecting out from the sides (see Greene Robertson 1985b: 26–31).

**HOUSE A WEST FOUNDATION *ALFARDAS:* DESCRIPTION (GR 58).** As outlined in the Palace Phase 2 discussion, it is likely that House A was originally designed to mirror House C and that the addition of the *alfardas* (GR 58), immediately to the left and right of the House A stairs, was undertaken to reinforce the symmetry expressed by

Table 4.15. *House A, west foundation* alfardas *dates*

| | | | | | |
|---|---|---|---|---|---|
| K'inich Janab' Pakal I | Left: 9.11.9.10.12 | 6 Eb' | 10 Woh | 25 March | 662 |
| | Right: 9.11.9.10.13 | 7 B'en | 11 Woh | 26 March | 662 |

*Note:* After Martin (2003: 111).

the two structures (see Section 4.3.2, The Buildings of the Palace Phase 2, House A: Description and Sculpture, this chapter). Furthermore, I have proposed that it was only after the closing of the East Court with House A-D that the seven remaining captive panels (GR 58) were added to the western foundation wall of House A.

The sculptural theme of the two House A west foundation *alfardas* is "Captive Display" (CapDis). The images on the monuments portray two kneeling captive figures plainly dressed, each wearing a cloth headband or turban, *ajaw* bone wristlets, a beaded necklace with mask pectorals, in addition to a loincloth portraying an inscription. The texts present on loin aprons display two dates that can be placed in the Long Count. The Long Count date present on the left figure's loin apron should read 9.11.9.10.12 6 Eb' 10 Woh, whereas the date on the right figure should read 9.11.9.10.13 7 B'en 11 Woh (see Table 4.15). These dates fall a day apart in the year A.D. 662, placing the carving of these monuments within the reign of K'inich Janab' Pakal I. Due to their proximity against the western foundation of House A, the completion of House A's platform can be loosely correlated with the *alfarda* dates. At position A3 on both texts an event is recorded that reads *nahwaj*, which refers to the dressing or display of captives. Following this event, recorded in both inscriptions, is the title *yahil* meaning "conquered one" (Martin 2003: 111).

House A's north foundation *alfarda* portrays a kneeling captive in a frontal orientation with his arms crossed in front of his body. The head of the individual is depicted in profile facing towards the viewer's right. The captive wears *ajaw* bone wristlets, a beaded necklace supporting an *ajaw* pectoral framed with small round beads, as well as pendulous earflares that display floral elements at the front of the ear. Around the head of the figure is a headband from which a piece of cloth hangs, covering the right shoulder of the individual. His hair has a typically male cut, short at the back and longer on top with an area shaved in between.

The costume, posture, and orientation of House A's southern *alfarda* are virtually identical to the northern *alfarda*. Some of the relatively minor differences between the figures (other than the more obvious reversal of the head orientation and the arm positions) are the slightly lower right forearm of the right *alfarda*, which exposes more of the *ajaw* pectoral. In addition (as previously mentioned), the content of the inscription on the right loincloth also varies. The fact that both the *alfardas* resemble each other so closely indicates that they were designed to be together and interpreted as a single sculptural composition. Combined, the House A *alfardas* compositionally mirror each other like those flanking the H.S. of House C on the opposite side of the East Court.

### 4.3.3. The buildings of the Palace Phase 3

Phase 3 of the Palenque Palace has been attributed to K'inich K'an Joy Chitam II (A.D. 702–721); among this ruler's contributions were the final extension to the Palace platform (see Figure 4.1) as well as the addition of Houses D and A-D. K'inich K'an Joy Chitam II was also responsible for erecting the monumental stairs located on the north and west sides of the Palace, as these massive stairways provide access to Houses D and A-D. In addition to the burial of the Northern Galleries (also referred to as the early subterranean structures), K'inich K'an Joy Chitam II's modifications to the Palace complex resulted in the closing off of the East Court and West Court of the Palace. With the closing of the Palace's East Court with House A-D, a number of curtain walls were added to House B to prevent through access to the Southeast Court (Schele 1986: 119, see also Greene Robertson 1985b: 50).

Houses D and A-D were constructed together as a single architectural unit as indicated by the absence of construction joins identifying where Houses D and A-D may have intersected. Furthermore, there is no position where an earlier northern wall to House D could have logically been positioned. With the construction of Houses D and A-D, a western entrance to the Palace was created between the southern end of House D and the northern end of House K (Andrews 1974). As noted earlier in this chapter, House K was the product of two construction phases described as the southern portion of House K and its northern extension. The purpose of House K's northern extension was to join the building to House D. It is probable that the northern extension to House K was added some time after the construction of House D, which resulted in the formation of a gap between the two buildings at the top of the western monumental stair (Figure 4.37). Fronted by two narrow piers, the six-metre span of this gap (combined with its position adjacent to the Great Plaza and House E's throne room) suggests that it was once both an elaborate and formal entrance to the Palace Group.

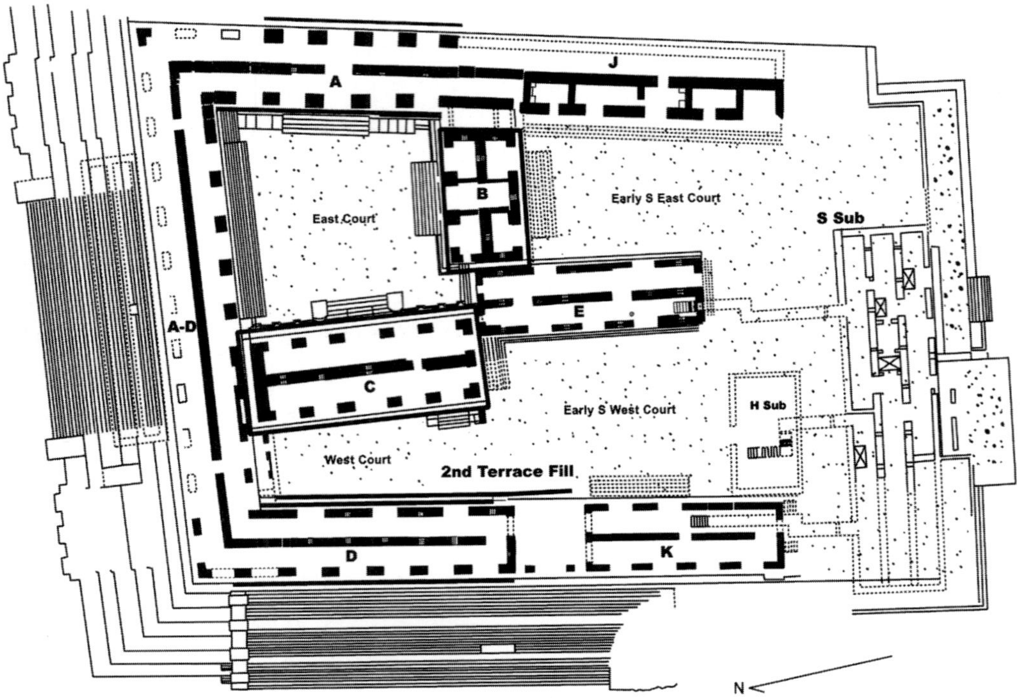


**Figure 4.37.** Map of the Palenque Palace Phase 3, modified after Greene Robertson (1985b: fig. 1).

**House A-D: Description and sculpture**

House A-D was constructed during the reign of K'inich K'an Joy Chitam II (Figure 4.38); this is indicated by the presence of the Palace Tablet (GR 57) adhered to the northern surface of House A-D's medial wall detailing his accession. In addition to the Palace Tablet, a glyph naming K'inich K'an Joy Chitam II was discovered on the remains of Pier J (see GR 56) of the same structure.

Because of the foundation giving way, little of House A-D survives today (Mauldslay 1889–1902a, IV: 19). Remaining architecture indicates that the building once comprised two long galleries separated by a continuous central wall (see Table 4.16 for gallery dimensions). At some later stage, the northern wall of House A was demolished so that the galleries of Houses A and A-D could be combined. Thirteen piers once defined the northern periphery of House A (if one counts the north-eastern and northwestern corners of the house), and ten piers defined the southern boundary. Unfortunately, few complete piers survive on the northern side of House A-D. According to the pattern of pier sculpture on other structures at Palenque, it is probable that the external surfaces of Piers 1 and 13 would have been textual, whereas the contents of Piers 2 through 12 were predominately iconographic (Greene Robertson 1985b: 51). The massive northern Palace stairway, which is approximately twenty-five metres wide, provided an impressive approach to the building. Access to the northern corridor of House

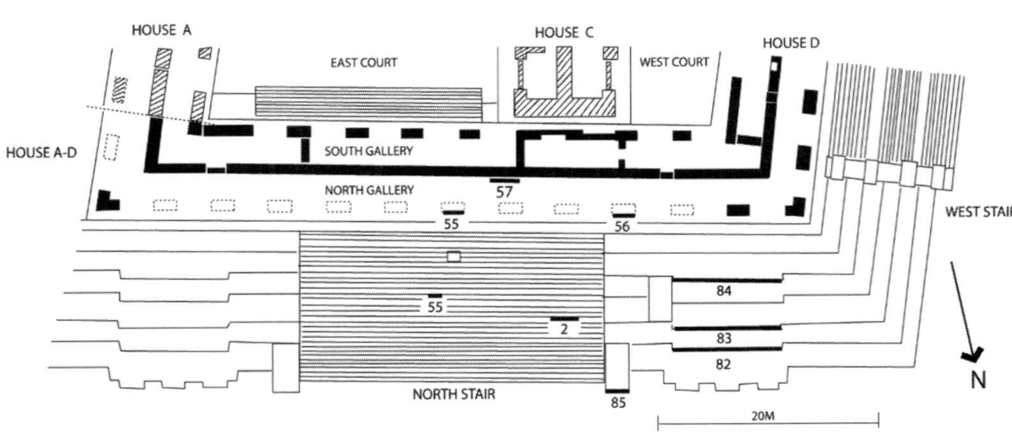


**Figure 4.38.** House A-D and Northern Palace Stair Plan, Palenque Palace.

Table 4.16. *House A-D, interior gallery dimensions*

| Gallery | Length | Width | Area |
|---|---|---|---|
| North | 56.70 m | 2.30 m | 130.41 $m^2$ |
| South | 49.80 m | 2.30 m | 114.54 $m^2$ |

*Note:* Dimensions in accordance with Andrews (1974).

A-D was achieved by passing through one of the twelve entrances between its piers. Admittance to the southern corridor of House A-D was given via two narrow doorways, one situated at the eastern end of the medial wall and one at its western end. The southern gallery of House A-D was designed to be long and unobstructed, providing easy access to the western corridor of House A and to the eastern corridor of House D. Later partitioning saw the southern corridor of House A-D divided into four rooms, two adjacent to the East Court and two contiguous with the West Court (see Greene Robertson 1985b: 50–51).

**HOUSE A-D, PIER G: DESCRIPTION (GR 55).** The sculptural theme of House A-D, Pier G, is tentatively defined as "Royal Portraiture Solo?" (RPSolo?). Little sculpture remains on what is left of Pier G; therefore, it is impossible to determine who was depicted on this pier, whether it was K'inich K'an Joy Chitam II himself or perhaps one of his predecessors. All that survives of the image once present on Pier G is the profile view of a large feathered headdress of the type generally worn by Palenque kings (e.g., the western piers of House C, see Section 4.3.2, The Buildings of the Palace Phase 2, House C: Description and Sculpture, this chapter, Figures 4.20, 4.21, and 4.22). The orientation of the headdress indicates that the head of the wearer was turned towards the viewer's left. It remains unclear what the body orientation of the figure wearing the headdress would have been, whether the individual held a profile or frontal stance. One of the more obvious elements that comprise the headdress is an array of large feathers emerging from a band of shorter cropped feathers. Three jaguar ears are attached to the outer edge of the cropped feathers, and a serpent wing is attached at their base. At the top of the headdress is a jaguar tail, as is indicated by the spots and a cross-hatched section at its end. At the base of the jaguar tail is a forelock of hair pulled through a bead that is just above a *k'in* sign that is embedded in the forehead of the god. It is probable that the *k'in* sign marks this deity as the Sun God. The god wears a cap consisting of a jade symbol and a bead that has a mirror embedded. Four tubular beads can also be seen emerging from this element, which have *ajaw* beads attached at their ends. "A modification of the Quadripartite Badge" (Greene Robertson 1985b: 52) is present just above the serpent wing. The badge comprises cross-bands within a border, a flattened U with *ajaw* bones attached to its sides, as well as plant foliation that emerges from its top (see Greene Robertson 1985b: 51–52).

**HOUSE A-D, PIER J: DESCRIPTION (GR 56).** As with Pier G, the sculptural theme of House A-D's Pier J has been tentatively defined as "Royal Portrait Solo?" (RPSolo?). Other than some modeled stucco believed to be associated with the upper portion of the pier, nothing else survives of this fragmented monument. The remaining Pier J sculpture portrays a representation of an inverted snaggle-tooth serpent that has a blade emerging from its gullet, as well as the remnants of a zoomorphic headdress. The serpent is stylistically consistent with those appearing on the House A Medallions (see Section 4.3.2, Buildings of the Palace Phase 2, House A: Description and Sculpture, this chapter). The creatures represented in both houses have a similar supraorbital plate and nose curl, a long upper jaw, a forward-positioned snaggle-tooth, as well as a *tau* tooth and incisors. The blade emerging from the snaggle-tooth monster probably represents the top section of a lance or spear.

The creature on the viewer's right of the Pier J fragment represents what is probably the upper forward portion of a headdress. Seen here in profile, it comprises the snout of a peccary terminating with the nose of a leaf-nosed bat. The creature has a prominent brow and heavy cheekbones; it wears oval-shaped earflares that have cropped ribbons hanging at their base. On top of the creature's head are the remains of a cap of some kind, most of which is now missing, with the exception of the lower portion. Small scrolls are visible running along the wavelike brim of the creature's hat.

Two glyphs survive from House A-D's Pier J – the glyph on the viewer's right names K'inich K'an Joy Chitam II, who acceded to the throne after his brother, K'inich Kan B'alam II in A.D. 702. The presence of K'inich K'an Joy Chitam II's name on Pier J may be viewed as circumstantial evidence that he was the ruler who oversaw the construction of House A-D. The glyph block to the right of K'inich K'an Joy Chitam II's name glyph displays two male heads set back-to-back; this glyph has yet to be transcribed (see Greene Robertson 1985b: 52–54).

**HOUSE A-D PALACE TABLET: DESCRIPTION (GR 57).** The principal themes of the Palace Tablet have been defined as "Royal Audience Accession" (RAudAcc) and "Text, Dynastic Historical" (TDynHist). Once positioned above

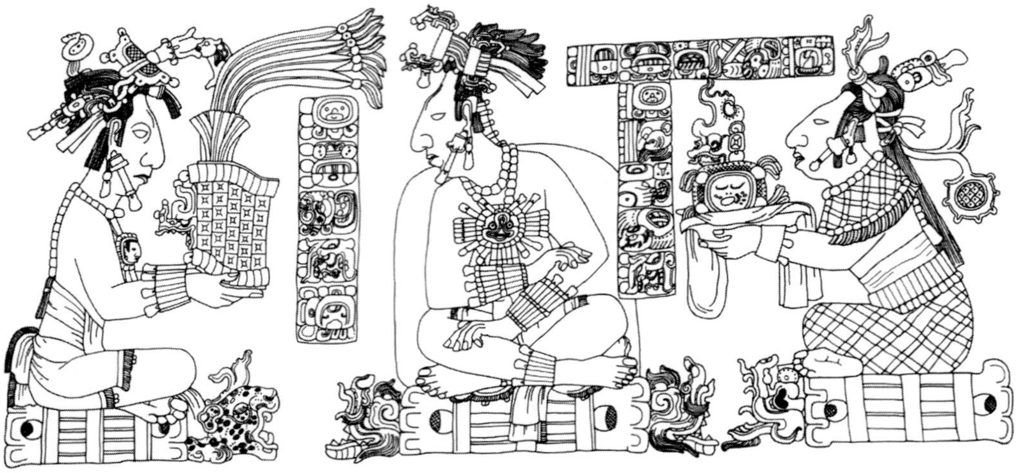

**Figure 4.39.** Palace Tablet Detail (GR 57). Drawing by Linda Schele (No. 124), © David Schele, courtesy of the Foundation for the Advancement of Mesoamerican Studies, Inc., www.famsi.org.

a throne, the Palace Tablet is located on the northern side of the medial wall of House A-D (Greene Robertson 1985a: 54). The Palace Tablet comprises a large body of text with an accompanying scene depicting the accession ceremony of K'inich K'an Joy Chitam II (Figure 4.39). Totaling 268 glyphs, the text also records historical events leading up to the accession of K'inich K'an Joy Chitam II. The latest dates recorded on the monument are 9.14.8.14.15, 9 Men 3 Yax (A.D. 720) and 9.14.8.15.18 6 Etz'nab' 6 Sak (A.D. 720), dates which both fall towards the end of K'inich K'an Joy Chitam II's reign (see Table 4.2, Palenque Rulers' Birth, Accession, and Death Dates). On the viewer's left of the composition, K'inich Janab' Pakal I is portrayed seated cross-legged and in profile on a water lily–jaguar (God III) throne fabricated from stacked mats. In his raised hands he holds a drum-major headdress, which he presents to K'inich K'an Joy Chitam II, the younger brother of K'inich Kan B'alam II. A Jestor God is attached to the front of the headdress as well as a spray feather that can be seen emerging from the top. Here K'inich Janab' Pakal I wears a simple hipcloth wound up around his waist, pendulous ear ornaments, and bracelets made from tubular beads and *ajaw* bones. Around his neck, K'inich Janab' Pakal I wears a beaded necklace supporting an *ajaw* pectoral. A single strand of beads can be seen hanging at the back of the necklace. K'inich Janab' Pakal I's headdress consists of a water lily pad and a fish nibbling a flower. Just below the water lily pad is a bead with a forelock of hair pulled through it; to the rear of the headdress, jade and plant elements are also visible.

K'inich K'an Joy Chitam II is depicted in a frontal orientation at the centre of the composition, flanked by his parents. He is seated with his arms crossed on a throne of stacked mats, and he is leaning against a large cushion. On the left and right of the throne is a *xoc* fish representative of God I (GI) of the Palenque Triad. Although he is simply dressed, the embellishment in his costume exceeds that in his father's (K'inich Janab' Pakal I's), comprising anklets and double-length wristlets with *ajaw* bones attached and pendulous ear ornaments. The pectoral that K'inich K'an Joy Chitam II wears is large and displays the head of a snarling jaguar at its centre. Framing the jaguar head are tubular beads with *ajaw* bones at their ends. Behind K'inich K'an Joy Chitam II's folded arms, a plain hipcloth is visible, similar to the one worn by K'inich Janab' Pakal I. The headband worn by K'inich K'an Joy Chitam II comprises cylinder elements from which a jawless serpent emerges at both ends.

On the viewer's right of the tablet, K'inich K'an Joy Chitam II's mother, Lady Tz'akb'u Ajaw, is portrayed seated in a pose identical to that of her husband, K'inich Janab' Pakal I. Instead of offering a headdress to K'inich K'an Joy Chitam II, however, she holds forth a bowl containing what appears to be a shield with a flayed human face and a flint god, which rests on a piece of cloth. Like K'inich Janab' Pakal I, Lady Tz'akb'u Ajaw is portrayed seated cross-legged and in profile on a throne of stacked mats. Attached to the left of the throne is the head of a serpent, signaling that Lady Tz'akb'u Ajaw is affiliated with God II of the Palenque Triad. In this scene, Lady Tz'akb'u Ajaw is portrayed wearing a shoulder piece, matching skirt made from long tubular beads, and smaller round beads configured in a netting pattern. Along the lower edge of the shoulder piece is a fringe of short feathers. Lady Tz'akb'u Ajaw also wears a necklace and bracelets similar to those worn by K'inich Janab' Pakal I, but with no *ajaw* pectoral attached to the front

Figure 4.40. Northern Monumental Staircase Masks, Palenque Palace (GR 82, 83, and 84). Drawing by Linda Schele (No. 125), © David Schele, courtesy of the Foundation for the Advancement of Mesoamerican Studies, Inc., www.famsi.org.

of the necklace. Around her waist is a wide belt fashioned from long, tubular beads and round beads attached to its lower edge. Beneath the belt some material can also be seen, wound around her torso. Although she does not wear a headdress as such, there are still a number of elements that have been incorporated to embellish her hairstyle; among these are ribbons used to tie the hair up as well as water lily foliation. Ribbons are also visible binding Lady Tz'akb'u Ajaw's hair at the rear so that a large ponytail falls neatly down her back. She also wears her name glyph in her hair (see Greene Robertson 1985b: 54–60).

**THE PALACE NORTH STAIR MASKS (GR 82, 83, 84) AND *ALFARDAS* (GR 85): DESCRIPTION.** The sculptural themes of the northern stairway masks (Figure 4.40) are "Royal Portraiture Solo" (RPSolo) and "Cosmological Celestial Underworld" (CosCU). Once flanking the northern monumental staircase of the Palace were multiple portraits of K'inich K'an Joy Chitam II (Schele 1998: 482) (see the Palace reconstruction by Greene Robertson 1985b: fig. 356). Once numbering four on each side, each portrait was the same, portraying a central portrait of K'inich K'an Joy Chitam II beneath a representation of the Cosmic Monster. The left head of this creature displays quadripartite symbolism that comprises a spondylus seashell, cross-bands, a stingray spine, and a bowl, and a sky-band represents the body of the creature. Combining celestial and Underworld symbolism, the forward head of a Cosmic Monster represents the heavens, "the front head being identified with Venus pulling behind it the fleshless head of the sun in the Underworld" (Miller and Taube 1993: 45). To the left and right of the central mask, a God K is portrayed emerging from the gaping mouth of the serpent. Adorning each portrait of K'inich K'an Joy Chitam II is a zoomorphic headdress that has feathers emerging from its sides; below these feathers hang massive ear ornaments.

As indicated by the discovery of a large tablet-like fragment set against the balustrade at the base of the northern Palace stairway, there were once two massive *alfardas* (GR 85) flanking the northern Palace staircase, each portraying standing portraits of a human figure. As K'inich K'an Joy Chitam II is purported to be responsible for the construction of the northern Palace monumental stairs, it is probable that these full-figured high-relief portraits once portrayed representations of K'inich K'an Joy Chitam II in full royal regalia (Greene Robertson 1985b: figures 354 and 356).

**House D: Description and sculpture**

Built together with House A-D, House D was commissioned by K'inich K'an Joy Chitam II. House D comprises a portion of the northwest quadrant of the Palace complex. The building runs in a north–south direction, defining the western boundary of the West Court (Figure 4.41). The plan of this building indicates that this structure was probably built at the same time as House A-D, only to be later connected at its southern end to House K. Both Mauldslay (1889–1902a, Vol. IV: 13) and Greene Robertson (1985b) have argued that, like House A, House D once stood independently of all other Palace structures. As Andrews (1974) points out, however, "there is simply no logical place where an earlier north end wall of House D could have been positioned and certainly none which could not have left some trace of its presence" (Andrews 1974). Access between the eastern and western galleries of House D is provided by a single entrance at the southern end of the medial wall of the building (see Table 4.17 for gallery dimensions). Admittance to the eastern gallery of House D was achieved by ascending a broad stairway rising from the level of the plaza directly adjacent to the Temple of Inscriptions.

Sometime late in the history of the Palenque Palace, the eastern side of House D was divided into four rooms with the addition of curtain walls (Greene Robertson 1985b: 32). On the western wall of one of the rooms are the remains of a painted mural (GR 54), although too little of the mural survives to distinguish any significant

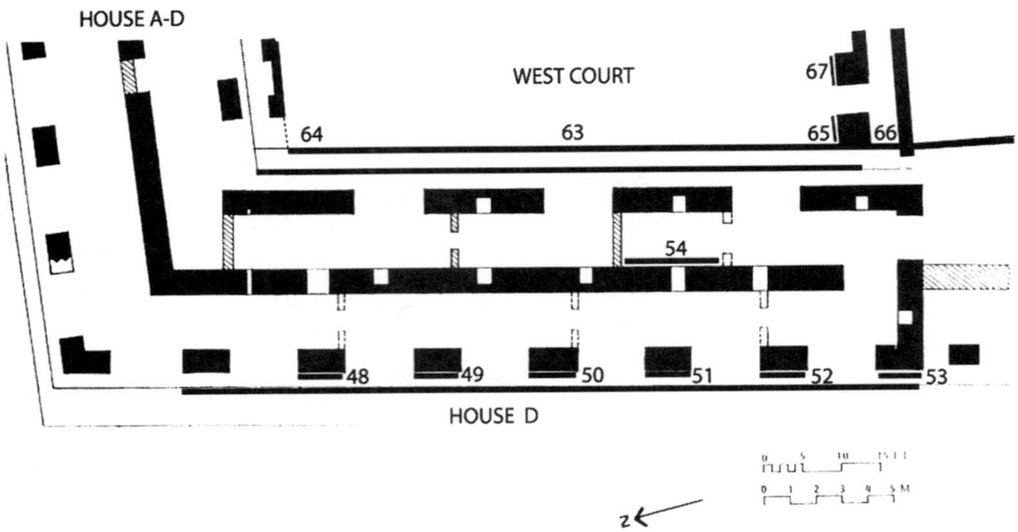


**Figure 4.41.** House D Plan, Palenque Palace. Modified after Andrews (1974). George F. and Geraldine D. Andrews papers. The Alexander Architectural Archive, the University of Texas Libraries, the University of Texas at Austin.

detail. There are eight *Ik'* openings present in the House D walls, four of which were at some time sealed up, and two that never completely penetrated the walls of the building (Greene Robertson 1985b: 33).

Sculpture survives on six of House D's western piers, described as Piers B, C, D, E, F, and G. With the exception of Pier G, two figures are represented on each of the House D piers, in contrast to the three individuals portrayed on the external piers of House A. Comparatively speaking, there is fluidness to the postures of the figures represented in each of the House D piers which is substantially different from the more static style of figural art appearing on the exterior of House A. According to Greene Robertson, the most prominent theme represented in the House D pier sculpture is sacrifice: "Pier B and C concerns persons about to be sacrificed. Pier F is concerned with the actual beheading. Pier D depicts a sacrifice about to be performed involving two people and a serpent" (Greene Robertson 1985b: 33). It remains possible that the House D piers were designed to be interpreted collectively, with each composition representing a different stage in a ritual sequence. At present, it cannot be confirmed whether the protagonist on each pier is a representation of K'inich K'an Joy Chitam II or one or more other rulers. Given, however, that K'inich K'an Joy Chitam II is the most likely ruler to have presided over the construction of House D (sometime late in his reign), it is argued here that the principal figure represented on each of the House D piers is in fact K'inich K'an Joy Chitam II.

Table 4.17. *House D, approximate interior gallery dimensions before partitioning*

| Gallery | Length | Width | Area |
|---|---|---|---|
| West | 32.0 m | 2.2 m | 70.4 $m^2$ |
| East | 29.0 m | 2.2 m | 63.8 $m^2$ |

*Note:* Dimensions in accordance with Andrews (1974).

**HOUSE D, PIER B: DESCRIPTION (GR 48).** The sculptural theme of House D, Pier B (Figure 4.42), has been defined as "Human Sacrifice" (HumSac); the scene portrays two male individuals, one seated and the other standing. In the scene, the seated figure (viewer's left) is about to be struck with an axe; he is portrayed in profile with his upper shoulders and torso turned in a frontal orientation. The left arm of the seated figure is raised (perhaps defensively), and his right arm hangs by his side. Remaining sculpture suggests that the individual was minimally dressed, wearing simply a hipcloth, a beaded necklace, and a pendant of some kind. The right figure stands in profile holding an eccentric flint axe in his raised right hand. The protagonist's forward-tilted posture suggests that he is threatening to strike the seated individual. In his right hand, the standing figure holds a narrow length of cloth. In addition to wearing wristlets, the standing individual wears sandals made from jaguar skin, as well as a beaded shoulder piece and a pectoral adorning his chest. The front panel of a loincloth is visible hanging to the front of the individual's body as well as a jaguar tail that hangs at the rear of the individual. The hair of the protagonist is pulled back into a ponytail and is bound at a number of points along its length. Just below his ponytail is a shaved area, below which is a band of shorter hair. In addition to the above elements, the figure also

**Figure 4.42.** House D, Pier B, Palenque Palace (GR 48). Maudslay (1889–1902a: Volume IV, plate 32).

wears a headband that supports a spray of feathers that emerges at the rear of his head. Rising vertically from this arrangement is an element resembling a Long-lipped God that is surrounded by shorter feathers.

On the lower register, on the viewer's left of the composition, the head of an *Ik'* God is portrayed. The god has a skeletal lower jaw and a *tau* symbol attached to its headdress symbolising the watery Underworld. The god also wears a round ear ornament that supports an inverted *ajaw* element as well as an *Ik'* symbol, which is attached to the brow of the god. The protagonist of the scene stands on a platform comprising three overlapping rectangular cartouches containing *tau* symbols and plant foliation. Surviving text on Pier B provides remnants of a Long Count date reading 9 Baktuns and 4 Katuns; this date is followed by a verb currently unidentified, and the remaining glyph block represents what was probably a title (see Greene Robertson 1985b: 34–36).

**Figure 4.43.** House D, Pier C, Palenque Palace (GR 49). Maudslay (1889–1902a: Volume IV, plate 32).

**HOUSE D, PIER C: DESCRIPTION (GR 49).** The sculptural theme of Pier D (Figure 4.43) has been defined as "Sacrificial Dance" (SacDan), as implied by the themes of the accompanying piers in addition to the raised right heel and the forward posture of the standing individual (Grube 1992: 201). The protagonist is portrayed in profile on the viewer's left of the composition, whereas the second individual (depicted on the viewer's right) kneels in a submissive pose. The kneeling figure is also depicted in profile; his gaze is turned up and fixed on the face of the individual standing before him. The relative status of the two individuals portrayed is implied by their submissive versus authoritative poses, as well as differences in the degree of costuming worn by each individual. For example, the subordinate figure portrays no emblems of authority. Besides a belt and loincloth, the individual wears no other costume elements. In contrast, the dancing figure is elaborately adorned wearing a feathered

**Figure 4.44.** House D, Pier D, Palenque Palace (GR 50). Maudslay (1889–1902a: Volume IV, plate 33).

headdress comprising a Long-lipped God with a second god that appears at the top of the headdress. The figure also wears a beaded necklace, a cross-banded belt, *ajaw* bone wristlets, and a beaded hip garment with a feathered hem. The dancing figure also cradles a Quadripartite God in his left arm which comprises a *k'in* sign, a stingray spine, a cross-band, and a spondylus seashell (see Greene Robertson 1985b: 39–43).

**HOUSE D, PIER D: DESCRIPTION (GR 50).** The sculptural theme of House D, Pier D (Figure 4.44), is "Sacrificial Dance" (SacDan). As noted by Grube (1992: 201), tilted posture and an upturned heel indicate the activity of dance in Maya iconography. The protagonist in the scene holds the serpent in his left hand whilst holding an eccentric flint axe in his right hand. He wears a short, beaded skirt with a feathered hem over a hipcloth made from tubular beads stitched together in a netting design. The front panel of the loincloth can be seen hanging at the front of the figure; attached to the top of the loincloth is a mat motif (Greene Robertson 1985b: 41). Around the protagonist's midriff, the top of the undergarment can be seen behind the beaded shoulder sash that supports a low-slung serpent bar pectoral. At both ends of the bar pectoral emerge the heads of Long-lipped Gods. The individual also wears jaguar skin sandals, wristlets, and feathered leg bands with bones attached. His hair is step cut at the front and bound up and forward with ribbons that are tied in bows. A lock of hair can also be seen pulled through a bead at the front of the individual's head. In addition, a spray of long feathers can also be seen protruding from the back of the figure's head, together with a leaf and flower element believed to represent a "personified tree" god (Schele in Greene Robertson 1985b: 41).

The stance of the female figure on the viewer's right of the scene suggests a more static activity; she is seen here grasping the serpent in both hands, controlling its movement. This actor's sex is signaled by gender-specific costume elements, by her long, tubular beaded skirt and the *xoc* fish emblem that is attached at the front of the skirt. In addition to these elements, she wears a matching beaded cape with a feathered hem and a quadripartite badge on her back comprising a stingray spine, a spondylus shell, and a *kimi* death sign as a substitute for cross-bands. The remains of her headdress display a beaded serpent at its top as well as an array of feathers emerging from the back. Hanging from her neck is a beaded necklace supporting an *ajaw* masket pectoral. The border around the composition consists of *yax* symbols and jade signs believed to be representative of the Underworld. The platform on which both characters are standing comprises two deeply incised stucco layers with rows of dots separated by a zigzagging feature; between the parallel layers is a feature that has a *k'in* sign at its centre. A text was once sculptured between the two figures represented on Pier D; the text today has all but completely eroded away (see Greene Robertson 1985b: 39–43).

**HOUSE D, PIER E: DESCRIPTION (GR 51).** Although little of the sculpture survives on Pier E of House D, it is likely that the theme of the sculpture once represented was consistent with the other House D Piers. Therefore, the sculptural theme of Pier E has been tentatively defined "Sacrifice?" (Sac?). The position of the headdress, combined with the orientation of the hand sculptured at the centre of the composition, indicates that the scene would have, at one time, portrayed two individuals. The headdress displayed on the upper left of the composition comprises the head of a Long-lipped God with a serpent wing attached at its rear. A circular cross-hatched element is also visible above the god's head, in addition to a T-shaped element protruding from the top. All that remains of the border that once framed the composition is a row of *k'an* and jade signs at the bottom of the pier, probably signifying blood (see Greene Robertson 1985b: 43–44).

**HOUSE D, PIER F: DESCRIPTION (GR 52).** The principal theme portrayed on House D, Pier F (Figure 4.45), is "Human Sacrifice" (HumSac). The scene portrays the instant just before the decapitation of a captive. In this scene, the protagonist (viewer's right) is portrayed elaborately dressed and in profile, whereas the seated victim (viewer's left) wears only a simple hipcloth over an undergarment. The standing figure holds an axe in his raised right hand while grasping the hair of the seated individual, pulling him forward with the intention of exposing the back of his neck. The forward lean of the right figure is indicated by the low slung bar pectoral, which hangs away from the protagonist's body. The costume worn by the standing figure includes a twin Jestor God diadem, an *ajaw* bone necklace supporting a bar pectoral, a wide tubular beaded belt, a beaded hipcloth in a net configuration, and jaguar skin sandals. Attached to the front of the belt is a jaguar head that represents the Underworld Jaguar Sun. Diagnostic elements that identify this god include a skeletal lower jaw, a *tau* tooth, and the three dots beneath the eye (Greene Robertson 1985b: 44). Below the jaguar head is a mat motif; below this feature hangs a celt with an embedded mirror. Attached to the base of the celt is a jade god, as signaled by the jade glyph, the supraorbital plate, and the presence of another jade sign just below the creature. Below this god hangs a bell-shaped feature similar to what is depicted hanging at the rear of the individual. Tubular beads and *ajaw* bones are visible protruding from the bell-shaped element. The standing figure's headdress comprises a large spray of feathers, ribbons, and some plant foliation that can be seen protruding from the rear of the figure's head.

The figure sculptured on the viewer's left of the composition is seated on what is purported to be an *Ik'* god throne. The identity of the god is signaled by the *Ik'* element present in the ear ornament of the creature just above the inverted *ajaw* (Greene Robertson 1985b: 46). The creature's head is portrayed in profile with its tongue hanging out; a square element is also present in the god's eye. According to Greene Robertson, the presence of the water lily god on Pier F, House D,

**Figure 4.45.** House D, Pier F, Palenque Palace (GR 52). Maudslay (1889–1902a: Volume IV, plate 33).

"definitely places the scene in the Underworld" (Greene Robertson 1985b: 46).

At the lowest register of Pier F, plant foliation (probably water lilies) emerges from the head of the god. In addition, seashells (two on each side) are depicted just behind the foliation flanking both sides of the water lily god. On the lower left of the pier, God of Number Six (as signaled by the axe eye) can be seen emerging from a flower. Framing the entire scene were alternating *ajaw* bones and jade signs. Water or blood symbolism is the likely message offered by the imagery of the border (see Greene Robertson 1985b: 44–47).

**HOUSE D, PIER G: DESCRIPTION (GR 53).** No theme could be determined for House D, Pier G; the monument was primarily textual, comprising eighteen glyph blocks arranged in two equal columns of nine glyphs. Only three of the glyph blocks survive today, at positions A1, A3, and A4 (Greene Robertson 1985b: 47). The glyph at position A4 records the name "Siyaj K'ak', leader of the Mexican *entrada*" (Martin and Grube 2000: 156). Some scholars have interpreted the reference to this individual on a seventh-century monument at Palenque as evidence of an earlier foray of Teotihuacan into the Maya lowlands (Martin and Grube 2000).

**HOUSE D EAST FOUNDATION *CASITA* WALL: DESCRIPTION (GR 63, 64, 65, 66, AND 69).** With the closing of the north and northeast end of the Palace with Houses D and A-D, both the northern and eastern boundaries of the West Court were established. Following this extensive architectural contribution by K'inich K'an Joy Chitam II, additional sculpture was assigned to the east foundation wall of House D, which is now referred to as the *Casita* Wall. Parts of this decorated wall (GR 66 and 69) were later covered with the construction of the Palace Tower and Toilets (Greene Robertson 1985b: 74, see also Section 4.3.4, Buildings of the Palace Phase 4, this chapter). The principal sculptural themes of the *Casita* Wall have been defined as "God Portraiture with Attendants" (GPWA) and "Cosmological Celestial" (CosC).

The sculpture of the *Casita* Wall comprises stones configured to resemble what Greene Robertson describes as "little stone houses" (Greene Robertson 1985b: 74). These "houses" form cartouches that run along the basement wall on the western side of the West Court just below House D. Within these cartouches, three bust-style portraits were sculptured: two peripheral human figures and a central figure resembling a god or a supernatural being of some kind. On one of the southern *casita* cartouches, two creatures with long snouts project from the left and right of the cartouche. In addition, square-nosed serpents emerge from the top left and right corners of the rectangular framing surrounding the bust-style portraits (Greene Robertson 1985b: 74). The central figure in the cartouche is portrayed in frontal orientation, whereas the two peripheral characters are depicted in profile. The central godlike figure appears to be emerging from the frame that surrounds all three individuals.

In between each of the "stone houses" are hourglass-shaped armatures that once supported additional stucco sculpture. On one of the hourglass armatures a peccary is portrayed, which has its front legs threaded through a Venus symbol. The peccary is sculptured in profile, facing the viewer's right. To the right of the peccary, on the next hourglass armature, is a sculptured representation of God K, which is also depicted with its limbs entwined in a Venus symbol (see Greene Robertson 1985b: 74–76).

**THE PALACE SOUTH SUBSTRUCTURE ALTAR: DESCRIPTION (GR 1).** The sculptural theme of Altar 1 is "Cosmological Celestial" (CosC). Elements of the sculpture

**Figure 4.46.** House A West Foundation Captives, Captive Panels North and South of Stairs (GR 58). Maudslay (1889–1902a: Volume IV, plate 13).

present on the southern substructure altar appear to be consistent with elements found on the east foundation wall of House D (Greene Robertson 1985b: 85). More specifically, the upper portion of Altar 1 portrays a God K entwined in two Venus symbols that is similar to the God K–Venus combination (see GR 65) present at the southwest corner of the West Court. The iconographic parallels between the *Casita* Wall and Altar 1 suggest that K'inich K'an Joy Chitam II constructed both at approximately the same time. Below the God K, on both the eastern and western sides of the altar, a Long-lipped God is portrayed. On the southern surface of the altar, a Sun God is also depicted (see Greene Robertson 1985b: 85).

### House A: Further description and sculptural additions

**HOUSE A WEST FOUNDATION CAPTIVE PANELS: DESCRIPTION (GR 58).** Captive Panels 4 and 5 (GR 58, see Figure 4.46), or the House A west foundation

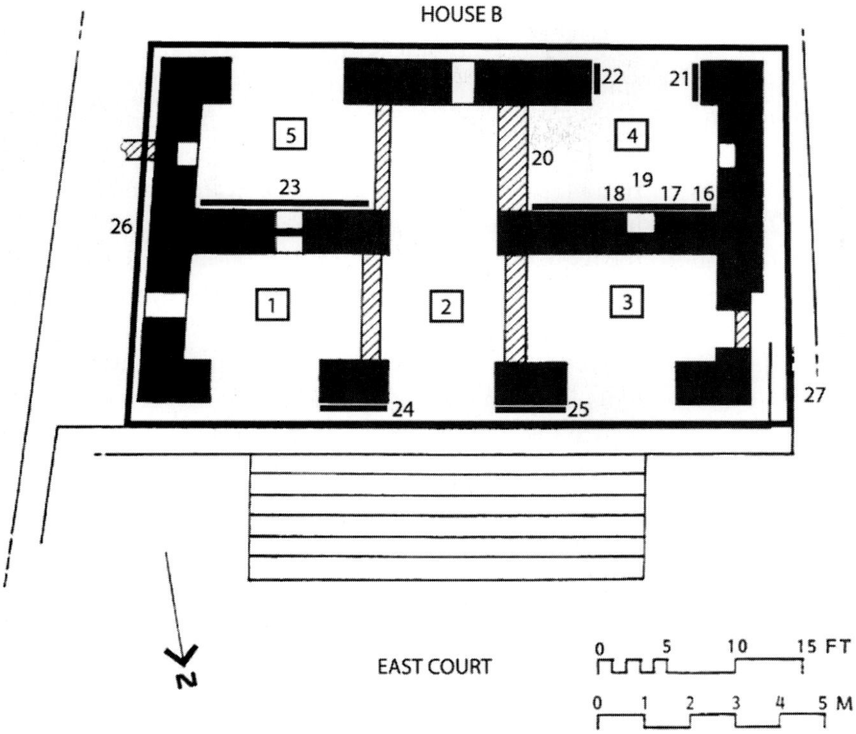


**Figure 4.47.** House B Plan, Palenque Palace. Modified after Andrews (1974). George F. and Geraldine D. Andrews papers. The Alexander Architectural Archive, the University of Texas Libraries, the University of Texas at Austin.

*alfardas* as they are referred to in this study, were installed during the reign of K'inich Janab' Pakal I as evidenced by the dates carved on the monuments (Martin 2003: 111). This installation probably occurred before the placement of the seven remaining captive panels (GR 58; Figure 4.46), which would have been undertaken after the closing of the northern end of the East Court with House A-D. As noted earlier in this chapter, the cropping of the captive panels suggests that their current positions do not represent their primary context and that they may have originated from elsewhere. The resizing of these monuments is especially evident when viewing Captive Panel 1, where the entire left half of the standing portrait appears to have been cut away to fit in the narrow space between Captive Panel 2 and the House A-D foundation wall. As the larger width of Captive Panel 3 implies, space was of little concern when it was positioned against the western foundation of House A, whereas space appears to have become more of a problem with the addition of Captive Panels 1 and 2, as suggested by the narrow appearance of both monuments.

The theme of all sculptures assigned to the western foundation wall of House A is "Captive Display" (CapDis). As stated previously, it remains possible that the seven captives referred to on the west foundation of House C may be those represented on the panels resting against the west foundation wall of House A (see Section 4.3.2, the Buildings of the Palace Phase 3, this chapter). Excluding the House A *alfardas*-style panels, a total of seven captives are carved in bas-relief relief on limestone panels on the eastern side of the East Court. Each of these figures is minimally dressed, wearing simply a loincloth, wristlets, flower and *ajaw* bone ear ornaments, as well as a headband and/or cloth turban. Additional elements adorning some of the captives include beaded necklaces and *ajaw* pectorals. The individual portrayed on Captive Panel 9 is the only captive stripped of all costume embellishments. The figure kneels in a frontal orientation with his arms held or bound behind his back. His enormous penis is exposed and has the front panel of his loincloth tucked behind it. The figure's posture, the absence of authoritative costume elements, and his exposed genitalia imply that this individual is marked for, or perhaps has participated in, sacrificial rites (see Greene Robertson 1985b: 64–66).

**House B: Further description and sculptural additions**

With the blocking of the northern end of the Palace with House A-D, the interior space of House B was modified to prevent access through the building into the Southeast Court (see Figure 4.47). The internal partitioning of House B postdates the closing of the north end of the Palace with House A-D (Schele 1986: 119), which places the refurbishment of House B, at its earliest, late in the reign of K'inich K'an Joy Chitam II (see Table 4.18 for room dimensions). As noted previously, it is likely K'inich K'an Joy Chitam II oversaw the completion of House A-D on or around 9.14.8.14.15 9 Men 3 Yax (A.D. 720) (see Section 4.3.3, Buildings of the Palace Phase 3, House A-D: Description and Sculpture, this chapter).

Table 4.18. *House B, interior room dimensions, rooms 1–5*

| Room | Length | Width | Area |
|---|---|---|---|
| 1 | 4.00 m | 2.40 m | 9.60 $m^2$ |
| 2 | 6.02 m | 2.44 m | 14.68 $m^2$ |
| 3 | 4.18 m | 2.39 m | 9.99 $m^2$ |
| 4 | 4.27 m | 2.41 m | 10.29 $m^2$ |
| 5 | 3.88 m | 2.74 m | 10.63 $m^2$ |

*Note:* Dimensions in accordance with Andrews (1974).

The partitioning of House B effectively broke up the northern and southern galleries of House B into a larger number of smaller rooms. Today, House B comprises five rooms, three of which are accessible only from the northern side of the structure (Rooms 1, 2, and 3) and two only from the southern side of the building (Rooms 4 and 5). Eight steps provide access to Rooms 1, 2, and 3 from the East Court. The entrances of Rooms 4 and 5 face the remnants of what was once a much larger Southeast Court. House F now occupies the space that once functioned as a courtyard shared by Houses B, E, G, and J (see Section 4.3.4, The Buildings of Palace Phase 4, this chapter).

The northern rooms of House B may have been used for public activities, such as receiving dignitaries, whereas the rooms on the south side of the structure were "private and possibly secret" (Greene Robertson 1985a: 47). Following the partitioning of House B's interior, additional sculpture and painting was commissioned in Rooms 4 and 5 of the building. Most if not all of the sculpture appearing in Rooms 4 and 5 postdate the partitioning of House B as suggested by the differences in the sculptural themes represented. The interior partitioning of House B, whilst defining the boundaries of each room, also marks the limits of the artwork's governance within these spaces. It becomes apparent when comparing the sculptural themes portrayed in Rooms 4 and 5 that they are markedly different, suggesting possible differences in the functions of the corresponding spaces. Given the disparate nature of the artistic themes represented in Rooms 4 and 5, it is probable that the secondary partitioning of House B occurred before the installation of the artworks. The iconography of the southwestern room (Room 4) indicates that some form of social activity occurred within the space, whereas the absence of human beings in the sculpture of the southeastern room (Room 5), combined with the presence of more esoteric imagery, suggests that the room functioned primarily as a symbolic space (see Greene Robertson 1985a: 46–48).

**HOUSE B, ROOM 4, NARRATIVE SCENE DESCRIPTION (GR 16, 17, 18, AND 19).** The principal theme of House B's narrative scene, located in Room 4 of House B, has tentatively been defined as "Royal Audience" (RAud). The northern wall of this room portrays the remains of a scene sculptured in stucco. The scene once depicted two figures seated or kneeling on a throne positioned above a bicephalic *Cauac* Monster flanked by two standing individuals (see Seler 1915). In Greene Robertson's more recent drawing (1985a: fig. 160), to the left of the throne a figure stands in profile on what appears to be a step, presenting a plate of unknown contents as an offering. This individual stands formally with his head wrapped in a turban-style headdress that has a large spray of feathers emerging from the back. Just above the throne, to the right of the standing figure, is an *Ik'*-shaped niche that is framed by a twisted stucco border. The presence of hair ribbons to the right of the niche indicates that an individual was once sculptured next to it; above the ribbons are the remnants of a painted text. An outline of a name caption is also visible above the offering held by the standing individual. Between the two heads of the *Cauac* Monster at the base of the scene, a small figure is portrayed seated cross-legged and in profile; he is bent forward facing the viewer's left of the composition wearing only a loincloth and belt. On his back sits what is either a representation of the Principal Bird Deity or a quetzal.

The presence of a throne impression on the rear wall of House B's southeast room may indicate that a bench once appeared in the room obstructing the narrative scene (Greene Robertson 1985a: 45. The installation of this bench suggests a change in the function of the space from one that housed esoteric activities to one that was used by the king to receive guests or members of his own court. Curtain holders present on opposing walls near the entrance of the room indicate that the space was at some time concealed from general view (see Greene Robertson 1985a: 43–46).

**HOUSE B, SOUTHEASTERN ROOM: DESCRIPTION (GR 23).** The sculptural theme of the double-stepped cleft fret pattern (Figure 4.48) located on the rear wall of the southeastern room of House B is "Cosmological Terrestrial" (CosT). This pattern represents "the mountain cleft in its most reduced form" (Schele 1998: 497). In this representation the pattern is combined with two *Ik'* symbols. The *Ik'* on the viewer's right of the composition

**Figure 4.48.** House B Fret Pattern, Southeastern Room (GR 23). Maudslay (1889–1902a: Volume IV, plate 18).

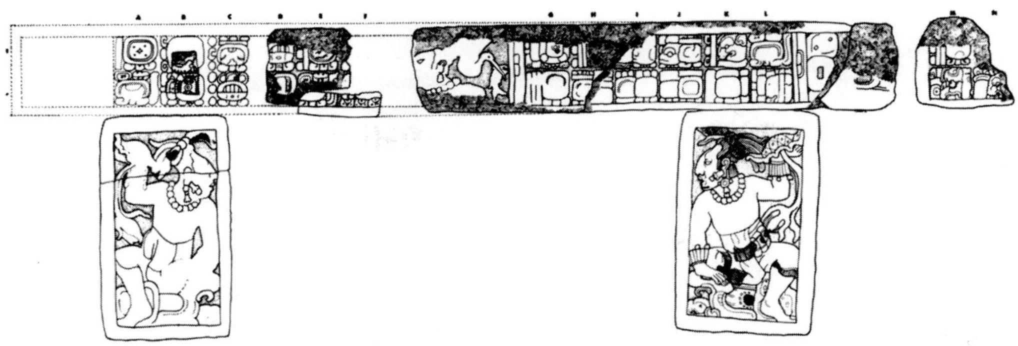

**Figure 4.49.** The Del Río Throne (GR 6). Drawing by Linda Schele (No. 128), © David Schele, courtesy of the Foundation for the Advancement of Mesoamerican Studies, Inc., www.famsi.org.

completely penetrates the wall providing a view into the northeast room of House B (Room 1). The *Ik'* symbol on the viewer's left has no recess at all other than a T-shaped scoring at the centre of the *Ik'*. Two smaller incised *Ik'* symbols are framed by the inward curling arms of the step-fret pattern. Serpent bands are visible emerging from the horizontal sections of the larger *Ik'* symbols, as well as from the curling arms of the step-fret pattern. Within these serpent bands are disk-shaped elements that alternate with cross-bands. From the ends of these bands emerge three-pronged plant elements. In addition to these elements, two serpent-birds with shell wings are depicted perched on Venus signs at the top of the larger *Ik'* symbols, and another two serpent-birds are portrayed resting on the lower serpent bands (see Greene Robertson 1985a: 46–48).

**House E: Further description and sculptural additions**

**HOUSE E DEL RÍO THRONE: DESCRIPTION (GR 6).** The principal theme of the Del Río Throne (Figure 4.49) has been defined as "Text Dynastic Historical" (TDynHist). Comprising a large slab of stone inscribed on three of its edges, the front legs of the throne portray two figures that reflect each other in both their costume and pose. More specifically, both figures appear in frontal orientation with their heads turned towards the other. Identified as *bacabs* by their netted headdresses, each individual holds a water lily in his or her outer raised hand and is sitting on an *Imix* Monster. Additional costume elements include pendulous ear ornaments, decorated hip- and loincloths, as well as beaded necklaces (Schele and Mathews 1979).

As stated earlier in this chapter (see Section 4.3.1, The Buildings of the Palace Phase 1, The Northern and Southern Subterranean Galleries and Passages, this chapter), the Del Río Throne was probably made to replace a preexisting throne commissioned by K'inich Janab' Pakal I (Bench 1, GR 80), which was moved from House E to the Subterraneos. The Del Río Throne (GR 6) was installed beneath House E's Oval Tablet by K'inich K'an Joy Chitam II to coincide with his accession on 9.13.10.6.8 5 Lamat 6 Xul. The throne text also records the accession dates of two other rulers, K'inich K'an Joy Chitam II's father (K'inich Janab' Pakal I) and older brother (K'inich Kan B'alam II).

*Del Río Throne Dates*

| | | | | |
|---|---|---|---|---|
| AI – BI | (9. 9. 2. 4. 8) | 5 Lamat | 1 Mol | |
| CI – C2 | *3.[9]. 8. 2* | | | DN, add |
| EI – FI | 9.12.11.12.10 | 8 Ok | 3 K'ayab' | |
| H1 – G2 | | 13? ?? | 12, 13 or 14 ?? | |
| JIa – I2 | 3. 4. 6 ?? | | | DN, add |
| J2b – KI | 9.13.10. 6. 8 | 5 Lamat | 6 Xul | |
| MI – NI | ? ?14 ? | | | DN, add |

(Schele and Mathews 1979: 140)

The discovery of the Del Río Throne with the Oval Tablet in the eastern corridor of House E provides evidence of the functional continuity of the building indicating that, for generations, House E was used to inaugurate kings (see Martin and Grube 2000: 163).

### 4.3.4. The buildings of the Palace Phase 4

Phase 4 of the Palace at Palenque (Figure 4.50) has been attributed to K'inich Akul Mo'Nab' III (A.D. 721–740?), whose main architectural contributions were the closing of the southern end of the Palace with Houses G and H, the construction of the Tower and Toilets, as well as the joining of Houses D and K with House K's northern addition (see Figure 4.51). The construction sequences proposed by Andrews (1974) and Greene Robertson (1985b) both agree that following the fabrication of Houses D, A-D, the Tower and Toilets, Houses G and H (Figures 4.51 and 4.52) were the next buildings to be erected (see also Table 4.1). Dates deciphered from the Tablet of the Scribe and the Tablet of the Orator, both of which once flanked the southern Tower stair, fall within

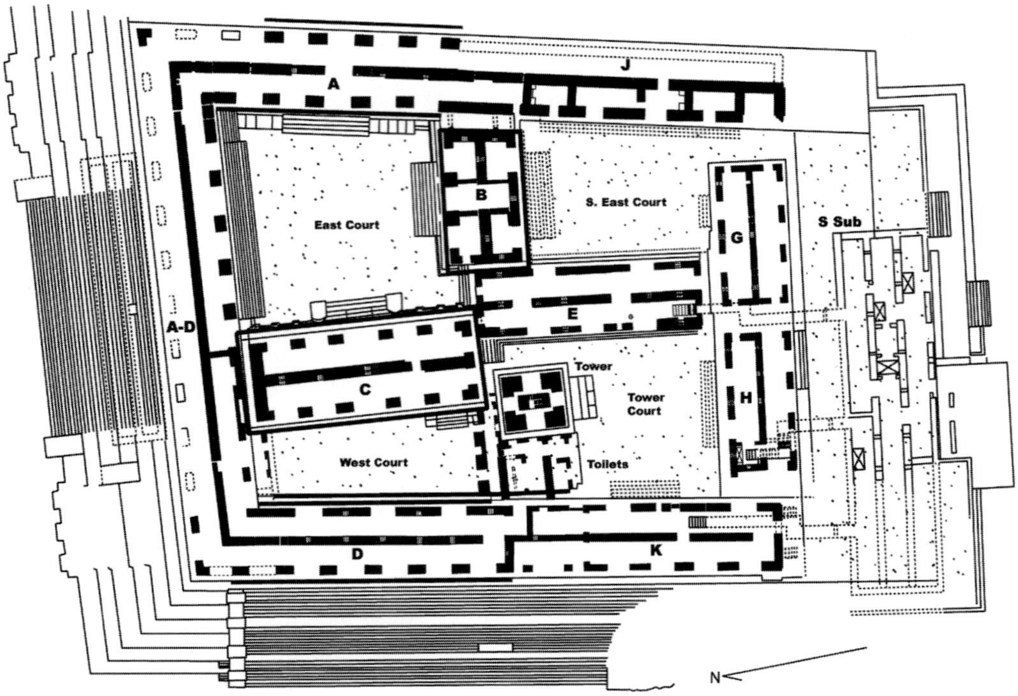


Figure 4.50. Map of the Palenque Palace Phase 4, modified after Greene Robertson (1985b: fig. 1).

the reign of K'inich Akul Mo'Nab' III. For the purposes of this investigation, the proximity of these monuments is viewed as circumstantial evidence that the completion of the Tower and Toilets occurred during K'inich Akul Mo'Nab' III's reign. It is acknowledged, however, that the presence of several ceramic caches that were found in association with the Toilets and construction west of the Tower, possibly dating from the Early *Balunte* period, may place the construction of these buildings slightly later (i.e., perhaps during the reign of K'inich Kan B'alam III and/or K'inich K'uk' B'alam II) (see Rands 1974: 5 and Table 4.4). Evidence in support of the view that the Toilets and Tower were erected together comes in the form of an acceptance that access to the Tower's interior was never provided from the level of the Tower Court. The interior of the Tower could only be accessed by crossing over the roof of the Toilets (now fallen) and passing through an entrance located on the first level on the western side of the building (Figures 4.53 and 4.54) (Andrews 1974).

The symmetry expressed by the Palace Phase 4 plan (Figure 4.50) suggests that it may represent the intended

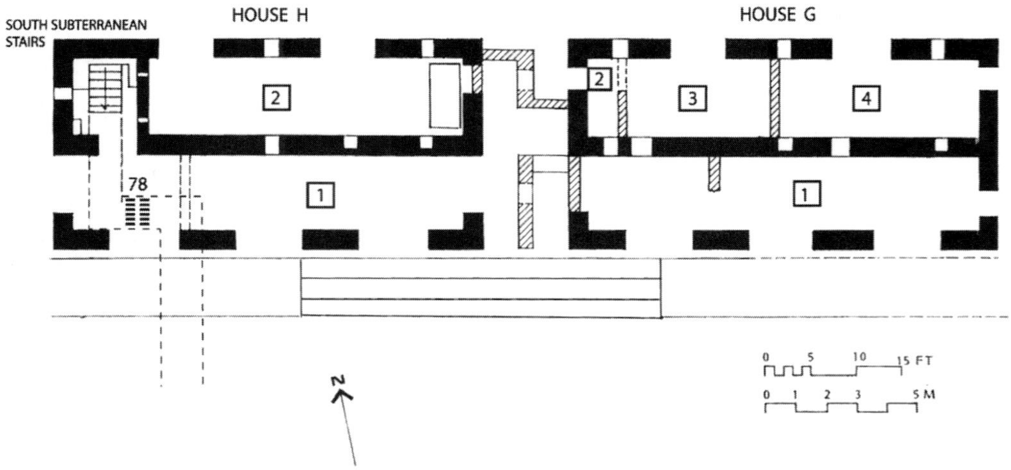


Figure 4.51. House G and H Plan, Palenque Palace. Modified after Andrews (1974). George F. and Geraldine D. Andrews papers. The Alexander Architectural Archive, the University of Texas Libraries, the University of Texas at Austin.

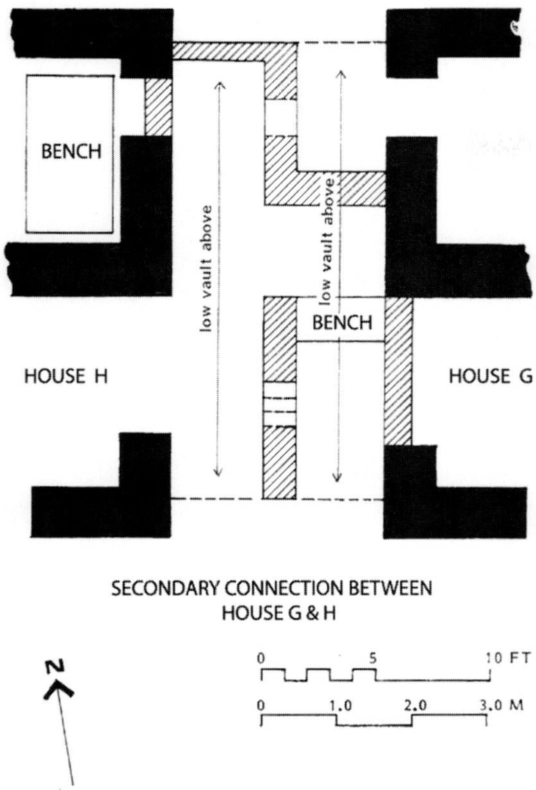


**Figure 4.52.** House G and H Secondary Extension Plan, Palenque Palace. Drawing by George Andrews (1974). George F. and Geraldine D. Andrews papers. The Alexander Architectural Archive, the University of Texas Libraries, the University of Texas at Austin.

and Tower Court probably housed meetings and/or rituals of a more private nature.

**Houses G and H: Description**

Houses G and H are similar in their general layout and configuration, an indication that both buildings were probably built at the same time (see Figure 4.51). Described as twin structures (Andrews 1974), neither building shows evidence of ever being decorated. Oriented in an east–west direction, both structures originally comprised two galleries divided by a medial wall (see Table 4.19 for gallery dimensions). Two entrances provided access to the northern galleries of Houses G and H, and three entrances provided access to the southern gallery in both buildings. Doorways were also once present in the eastern and western walls of both structures that were later blocked. Houses G and H each measured 14.37 m by 7.01 m from end to end and were separated by a 2.74-m gap (see Table 4.19; Andrews 1974). The most obvious differences between the two structures is the presence of an extra room in the northeast corner of House H that has a stairway leading down into the South Subterranean Passage of the Palace and on into the Subterraneos. At some later stage a bench was placed against the rear wall of this space, and several transverse walls were added to both structures inhibiting access through

design of the building group before the addition of Houses F and I (see Section 4.3.4, The Buildings of Palace Phase 4, this chapter) and the intensive partitioning of the Palace interior by subsequent rulers (see Section 4.3.6, The Buildings of Palace Phase 6, this chapter). The effect of K'inich Akul Mo'Nab' III's architectural amendments on the plan of the Palace was the complete closing off of the Palace interior as well as the partitioning of the building group into quadrants of varying accessibility. The closing of the gap between Houses D and K with House K's northern addition (see Section 4.3.3, The Buildings of the Palace Phase 3, this chapter) resulted in House A's medial entrance becoming the main point of entry into the Palace complex. Consequently, whereas the more accessible East Court would have functioned to receive nobles, dignitaries, and captives, the deeper, more restricted Southeast Court, West Court,

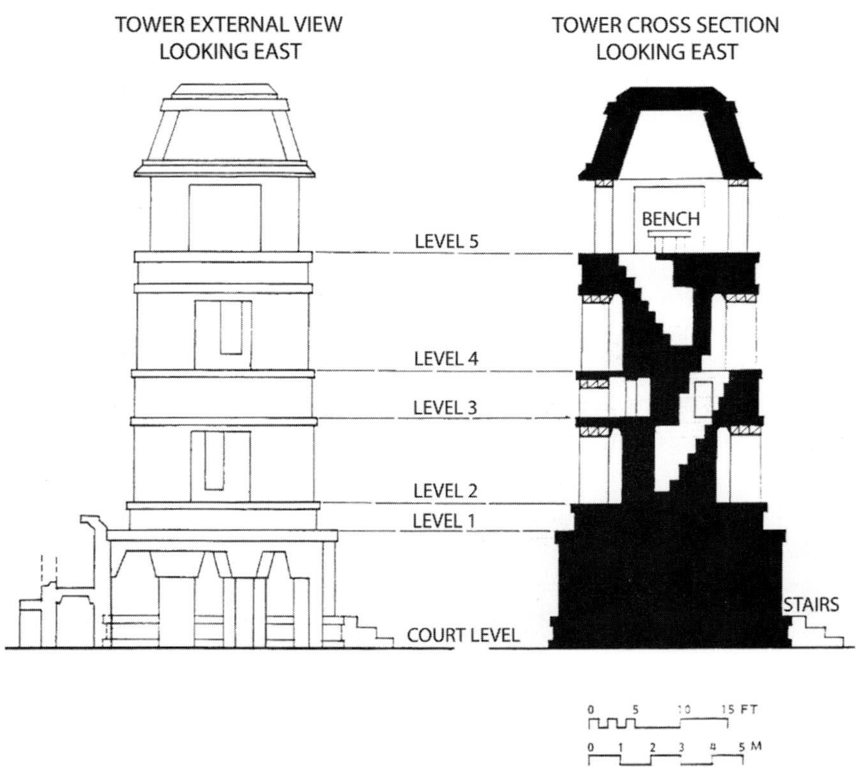


**Figure 4.53.** Palace Tower Cross Section Showing Principal Levels. Modified after Andrews (1974). George F. and Geraldine D. Andrews papers. The Alexander Architectural Archive, the University of Texas Libraries, the University of Texas at Austin.

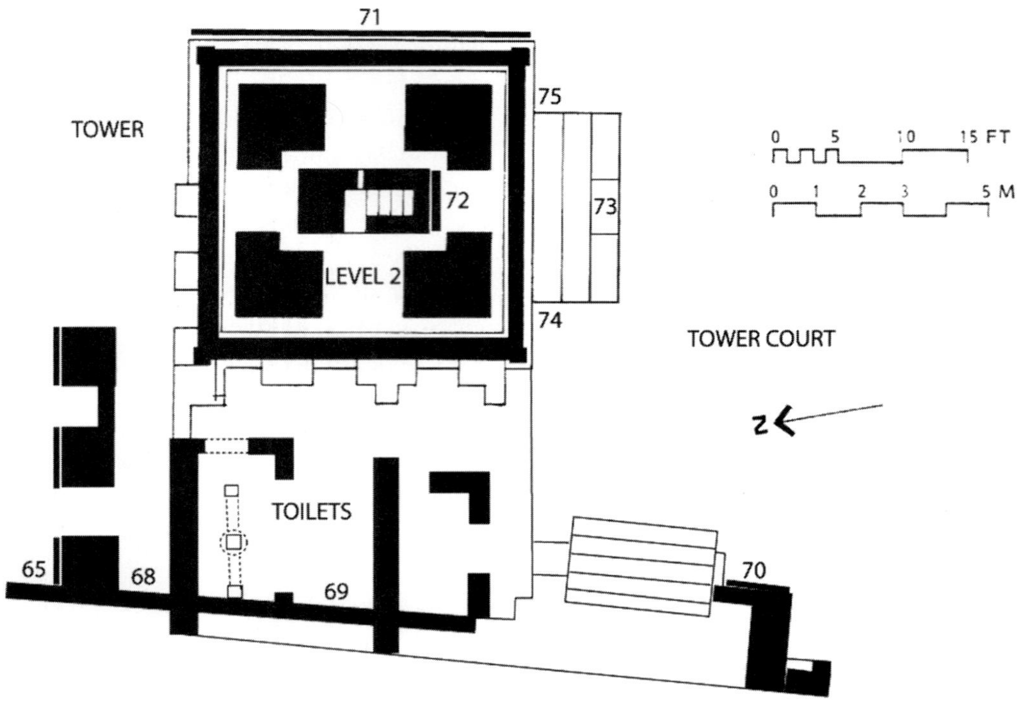


**Figure 4.54.** Tower and Toilets Plan, Palenque Palace. Modified after Andrews (1974). George F. and Geraldine D. Andrews papers. The Alexander Architectural Archive, the University of Texas Libraries, the University of Texas at Austin.

the buildings and dividing the interior of House G into several smaller rooms. Houses G and H were also later joined by two, low half-vaults and additional transverse walls that, combined with a crude bench, prevented access into the Palace interior by passing between the two buildings (see Figure 4.52).

**The Tower and Toilets: Description and their sculpture**

Positioned on the northern side of the Tower Court is the Palace Tower, a structure that is unique to Palenque. Immediately to the south of the Tower are the Palace Toilets; together these two structures formed the boundary between the West Court and the Tower Court (Figure 4.50). The Palace Tower comprises three principal levels: On the Tower's northern side are a series of vaulted niches two storeys high (see Figure 4.53). Maudslay (1889–1902b: 21) described the structure as a shell enclosing a narrow inner stairway with two main runs – one oriented southward and the other oriented northward. From the level of the courtyard to the uppermost level of the building, the structure measures 13.0 metres (Andrews 1974). As stated earlier, access to the interior of the Tower would have first required crossing over the once-roofed Toilets (see Figures 4.53 and 4.54), as there was never any entrance at the level of the Tower Court. Currently it remains unclear for what precisely the Tower may have been used. The presence of a bench, in addition to the wide open windows located at the top level of the structure, suggests that it may have functioned either as an observatory (Greene Robertson 1985b: 80) or perhaps as a lookout used for defensive purposes.

At the base of the Tower, on its eastern side, is a row of captives sculptured in stucco, a sculptural theme that was repeated on the south of the structure where additional captive sculptures once flanked a low staircase; furthermore, just northwest of the Tower in front of the Toilets are three sculptured piers portraying additional figures.

Table 4.19. *Houses G and H, approximate interior gallery dimensions*

| | Length | Width | Area |
|---|---|---|---|
| House G | | | |
| Northern Gallery | 13.12 m | 2.50 m | 32.80 m$^2$ |
| Southern Gallery | 13.12 m | 2.50 m | 32.80 m$^2$ |
| House H | | | |
| Northern Gallery | 11.37 m | 2.50 m | 28.42 m$^2$ |
| Southern Gallery | 13.12 m | 2.50 m | 32.80 m$^2$ |

*Note:* Dimensions in accordance with Andrews (1974).

**THE WESTERN COURT SCULPTURE PIERS: DESCRIPTION (GR 67).** The sculptural theme of the West Court piers has been tentatively defined as "Ritual Dance?" (RitDan?). Forming the southern boundary of the Western Court are three short sculptured piers; their placement at the southern boundary effectively blocked access from

**Figure 4.55.** Tablet of the Orator (GR 74). Drawing by Linda Schele (No. 145), © David Schele, courtesy of the Foundation for the Advancement of Mesoamerican Studies, Inc., www.famsi.org.

the Western Court to what is presently referred to as the Tower Court. On the northern surfaces of these piers the remains of five figures are portrayed; two are on the easternmost pier, two are on the central pier, and one is on the westernmost pier. Although only the legs of the figures survive today, remaining sculpture indicates that figures were portrayed in frontal orientation. The figure to the viewer's left of the central pier stands with his right leg bent as though he is dancing, as does the lone figure on the westernmost pier. As pointed out by Grube (1992: 201), a bent knee and lifted heel represent dancing in Maya iconography. In terms of costume, all that is discernible on the easternmost pier is the outline of the sandals worn by the figure on the viewer's right. Pompoms are visible on the tops of the sandals worn by the figures on the western and central piers. The western pier figure also wears a hipcloth with vertical slits as well as a loincloth, the front panel of which can be seen hanging in front of the figure.

**THE EAST WALL TOWER FIGURES: DESCRIPTION (GR 71).** The principal theme of the Tower East Wall Figures is "Captive Display" (CapDis). Eight individuals are sculptured at the base of the eastern wall of the Palace Tower, seven of which are portrayed seated in a frontal orientation and with their heads turned south towards the Tower Court. Stripped of all emblems of authority, the six captives wear only loincloths, knotted cloth armbands, and a length of material that is pulled through the perforations in their ears, an element marking them for sacrifice (P. Mathews 2000: personal communication). The eighth figure pictured on the far right of the scene appears to be dancing.

**TABLET OF THE ORATOR (GR 74) AND THE TABLET OF THE SCRIBE (GR 75): DESCRIPTION.** Once flanking the southern Tower stairs, the sculptural theme for both the Tablet of the Orator and the Tablet of the Scribe is "Captive Display" (CapDis) (see Figures 4.55 and 4.56). The ages of both monuments are thought to be approximately the same with the events depicted on the tablets occurring between 9.14.13.5.0 and 9.14.14.5.4? (A.D. 725?) – dates that fall within the reign of K'inich Akul Mo'Nab'

**Figure 4.56.** Tablet of the Scribe (GR 75). Drawing by Linda Schele (No. 147), © David Schele, courtesy of the Foundation for the Advancement of Mesoamerican Studies, Inc., www.famsi.org.

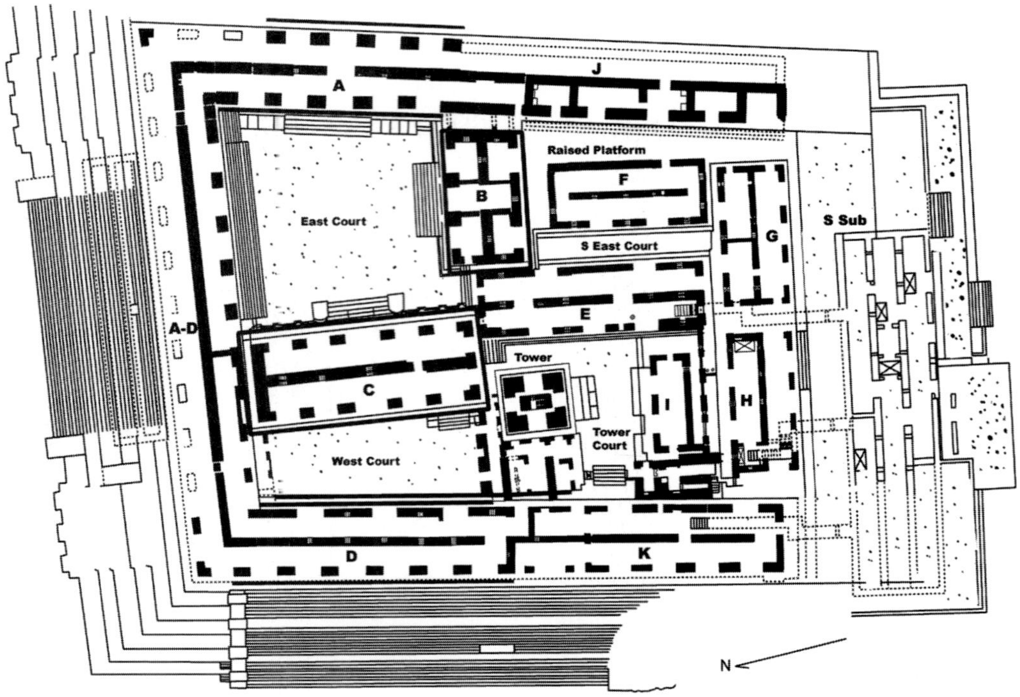


**Figure 4.57.** Map of the Palenque Palace Phase 5, modified after Greene Robertson (1985b: fig: 1).

III (A.D. 721–740?). The individuals portrayed on the tablets are depicted stripped of all authoritative symbols and marked for sacrifice. In addition to kneeling in subordinate poses, both figures are sparsely dressed and have cloth strips pulled through their ears. The captives also carry banners, and each has a piece of material draped over one arm. Originally thought to be rulers undergoing blood rites (Schele 1984: 30), the two individuals depicted are now thought to be trophies of war captured during times of conflict between Palenque and other neighbouring centres (P. Mathews 2000: personal communication). The figure represented on the Tablet of the Orator (Figure 4.55) is Ni-Sak-Kabay, *u sajal* of Piedras Negras.

### 4.3.5. The buildings of the Palace Phase 5

The construction of House F, House I, and the Sweat Baths within the confines of two of the Palace courtyards characterise Phase 5 of the Palace at Palenque (see Figure 4.57). Although it remains unclear who erected these buildings, it is likely that their fabrication occurred towards the end of the Palace's evolution, estimated to have occurred between A.D. 742 and 799. The construction of these buildings places Phase 5 of the Palace within the reigns of U Pakal K'inich Janab' Pakal II, K'inich Kan B'alam III, and K'inich K'uk' B'alam II (see Table 4.2). Built within the confines of the Southeast Court, the construction of House F included the installation of a supporting platform that raised the level of the structure (Figure 4.58). The addition of this building reduced the southeast courtyard into a series of narrow corridors on the northern, eastern, and western sides of House F. Also constructed on a low platform was House I (Figure 4.59), which was built on the southern side of the Tower Court. Directly behind House I is the structure referred to as the Sweat Baths where Palenque kings, and perhaps other nobles, undertook or participated in purification rituals (Schele 1986: 122).

Why the architects and builders chose to fill the Palace courtyards with architecture rather than extend the Palace over the Southern Palace Terrace remains unknown. One reason may have been a concern for placing too much weight over the southern Subterraneos, fearing perhaps that it may have collapsed. Interestingly, the rubble fill may have been placed in the corridor at the southeast corner of the substructure (see Section 4.3.1, The Buildings of the Palace Phase 1, The Northern and Southern Subterranean Galleries and Passages, this chapter), directly beneath the southern limits of House L, to prevent just such a disaster. Sculpture appearing during Palace Phase 5 includes the House I Sculptured Figure (GR 77), the House I Serpent Door (GR 76), and the Tablet of the 96 Glyphs (GR 73).

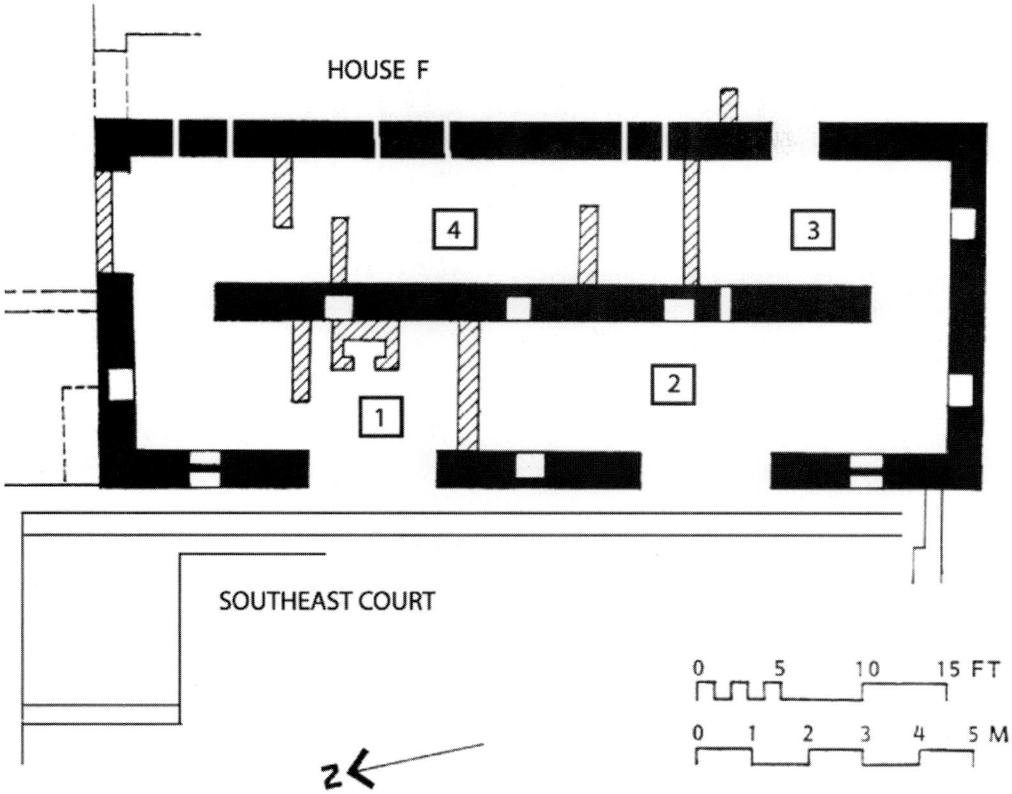


**Figure 4.58.** House F Plan, Palenque Palace. Modified after Andrews (1974). George F. and Geraldine D. Andrews papers. The Alexander Architectural Archive, the University of Texas Libraries, the University of Texas at Austin.

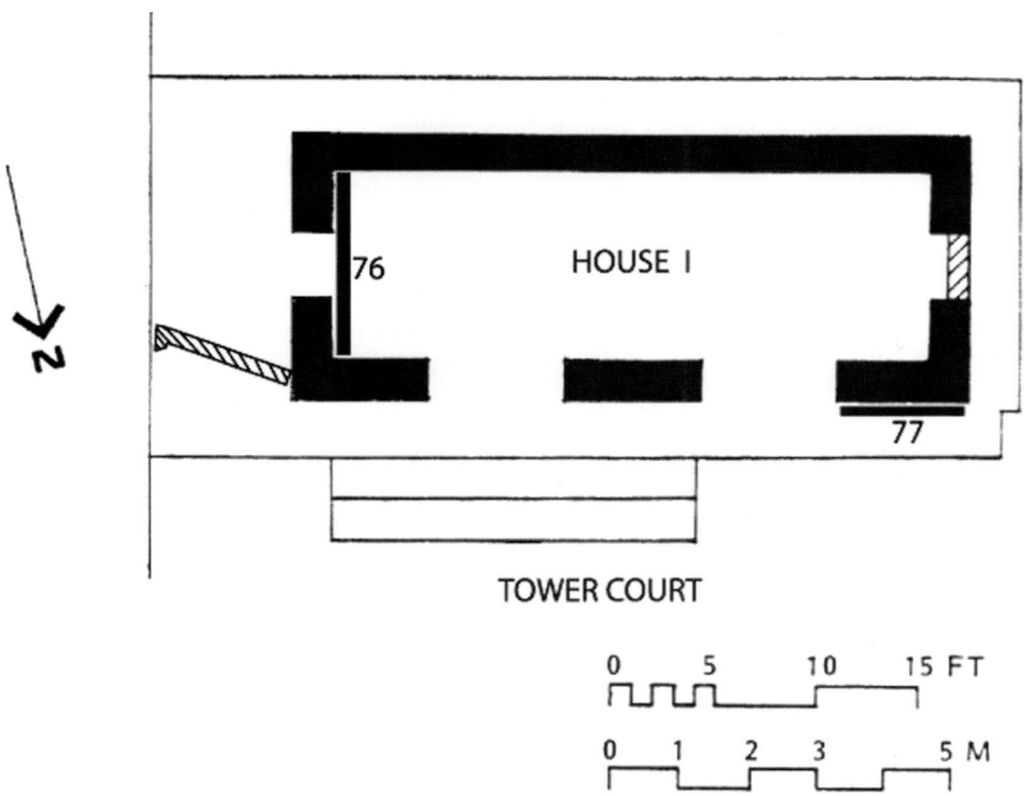


**Figure 4.59.** House I Plan, Palenque Palace. Modified after Andrews (1974). George F. and Geraldine D. Andrews papers. The Alexander Architectural Archive, the University of Texas Libraries, the University of Texas at Austin.

Table 4.20. *House F, approximate interior gallery dimensions*

| Gallery | Length | Width | Area |
|---|---|---|---|
| East | 15.00 m | 3.50 m | 52.50 $m^2$ |
| West | 15.00 m | 3.50 m | 52.50 $m^2$ |

### House F: Description

Oriented north to south, House F originally comprised two galleries separated by a medial wall (see Figure 4.58, see Table 4.20 for gallery dimensions); at the northern and southern ends of the central dividing wall were vaulted doorways that provided access between the eastern and western galleries of the building. Two wide entrances were also constructed in the western wall of the house, and at least one narrow doorway in the eastern wall of the building. An entrance was also constructed piercing the northern wall of the building at the end of the eastern gallery. Some time later, transverse walls were added to break the interior space of House F into a series of smaller rooms. In one of the rooms (Room 1, Figure 4.58), a small sanctuary was constructed against the medial wall of the building directly adjacent to the northwest entrance to the building (Andrews 1974). Unfortunately, today little survives of House F, and there is no evidence to suggest that the structure was ever decorated with stucco sculptures or paintings.

### House I: Description and sculpture

Located on the southern side of the Tower Court, presently little remains of House I. The structure in its entirety measured 10.08 m by 3.90 m (Andrews 1975). Comprising a single room (see Figure 4.59), the importance of this building is implied by its close proximity to House E and by some evidence of sculptural embellishment. More specifically, there are remains of a stucco figure on the northwestern corner of the building as well as a serpent sculptured around the interior of House I's northern entrance. Sometime after the construction of House I, a wall was erected at the northeastern corner of the building. The purpose of this wall was probably to bar or restrict access to the eastern entrance of the building and/or the corridor leading to the Sweat Baths that are located behind House I (see Figure 4.59).

**HOUSE I SCULPTURED FIGURE (GR 77) AND SERPENT DOOR (GR 76): DESCRIPTION.** The theme of the sculptured figure appearing on the exterior of House I (GR 77) is tentatively defined as "Royal Portraiture Solo" (RPSolo). The figure stands in profile facing towards the northeast entrance of House I. Holding a static pose, the individual is portrayed wearing tubular bead anklets, a plain hipcloth, and a cross-band belt. Just above the figure's belt is a wrap of cloth that displays a knot at the back. Due to the Classic Maya's fixation with compositional symmetry, it is highly likely that there was a similar figure sculptured on the opposite side of House I's northern entrance.

The House I Serpent Door has been defined as "Cosmological?" (Cos?). "In religious terms, serpents may have been the most important fauna of Mesoamerica . . . [displaying] the most varied roles in religion and religious symbolism" (Miller 1998: 148–149). Sculptured above the eastern entrance of the building is a boa constrictor. The remains of both the head and tail of the serpent can still be seen as if meandering over the top of the doorway. As suggested by the scale pattern tracing the top right edge of the entrance, it appears that a snake once framed the entire doorway (see Greene Robertson 1985b: 84).

**TABLET OF THE 96 GLYPHS (GR 73): DESCRIPTION.** The sculptural theme of the Tablet of the 96 Glyphs (Figure 4.60) is "Text, Dynastic Historical" (TDynHist). Discovered within the confines of the Tower Court, the Tablet of the 96 Glyphs represents one of the finest examples of Maya calligraphic art, where the carving of each of the glyphs replicates the strokes of a painter's brush. Once positioned on the stairs on the southern side of the Tower, it is probable that the Tablet of 96 Glyphs functioned as a throne bench (P. Mathews 2004: personal communication). In addition to recording the dedication of House E during the reign of K'inich Janab' Pakal I, the text also documents the accessions of K'inich K'an Joy Chitam II, K'inich Akul Mo'Nab' III, and K'inich K'uk' B'alam II within the building. The monument was commissioned by K'inich K'uk' B'alam II to celebrate his first Katun ending, which occurred in A.D. 783 (Martin and Grube 2000: 174–175).

### Sweat Baths: Description

Immediately behind House I are the Palace Sweat Baths (see Figure 4.61), a structure believed to have been used by rulers at Palenque for ritual purification (Schele 1986: 122). Three rooms make up the structure's interior, and together they measure approximately 13.0 m by 1.20 m. The configuration of the Sweat Baths has been described simply as three narrow rooms covered by a low vaulted roof barely high enough for the modern visitor to stand (Andrews 1974). Three doorways in the northern wall of the structure provide access to the interior of the building. It is argued that holes present in the floor of the structure (see Room 2, Figure 4.61) vented steam when the stones

**Figure 4.60.** The Tablet of the 96 Glyphs (GR 73). Drawing by Linda Schele (No. 136), © David Schele, courtesy of the Foundation for the Advancement of Mesoamerican Studies, Inc., www.famsi.org.

present in the holes were heated and water was poured over them (Andrews 1975).

### 4.3.6. The buildings of the Palace Phase 6

I have placed Phase 6 of the Palace (Figure 4.62) after the reign of K'inich K'uk' B'alam II and under the patronage of the last king of Palenque: Wak Kimi Janab' Pakal III, who acceded to the throne in 9.18.9.4.4 (A.D. 799). The buildings that I have assigned to this phase are two structures that fall last in both Andrews (1974) and Greene Robertson's (1985b: table IV) construction sequences; these are House L and the rooms at the southwest corner of the Palace at plaza level (see Figure 4.63). The only other contribution to the Palace I attributed to Wak Kimi Janab' Pakal III is the bulk of the masonry partitioning still visible throughout the Palace today.

To date, the only reference to Wak Kimi Janab' Pakal III is found on an incised black vessel recovered in a residential area of the city (Martin and Grube 2000: 175).

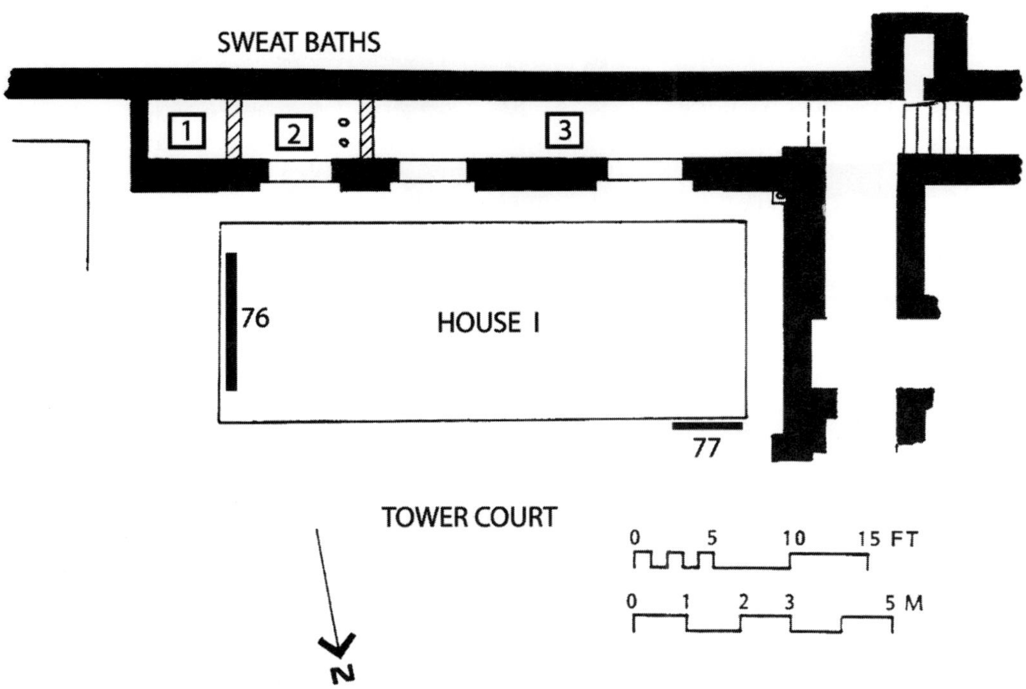


**Figure 4.61.** Sweat Baths Plan, Palenque Palace. Modified after Andrews (1974). George F. and Geraldine D. Andrews papers. The Alexander Architectural Archive, the University of Texas Libraries, the University of Texas at Austin.

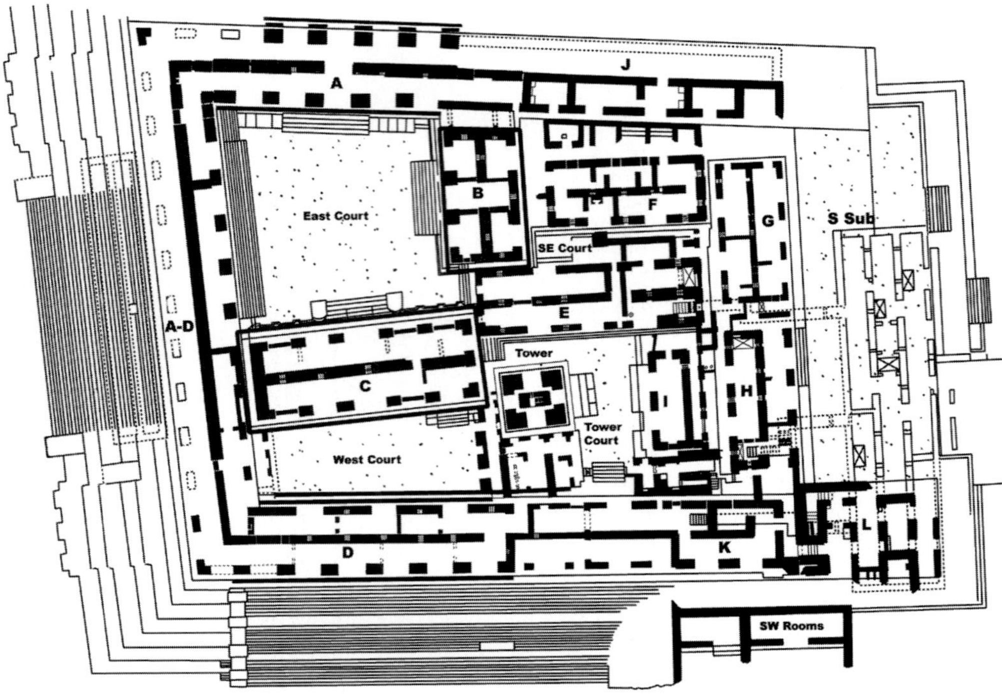


**Figure 4.62.** Map of the Palenque Palace Phase 6, modified after Greene Robertson (1985b: fig. 1).

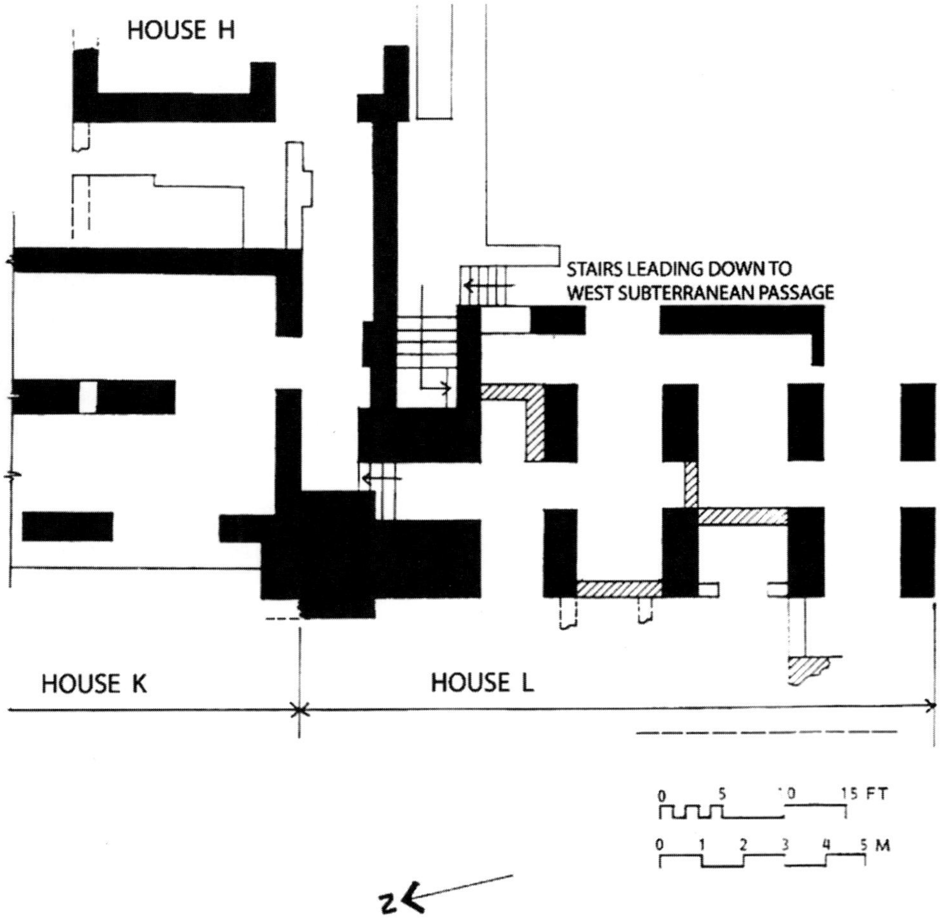


**Figure 4.63.** House L Plan, Palenque Palace. Drawing by George Andrews (1974). George F. and Geraldine D. Andrews papers. The Alexander Architectural Archive, the University of Texas Libraries, the University of Texas at Austin.

The absence of any monuments documenting this ruler may be viewed as an indication that Wak Kimi Janab' Pakal III lacked the necessary authority, infrastructure, and/or prosperity to commission monuments in his own honour. The intensive partitioning of the Palace complex probably occurred late in the history of Palenque, just before its collapse (Greene Robertson 1985a: 63–64). The shift away from monumental construction to the blockage and segmenting of existing Palace buildings was probably due to one or a number of influencing factors, perhaps diminishing resources, such as trees and limestone; an absence or loss of the necessary engineering and artistic skills; and the protection of an ever-growing population of Palace residents from outsiders. It remains possible that, as a consequence of increased partitioning within the Palace proper, the building of House L and the southwest building on the periphery of the building group represents an attempt to ease further congestion within the Palace's interior.

**House L: Description**

House L (Figure 4.63) is located at the southeast corner of the Palace directly adjacent to the southern end of House K. Formulated from eight piers roofed over by a low vault, the structure comprises a confusing array of short passages oriented east–west, positioned over the eastern quarter of the Subterraneos (Andrews 1974). At the northwestern corner of the structure is a narrow stair that first runs north–south, then turns east–west, and finishes in a dead end, perhaps once leading down to the West Subterranean Passage. It remains unclear what the purpose of this structure might have been; the addition of secondary walls breaking up the interior of the building into a series of compartments suggests that the structure may have been used for storage. Today, little remains of House L, the roof having long ago collapsed. Furthermore, there is no evidence to suggest that the building was ever decorated with sculpture or paintings.

5

# Access analysis of the architectural and sculptural programs of the Palenque Palace Group

## 5.1. Access analysis of the Palace Phases 1–6

As outlined at the beginning of Chapter 4, the first stage of the Palace Group investigation required that I establish a relative chronology of Palace buildings at Palenque, followed by the breaking up of the proposed sequence of construction into phases that correspond to the reigns of the kings. My objective in doing this was to more clearly articulate the various architectural and sculptural programs of the Palace as well as to provide a platform for inferring the intent of different rulers in the placement of certain sculpture within specific spatial contexts. In Chapter 4, six maps of the Palenque Palace were presented, each representing different stages in the evolution of the building group. These maps were used to construct a series of "access diagrams" that were implemented to hierarchically order and interpret the "spaces" and corresponding sculpture of each Palace phase (see Appendix 3, Palenque Palace Sculpture Phases 1–6). Using various techniques in "access analysis," in addition to the "thematic–analytical" inquiry of monumental art, my goal was to detect shifts in the content and placement of Palace Group sculpture over time and through space, as prescribed by changes in corresponding "carrier depths," "relative asymmetry" (RA), and "control value" calculations. General questions that I address in this chapter are: (1) Is there specific sculpture marking certain types of spaces? (2) What may have motivated the apparent strategic placement of fixed sculpture media in Maya city centres? and (3) Are there clues embedded within compositions as to how certain space was used and socially demarcated?

To summarise, according to the methodological procedures implemented in Chapter 5, all Palace Group sculpture was first classified thematically and to the reigns of kings. Then, their distribution was interpreted on access diagrams. Sculptural themes were determined through compositional inquiry in addition to the interpretation of hieroglyphic texts, where present. Elements such as figural position, perspective, costume, furniture, and texts were considered in determining the behaviour and symbolism represented on monuments and other artworks and to surmise behaviours associated with corresponding spatial contexts (see Chapter 2, Section 2.4.7, Some Categories for Thematic Analytical Investigation). All access diagrams derived from the Palace phase maps were "justified" from the perspective of the "carrier" position or, more specifically, from the viewpoint of the "stranger" and/or "visitor" standing externally to the spatial complex (Foster 1989a: 41). Justified access maps were used to determine depth from the perspective of the "outsider," in addition to the calculation of indices that were used to describe patterns in the relationships between "inhabitants" and between inhabitants and "visitors" (i.e., persons permitted to temporarily occupy the Palace complex). These indices are referred to as the RA index (or Relative Depth Index) and the Control Values Index (refer to Appendix 7 for Path Matrix tables, Palace Phases 1–6). As described in Chapter 2, a control value is a measure of the amount of control that a space possesses over access to other spatial units in its local proximity: The more units that a particular space controls access to, the higher the level of control that that space is said to have. In this calculation, any space assigned a value of one or greater is considered to be a "controlling" space; conversely, any spatial unit assigned a value less than one is considered to be a "controlled" space. Values derived from RA calculations operate as indicators of "relative depth," specifying how deep or shallow a unit of space is from the rest of the spatial system to which it belongs (see Chapter 2, Section 2.4.5, Measuring Control Values and RA Analysis). RA calculations may be interpreted as measures of "integration": the higher the value derived from the calculation (i.e., the closer it is to a value of one) the more "segregated" (or private) that space is from the rest of the spatial system; the lower the resulting value, the more "integrated" (or accessible) the space is said to be (see Hillier and Hanson 1984: 109).

To assist in the following discussion, I have tabulated by phase the data derived from the Palace access maps, focusing specifically on spaces found to display sculpture as well as other artistic media. In the far left column of each table are reference numbers that have been assigned to all spatial units ("nodes") on both the unjustified and justified access maps. The next column in the tables lists either the depth of each space from the perspective of the carrier position (i.e., Carrier Rank), the RA, and the control values for each unit of space. These columns are followed by Greene Robertson (GR) Numbers that have long been assigned as a system of reference for Palenque Palace artworks (see Greene Robertson 1985b: table 1). The remaining columns in each Palace phase table list the sculptural themes of the sculpture designated to each space. Any GR number and sculptural theme in bold lettering indicates that it corresponds temporally to the phase table in which it is listed. Conversely, all GR numbers and themes not in bold lettering indicate that they correspond to one of the earlier Palace phase tables. To assist in the interpretation of the tabulated data, I have provided a legend for all table headings and for the various themes (which appear individually or in combination) listed in the far right column of each table (see Tables 5.1 and 5.2).

### 5.1.1. Analysis of the Palace Phase 1

As outlined in Chapter 4, Palace Phases 1 and 2 have both been attributed to K'inich Janab' Pakal I, who ruled at Palenque for just over 68 years (A.D. 615–683). During the first half of his reign (circa A.D. 615–650), Houses E, B, J, K, and H-Sub were built, followed by Houses C and A, which were constructed during the second half of his reign (circa A.D. 650–683). (Note: As stated in Chapter 4, it remains possible that House A was commissioned a little later by K'inich Kan B'alam II [see Chapter 4, Section 4.3.2, The Buildings of the Palace Phase 2, House A: Description and Sculpture]). Additional structures used during the first two Palace phases, although probably built before K'inich Janab' Pakal I's reign, include the northern and the Southern Subterranean Structures (see Figure 5.1).

Palenque Palace Phase 1 comprises fifty-three units of space (or nodes) if one counts the carrier position, as indicated by the corresponding access diagram: Two spaces are represented at Level 1; five spaces at Levels 2, 3, 4, and 8; seven spaces at Level 5; six spaces at Level 6; seven spaces at Level 7; eight spaces at Level 9; and two spaces at Level 10 (see Figure 5.2). Of the fifty-three spaces represented, only seven spatial units were found to display sculptural residues that could be sorted into thematic categories; these spaces were House B's northern room (node 5b), House B's southern room (node 6a), House E's eastern room (node 6b), House E's western room (node 6c), the East Subterranean Passage's northern corridor (node 7c), the South Subterranean Passage bend (node 8b), and the West Subterranean Passage's southern corridor (node 9g). In addition to those spaces that display sculpture, nineteen spaces were defined as "transitional" (i.e., their primary function was to provide access between other spatial units within the system). Two courtyards are proposed at Palace Phase 1, both of which

Table 5.1. *Headers for Palace phase tables*

| Table header | Meaning of header |
|---|---|
| Carrier R | = Carrier Rank |
| Cont. Val | = Control Value |
| X | = Controlled Space |
| GR | = Greene Robertson number for Palace artworks |
| Node | = Reference number of spatial unit or cell |
| RA Val | = Relative Asymmetry Value |
| Theme | = Designated sculptural theme |

Table 5.2. *Sculptural themes*

| Abbreviated sculptural theme | Sculptural theme |
|---|---|
| APSolo | = Ancestor Portraiture Solo |
| APWA | = Ancestor Portraiture with Attendants |
| Acc | = Accession |
| CapDis | = Captive Display |
| TCapDis | = Text, Captive Display |
| CosU | = Cosmological Underworld |
| CosT | = Cosmological Terrestrial |
| CosC | = Cosmological Celestial |
| GPSolo | = God Portraiture Solo |
| GPWA | = God Portraiture with Attendants |
| HumSac | = Human Sacrifice |
| MMot | = Mat Motif |
| PSolo | = Portraiture Solo |
| SacDan | = Sacrificial Dance |
| TCos | = Text, Cosmological |
| THist | = Text, Historical |
| TMyth | = Text, Mythological |
| TDed | = Text, Dedication |
| RAud | = Royal Audience |
| RitDan | = Ritual Dance |
| RPSolo | = Royal Portraiture Solo |

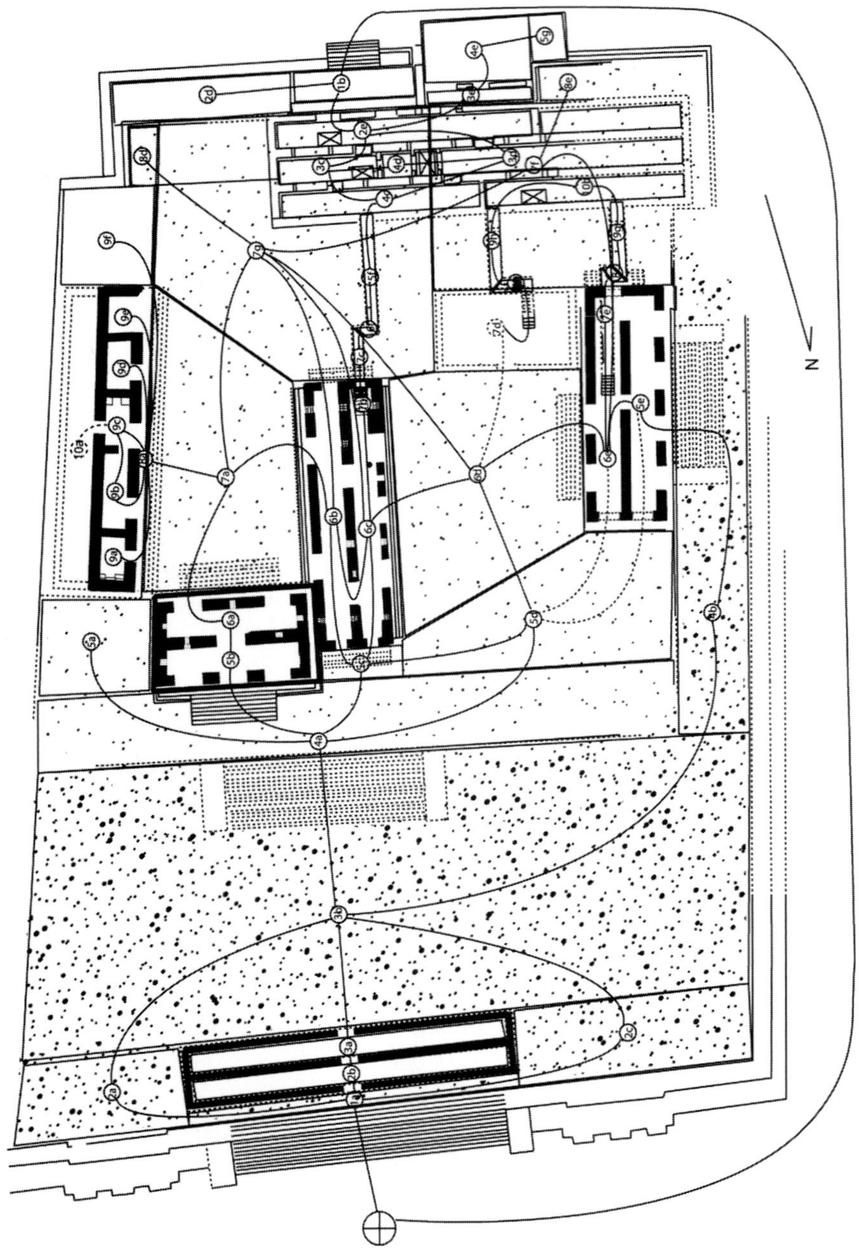

**Figure 5.1.** Palenque Palace Phase 1, Combined Convex Map and Access Diagram. Palenque Palace modified after Greene Robertson (1985b: fig. 1).

occupy the level of the Second Palace Terrace platform. These courtyards later evolved into the Southeast Court and the Tower Court of the Palace (see Chapter 4, Section 4.3.4, The Buildings of the Palace Phase 4). In the following discussion, they are referred to as the Early Southeast Court (node 7a) and the Southwest Court (node 6d).

Phase 1 of the Palace comprised two principal levels: the then-exposed Northern Galleries and the Southern Subterranean Structures, occupying the level of the First Palace Terrace; and all other buildings that occupy the level of the Second Palace Terrace (i.e., Houses K, H-Sub, E, B, and J) (see Figure 5.1). From the perspective of carrier position (i.e., the space external to the Palace complex), formal access to the interior of the building group was provided via the Northern Galleries (node 2d), which I have previously argued functioned as a form of gateway structure controlling access into the Palace Group interior. As outlined in Chapter 4, at this phase of the Palace's construction, the northern half of the First Palace Terrace platform was not yet covered by the Second Palace Terrace extensions A and B, which left the earlier northern subterranean structures exposed. The combined effect of these platforms was to provide a progressive change in elevation from the perspective of someone entering the Palace Group complex from the plaza fronting the earlier northern Palace stairs (see Chapter 4, Section 4.3.1, The Northern and Southern Subterranean Galleries and Passages). Once at the level of the First Palace Terrace platform (node 3b), accessing structures resting on the Second Palace Terrace platform would have necessitated climbing an additional staircase positioned somewhere between the Northern Galleries of the Palace and House B.

Although it remains unclear what structures, if any, were erected just beyond the then-exposed northern substructure at the level of the First Palace Terrace (see Chapter 4, Section 4.3.1, The Northern and Southern Subterranean Galleries and Passages), a stairway in this proximity would have been necessary to access the upper houses of the Palace. Access to the upper Palace platform was also possible by passing through the Subterraneos (the Southern Subterranean Structures). As this entrance is found to the rear of the Palace Group (node 2e), and access to the upper Palace houses requires passing through narrow underground passages (one of which provides direct access to the Palace's throne room [node 6c]), this route into the Palace interior was likely to have been highly restricted. (Note: At Palace Phase 1, it is likely that the stairway that today provides access to the Southern Palace Terrace [node 7g] did not exist at this time [P. Mathews

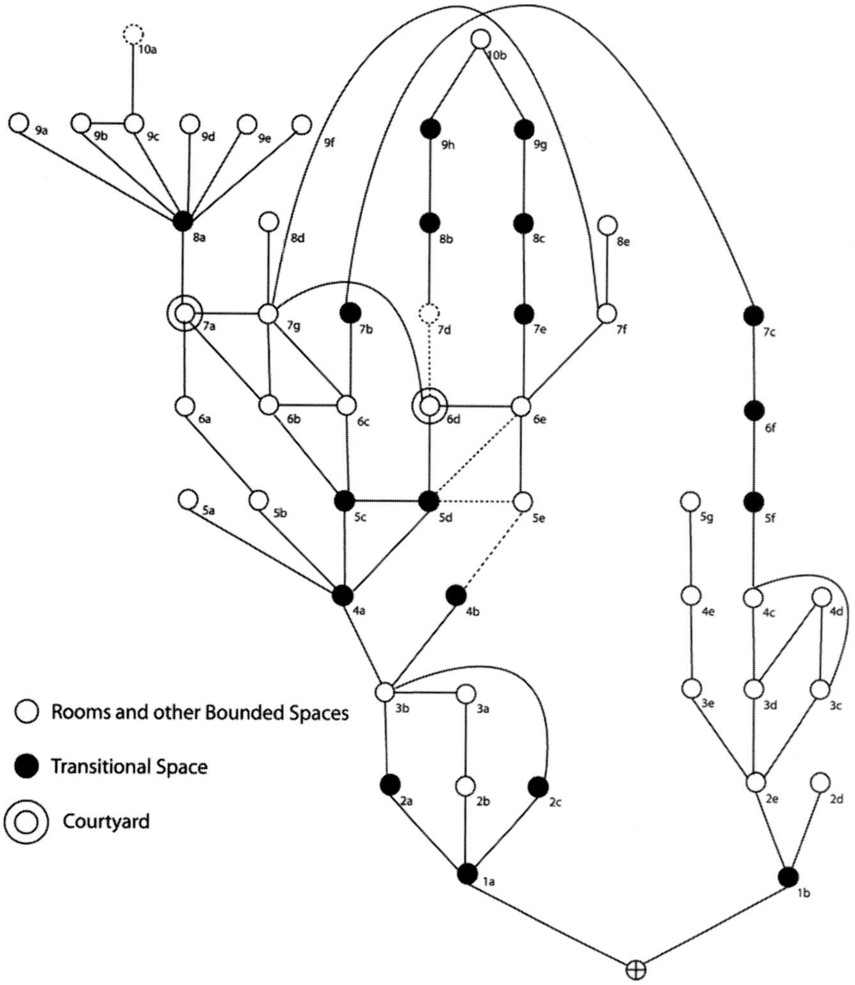


**Figure 5.2.** Palenque Palace Phase 1 "Justified" Access Diagram.

2005: personal communication]. It appeared only after the construction of Houses G and H; see Chapter 4, Section 4.3.4, The Buildings of the Palace Phase 4). Access to the upper Palace platform also may have been possible via another earlier stairway on the east side of House K (see Andrews 1975), which was later buried beneath a much larger staircase with the construction of House D (see Figure 5.5).

Although the Palace Phase 1 access diagram is relatively deep, comprising ten levels of permeation, many of the spaces represented may be described as "distributed" (i.e., there is more than one route of access to many of the spaces). As noted in Chapter 2, a space or system of spaces is described as distributed if there is more than one route of access between spatial units, including passing through an additional space to get to the target space. Conversely, a space or system of spaces is categorised as "nondistributed" when there is no more than one route to a given unit of space (Hillier and Hanson 1984: 108). Distributed and nondistributed relationships between spatial units function as indicators of how well a system of spaces is integrated in addition to how effectively it may be controlled. As outlined previously, a highly nondistributed access map will appear treelike or "dendritic" in shape, displaying fewer or no redundant connectors ("edges") between spatial units ("nodes"). Consequently the spatial configuration will be less integrated and more easily controlled. In contrast, an access map showing a similar number of spaces, but a greater number of connections, will often appear "shallower," comprising fewer permeation levels. In such a system, accessibility between spaces is likely to be improved and less easily controlled as the spatial configuration displays a greater number of route options between the spatial units represented (see Chapter 2, Section 2.4.4, Principals of Distributedness–Nondistributedness and Symmetry–Asymmetry).

In addition to the fifty-three nodes that make up the Palace Phase 1 access map are the seventy edges that connect them. The more distributed spaces represented (i.e., those displaying the most number of connections to neighbouring spaces) are the northern terrace of House B (node 4a), the eastern and western galleries of House E (nodes 6b and 6c), the upper southeastern terrace of the Palace (node 7g), the Early Southwest Court and Tower Court (nodes 6d and 7a), and the platform providing access to the interior of House J (node 8a). Conversely, some of the nondistributed spaces represented (i.e., those displaying fewer connections between nodes) include the spaces that comprise the eastern, western, and southern subterranean passages (nodes 5f, 6f, 7c, 7e, 8b, 8c, 9g, and 9h), the northern and southern galleries of House B (nodes 5b and 6a), three of House J's outer rooms (nodes 9a, 9d, and 9e), and the northwestern and central rooms found in the Southern Substructure (nodes 10b and 4d).

A total of seven structures are represented at Palenque Palace Phase 1, all of which comprise spaces of varying "depth," integration, and control. To gain some clearer sense of accessibility at the building rather than the "unit-space" level, the RA values of all seven Palace structures were collated and averaged, revealing which buildings, relatively speaking, are more or less integrated with respect to the rest of the Palace complex (see D. Mathews 2004: 58–59 for an equivalent procedure

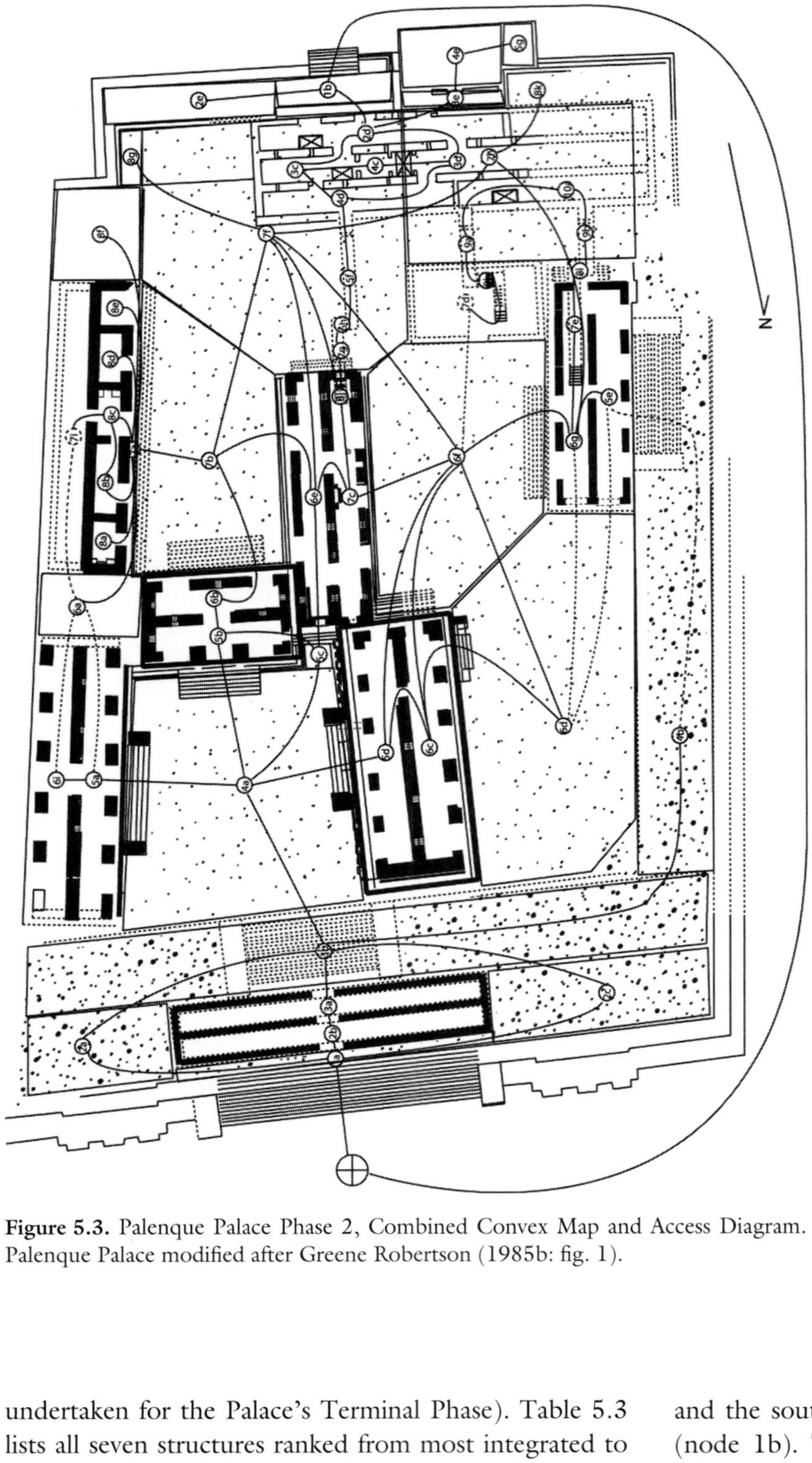

**Figure 5.3.** Palenque Palace Phase 2, Combined Convex Map and Access Diagram. Palenque Palace modified after Greene Robertson (1985b: fig. 1).

undertaken for the Palace's Terminal Phase). Table 5.3 lists all seven structures ranked from most integrated to least integrated; the most integrated structure at Phase 1 of the Palace was House E, and the least integrated structures were the southern subterranean galleries and House J.

House E's relatively high level of integration may be explained in terms of the structure's function as this building represents what was probably the primary interface between the "principal inhabitant" of the Palace (i.e., K'inich Janab' Pakal I) and other inhabitants, as well as the principal-inhabitant and visitors permitted to access the Palace interior. The structure's high level of integration may also be due to issues of convenience, permitting persons who once occupied this building easier access to the rest of the Palace complex. The more restricted character of House J and the Southern Subterranean Substructures and passages suggests that more private activities were undertaken within the confines of these spaces, such as K'inich Janab' Pakal I's ritualised ventures into the Underworld (in the case of accessing the Southern Subterranean Substructures), or perhaps as a place to retire for persons who once occupied the small western rooms of House J.

It is reasonable to argue that the higher the level of integration that a unit of space possesses, the more difficult the space would be to control. An effect of high integration may be an increase in the potential for uninhibited social encounters to take place within that space. In contrast, the lower the level of integration that a unit of space exhibits, the easier it would be to control, effectively reducing the likelihood of arbitrary social encounters taking place (see Grahame 1997: 150). Of all the spaces represented in the Palace Phase 1 access diagram, spaces displaying the greatest level of control over localised access to other spaces include the narrow terrace providing access into House J's interior (node 8a), the upper southeastern terrace of the Palace (node 7g), the northern substructure's southern terrace (node 3b), the northern substructure's northern terrace (node 1a), and the southern substructure's lower southern terrace (node 1b). Three of these spaces (nodes 1a, 1b, and 3b) are located around the entry points of the building group. Beyond these zones, levels of control among the remaining Palace spaces become more dilute due to the higher number of distributed spaces that characterise the Palace interior. As noted by Hillier and Hanson, when an array of spaces is found to be highly distributed, "the more there will be a tendency towards the diffusion of spatial control, while nondistributedness will indicate a tendency towards a unitary, superordinate

Table 5.3. *Average RA values for Palace Phase 1 houses and other structures*

| Structure | Average RA value | Rank |
|---|---|---|
| House E | 0.1322 | 1 |
| House K | 0.1410 | 2 |
| House B | 0.1463 | 3 |
| House H-Sub | 0.1584 | 4 |
| North Substructure (exposed) | 0.1689 | 5 |
| House J | 0.2086 | 6 |
| South Substructure | 0.2370 | 7 |

*Note:* Ranked from most integrated to least integrated.

Table 5.5. *Palace Phase 1 Court carrier depths, RA values, and control values*

| Palace Courts | Node | Carrier rank | RA value | Control value |
|---|---|---|---|---|
| Southwest Court (early Tower Court) | 6d | 6 | 0.1267 | 0.8666 |
| Southeast Court (early) | 7a | 7 | 0.1388 | 1.0595 |

control" (Hillier and Hanson 1984: 97). Of the few Palace spaces to still display some sculpture, the western gallery of House E (node 6c) and the three subterranean passages (nodes 7c, 8b, and 9g) can be described as controlling spaces, whereas House B's northern and southern galleries (nodes 5b and 6a), as well as the eastern gallery of House E, would more accurately be described as controlled spaces (see Table 5.4).

The size and distributed character of spaces such as the Palace Phase 1 courtyards (nodes 6d and 7a), which correspondingly show relatively high levels of integration (see Tables 5.5 and 5.6), relate directly to their function as areas for group interaction. In contrast, more confined spaces that display fewer access pathways suggest that activities once undertaken within their confines were probably of a more restricted nature. From the perspective of the carrier position, it follows that more deeply distributed spaces of the Palace, such as the Early Southeast Court (node 7a), were areas that functioned as inhabitant–inhabitant interfaces, whereas distributed spaces occupying shallower positions with respect to the carrier, such as the Southwest Court (node 6d), functioned as inhabitant–visitor interfaces. The relatively shallow proximity of the Southwest Court, combined with its position directly adjacent to the House E throne room (node 6c), suggests that the space was probably accessible to those Palace visitors and inhabitants who sought audience with K'inich Janab' Pakal I, such as those who once participated in rituals connected with the ruler's accession. One of the principal functions of the western gallery of House E was the inauguration of kings (see Martin and Grube 2000: 163). In contrast, the deeper Southeast Court (node 7a) was probably reserved for more private ceremonies and meetings between members of K'inich Janab' Pakal I's inner circle. In support of this proposal, directly adjacent to the Southeast Court is House J, which, in addition to the Subterraneos, contains some of the deepest nondistributed spaces represented in the Palace Phase 1 access diagram. The relatively small sizes and restricted nature of House J's western rooms (see nodes 9a, 9b, 9c, 9d, and 9e), suggest that they functioned as individual–private domains where perhaps inhabitants or members of Palenque's nobility withdrew to avoid public scrutiny (see Chapter 4, Section 4.3.1, The Buildings of the Palace Phase 1, for a more detailed description of House J). The fact that House B defines the northern boundary of the Southeast Court, access through which is highly controlled, also signals the exclusive character of the southeast quarter of the Palace. It has long been argued that, due to the presence of a specific iconographic element, House B probably functioned as a

Table 5.4. *Palenque Palace Phase 1 sculpture ranked according to control values, K'inich Janab' Pakal I (circa* A.D. *615–650)*

| Node | Control value | Rank | Sculpture GR No. | Theme |
|---|---|---|---|---|
| 6c | 1.1666 | 1 | **3, 4, 5, 13, 11, 80** (Bench) | **RAud, Acc, CosTU, THist, TCosCU** |
| 7c | 1.0000 | 2 | 7 | **CosCU** |
| 8b | 1.0000 | 2 | **78** | **CosU** |
| 9g | 1.0000 | 2 | **79** | **CosU** |
| 6b | 0.9166 | X | **8** | **CosCU** |
| 6a | 0.7500 | X | **21, 22, 26, 27, 28** | **PSolo, RitDan, MMot, CosU** |
| 5b | 0.7000 | X | **24, 25, 26, 27** | **RPSolo, MMot, CosU, CosCU** |

*Note:* See Appendix 6 for building and location information for nodes.

Table 5.6. *Palenque Palace Phase 1 sculpture ranked according to RA values, K'inich Janab' Pakal I (circa* A.D. *615–650)*

| Node | RA value | Rank | Sculpture GR No. | Theme |
|---|---|---|---|---|
| 9g | 0.2330 | 1 | **79** | **CosU** |
| 8b | 0.1916 | 2 | **78** | **CosU** |
| 7c | 0.1637 | 3 | 7 | **CosCU** |
| 6a | 0.1501 | 4 | **21, 22, 26, 27, 28** | **PSolo, RitDan, MMot, CosU** |
| 5b | 0.1425 | 5 | **24, 25, 26, 27** | **RPSolo, MMot, CosU, CosCU** |
| 6b | 0.1282 | 6 | **8** | **CosCU** |
| 6c | 0.1244 | 7 | **3, 4, 5, 13, 11, 80** (Bench) | **RAud, Acc, CosTU, THist, TCosCU** |

*Notes:* Ranked from least integrated to most integrated. See Appendix 5 for building and location information for nodes.

*popol nah* or "mat house" (see Miller 1998: 199, 1999: 38). As pointed out in Chapter 4, Maya buildings displaying "mat" designs may have functioned as council houses, where the "Council of Lords" congregated to undertake various political and religious activities (Schele and Mathews 1998: 44; see also Sharer 1994: 329). The only other controlled space that displays sculpture at Palace Phase 1 is located in the eastern room of House E (node 6b), a space housing Cosmological Celestial Underworld themes which delineates the eastern boundary of the Southeast Court.

When considering the seven spatial units that display sculpture, and ranking each according to its distance from the carrier position, the three carved vaults (GR 7, 78, and 79; see Figures 4.7 and 4.8) located in the subterranean passages (nodes 7c, 8b, and 9g) are the most restricted sculptures of the Palace (see Table 5.7), positioned at Levels 7, 8, and 9 on the Phase 1 access diagram (Figure 5.2). Located in transitional spaces, the eastern, western, and southern subterranean passage vaults are also the least integrated (or most segregated) sculptures of the Palace (see Table 5.6), displaying predominantly Cosmological Underworld themes. As noted by Schele, the presence of these sculptures in the underground passages leading to the Southern subterranean structure marks these corridors as entrances into the Underworld (Schele 1986: 122).

Table 5.7. *Palenque Palace Phase 1 sculpture ranked according to depth from the carrier, K'inich Janab' Pakal I (circa* A.D. *650–683)*

| Node | Carrier rank | Sculpture GR No. | Theme |
|---|---|---|---|
| 5b | 5 | **24, 25, 26, 27** | **RPSolo, MMot, CosU, CosCU** |
| 6c | 6 | **3, 4, 5, 13, 11, 80** (Bench) | **RAud, Acc, CosTU, THist, TCosCU** |
| 6b | 6 | **8** | **CosCU** |
| 6a | 6 | **21, 22, 26, 27, 28** | **PSolo, RitDan, MMot, CosU** |
| 7c | 7 | 7 | **CosCU** |
| 8b | 8 | **78** | **CosU** |
| 9g | 9 | **79** | **CosU** |

*Note:* See Appendix 4 for building and location information for nodes.

The shallowest sculpture at Palace Phase 1 appears at a depth of Level 5 from the carrier position and encompasses all sculpture appearing on the roof and the northern piers of House B (see Table 5.7) – a structure commissioned by K'inich Janab' Pakal I. As mentioned previously, the Mat Motif present on the entablature of the building marks the function of House B (nodes 5b and 6a) as a place once used for important ritual and administrative gatherings. In addition to this mat pattern there is also cosmological imagery decorating the roof of the structure, which portrays Celestial, Terrestrial, and Underworld themes (GR 26 and 27). Just below the roof, on the northern piers of House B, are portraits of K'inich Janab' Pakal I standing authoritatively in royal costume (GR 24 and 25). Most notably, residual sculpture on Pier B (GR 24) portrays a royal figure wearing a feathered headdress comprising a stingray spine and a Jestor God diadem; combined, the sculpture of House B's roof and northern façade signal K'inich Janab' Pakal I's dominion over the cosmological realms.

Level 6 sculpture of Palace Phase 1 appears in the eastern and western galleries of House E (nodes 6b and 6c) as well as in the southern gallery of House B (node 6a). The southern gallery jambs of House B also display remains of a solo portrait (GR 21) in addition to the lower portion of a dancing figure (GR 22). The sculpture present in the western gallery or throne room of House E displays Royal Audience and Accession themes as seen on the Oval Tablet (GR 5; Figure 4.13), as well as a throne text (Figure 4.6) containing historical

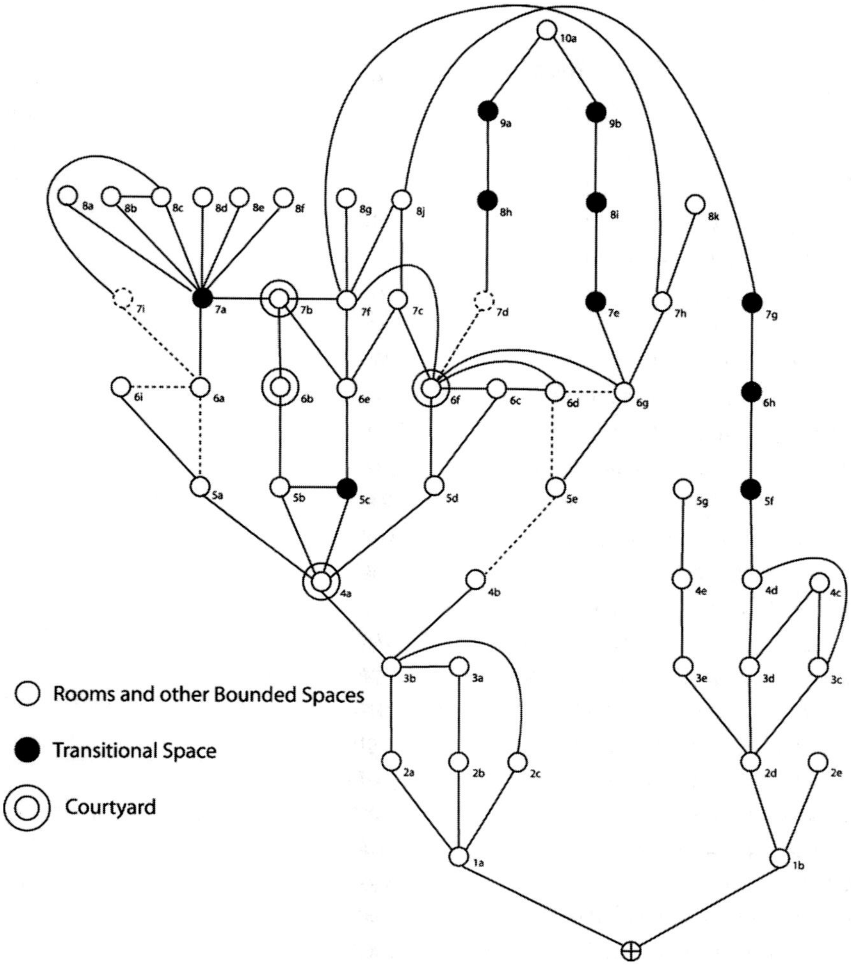


**Figure 5.4.** Palenque Palace Phase 2, "Justified" Access Diagram.

data (i.e., K'inich Janab' Pakal I's accession) and several Celestial and Underworld references. The western façade of House E portrays Cosmological Terrestrial Underworld themes in the form of flower motifs (GR 13; see Figures 4.10 and 4.11), whereas the eastern gallery of the same structure houses Cosmological Underworld imagery (GR 8; Figure 4.14), thematically connecting the structure with the East Subterranean Passage and southern substructures (see Schele 1986: 122).

## 5.1.2. Analysis of the Palace Phase 2

Phase 2 of the Palace differs from Palace Phase 1 due to the addition of Houses A and C to the complex (see Figure 5.3). As noted in Chapter 2, the construction of these two buildings required that the Second Palace Terrace platform be extended towards the north. K'inich Janab' Pakal I was the most likely ruler to have presided over the northward expansion of the Palace Group, which occurred in the last half of his reign (circa A.D. 650–683). The effect of these structural additions was the formulation of an early version of the East Court (node 4a), where Houses A and C define its east and west perimeter (see Chapter 4, Section 4.3.2, The Buildings of the Palace Phase 2). In addition, with the extension of the Second Palace Terrace platform came the beginnings of what would later become the West Court of the Palace (node 6d; see Chapter 4, Section 4.3.4, The Buildings of the Palace Phase 4, Section 4.3.5, The Buildings of the Palace Phase 5, and Section 4.3.6, The Buildings of the Palace Phase 6). It should be noted here that, at this stage in the Palace's evolution, Houses D and A-D did not yet define the northern periphery of the Palace, which (as at Phase 1) left the northern Palace substructure exposed. It is argued here that the Northern Galleries remained the principal point of entry into the Palace complex until the construction of Houses D and A-D and that the northern monumental stairway buried the Northern Galleries, shifting primary access into the Palace interior through either House A or between Houses D and K (see Chapter 4, Section 4.3.3, The Buildings of the Palace Phase 3).

As one would expect, with the addition of Houses A and C in the northeast quarter of the Palace, the access diagrams of both Phase 1 and Phase 2 display some differences when compared. For example, Palace Phase 2 comprises four more units of space than does Palace Phase 1, comprising a total of fifty-seven nodes. In addition, although Phases 1 and 2 both have a total of ten levels of depth from the perspective of the carrier position, at Palace Phase 2 there is a shift in the number of spaces assigned to the deeper levels of the access diagram: At Palace Phase 2, two spaces are represented at Levels 1 and 9; five spaces at Levels 2, 3, and 4; seven spaces at Level 5; nine spaces at Levels 6 and 7; eleven spaces at Level 8; and one space at Level 10 (see Figure 5.4). In addition, with the construction of Houses A and C, another five units of space display sculpture; these spaces are the East Court (node 4a), the eastern room of House C, (node 5d), House C's western gallery (node 6c), the west foundation of House C (node 6d), and House A's eastern gallery (node 6i). Seventeen spaces are defined as transitional at Palace Phase 2, compared to the total of

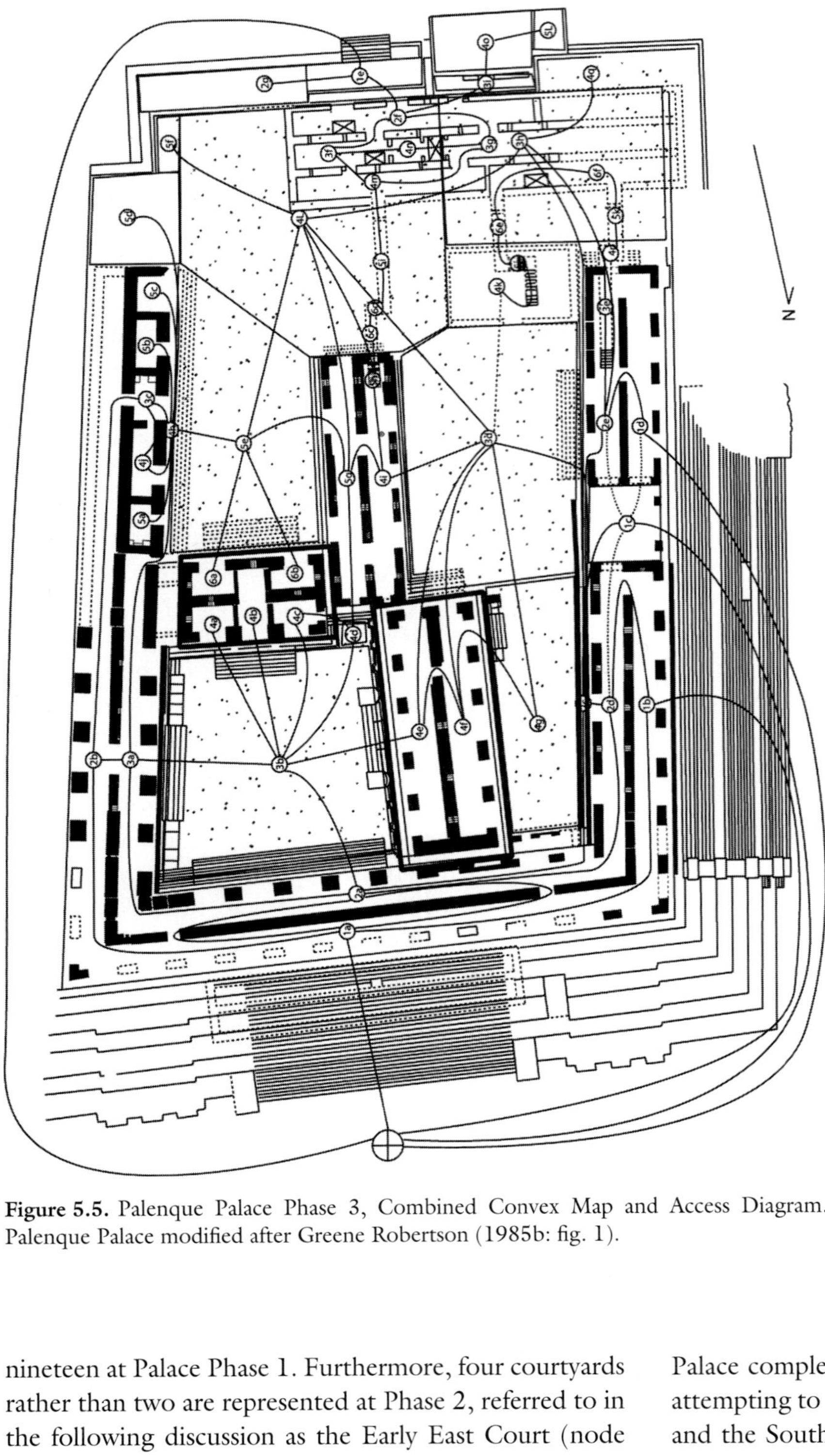

Figure 5.5. Palenque Palace Phase 3, Combined Convex Map and Access Diagram. Palenque Palace modified after Greene Robertson (1985b: fig. 1).

H-Sub and the then-exposed northern substructure, whereas House C falls between House E and House K. As Table 5.8 illustrates, the least integrated (and subsequently more inaccessible) buildings of the Palace at Phase 2 are still the South Subterranean Substructures and House J, whereas Houses C and E are the most integrated (accessible) of the Palace Phase 2 buildings.

It appears that the primary purpose in constructing Houses A and C, together with the early version of the East Court (see Figure 4.31), was to replace the Southwest Court and the western gallery of House E as the principal inhabitant–visitor interface of the Palenque Palace, where persons invited to enter could interact with inhabitants in a space designed to be separated from other areas in the Palace complex. At Palace Phase 2, formal access to the Palace interior was still provided via the Northern Galleries (node 2b and 3a). With the addition of the Early East Court (node 4a; a shallower reception area with respect to its distance from the carrier), however, there was a shift in the proximity of the Palace's primary inhabitant–visitor interface away from the Southwest Court and throne area of House E to a more intermediate area adjacent to Houses A, B, and C. It appears that the formulation of the Early East Court may have been a strategy used by K'inich Janab' Pakal I to increase the level of privacy associated with the Early Southwest Court and House E, as well as to more effectively control access to other areas of the Palace complex. For example, at Palace Phase 2, visitors attempting to access the House E throne room (node 7c) and the Southwest Court of the Palace (node 6f) from the East Court (node 4a) would have had to pass through House B, House C, or the eastern gallery of House E. (Note: At Palace Phase 2, more restrictive and informal access into the Palace remained possible via the subterranean galleries and passages and through the southern portion of House K.)

nineteen at Palace Phase 1. Furthermore, four courtyards rather than two are represented at Phase 2, referred to in the following discussion as the Early East Court (node 4a), the Early West Court (node 6d), the Early Southeast Court (node 7b), and the Southwest Court (node 6f).

Nine buildings comprise the Palace at Phase 2, compared to seven structures that collectively make up Palace Phase 1. On ranking these structures from most integrated to least integrated, and comparing their relative positioning with that of Phase 1, the order of the existing Palace Phase 2 structures remains unchanged. The RA value calculated for House A places it between House

The shallow positioning of the East Court from the carrier with its Captive Display themes (GR 58, 59, 60, and 61; see Figures 4.28 and 4.46), in addition to its

Table 5.8. *Average RA values for Palace Phase 2 houses and other structures*

| Structure | Average RA values | Rank |
|---|---|---|
| House E | 0.1177 | 1 |
| House C | 0.1214 | 2 |
| House K | 0.1292 | 3 |
| House B | 0.1299 | 4 |
| House H-Sub | 0.1383 | 5 |
| House A | 0.1419 | 6 |
| North Substructure | 0.1500 | 7 |
| House J | 0.1679 | 8 |
| South Substructure | 0.2101 | 9 |

*Note:* Ranked from most integrated to least integrated.

relatively high level of integration (see Table 5.9), is consistent with the proposal that the northeastern quadrant of the Palace functioned as the principal meeting area used by Palace inhabitants and outsiders permitted to enter the complex (Liendo Stuardo 2003: 197). In support of this view, House C (the building that defines the eastern boundary of the northeastern quadrant of the Palace) is described as the *yet nah* ("companion house") on the Hieroglyphic Stairs (H.S.; GR 61; Figure 4.27), which provide access from the East Court to the eastern gallery of the structure (Martin 2003: 94–95).

In order from most integrated to least integrated, all spatial units that display residual sculpture at Palace Phases 1 and 2 are as follows: the East Court (node 4a) and the eastern gallery of House C (node 5d); the eastern gallery of House E (node 6e); the Early West Court (node 6d); the western gallery of House C (node 6c); northern and southern rooms of House B (nodes 5b and 6b); the East Subterranean Passage's northern corridor (node 7g); the western gallery of House A (node 6i); the South Subterranean Passage bend (node 8h); and the West Subterranean Passage's southern corridor (node 9b). As at Palace Phase 1, at Phase 2 of the Palace, spaces displaying some of the lowest levels of integration are those spaces within the subterranean passages containing the subterranean vaults (GR 7, 78, and 79; Figures 4.7 and 4.8). At Palace Phase 2, however, the Ancestor Portraiture theme (GR 43, 43, 44, 45, 46, and 47; Figures 4.32, 4.33, 4.34, 4.35, and 4.36) associated with the eastern gallery of House A (node 6i) is also found at a comparably low level of integration. As just mentioned, some of the least integrated spaces of the Palace at Phase 2, and subsequently the most segregated from the rest of the complex, are those spatial units that collectively comprise Southern Subterranean Structures, the underground passages of the Palace, and House J. Among the least integrated spaces associated with the Subterraneos are the south substructure patios (nodes 5g and 4e), the south substructure central room (node 4c), and the South Substructure northwestern room (node 10a). Defined as transitional units of space, the least integrated spaces of the Palace's underground passages are the southern corridors of the West, South, and the East Subterranean Passages (nodes 5e, 9b, and 9a), whereas some of the most segregated spaces comprising House J include the northwestern room (node 8a), the west south-central room (node 8d), the west north-central room (node 8b), and the west central room (node 8c; see Chapter 4, Section 4.3.2, The Buildings of the Palace Phase 2).

As at Palace Phase 1, the bulk of the indoor and outdoor spaces that comprise Phase 2 of the Palace are distributed, displaying a relatively high number of redundant

Table 5.9. *Palenque Palace Phase 2 sculpture ranked according to RA values, K'inich Janab' Pakal I (circa* A.D. *650–683)*

| Node | RA value | Rank | Sculpture GR No. | Theme |
|---|---|---|---|---|
| 9b | 0.2169 | 1 | 79 | CosU |
| 8h | 0.1695 | 2 | 78 | CosU |
| 6i | 0.1539 | 3 | **42, 43, 44, 45, 46, 47** | **THist, APWA, APSolo** |
| 7g | 0.1396 | 4 | 7 | CosCU |
| 6b | 0.1312 | 5 | 21, 22, 26, 27, 28 | PSolo, RitDan, MMot, CosU, CosCU |
| 5b | 0.1286 | 6 | 24, 25, 26, 27 | RPSolo, MMot, CosU, CosCU |
| 6c | 0.1279 | 7 | **30, 31, 32, 33, 34, 35, 36, 37, 39, 40, 41** | **RAud, APSolo, PSolo, CosT** |
| 6d | 0.1273 | 8 | **68** | **TCapSac** |
| 6e | 0.1156 | 9 | 8 | CosCU |
| 5d | 0.1149 | 10 | **29, 38, 40, 41, 62** | **GPSolo, CosT, TDed?** |
| 7c | 0.1149 | 10 | 3, 4, 5, 13, 11, 80 (Bench) | RAud, Acc, CosTU, THist, TCosCU |
| 4a | 0.1078 | 11 | **58** (*alfardas* only), **59, 60, 61** (H.S.) | **CapDis, THist** |

*Notes:* Ranked from least integrated to most integrated. See Appendix 5 for building and location information for nodes.

Table 5.10. *Palace Phase 2 court carrier depths, RA values, and control values*

| Palace courts | Node | Carrier rank | RA value | Control value |
|---|---|---|---|---|
| Southwest Court (early Tower Court) | 6f | 6 | 0.1071 | 2.1166 |
| East Court (early) | 4a | 4 | 0.1078 | 1.5333 |
| Southeast Court (early) | 7b | 7 | 0.1162 | 1.0416 |
| West Court (early) | 6d | 6 | 0.1273 | 1.0090 |

pathways between spatial units: a total of fifty-seven nodes compared to eighty-three edges. Some of the more distributed spaces represented at Phase 2 include the patio space situated between Houses A and J (node 6a), the eastern room of House K (node 6g), the platform area providing access to the House J interior (node 7a), and the upper southeastern terrace of the Palace (node 7f). The Palace courtyards (nodes 4a, 6f, 6d, and 7b) are some of the most highly distributed spaces found at Palace Phase 2 (see Table 5.10). Again (as noted in Chapter 4, Section 4.3.1, The Buildings of the Palace Phase 1), this configuration is probably a reflection of their function as reception spaces and as venues for various semi-public and semi-private activities. In terms of depth from the carrier position, the East Court (node 4a) is the shallowest courtyard of the Palace, followed by the Early West Court (node 6d) and the Southwest Court (node 6f), which are at equal depths. As was the case in Palace Phase 1, the deepest courtyard of the Palace at Phase 2 is the Early Southeast Court. Some of the nondistributed spaces appearing on the Palace Phase 2 access diagram (indicating their more limited accessibility) are again those that comprise the east, west, and south subterranean passages (nodes 5f, 6h, 7g, 7e, 8h, 8i, 9a, and 9b); the northern and southern galleries of House B (nodes 5b and 6b); three of House J's outer rooms (nodes 8a, 8d, and 8e); in addition to the northwest and central rooms located in the Subterraneos (nodes 10a and 4c).

Of the fifty-seven spaces represented in the Palace Phase 2 access diagram, those that display the greatest level of control over localised access to surrounding spaces (in descending order) are as follows: House J's western terrace (node 7a); the Palace's upper southeastern terrace (node 7f); the Southwest Court (node 6f); the northern substructure's southern terrace (node 3b); the northern substructure's northern terrace (node 1a); the northern corridor of the East Subterranean Passage (node 7g); the eastern room of House K (node 6g); the East Court (node 4a); the southern substructure patio (node 4e); and the southern corridor of the Subterraneos (node 2d). At Palace Phase 2, the relatively high level of control that the Early East Court exhibits relates specifically to the inability of Houses A, B, C, and E to be accessed from the direction of the Northern Galleries without first passing through the East Court.

The controlling spaces that display sculpture at Phase 2 of the Palace (see Table 5.11) are the Early East Court and the West Court (nodes 4a and 6d), both of which portray predominantly Captive themes (GR 58, 59, 60, 61, and 68; see Figures 4.28, 4.29, and 4.46). Conversely,

Table 5.11. *Palenque Palace Phase 2 sculpture ranked according to control values, K'inich Janab' Pakal I (circa* A.D. *650–683)*

| Node | Control value | Rank | Sculpture GR No. | Theme |
|---|---|---|---|---|
| 7g | 1.8333 | 1 | 7 | CosCU |
| 4a | 1.5333 | 2 | **58** (*alfardas* only), **59, 60, 61** (Hieroglyphic Stairs) | **CapDis, THist** |
| 5b | 1.0333 | 3 | 24, 25, 26, 27 | RPSolo, MMot, CosU, CosCU |
| 6d | 1.0090 | 4 | **68** | **TCapSac** |
| 8h | 1.0000 | 5 | 78 | CosU |
| 9b | 1.0000 | 5 | 79 | CosU |
| 6e | 0.7499 | X | 8 | CosCU |
| 7c | 0.7261 | X | 3, 4, 5, 13, 11, 80 (Bench) | RAud, Acc, CosTU, THist, TCosCU |
| 6c | 0.7261 | X | **30, 31, 32, 33, 34, 35, 36, 37, 39, 40, 41** | **RAud, APSolo, PSolo, CosT** |
| 5d | 0.6761 | X | **29, 38, 40, 41, 62** | **GPSolo, CosT, TDed?** |
| 6b | 0.5833 | X | 21, 22, 26, 27, 28 | PSolo, RitDan, MMot, CosU, CosCU |
| 6i | 0.5833 | X | **42, 43, 44, 45, 46, 47** | **THist, APWA, APSolo** |

*Note:* See Appendix 6 for building and location information for nodes.

Table 5.12. *Palenque Palace Phase 2 sculpture ranked according to depth from the carrier, K'inich Janab' Pakal I (circa* A.D. *650–683)*

| Node | Carrier R | Sculpture GR No. | Theme |
|---|---|---|---|
| 4a | 4 | **58** (*alfardas* only), **59, 60, 61** (Hieroglyphic Stairs) | **CapDis, THist** |
| 5b | 5 | 24, 25, 26, 27 | RPSolo, MMot, CosU, CosCU |
| 5d | 5 | **29, 38, 40, 41, 62** | **GPSolo, CosT, TDed?** |
| 6b | 6 | 21, 22, 26, 27, 28 | PSolo, RitDan, MMot, CosU, CosCU |
| 6c | 6 | **30, 31, 32, 33, 34, 35, 36, 37, 39, 40, 41** | **RAud, APSolo, PSolo, CosT** |
| 6d | 6 | **68** | **TCapSac** |
| 6e | 6 | 8 | CosCU |
| 6i | 6 | **42, 43, 44, 45, 46, 47** | **THist, APWA, APSolo** |
| 7c | 7 | 3, 4, 5, 13, 11, 80 (Bench) | RAud, Acc, CosTU, THist, TCosCU |
| 7g | 7 | 7 | CosCU |
| 8h | 8 | 78 | CosU |
| 9b | 9 | 79 | CosU |

*Note:* See Appendix 4 for building and location information for nodes.

the controlled spaces where sculpture appears are the eastern and western galleries of House C (nodes 5d and 6c) as well as the eastern gallery of House A (node 6i). Sculptural themes represented in the controlled spatial contexts include God Portraiture Solo (GR 29; Figures 4.23, 4.24, 4.25, and 4.26); Cosmological Terrestrial (GR 38, 40, and 41); Ancestor Portraiture Solo (GR 32, 34, 35, and 36; Figures 4.20, 4.21, and 4.22); Ancestor Portraiture with Attendants (GR 43, 44, 45, and 46; Figures 4.32, 4.33, 4.34, and 4.35); and Royal Audience (GR 30).

When considering the five spatial units that display sculpture in terms of their relative depth from the carrier (see Table 5.12), the shallowest sculpture of Palace Phase 2 appears at Level 4 in the area of the Early East Court (node 4a). As already mentioned earlier in text, the principal theme of the monuments confined to the area of the East Court is Capture Display, as seen on the eastern foundation wall of House C (GR 59 and 60; Figure 4.28) and on either side of the foundation stair of House A (GR 58 [*alfardas* only]; Figure 4.46), where several bound captives are named and represented. Also in the proximity of the East Court are the H.S. (GR 61; Figure 4.27) assigned to the eastern foundation of House C, which, in addition to recording the *ch'ak* ("axing") of *Lakam Ha'* (Palenque) at the hands of Calakmul (Martin 2003: 88), document the *chukaj* ("capture") of four vassals who are also named and portrayed on the east basement wall of the building (GR 59; Figure 4.28). K'inich Janab' Pakal I is recorded as the protagonist of the action, and the captures are purported to have occurred on 9.11.6.16.11 (7 August 659), four years after the attack on Palenque (Martin 2003: 90–93, 95). Also recorded on the H.S. of House C is the name of the structure itself reading *yet nah* ("companion house"; Martin 2003: 94–95), in addition to the "opening" of the house: *i-pasaj* ("and then it was opened"; Martin 2003: 89). The next sculptural theme to appear on the Palace Phase 2 access diagram is at Level 5, where God Portraiture Solo is the principal theme appearing in the eastern gallery of House C (GR 29; Figures 4.23, 4.24, 4.25, and 4.26). Set in a horizontal line on the eastern medial wall of the house are the remains of seven portrait masks of gods, adhered to the medial wall of House C (see Chapter 4, Section 4.3.2, The Buildings of the Palace Phase 2, House C: Description and Sculpture).

At Level 6, the Royal Audience (GR 30) and Ancestor Portraiture Solo (GR 32, 34, 35, and 36; Figures 4.20, 4.21, and 4.22) themes appear in the western gallery of House C (node 6c), where the remnants of a painted narrative scene cover the southern portion of the medial wall on the western side of House C. The mural once portrayed two seated individuals making offerings to an elevated figure seated centrally on a throne. To the left of the scene, overlying the earlier mural, a human figure is sculptured. The profile stance and proximity of the individual (left of centre) suggest that the figure was subordinate to whatever once decorated the central area of the wall. As stated in Chapter 4, it is my view that this figure is compositionally consistent with an attendant standing in the company of something or someone of greater authority. The surviving sculpture on the four western piers of House C (GR 32, 34, 35, and 36; Figures 4.20, 4.21, and 4.22) all portray the remains of individuals seated on thrones, in frontal orientation, and in semi-elaborate royal attire. It is probable that the pier sculpture portrays K'inich Janab' Pakal I (GR 35; Figure 4.22) in the company of several of his ancestors who are depicted on the other piers. In support of this proposal, variations are evident in sculptured costumes and throne designs associated with the individuals represented, in addition to the representation of a woman that appears on Pier F (GR 36) of the house (see Chapter 4, Section 4.3.2, The Buildings of the Palace Phase 2, House C: Description and Sculpture). The eastern piers of House C, which face in on the East Court, also once portrayed sculpture of some kind – most likely standing royal portraits similar to those decorating the northern façade of House B, as this side of the structure represents the formal entrance to the building.

Also appearing at a depth of Level 6 from the carrier are the ancestral portraits assigned to the interior and exterior of House A's eastern gallery (node 6i) and texts recording human capture and sacrifice located on the western foundation of House C (GR 68; Figure 4.29) in the West Court (node 6d). The theme of the text located on the western foundation of House C is Text, Captive Sacrifice, which is perhaps connected with activities once undertaken within the space. The seven texts carved on the western foundation of House C refer to six captive lords from Santa Elena and one from Pomona, one of whom was sacrificed (Martin 2003: 114–116). The first glyph in each of the seven inscriptions records the name of the captive, whereas the remaining text in each of the columns records his place of origin and title, respectively.

The principal theme of House A's eastern piers is Ancestor Portraiture with Attendants (GR 43, 44, 45, and 46; see Figures 4.32, 4.33, 4.34, and 4.35); each sculpture depicts three individuals – one who stands centrally and two who are seated peripherally. When comparing the figural scenes, it becomes apparent that, although there are parallels in the way that the individuals are represented, each pier portrays a different protagonist, implying that the piers of House A are most likely to be commemorative portraits of different rulers undertaking the same ritual activity. In the eastern gallery of House A, spaced in a row on the medial wall, is a row of portrait busts referred to as the Palenque Medallions (GR 47; Figure 4.36). These Medallions (like the House A piers) portray what were probably ancestral portraits of the Palenque kings, which in these examples are depicted emerging from *k'in* signs (Greene Robertson 1985b: 26; see Chapter 4, Section 4.3.2, The Buildings of the Palace Phase 2, House A: Description and Sculpture).

### 5.1.3. Analysis of the Palace Phase 3

As outlined in Chapter 4, K'inich K'an Joy Chitam II (A.D. 702–721?) presided over the structural additions and modifications that collectively comprise Phase 3 of the Palenque Palace (see Figure 5.5). The main architectural contribution implemented by K'inich K'an Joy Chitam II was the final extension of the Second Palace Terrace platform (i.e., Extension B), followed by the construction of Houses D and A-D and the monumental staircases erected on the northern and western sides of the complex (see Chapter 4, Section 4.3.3, The Buildings of the Palace Phase 3). As a consequence of these structural additions, the Northern Galleries of the Palace (appearing at Phases 1 and 2) were finally buried beneath the northern monumental stairway, becoming what is currently known as the northern subterranean structure of the Palace. K'inich K'an Joy Chitam II's amendments dramatically change the general appearance and spatial configuration of the Palace Group. More specifically, towards the end of K'inich K'an Joy Chitam II's reign, the Palace complex shifted from what was a multilevel plan implemented by predecessor K'inich Janab' Pakal I (comprising transitions in elevation as one moved progressively deeper into the complex) to one that consisted of a single elevated platform approximately 10 metres high that was enclosed on three of its sides by "range"-type structures. It appears that one of K'inich K'an Joy Chitam II's objectives in building Houses D and A-D was to limit access to the East Court as well as to define the northern and western limits of the Early West Court. In my view, the presence of these massive stairs (see Table 5.13), whilst providing an impressive approach for persons permitted to ascend them, were constructed primarily to function as boundaries to more effectively segregate Palace occupants from the general population at Palenque.

Table 5.13. *Counts, widths, and elevations of the North, East, and West Palace stairs*

| Palace staircase | Number of stairs | Approximate width | Approximate height |
|---|---|---|---|
| North Stair | 33 | 25.0 m | 10.9 m |
| West Stair | 28 | 38.0 m | 8.1 m |
| East Stair (incomplete) | 18 | 35.0 m | 6.0 m |

As outlined in Chapter 4 (see Chapter 4, Section 4.3.3, The Buildings of the Palace Phase 3), it is probable that Palace Houses D and A-D were constructed together rather than separately, as suggested by the absence of construction joins indicating where the two structures were connected. A western entrance to the Palace appeared with the construction of Houses D and A-D, between the southern end of House D and the northern extremity of an earlier version of House K. As discussed in Chapter 4, the current plan of House K is the product of two stages of construction that Andrews (1974) described as the

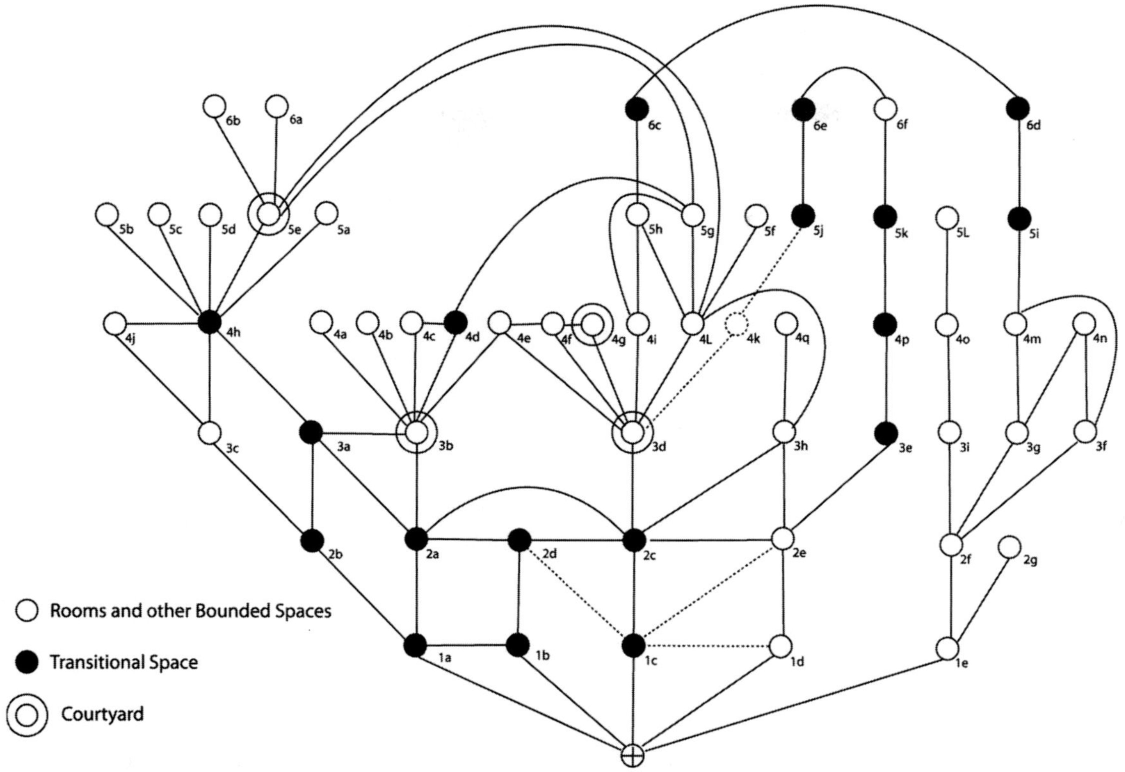


**Figure 5.6.** Palenque Palace Phase 3, "Justified" Access Diagram.

"southern portion of House K" and its "northern extension." House K's northern extension was added some time after the construction of House D, resulting in the formation of a gap between Houses D and K to which access was provided via the western monumental stairway (see Figures 4.37 and 5.5). The appearance of this entrance suggests that primary access into the building group may have shifted from the north of the Palace via the East Court to either the eastern gallery of House A or between Houses D and K. (Note: As pointed out in Chapter 4, Section 4.3.3, The Buildings of the Palace Phase 3, House D: Description and Sculture, the northern and western Palace monumental stairs provided formal access to the upper Palace houses, as the stairway on the eastern side of the Palace never reached the level of the Second Place Terrace platform [see Maudslay 1889–1902b: 12]). Fronted by two narrow piers (which probably supported a roof at one time), the proximity of the eastern entrance (directly adjacent to House E's western gallery) indicates that it may have once been an elaborate and formal point of entry into the Palace complex from the plaza fronting the Temple of Inscriptions.

As a consequence of K'inich K'an Joy Chitam II's modifications to the Palace, the Phase 3 access diagram comprises the same number of nodes as Phase 2 but has a higher number of edges connecting them; Phase 3 of the Palace comprises eighty-three edges versus Palace Phase 2's total of seventy-seven. Due to this increased interconnectedness between nodes at Palace Phase 3, the access diagram is more distributed than that of Phase 2, displaying a higher degree of spatial integration overall. Consequently, the Phase 3 access diagram appears shallower, comprising fewer levels of depth when compared to the first two phases of the Palace. Whereas Phases 1 and 2 comprise ten levels of depth, Phase 3 portrays only six from the perspective of the carrier position.

Palace Phase 3 consists of fifty-seven units of space, including the carrier, as indicated by the corresponding access diagram (see Figure 5.6): Five spaces are represented at Level 1, seven spaces at Level 2, nine spaces at Level 3, seventeen spaces at Level 4, twelve spaces at level 5, and six spaces at Level 6. At Palace Phase 3, sculpture was assigned to an additional eight units of space within the Palace complex; among these spaces were the highly exposed northern monumental stairs (visible from the carrier position), House A-D's northern gallery (node 1a), the exterior of House D's western gallery (node 1b), the East Court (node 3b), and the southeastern and southwestern rooms of House B (nodes 6a and 6b). In addition to the increase in the number of sculptures at Palace Phase 3, nineteen spaces were defined as transitional. Four courtyards also proposed at Palace Phase 3 are referred to in the following discussion as the completed East Court (node 3b), Early West Court (node 4g), the Early Southeast Court (node 5e), and the Early Southwest Court (node 3d).

Table 5.14. *Average RA values for Palace Phase 3 houses and other structures*

| Structure | Average RA values | Rank |
|---|---|---|
| House A-D | 0.0929 | 1 |
| House K | 0.1019 | 2 |
| House D | 0.1042 | 3 |
| House E | 0.1075 | 4 |
| House A | 0.1081 | 5 |
| House C | 0.1100 | 6 |
| House H-Sub | 0.1156 | 7 |
| House B | 0.1331 | 8 |
| House J | 0.1356 | 9 |
| South Substructure | 0.1749 | 10 |

*Note:* Ranked from most integrated to least integrated.

Ten structures collectively comprise Phase 3 of the Palace at Palenque. On ranking these buildings from most integrated to least integrated, there is a notable shift in the relative ordering of the existing structures when compared to Palace Phase 2 (see Table 5.14). For example, with the addition of Houses D and A-D to the Palace complex, House E has become less integrated with respect to House K, and House A has become more integrated with respect to Houses B and C. Shifts in the relative depth of existing Palace buildings suggest that one of the objectives of K'inich K'an Joy Chitam II in adding Houses D and A-D to the Palace complex may have been to reconfigure the building group in terms of accessibility and control or, more specifically, to make certain spaces within the Palace Group more or less accessible to inhabitants occupying the building group and to visitors permitted to enter the complex. Changes that occur to the level of integration and control of a Palace house or unit of space over time imply that activities once undertaken within the confines of these spaces may have become either increasingly (1) hidden from, or (2) exposed to the observances of others occupying the Palace complex. Such shifts would have necessitated, at least in some instances, changes in the proximity of activities that were intended to remain private. Whereas Houses D, A-D, and K replaced House E as the most integrated structures of the Palace, House J as well as the south substructure remained the least integrated (most private) buildings of the Palace complex during K'inich K'an Joy Chitam II's reign.

Due to their function as arenas for public or semi-public activity, again several of the Palace courtyards stand out as some of the most distributed spaces represented at Palace Phase 3 (see Table 5.15); these spaces are the East Court (node 3b), the Early Southwest Court (node 3d), and the Early Southeast Court (node 5e). In terms of their depth from the carrier position, the East Court and the Southeast Court are the shallowest courtyards of the Palace, located at a depth of Level 3 from the carrier. The Early West Court is the next deepest court space positioned at Level 4 (see node 4g), followed by the Southeast Court, which is located at a depth of Level 5. Although the Early West Court is shallower than the Southeast Court, it is a relatively nondistributed space displaying a low level of integration when compared to the other court spaces represented. In addition, of all the Palace Phase 3 courtyards, the West Court is the only one calculated as being a controlled rather than a controlling space. Other nondistributed spaces indicating their relatively limited accessibility at Palace Phase 3 remain those spaces that comprise the southern substructure (nodes 3f, 3g, 4n, 4m, and 6f); the east, west, and south subterranean passages (nodes 4p, 5k, 5i, 5j, 6d, and 6e); and the rooms that make up the interior of House J (nodes 5a, 5b, 5c, and 5d).

Increased connectivity between nodes at Palace Phase 3 (see Figure 5.6) indicate that fewer boundaries needed to be crossed to reach any one point in the Palace complex. A consequence of this increased interconnectivity between nodes is a reduction in the overall depth of Palace Phase 3's access diagram from the perspective of the carrier. Interestingly, even with a comparatively shallow access diagram, most of the spaces that comprise the southern substructure, the subterranean passages, and House J at Palace Phase 3 remain the most segregated of the Palace. Furthermore, the East Court (node 3b) at Palace Phase 3 remains one of the more integrated spaces of the complex. The assigning of additional sculpture to this semi-public space may be viewed as an indication that this already highly embellished courtyard remained a hub of semi-public activity during the reign of K'inich K'an Joy Chitam II.

In order from most integrated to least integrated, other Phase 3 spaces displaying sculpture are as follows: the northern corridor of House A-D (node 1a); the House E western room (node 4i); the western gallery of

Table 5.15. *Palace Phase 3 court carrier depths, RA values, and control values*

| Palace courts | Node | Carrier rank | RA value | Control value |
|---|---|---|---|---|
| East Court (completed) | 3b | 3 | 0.0961 | 3.6166 |
| Southwest (early Tower Court) | 3d | 3 | 0.0851 | 2.3333 |
| West Court (early) | 4g | 4 | 0.1201 | 0.4761 |
| Southeast Court (early) | 5e | 5 | 0.1032 | 2.4916 |

Table 5.16. *Palenque Palace Phase 3 sculpture ranked according to RA values, K'inich K'an Joy Chitam II* (A.D. *702–721?)*

| Node | RA value | Rank | Sculpture GR No. | Theme |
|---|---|---|---|---|
| 5k | 0.1864 | 1 | 79 | CosU |
| 5j | 0.1461 | 2 | 78 | CosU |
| 2f | 0.1429 | 3 | **80** (bench moved to South Substructure) | **TCosCU** |
| 6a | 0.1390 | 4 | **23**, 26, 27, 28 | **CosT?,** MMot, CosU, CosCU |
| 6b | 0.1390 | 4 | **16, 17, 18, 19, 20,** 21, 22, 26, 27, 28 | **RAud, Text?,** PSolo, RitDan, MMot, CosU |
| 6c | 0.1331 | 5 | 7 | CosU |
| 4a | 0.1318 | 6 | 24, 25, 26, 27 | RPSolo, MMot, CosU, CosCU |
| 4b | 0.1318 | 6 | 24, 25, 26, 27 | RPSolo, MMot, CosU, CosCU |
| 4c | 0.1240 | 7 | 24, 25, 26, 27 | RPSolo, MMot, CosU, CosCU |
| 4g | 0.1201 | 8 | **63, 64, 65, 66,** 68, **69** | **GPWA, CosC,** TCapSac |
| 2b | 0.1149 | 9 | 42, 43, 44, 45, 46, 47 | THist, APWA, APSolo |
| 4f | 0.1149 | 9 | 30, 31, 32, 33, 34, 35, 36, 37, 39, 40, 41 | RAud, APSolo, PSolo, CosT |
| 1b | 0.1117 | 10 | **48, 49, 50, 51, 52, 53** | **HumSac, SacDan, Text?** |
| 4e | 0.1052 | 11 | 29, 38, 40, 41, 62 | GPSolo, CosT, TDed? |
| 4i | 0.1026 | 12 | 3, 4, 5, **6** (Del Río Throne), 13, 11 | RAud, Acc, CosTU, THist, **TDynHist** |
| 1a | 0.0994 | 14 | **55, 56, 57** (Throne?) | **RPSolo, RAudAccTDynHist** |
| 3b | 0.0961 | 15 | **58** (remaining Panels), 59, 60, 61 (Hieroglyphic Stairs) | **CapDis,** THist |

*Notes:* Ranked from least integrated to most integrated. See Appendix 5 for building and location information for nodes.

House D (node 1b); the Early West Court (node 4g); the southwest room of House B (node 6b); the southeastern room of House B (node 6a); and the southern corridor of the Subterraneos (node 2f). From the perspective of all other points in the Palace, on average, the East, West, and South Subterranean Passage vaults (GR 7, 78, and 79; see Figures 4.7 and 4.8) and their corresponding Underworld themes are the least accessible sculptures of the Palace Group at Phase 3. Also at comparably low levels of integration is the space now housing K'inich Janab' Pakal I's throne (Bench 1, GR 80; Figure 4.6) as well as the southeastern and southwestern sanctuaries of House B. Facing in on the relatively private Southeast Court, the House B sculptural additions (GR 23; Figure 4.48) – at a level of integration similar to that of the subterranean passage vaults – mark a shift in the exclusivity of the southern portion of House B and, correspondingly, activities once associated with those spaces. In contrast to K'inich K'an Joy Chitam II's Palace design, during K'inich Janab' Pakal I's reign (Palace Phases 1 and 2), the southern portion of House B was more integrated with the rest of the Palace complex and subsequently more accessible from the perspective of persons occupying different areas of the Palace complex.

The most integrated sculpture at Palace Phase 3 is located in the East Court (see Table 5.16), which, as previously mentioned, predominantly portrays capture themes (GR 58, 59, and 61; Figures 4.28, 4.27 and 4.46). The next most integrated sculpture is the Palace Tablet, which depicts K'inich K'an Joy Chitam II's accession scene and a dynastic–historical text in the northern gallery of House A-D (GR 57; Figure 4.39), followed by the dynastic–historical throne text appearing in the western gallery of House E (GR 6). Palace Phase 3 sculpture holding more intermediate positions on the RA order of ranking (Table 5.16) are the sacrificial themes found on the western gallery piers of House D (GR 48, 49, 50, and 52; Figures 4.42, 4.43, 4.44, and 4.45) as well as the God Portraiture and Cosmological Celestial themes appearing on the *Casita* Wall in the Early West Court of the Palace.

At Palace Phase 3, fifty-seven nodes are represented in the corresponding access diagram, twenty of which were calculated as being controlling spaces. Among these spaces, those that display the highest level of control over localised access to surrounding spaces (in descending order of control) are as follows: House J's western terrace (node 4h); the East Court (node 3b); the Southeast Court (node 5e); the Southwest Court (node 3d); the upper southeastern terrace of the Palace (node 4L); the upper southwestern terrace of the Palace (node 3h); the southern substructure's southern corridor; and the eastern room of House K (node 2e). The high level of control that the East Court exhibits at Palace Phase 3 relates specifically to the inability to access the northern rooms of House B and the eastern gallery of House C without first passing through the East Court.

At Phase 3 of the Palace, the East Court (node 3b) and House A-D's northern gallery (node 1a) are the only two controlling spaces that were found to display sculpture (see Table 5.17). All other artwork appearing at Phase 3 of the Palace was assigned to controlled spaces. Sculptural

Table 5.17. *Palenque Palace Phase 3 sculpture ranked according to control values, K'inich K'an Joy Chitam II* (A.D. *702–721)*

| Node | Control value | Rank | Sculpture GR No. | Theme |
|---|---|---|---|---|
| 3b | 3.6166 | 1 | **58** (remaining panels), 59, 60, 61 (Hieroglyphic Stairs) | **CapDis,** THist |
| 2f | 1.4999 | 2 | **80** (bench moved to South Substructure) | **TCosCU** |
| 1a | 1.0666 | 3 | **55, 56, 57** (Throne?) | **RPSolo, RAudAccTDynHist** |
| 5j | 1.0000 | 4 | 78 | CosU |
| 5k | 1.0000 | 4 | 79 | CosU |
| 4f | 0.9761 | X | 30, 31, 32, 33, 34, 35, 36, 37, 39, 40, 41 | RAud, APSolo, PSolo, CosT |
| 2b | 0.8333 | X | 42, 43, 44, 45, 46, 47 | THist, APWA, APSolo |
| 6c | 0.8333 | X | 7 | CosU |
| 4i | 0.7261 | X | 3, 4, 5, **6** (Del Río Throne), 13, 11 | RAud, Acc, CosTU, THist, **TDynHist** |
| 1b | 0.7000 | X | **48, 49, 50, 51, 52, 53** | **HumSac, SacDan, Text?** |
| 4e | 0.6190 | X | 29, 38, 40, 41, 62 | GPSolo, CosT, TDed? |
| 4g | 0.4761 | X | **63, 64, 65, 66,** 68, **69** | **GPWA, CosC,** TCapSac |
| 4c | 0.4761 | X | 24, 25, 26, 27 | RPSolo, MMot, CosU, CosCU |
| 6a | 0.2000 | X | **23,** 26, 27, 28 | **CosT?;** MMot, CosU, CosCU |
| 6b | 0.2000 | X | **16, 17, 18, 19, 20,** 21, 22, 26, 27, 28 | **RAud, Text?,** PSolo, RitDan, MMot, CosU |
| 4a | 0.1428 | X | 24, 25, 26, 27 | RPSolo, MMot, CosU, CosCU |
| 4b | 0.1428 | X | 24, 25, 26, 27 | RPSolo, MMot, CosU, CosCU |

*Note:* See Appendix 6 for building and location information for nodes.

themes appearing in controlling spaces of the Palace during the reign of K'inich K'an Joy Chitam II include Royal Audience–Accession and Dynastic Historical text, as seen on the Palace Tablet (GR 57; Figure 4.39), in addition to the Captive Display imagery located on the western foundation of House A and the eastern foundation of House C (GR 58, 59, and 60; Figures 4.46 and 4.28). Conversely, sculptural themes located in controlled spatial contexts at Palace Phase 3 include the Dynastic Historical text appearing on the Del Río Throne (GR 6; Figure 4.49), which was once assigned to the western medial wall of House E, as well as Human Sacrifice and Sacrificial Dance themes, which are displayed on the western gallery piers of House D (GR 48, 49, 50, 51, and 52; see Figures 4.42, 4.43, 4.44, and 4.45). Additional sculptural themes occupying controlled spatial contexts within the confines of the Palace are God Portraiture with Attendants and Cosmological Celestial that appear in the Early West Court (GR 63, 64, 65, 66, and 69; see Chapter 4, Section 4.3.3, The Buildings of the Palace Phase 3, House D East Foundation *Casita* Wall: Description).

When considering the eight spatial units that display sculpture in terms of their relative depth from the carrier position, the most public artwork to appear at Palace Phase 3 is found at the base of the northern monumental stairway (see Table 5.18). Once flanking the northern Palace monumental staircase were multiple mask portraits of K'inich K'an Joy Chitam II (Schele 1998: 482). The sculptural theme of these masks has been designated as Royal Portraiture Solo and Cosmological Celestial Underworld, as each mask portrayed K'inich K'an Joy Chitam II beneath a representation of the Cosmic Monster (GR 82, 83, and 84; Figure 4.40) comprising elements symbolic of both the celestial realm and the Underworld (see Miller and Taube 1993: 45). In addition to these masks, once set against the balustrades flanking the base of the northern Palace stairs were two large *alfardas* (GR 85) portraying standing solo portraits of K'inich K'an Joy Chitam II in full royal regalia (see Greene Robertson 1985b: figs. 354 and 356).

Sculpture located at a depth of Level 1 from the carrier includes the Palace Tablet, which is centrally positioned on the northern side of the medial wall of House A-D (node 1a), and the western gallery piers of House D (node 1b). Facing north, the Palace Tablet (GR 57; Figure 4.39) comprises a large Dynastic Historical text with an accompanying scene depicting the accession ceremony of K'inich K'an Joy Chitam II (see Chapter 4, Section 4.3.3, The Buildings of the Palace Phase 3, House A-D: Description and Sculpture). Compositionally emulating the Oval Tablet of House E (GR 5; Figure 4.13), the monument is believed to have once been positioned above a throne (Greene Robertson 1985b: 54). The exterior piers of House D (GR 48, 49, 50, 51, and 52; Figures 4.42, 4.43, 4.44, and 4.45) represent what is probably a sequence of stills connected with a single ritual event; each pier portrays two human figures participating in various sacrificial activities (Greene Robertson 1985b: 33). As

Table 5.18. *Palenque Palace Phase 3 sculpture ranked according to depth from the carrier, K'inich K'an Joy Chitam II* (A.D. *702–721)*

| Node | Carrier R | Sculpture GR No. | Theme |
|---|---|---|---|
| C | 0 | **1, 2, 82, 83, 84, 85** | **CosC, RPSoloCosCU, RPSolo** |
| 1a | 1 | **55, 56, 57** (Throne?) | **RPSolo, RAudAccTDynHist** |
| 1b | 1 | **48, 49, 50, 51, 52, 53** | **HumSac, SacDan, Text?** |
| 2b | 2 | 42, 43, 44, 45, 46, 47 | THist, APWA, APSolo |
| 2f | 2 | **80** (bench moved to South Substructure) | **TCosCU** |
| 3b | 3 | **58** (remaining panels), 59, 60, 61 (Hieroglyphic Stairs) | **CapDis,** THist |
| 4a | 4 | 24, 25, 26, 27 | RPSolo, MMot, CosU, CosCU |
| 4b | 4 | 24, 25, 26, 27 | RPSolo, MMot, CosU, CosCU |
| 4c | 4 | 24, 25, 26, 27 | RPSolo, MMot, CosU, CosCU |
| 4e | 4 | 29, 38, 40, 41, 62 | GPSolo, CosT, TDed? |
| 4f | 4 | 30, 31, 32, 33, 34, 35, 36, 37, 39, 40, 41 | RAud, APSolo, PSolo, CosT |
| 4g | 4 | **63, 64, 65, 66,** 68, **69** | **GPWA, CosC,** TCapSac |
| 4i | 4 | 3, 4, 5, **6** (Del Río Throne), 13, 11 | RAud, Acc, CosTU, THist, **TDynHist** |
| 5j | 5 | 78 | CosU |
| 5k | 5 | 79 | CosU |
| 6a | 6 | **23,** 26, 27, 28 | **CosT?,** MMot, CosU, CosCU |
| 6b | 6 | **16, 17, 18, 19, 20,** 21, 22, 26, 27, 28 | **RAud, Text?,** PSolo, RitDan, MMot, CosU |
| 6c | 6 | 7 | CosU |

*Note:* See Appendix 4 for building and location information for nodes.

outlined in Chapter 1, there were three instances within a ritual sequence of events that the Maya focused on recording with sculpture and other artistic media; these have been referred to as the "inceptive," the "progressive," and the "completive" moments (see Chapter 1, Section 1.5, Text, Image, and Literacy among the Classic Maya). The Maya often combined these instances into a series of images suggesting a progression of action – the separate scenes depict different episodes of the same event (Schele and Miller 1986: 38). As K'inich K'an Joy Chitam II has been attributed with the construction of House D, the protagonist once represented in each of the pier scenes has been interpreted as K'inich K'an Joy Chitam II himself (see Chapter 4, Section 4.3.3, The Buildings of the Palace Phase 3, House D: Description and Sculpture). The two principal themes represented on the House D piers are Human Sacrifice (GR 48 and 52; Figures 4.42 and 4.45) and Sacrificial Dance (GR 49, 50, and 51; Figures 4.43 and 4.44). Piers B and F of House D each portray two male individuals, one of whom is about to be beheaded with an axe. Piers C and D of the same structure depict an individual dancing. In one example, the figure dances with a serpent (GR 50); in the other, the figure dances in the company of a kneeling subordinate (GR 49; Figure 4.43).

The next sculpture to appear is at Level 3 in the East Court. As already outlined in the Palace Phase 2 discussion, the principal theme of the monuments confined to the East Court (node 3b) is Capture Display. Consistent with the existing sculptural theme portrayed in the space, soon after the closing of the East Court an additional seven panels (GR 58; Figure 4.46) were assigned to the western foundation of House A. They display captive individuals, each of whom may be identified by the texts appearing on the western foundation of House C (see Chapter 4, Section 4.3.3, The Buildings of the Palace Phase 3, House A: Further Description and Sculptural Additions).

Sculptures appearing at a depth of Level 4 on the Phase 3 access diagram include those located within the West Court (node 4g) and the throne room of House E (node 4i). Referred to as the *Casita* Wall, the principal themes of the sculpture appearing on the eastern foundation of House D in the West Court have been defined as God Portraiture with Attendants (GR 63, 64, and 66) and Cosmological Celestial (see GR 65 and 69). On the eastern foundation of House D are three cartouches that contain bust-style portraits of three individuals. Two of the figures depicted peripherally in each of the cartouches are human, whereas the third individual (who is centrally positioned and frontally oriented) represents a supernatural being or god of some kind (see Chapter 4, Section 4.3.3, The Buildings of the Palace Phase 3, House D East Foundation *Casita* Wall: Description). Located in between these cartouches are hourglass-shaped armatures that support the stucco remains of a peccary and a

God K whose arms and legs are threaded through Venus symbols. Similar imagery appears on the south substructure altar of the Palace (GR 1; see Chapter 4, Section 4.3.3, The Buildings of the Palace Phase 3, House D: Description and Sculpture). Also at a depth of Level 4 is the House E throne room (node 4i), where, at Phase 3 of the Palace, K'inich Janab' Pakal I's throne (Bench 1, GR 80; Figure 4.6) was substituted with the Del Río Throne (GR 6; Figure 4.49). Displaying a Dynastic Historical text carved on the outer edge of the throne, K'inich K'an Joy Chitam II's accession is the latest date recorded on the monument (Schele and Mathews 1979: 140). As I have argued previously, Bench 1 (GR 80) was originally placed in the northwestern room of House E below K'inich Janab' Pakal I's coronation scene (i.e., the Oval Tablet). Functioning then as K'inich Janab' Pakal I's principal throne, it was later moved to the Subterraneos (node 2f) by K'inich K'an Joy Chitam II and replaced with the Del Río throne (GR 6) to coincide with the later king's inauguration in 9.13.10.6.8 (A.D. 702).

The deepest artworks appearing at Phase 3 of the Palace (see Table 5.18) are located in House B's southeastern and southwestern rooms (nodes 6a and 6b). With the closing of the Palace's East Court with House A-D, a number of curtain walls were added to House B to prevent through access to the Southeast Court (Schele 1986: 119; see also Greene Robertson 1985b: 50). The effect of this architectural amendment was to break up the north and south galleries of House B into five smaller, sanctuary-type spaces (Figure 4.47), three of which face onto the East Court (Rooms 1, 2, and 3) and two of which are adjacent to the Southeast Court (Rooms 4 and 5). In the southeast room of House B (Room 5) is a step-fret pattern sculptured on the rear wall of the space (GR 23; Figure 4.48); the theme of the sculpture has been defined as Cosmological Terrestrial. The pattern represents "the mountain cleft in its most reduced form" (Schele 1998: 497). The southwest room of House B (Room 4) contains a narrative scene that has been tentatively argued to portray a Royal Audience scene (GR 16, 17, 18, and 19; see Chapter 4, Section 4.3.3, Buildings of the Palace Phase 3, House B, Room 4, Narrative Scene Description).

### 5.1.4. Analysis of the Palace Phase 4

The appearance of the Palenque Palace at Phase 4 (Figures 4.50 and 5.7) is the result of architectural additions and modifications implemented during the reign of K'inich Akul Mo'Nab' III (A.D. 721–740?). In Chapter 4, I proposed that K'inich Akul Mo'Nab' III presided over the construction of Houses G and H, the Tower and Toilets, as well as the extension of House K (see Chapter 4, Section 4.3.4, The Buildings of the Palace Phase 4). The effect that these structural modifications had on the general configuration of the Palace was to completely enclose the complex, as well as to segregate the four courtyards confined to the Palace interior. Following the closure of the Palace group with Houses A, J, G, H, D, K, and A-D, the Palace Tower was erected as a lookout – as a vantage point from which select inhabitants could observe events (whether celestial or societal in nature) occurring beyond the confines of the Palace, without leaving the comfort or security of the complex. The joining of Houses D and K into one long continuous structure resulted in the closing of the Palace's western entrance (see Chapter 4, Section 4.3.3, The Buildings of the Palace Phase 3). Consequently, House A's "ogee-trefoil arch" (see Greene Robertson 1985b: 7) became the main point of entry into the building group at Palace Phase 4. More restricted access was possible, however, via the southern subterranean substructures and passages in addition to a narrow stair at the southeastern corner of the complex. (Note: I have previously argued that the southeastern stairway [node 2e], which today provides access to the southern Palace terrace [node 3d], was built to coincide with the construction of Houses G and H [see Section 5.1.1, Analysis of Palace Phase 1, this chapter].) The symmetry expressed by the Palace Phase 4 plan (Figure 5.7) implies design intent rather than simply a gradual and arbitrary accretion in architecture. Probably instigated with the construction of Houses D and A-D by K'inich K'an Joy Chitam II, the Phase 4 plan would have been completed following his death, sometime during the reign of K'inich Akul Mo'Nab' III.

Comprising seventy-three nodes and one hundred three edges, the general configuration of the Phase 4 access diagram may be described as distributed, as the number of nodes is substantially less than the number of edges connecting them (see Figure 5.8). As noted in the Palace Phase 3 discussion, the effect that a higher number of edges has on a spatial system comprising fewer nodes is to increase the number of route options between nodes and, correspondingly, the overall level of integration expressed by the spatial configuration. Some of the more distributed spaces represented at Phase 4 include the Palace's southern terrace (node 3d), the East Court (node 4b), the western terrace of House J (node 4d), the Tower Court (node 4n), the Southeast Court (node 5n), and the northern terrace shared by Houses G and H (node 5p). Conversely, spaces that are characteristically nondistributed include, among others, the West Court (node 7d) and the western gallery of House C (node 6i), as well as spaces that comprise the interior of the Tower

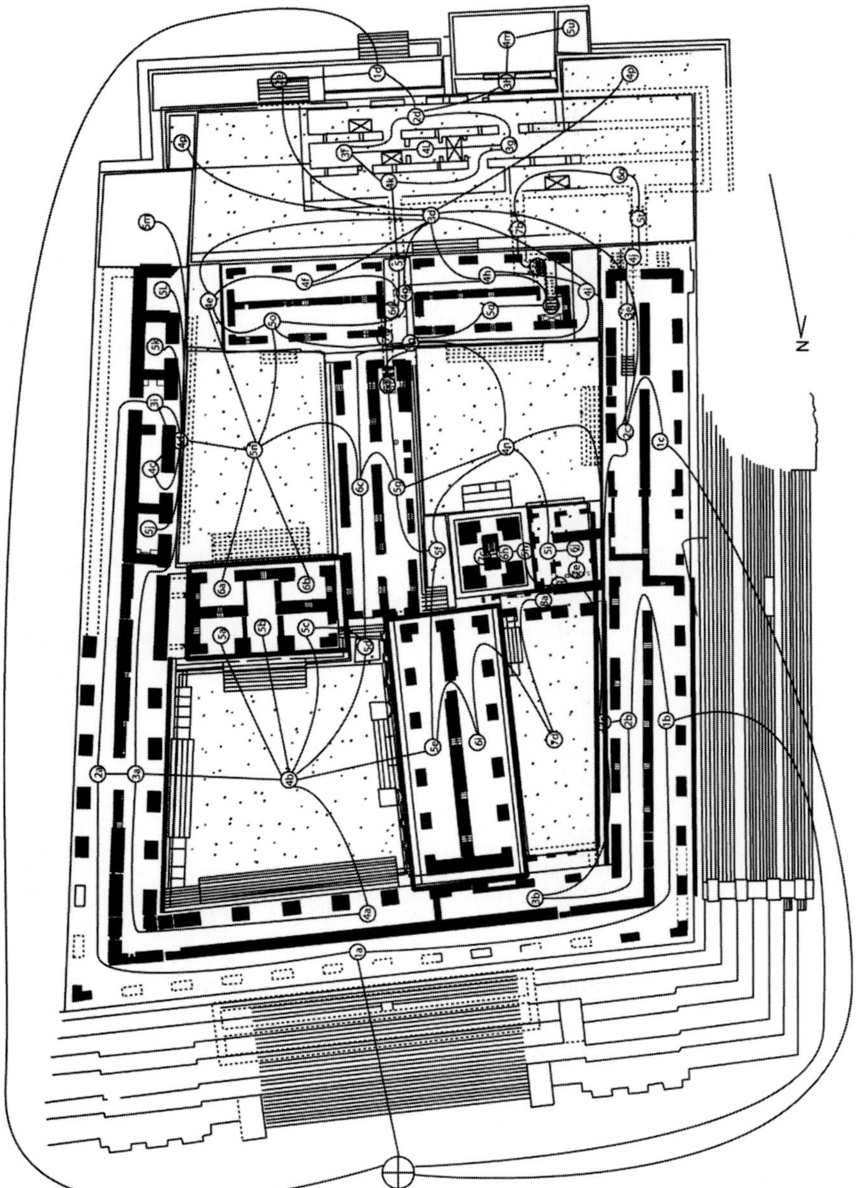

**Figure 5.7.** Palenque Palace Phase 4, Combined Convex Map and Access Diagram. Palenque Palace modified after Greene Robertson (1985b: fig. 1).

(nodes 5h, 6h, and 7c), the Toilets (nodes 5i, 6j, and 7e), and the southern subterranean substructure and passages (nodes 4j, 5s, 5t, 6d, 6e, 6f, 6g, 7a, and 7b).

As a result of K'inich Akul Mo'Nab' III's architectural contributions, Palace Phase 4 comprises sixteen more spatial units than the previous architectural phase, or more specifically, seventy-three units of space to Phase 3's total of fifty-seven. In addition, due to the closing of the Palace's western entrance, Phase 4's access diagram displays an increase in its overall depth when compared to the previous phase; whereas Phase 4 comprises eight levels of depth, Phase 3 shows six levels. The justified access diagram for Palace Phase 4 shows four spaces at Level 1, five spaces at Levels 2 and 7, nine spaces at Level 3, seventeen spaces at Level 4, twenty-one spaces at Level 5, ten spaces at Level 6, and one space at Level 8. Among the seventy-three spaces represented, twenty-two spaces are transitional; that is to say, their principal function was to provide access between other spaces in the system.

Thirteen structures comprise Phase 4 of the Palenque Palace, compared to a total of ten at Palace Phase 3. When comparing the averaged integration values for both phases, there is a notable shift in the relative depths of the existing structures at Phase 4. For example, with the addition of Houses G and H, as well as the Toilets and Tower, House E reappears as one of the more integrated buildings of the Palace Group. In contrast, House A-D drops from being the most integrated structure at Palace Phase 3 to one that holds a more intermediate position in Palace Phase 4's order of ranking (see Table 5.19). On average, the most integrated building at Phase 4 of the Palace (and subsequently the most accessible from the perspective of Palace inhabitants) is House G, whereas the least integrated existing structures remain Houses B and J as well as the southern substructures. The most segregated structures appearing at Palace Phase 4 are the Tower and Toilets. Forming the boundary between the West Court and Tower Court, it is highly likely that the Tower and Toilets were constructed simultaneously, as accessing the interior of the Tower necessitated crossing the roof of the Toilets and entering via a doorway on the western side of the structure (see Andrews 1974).

Architectural embellishment was not of primary importance to K'inich Akul Mo'Nab' III at Palace Phase 4, as relatively few sculptural additions can be attributed to his reign. In addition, Houses G and H (structures commissioned by K'inich Akul Mo'Nab' III) show no evidence of ever being decorated. The few spaces in the Palace where sculpture and other artwork do appear are the central area and northeastern corner of the Tower Court (nodes 4n and 5f), the second floor of the Tower

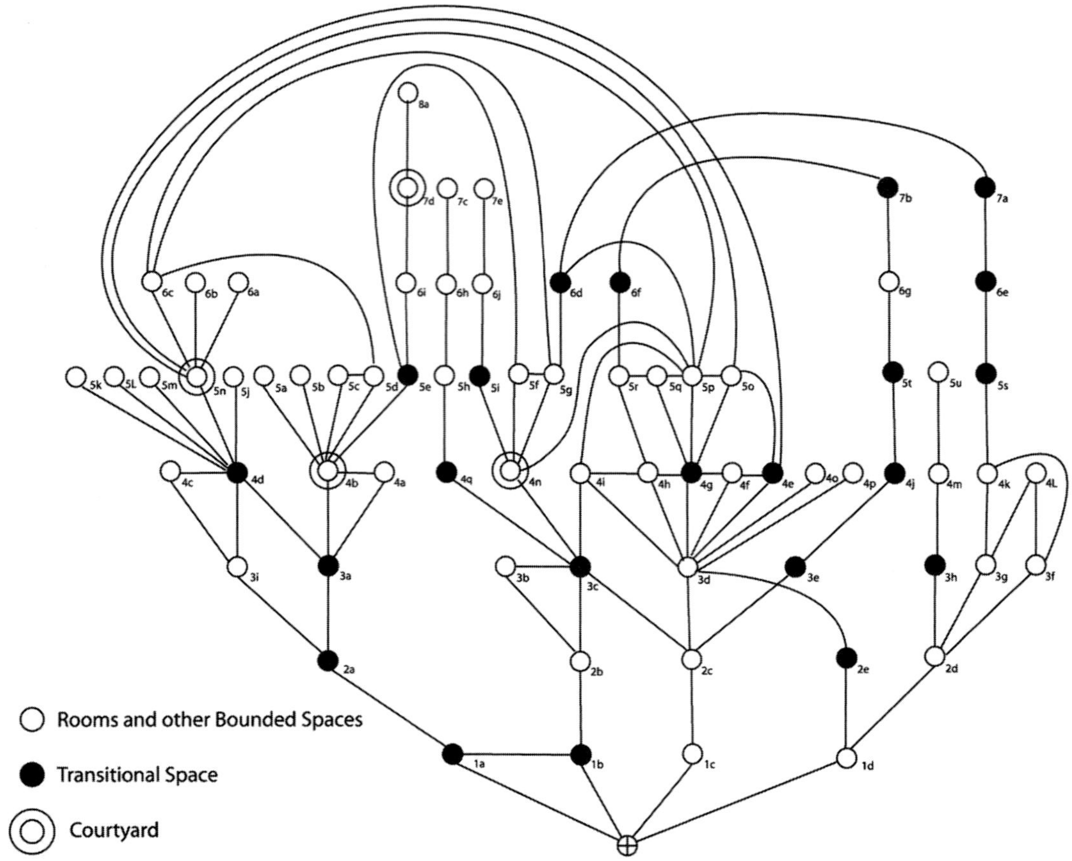


**Figure 5.8.** Palenque Palace Phase 4, "Justified" Access Diagram.

(node 6n), and the West Court of the Palace (node 7d). The least integrated and, subsequently, the most segregated of the Phase 4 Palace artworks is the Venus sign (GR 72) painted on the interior southern wall of the Tower, on the second level of the building. In this investigation,

Table 5.19. *Average RA values for Palace Phase 4 houses and other structures*

| Structure | Average RA values | Rank |
|---|---|---|
| House G | 0.0849 | 1 |
| House E | 0.0873 | 2 |
| House K | 0.0876 | 3 |
| House H | 0.0940 | 4 |
| House D | 0.0964 | 5 |
| House A | 0.1019 | 6 |
| House A-D | 0.1060 | 7 |
| House C | 0.1140 | 8 |
| House B | 0.1190 | 9 |
| House J | 0.1201 | 10 |
| Toilet | 0.1336 | 11 |
| South Substructure | 0.1501 | 12 |
| Tower | 0.1602 | 13 |

*Notes:* Ranked from most integrated to least integrated.

I argue that the principal theme of the Venus sign of the Tower is Cosmological Celestial. Referred to as *Noh Ek'* ("great star") or *Xux Ek'* ("wasp star") (Fuson 1969: 502), Venus was an important celestial object among the ancient Maya, as both Venus and the Sun were believed to be manifestations of the Hero Twins as described in the creation mythology of the *Popol Wuj* (Schele and Freidel 1990: 115; Sharer 1994: 288). Also at a relatively low level of integration is the theme Ritual Dance, as represented by five stucco figures decorating the three pillars that define the southern boundary of the West Court (GR 67). The portrayal of dancers in the West Court may be viewed as evidence of the type of activity once undertaken within the confines of the space (see Chapter 4, Section 4.3.4, The Buildings of the Palace Phase 4).

Palace Phase 4 sculpture appearing at the highest level of integration are those that occupy the Tower Court (i.e., GR 71, 74, and 75) (Table 5.20). Once flanking both sides of a small staircase on the south side of the Tower were two bas-relief panels referred to as the Tablet of the Orator (GR 74; Figure 4.55) and the Tablet of the Scribe (GR 75; Figure 4.56). Both tablets portray images thematically categorised as Captive Display, which were

Table 5.20. *Palenque Palace Phase 4 sculpture ranked according to RA values, K'inich Akul Mo'Nab' III* (A.D. *721–740?*)

| Node | RA value | Rank | Sculpture GR No. | Theme |
|---|---|---|---|---|
| 6h | 0.1600 | 1 | **72** | **CosC** |
| 5t | 0.1569 | 2 | 79 | CosU |
| 7d | 0.1557 | 3 | 63, 64, 65, 66, **67**, 68,69 | **RitDan?**, GPWA, CosC, TCapSac |
| 6f | 0.1303 | 4 | 78 | CosU |
| 5a | 0.1279 | 5 | 24, 25, 26, 27 | RPSolo, MMot, CosU, CosCU |
| 5b | 0.1279 | 5 | 24, 25, 26, 27 | RPSolo, MMot, CosU, CosCU |
| 6i | 0.1268 | 6 | 30, 31, 32, 33, 34, 35, 36, 37, 39, 40, 41 | RAud, APSolo, PSolo, CosT |
| 2d | 0.1193 | 7 | 80 (Bench) | TCosCU |
| 5c | 0.1174 | 8 | 24, 25, 26, 27 | RPSolo, MMot, CosU, CosCU |
| 7a | 0.1146 | 9 | 7 | CosU |
| 6a | 0.1111 | 10 | 23, 26, 27, 28 | CosT?, MMot, CosU, CosCU |
| 6b | 0.1111 | 10 | 16, 17, 18, 19, 20, 21, 22, 26, 27, 28 | RAud, Text?, PSolo, RitDan, MMot, CosU |
| 2a | 0.1041 | 11 | 42, 43, 44, 45, 46, 47 | THist, APWA, APSolo |
| 4b | 0.1013 | 12 | 58, 59, 60, 61 (Hieroglyphic Stairs) | CapDis, THist |
| 5e | 0.1013 | 12 | 29, 38, 40, 41, 62 | GPSolo, CosT, TDed? |
| 1b | 0.0986 | 13 | 48, 49, 50, 51, 52, 53 | HumSac, SacDan, Text? |
| 1a | 0.0982 | 14 | 55, 56, 57 (Bench) | RPSolo, RAudAccTDynHist |
| 5f | 0.0908 | 16 | **71** | **CapDis** |
| 5g | 0.0872 | 17 | 3, (Bench?), 4, 5, 6 (Del Río Throne), 11, 13 | RAud, Acc, CosTU, THist, TDynHist |
| 6c | 0.0829 | 18 | 8 | CosCU |
| 4n | 0.0794 | 19 | **74, 75** | **CapDis** |

*Notes:* Ranked from least integrated to most integrated. See Appendix 5 for building and location information for nodes.

probably commissioned during the reign of K'inich Akul Mo'Nab' III (see Chapter 4, Section 4.3.4, The Buildings of the Palace Phase 4, The Tower and Toilets: Description and Their Sculpture). In the northeastern corner of the Tower Court, portrayed at the base of the Tower's eastern wall, is an additional captive scene (GR 71; see Chapter 4, Section 4.3.4, The Buildings of the Palace Phase 4, The East Wall Tower Figures: Description). As with the Tablets of the Scribe and the Orator, the absence of clothing and authoritative emblems (in addition to their subordinate postures and the presence of cloth pulled through the perforations in their ears) marks six of the seven eastern tower figures as captives. All of the captives are portrayed seated, in frontal orientation, and in subordinate pose, with their heads turned to the viewer's left of the composition. A seventh figure on the viewer's right appears to be dancing while facing the direction of the western entrance of House E. Again, the appearances of capture themes in the Tower Court directly adjacent to House E's throne may be viewed as examples of the sorts of activities once undertaken in the presence of the ruler of the day. For example, the Tower Court may have been used to receive, parade, and/or humiliate captives in the presence of the enthroned king prior to their sacrificial disposal.

Due to the architectural changes implemented at Palace Phase 4, there is a notable shift in the relative depths of the deeper sculpture when compared to Phase 3, suggesting that the exclusivity of the earlier sculptural features of the Palace may have diminished over time. For example, several of the subterranean vaults that mark the function of the underground passages as symbolic conduits to the Underworld (Schele 1986: 122) appear at a higher level of integration relative to other Palace sculpture. From the perspective of all other spaces in the complex, this increase in integration implies that the vaults were becoming more and more accessible relative to other artworks. At Palace Phase 4, the East Subterranean Passage vault ranks at a lower level of integration than the sculpture of the western gallery of House C, the northern façade of House B, as well as the West Court. Among the sculptural themes appearing on these structures are Ancestor Portraiture (see Chapter 4, Section 4.3.2, The Buildings of the Palace Phase 2, House C: Description and Sculpture), God Portraiture with Attendants, Cosmological Celestial (as seen in the West Court), Mat Motif, and Royal Portraiture (see Chapter 4, Section 4.3.1, The Buildings of the Palace Phase 1, House B: Description and Sculpture). At Palace Phase 4, the South Subterranean Passage vault was also calculated as being more integrated with respect to the sculpture of the West Court.

As with the Palace buildings at Palace Phase 4, we also see a shift in the ranking of the Palace courtyards depending on whether they are ordered according to corresponding RA or control values or to their relative depth from

Table 5.21. *Palace Phase 4 court carrier depths, RA values, and control values*

| Palace courts | Node | Carrier rank | RA value | Control value |
|---|---|---|---|---|
| East Court | 4b | 4 | 0.1013 | 3.9166 |
| Tower Court | 4n | 4 | 0.0794 | 1.3928 |
| Southeast Court | 5n | 5 | 0.0841 | 2.8750 |
| West Court | 7d | 7 | 0.1557 | 1.5000 |

the carrier position. At Phase 4, the West Court replaces the Southeast Court as the most restricted court space from the perspective of both visitors and Palace inhabitants, as this courtyard is furthest from the carrier position and displays the highest RA value (see Table 5.21). From the perspective of the visitor (i.e., someone wishing to enter the spatial system from the carrier position), the East Court (node 4b) and the Tower Court (node 4n) both appear at a depth of Level 4, followed by the Southeast Court (node 5g), which is found at a depth of Level 5. Positioned at a depth of Level 7, the West Court (node 7d) represents the most restricted court space of the Palace at Phase 4, as access to this space is provided exclusively from the western gallery of House C. Consistent with the Palace Phase 3 analysis, the East Court maintains the highest level of control over localised access to neighbouring spaces, followed by the Southeast Court and West Court of the Palace. As indicated by its corresponding control value, at Palace Phase 4 the Tower Court displays the lowest level of control of all four of the court spaces.

Of all the seventy-three spaces that collectively represent Palace Phase 4, twenty-seven were calculated to be controlling spaces whereas the remaining forty-six spaces (excluding the carrier position) were assessed as being controlled spaces. Spatial units that display the greatest level of control over localised access to other spaces are as follows: House J's western terrace (node 4d); the upper southeastern terrace of the Palace (node 3d); the East Court (node 4b); the Southeast Court (node 5n); the pathway fronting Houses D and K (node 3c); the northern terrace of Houses G and H (node 5p); the south substructure patio (node 4m); the second level of the Tower (node 6h); the West Court (node 7d); the southern corridor of the Subterraneos (node 2d); the corridor between Houses G and H (node 4g); and the Tower Court (node 4n). Of all the spatial units represented at Palace Phase 4, the second floor of the Tower and the Tower Court are the only two controlling spaces that display artwork (see Table 5.22). The relatively low level of control that the Tower Court exhibits, however, relates primarily to the inability to access the Palace Toilets (nodes 5i, 6j, and 7e) without first passing through the courtyard. At Palace Phase 4, the only controlled space that displays artwork

Table 5.22. *Palenque Palace Phase 4 sculpture ranked according to control values, K'inich Akul Mo'Nab' III* (A.D. *721–740?)*

| Node | Control value | Rank | Sculpture GR No. | Theme |
|---|---|---|---|---|
| 4b | 3.9166 | 1 | 58, 59, 60, 61 (Hieroglyphic Stairs) | CapDis, THist |
| 7d | 1.5000 | 2 | 63, 64, 65, 66, **67**, 68,69 | **RitDan?**, GPWA, CosC, TCapSac |
| 6h | 1.5000 | 2 | **72** | **CosC** |
| 2d | 1.4999 | 3 | 80 (Bench) | TCosCU |
| 4n | 1.3928 | 4 | **74, 75** | **CapDis** |
| 5g | 1.1166 | 5 | 3 (Bench?), 4, 5, 6 (Del Río Throne), 11, 13 | RAud, Acc, CosTU, THist, TDynHist |
| 5t | 1.0000 | 6 | 79 | CosU |
| 5e | 0.9761 | X | 29, 38, 40, 41, 62 | GPSolo, CosT, TDed? |
| 1a | 0.9166 | X | 55, 56, 57 (Bench) | RPSolo, RAudAcc, TDynHist |
| 1b | 0.9166 | X | 48, 49, 50, 51, 52, 53 | HumSac, SacDan, Text? |
| 2a | 0.9166 | X | 42, 43, 44, 45, 46, 47 | THist, APWA, APSolo |
| 6c | 0.8928 | X | 8 | CosCU |
| 7a | 0.8333 | X | 7 | CosU |
| 6i | 0.8333 | X | 30, 31, 32, 33, 34, 35, 36, 37, 39, 40, 41 | RAud, APSolo, PSolo, CosT |
| 6f | 0.8333 | X | 78 | CosU |
| 5f | 0.7833 | X | **71** | **CapDis** |
| 5c | 0.4761 | X | 24, 25, 26, 27 | RPSolo, MMot, CosU, CosCU |
| 6a | 0.1666 | X | 23, 26, 27, 28 | CosT?, MMot, CosU, CosCU |
| 6b | 0.1666 | X | 16, 17, 18, 19, 20, 21, 22, 26, 27, 28 | RAud, Text?, PSolo, RitDan, MMot, CosU |
| 5a | 0.1428 | X | 24, 25, 26, 27 | RPSolo, MMot, CosU, CosCU |
| 5b | 0.1428 | X | 24, 25, 26, 27 | RPSolo, MMot, CosU, CosCU |

*Note:* See Appendix 6 for building and location information for nodes.

Table 5.23. *Palenque Palace Phase 4 sculpture ranked according to depth from the carrier, K'inich Akul Mo'Nab' III* (A.D. *721–740?)*

| Node | Carrier R | Sculpture GR No. | Theme |
|---|---|---|---|
| C | 0 | 1, 2, 82, 83, 84, 85 | CosC, RPSoloCosCU, RPSolo |
| 1a | 1 | 55, 56, 57 (Bench) | RPSolo, RAudAccTDynHist |
| 1b | 1 | 48, 49, 50, 51, 52, 53 | HumSac, SacDan, Text? |
| 2a | 2 | 42, 43, 44, 45, 46, 47 | THist, APWA, APSolo |
| 2d | 2 | 80 (Bench) | TCosCU |
| 4b | 4 | 58, 59, 60, 61 (Hieroglyphic Stairs) | CapDis, THist |
| 4n | 4 | **74, 75** | **CapDis** |
| 5a | 5 | 24, 25, 26, 27 | RPSolo, MMot, CosU, CosCU |
| 5b | 5 | 24, 25, 26, 27 | RPSolo, MMot, CosU, CosCU |
| 5c | 5 | 24, 25, 26, 27 | RPSolo, MMot, CosU, CosCU |
| 5e | 5 | 29, 38, 40, 41, 62 | GPSolo, CosT, TDed? |
| 5f | 5 | **71** | **CapDis** |
| 5g | 5 | 3 (Bench?), 4, 5, 6 (Del Río Throne), 11, 13 | RAud, Acc, CosTU, THist, TDynHist |
| 5t | 5 | 79 | CosU |
| 6a | 6 | 23, 26, 27, 28 | CosT?, MMot, CosU, CosCU |
| 6b | 6 | 16, 17, 18, 19, 20, 21, 22, 26, 27, 28 | RAud, Text?, PSolo, RitDan, MMot, CosU |
| 6c | 6 | 8 | CosCU |
| 6f | 6 | 78 | CosU |
| 6h | 6 | **72** | **CosC** |
| 6i | 6 | 30, 31, 32, 33, 34, 35, 36, 37, 39, 40, 41 | RAud, APSolo, PSolo, CosT |
| 7a | 7 | 7 | CosU |
| 7d | 7 | 63, 64, 65, 66, **67**, 68,69 | **RitDan?**, GPWA, CosC, TCapSac |

*Note:* See Appendix 4 for building and location information for nodes.

(GR 71) is the transitional space located in the northeastern corner of the Tower Court (node 5f; see Chapter 4, Section 4.3.4, The Buildings of the Palace Phase 4, The East Wall Tower Figures: Description).

The architectural modifications implemented by K'inich Akul Mo'Nab' III, resulted in no change in the distribution of existing sculpture (as it correlates with Palace Phase 4's privacy gradient) until a depth of Level 5 from the carrier position (see Table 5.23). More specifically, from the perspective of the visitor, changes in the ranking of the earlier Palace sculpture occur at the deeper levels of the Palace complex. Most notably, the subterranean vaults and their corresponding Underworld themes (GR 7, 78, and 79; see Figures 4.7 and 4.8), which once occupied the deepest areas of the Palace complex (see Chapter 4, Section 4.3.1, The Buildings of the Palace Phase 1, and Section 4.3.2, The Buildings of the Palace Phase 2), now hold shallower positions when compared to the Ancestor Portraiture of the western gallery of House C (GR 32, 34, 35, and 36; Figures 4.20, 4.21, and 4.22), the Royal Audience and Cosmological Terrestrial imagery of the House B sanctuaries (GR 16, 17, 18, 19, 20, and 23; Figure 4.47), as well as the Ritual Dance, God Portraiture with Attendants, and Cosmological Celestial themes of the West Court (GR 63, 64, 65, 67, and 69; see Chapter 4, Section 4.3.3, The Buildings of the Palace Phase 3, House D East Foundation *Casita* Wall: Description). As at Palace Phase 3, the sculpture appearing at a depth of Level 1 at Palace Phase 4 includes the Palace Tablet (GR 57; Figure 4.39), a Dynastic Historical text depicting the accession ceremony of K'inich K'an Joy Chitam II, as well as the exterior piers of House D (GR 48, 49, 50, 51, and 52; Figures 4.42, 4.43, 4.44, and 4.45), which display themes such as Human Sacrifice and Sacrificial Dance. At Level 2 are the ancestor portraits sculptured on the piers of the eastern gallery of House A (GR 43, 44, 45, and 46; Figures 4.32, 4.33, 4.34, and 4.35), followed by the Capture Display themes portrayed at Level 4 in the East Court (GR 58, 59, and 60; Figures 4.46, and 4.28; see also Chapter 4, Section 4.3.2, The Buildings of the Palace Phase 2, House C, East Foundation *Alfardas*). Appearing at the same depth as the East Court monuments, at Palace Phase 4, is the Tablet of the Orator (GR 74) and the Tablet of the Scribe (GR 75; Figures 4.55 and 4.56), both of which also portray Captive Display themes.

The shallowest sculpture appearing at Phase 4 of the Palace are the Tablet of the Scribe and the Tablet of the Orator, which were arguably placed in the Tower Court (node 4n) by K'inich Akul Mo'Nab' III as a form of self-assertion. The two individuals depicted on the tablets were probably trophies of war, captured during times

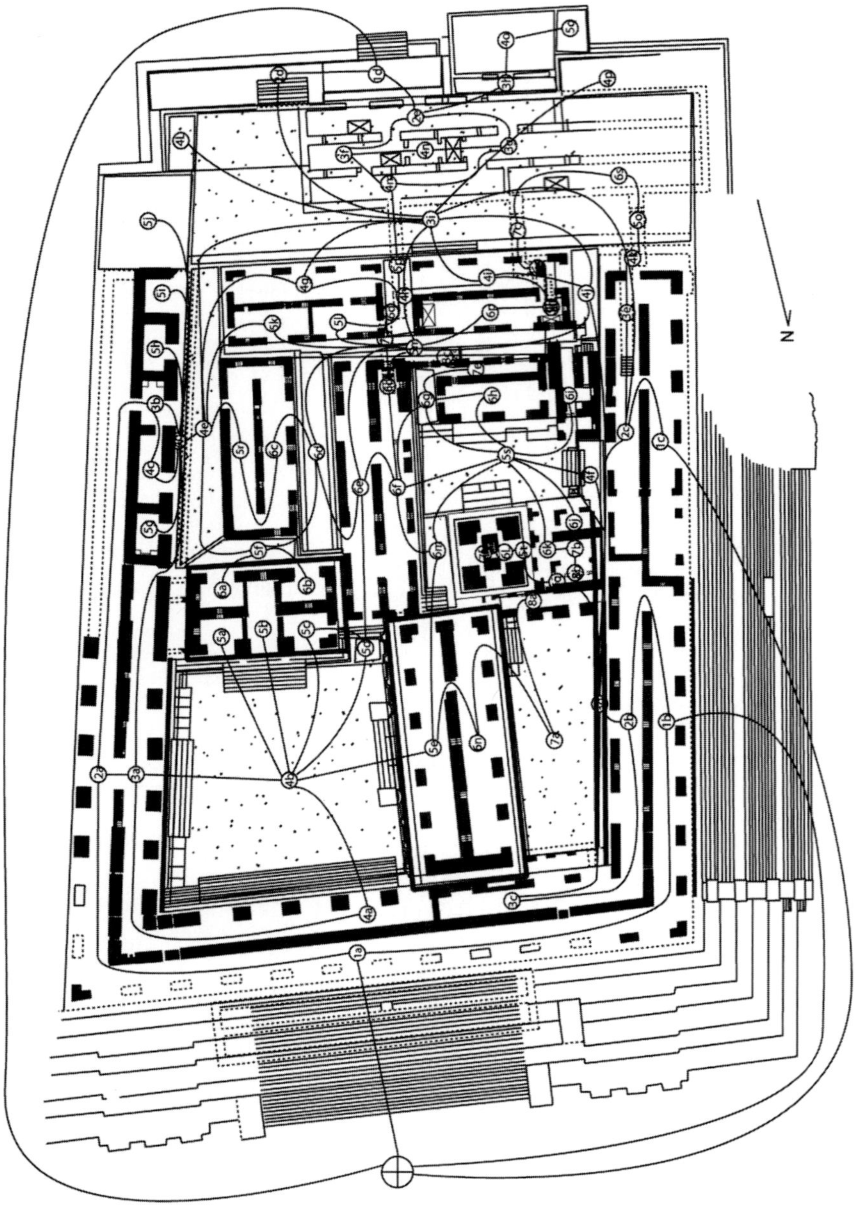


**Figure 5.9.** Palenque Palace Phase 5, Combined Convex Map and Access Diagram. Palenque Palace modified after Greene Robertson (1985b: fig. 1).

of conflict between Palenque and another neighbouring city (P. Mathews 2000: personal communication). At a slightly deeper level, still within the confines of the Tower Court, are the "east-tower captives" (GR 71), followed by the Venus painting (GR 72), which is confined to the second floor of the Tower at a depth of Level 6 from the carrier.

### 5.1.5. Analysis of the Palace Phase 5

Palenque Palace Phase 5 is characterised by buildings assigned to the interior of the Palace complex rather than to the peripheral extension of the building group (see Figure 5.9). The main architectural contribution appearing at Phase 5 was the construction of Houses F and I, in addition to that of the Palace Sweat Baths. In this study, I argue that these structures were built sometime during the reigns of U Pakal K'inich Janab' Pakal II, K'inich Kan B'alam III, and K'inich K'uk' B'alam II between A.D. 742 and A.D. 799. Although no precise date can be allocated to the Phase 5 additions, it is likely that their appearance postdated the construction of the Tower and Toilets as well as that of Houses G and H, and predated the intensive partitioning that occurred at Palace Phase 6 (see Chapter 4, Section 4.3.6, The Buildings of the Palace Phase 6). House F, the remains of which are confined to the area of the Southeast Court, included a supporting platform that was intended to raise the building to a height similar to that of the surrounding structures. House I, which is located in the southern half of the Tower Court, was also erected on a low platform elevating the building to a lesser extent. Behind House I are the Palace Sweat Baths where Palenque rulers are believed to have participated in purification rituals (Schele 1986: 122).

Supplementary construction inside rather than outside the confines of Palace Phase 4 was probably undertaken to maintain certain levels of privacy and protection for a growing population of elites once occupying the Palace complex. An alternative reason for the proximity of the Phase 5 architectural additions, however, may have been to prevent any overburdening of the Subterraneos with additional architecture that may have resulted in their collapse. (Note: It is clear that the Palace's southern terrace [node 3i] would have been the only space where additional buildings could have been erected to further extend the Palace complex at the level of the Second Palace Terrace Platform.) Due to the confinement of Houses F and I within the Southeast Court and the Tower Court of the Palace, both court spaces became substantially smaller, an indication perhaps that social encounters and activities occurring within these spaces were becoming increasingly intimate and private, as both court spaces

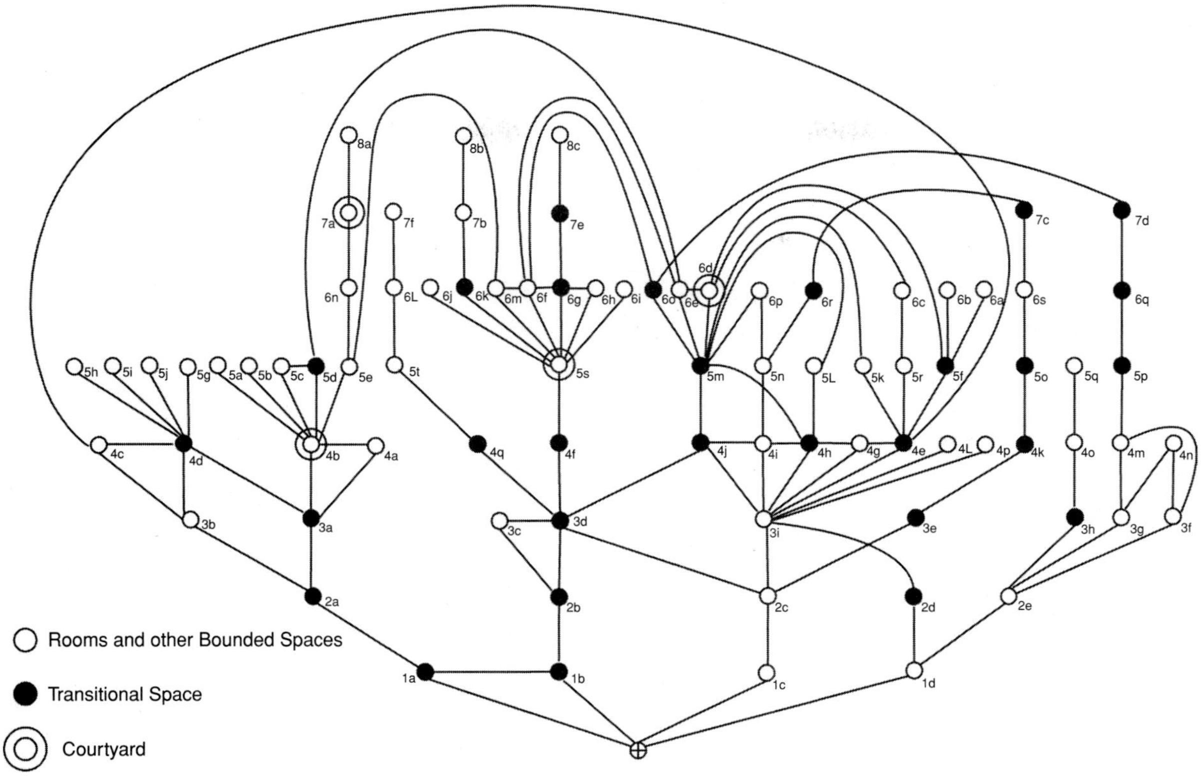


**Figure 5.10.** Palenque Palace Phase 5, "Justified" Access Diagram.

would have held fewer individuals in closer proximity to one another.

Fifteen buildings collectively comprise the Palace at Phase 5 compared to the thirteen structures that make up Palace Phase 4. Due to structural changes implemented at Palace Phase 5, the corresponding access diagram displays eighty-four nodes (including the carrier) compared to seventy-three as represented in the previous architectural phase, which is an overall increase of eleven spatial units. From the perspective of the carrier position, Palace Phase 5 shows four spaces at Level 1, five spaces at Level 2, nine spaces at Level 3, seventeen spaces at Level 4, twenty spaces at Level 5, nineteen spaces at Level 6, six spaces at Level 7, and three spaces at Level 8 (see Figure 5.10). Among the eighty-four spaces represented at Palace Phase 5, twenty-nine are transitional, functioning to afford access between other spaces in the system. At Palace Phase 5, there is no apparent increase in the overall depth when compared to Palace Phase 4, but the corresponding access diagram does show two additional spaces at the depth of Level 8, whereas Phase 4 displays only one. With the structural changes implemented at this stage of the evolution of the southern room of the West Court (node 8a), the northern room of the Toilets (node 8b) and the Palace Sweat Baths (node 8c) were the deepest spaces of the complex.

On ranking the Palace Phase 5 structures from most integrated to least integrated (Table 5.24), and comparing their order with Palace Phase 4, the relative positioning of existing Palace structures remains essentially unchanged. One minor variation between the two phases is in the positioning of House J: At Palace Phase 5, the structure becomes more integrated with respect to House C. With the structural modifications implemented at the Palace Phase 5, House I replaces House J's earlier position between House B and the Toilets, whilst the RA

Table 5.24. *Average RA values for Palace Phase 5 houses and other structures*

| Structure | Average RA values | Rank |
|---|---|---|
| House E | 0.0760 | 1 |
| House G | 0.0788 | 2 |
| House K | 0.0820 | 3 |
| House H | 0.0846 | 4 |
| House D | 0.0875 | 5 |
| House A | 0.0889 | 6 |
| House F | 0.0921 | 7 |
| House A-D | 0.0936 | 8 |
| House J | 0.1015 | 9 |
| House C | 0.1039 | 0 |
| House B | 0.1066 | 11 |
| House I | 0.1101 | 12 |
| Toilet | 0.1224 | 13 |
| South Substructure | 0.1318 | 14 |
| Tower | 0.1414 | 15 |

*Note:* Ranked from most integrated to least integrated.

Table 5.25. *Palenque Palace Phase 5 sculpture ranked according to RA values, U Pakal K'inich Janab' Pakal II, K'inich Kan B'alam III, and K'inich K'uk' B'alam II (circa* A.D. *742 to 799)*

| Node | RA value | Rank | Sculpture GR No. | Theme |
|---|---|---|---|---|
| 8a | 0.1628 | 1 | 66 | GPWA |
| 6L | 0.1411 | 2 | 72 | CosC |
| 7a | 0.1387 | 3 | 63, 64, 65, 67, 68 | RitDan?, GPWA, CosC, TCapSac |
| 5o | 0.1364 | 4 | 79 | CosU |
| 7b | 0.1278 | 5 | 69 | CosC |
| 6n | 0.1152 | 6 | 30, 31, 32, 33, 34, 35, 36, 37, 39, 40, 41 | RAud, APSolo, PSolo, CosT |
| 6r | 0.1128 | 7 | 78 | CosU |
| 5a | 0.1108 | 8 | 24, 25, 26, 27 | RPSolo, MMot, CosU, CosCU |
| 5b | 0.1108 | 8 | 24, 25, 26, 27 | RPSolo, MMot, CosU, CosCU |
| 2e | 0.1096 | 9 | 80 (Bench) | TCosCU |
| 6a | 0.1049 | 10 | 23, 26, 27, 28 | CosT?, MMot, CosU, CosCU |
| 6b | 0.1049 | 10 | 16, 17, 18, 19, 20, 21, 22, 26, 27, 28 | RAud, Text?, PSolo, RitDan, MMot, CosU |
| 6h | 0.1049 | 10 | **76, 77** | **Cos?, RPSolo** |
| 5c | 0.1020 | 11 | 24, 25, 26, 27 | RPSolo, MMot, CosU, CosCU |
| 2a | 0.0946 | 12 | 42, 43, 44, 45, 46, 47 | THist, APWA, APSolo |
| 7d | 0.0937 | 13 | 7 | CosU |
| 5e | 0.0926 | 14 | 29, 38, 40, 41, 62 | GPSolo, CosT, TDed? |
| 1a | 0.0923 | 15 | 55, 56, 57 (Bench) | RPSolo, RAudAccTDynHist |
| 1b | 0.0911 | 16 | 48, 49, 50, 51, 52, 53 | HumSac, SacDan, Text? |
| 6m | 0.0864 | 17 | 71 | CapDis |
| 4b | 0.0861 | 18 | 58, 59, 60, 61 (Hieroglyphic Stairs) | CapDis, THist |
| 5s | 0.0802 | 19 | **70, 73** (Tablet 96 Glyphs), 74, 75 | **?(Stucco fret), TDynHist,** CapDis |
| 6f | 0.0802 | 19 | 3 (Bench?), 4, 5, 6 (Del Río Throne), 11, 13 | RAud, Acc, CosTU, THist, TDynHist |
| 6e | 0.0726 | 20 | 8 | CosCU |

*Notes:* Ranked from least integrated to most integrated. See Appendix 5 for building and location information for nodes.

value calculated for House F places it between Houses A and A-D in the order of ranking. The least integrated and, subsequently, more inaccessible buildings from the perspective of persons occupying all other structures are House I, the Toilets, the south substructure, and the Tower, whereas Houses G, E, and K remain the most integrated (accessible) of the Palace structures.

The Palace Phase 5 access diagram displays eighty-four nodes and one hundred fifteen edges, indicating a further increase overall in the number of access routes between spatial units. Consequently, as with the four Palace Phases previously discussed, the spatial configuration of Palace Phase 5 is also highly distributed. Some of the more distributed and, subsequently, more integrated spaces represented at Phase 5 include the Palace's southern terrace (node 3i), the East Court (node 4b), the eastern (node 4e) and western (node 6d) sides of the Southeast Court, the western terrace of House J (node 4d), the northern terrace shared by Houses G and H (node 5m), and the Tower Court (node 5s). In contrast, some of the nondistributed spaces represented are the West Court (node 7a), the western gallery of House C (node 6n), and the eastern room of House F (node 5r). Other nondistributed spaces include many of those that collectively comprise the interior of the Tower (nodes 5t, 6l and 7f), the Toilets (nodes 6k, 7b, and 8b), House J (nodes 5g, 5h, and 5i), House B (nodes 5a, 5b, and 5c), House I (nodes 6h and 7e), as well as the Sweat Baths (node 8c). As noted in the discussion of the earlier Palace phases, many of the spaces that make up southern subterranean substructures and passages are also nondistributed (see nodes 3h, 3e, 4k, 4h, 5o, 5p, 6q, 6r, 6s, 7c, and 7d).

The few spaces where sculpture and other artwork do appear at Palace Phase 5 are in the area of the Tower Court; among these sculptural embellishments are the remains of a standing portrait on the northwestern wall of House I (GR 77) and a serpent framing the interior of the eastern door of the same structure (GR 76; see Chapter 4, Section 4.3.5, The Buildings of the Palace Phase 5, House I Sculptured Figure and Serpent Door: Description). In addition to the sculpture present on House I are the remains of a thematically indistinct fret pattern (GR 70) located on the southern side of the western Tower Court stair and an all-glyphic tablet referred to as the Tablet of the 96 Glyphs (GR 73; Figure 4.60). The Tablet of the 96 Glyphs is a beautifully incised monument that was once positioned on the lowest riser of the southern stair of the Tower. The tablet probably functioned as the top of a throne used in the area of the Tower Court (P. Mathews 2004: personal communication).

The decision to assign additional sculpture at relatively high levels of integration (Table 5.25) implies a concern for addressing Palace inhabitants, and presumably those more important people invited to enter the deeper areas of the complex. Of the few sculptures identified at Palace Phase 5, the monument located at the highest level of integration is the Tablet of the 96 Glyphs. As outlined in Chapter 4, the sculptural theme of this monument has been defined as Dynastic Historical text. The monument was commissioned by K'inich K'uk' B'alam II to celebrate his first *k'atun* ending, which occurred in A.D. 783 (Martin and Grube 2000: 174). As well as recording the dedicatory date of House E (9.11.2.1.11) under the auspices of K'inich Janab' Pakal I, the inscription also documents the accessions of K'inich K'an Joy Chitam II, K'inich Akul Mo'Nab' III, and K'inich K'uk' B'alam II. Located at more of an intermediate level of integration are the two sculptures associated with House I, defined tentatively in this study as Royal Portraiture Solo (GR 77) and Cosmological ? (GR 76; see Chapter 4, Section 4.3.5, The Buildings of the Palace Phase 5, House I Sculptured Figure (GR 77) and Serpent Door: Description). Framing the interior eastern entrance of House I is a sculpture of a boa constrictor, as indicated by the scale pattern tracing the top right corner of the entrance. On the exterior of the same structure, the figure of a king or noble stands in profile and in static pose facing in towards the northeastern entrance of House I. Probably once mirroring a similar figure at the opposite end of the structure, the surviving sculpture suggests that the individual's costume was relatively understated, comprising a plain hipcloth, a cross-banded belt, and a wrap of cloth that appears knotted at the back.

As at Palace Phase 4, there is a notable shift in the order of the Palace courtyards at Palace Phase 5, depending on whether they are ranked according to corresponding carrier depths, control values, or RA calculations. In terms of their respective depths from the carrier position, the only change in the ordering of the Palace courtyards at Phase 5 relates to the Tower Court. Once displaying the same level of depth as the East Court (see Chapter 4, Section 4.3.4, The Buildings of the Palace Phase 4), the Tower Court is now deeper than the East Court by one level of depth. From the perspective of the carrier, the shallowest and, subsequently, the most accessible of the Palace courtyards is the East Court (node 4b), followed by the Tower Court (node 5s), the Southeast Court (node 6d), and finally the West Court (node 7a). Most notably, at Phase 5 the West Court persists as the most restricted of the Palace courtyards from the perspective of both visitors and Palace inhabitants, as this courtyard remains the deepest from the carrier whilst displaying the highest RA value (see Table 5.26). In contrast, the Southeast Court replaces the Tower Court as the most integrated (i.e., accessible) court space from the perspective of all other Palace spaces. At Phase 5 of the Palace Group the Tower Court displays the highest level of control over localised access to neighbouring spaces, followed by the East Court, the West Court, and then the Southeast Court. The construction of House I within the confines of the Tower Court (in addition to the Sweat Baths located immediately behind the building) effectively increased the number of spaces to which the Tower Court controlled access, subsequently raising the overall level of control that the Tower Court possessed at Palace Phase 5.

Table 5.26. *Palace Phase 5 Court carrier depths, RA values, and control values*

| Palace courts | Node | Carrier rank | RA value | Control value |
|---|---|---|---|---|
| East Court | 4b | 4 | 0.0861 | 3.9999 |
| Tower Court | 5s | 5 | 0.0802 | 4.2833 |
| Southeast Court | 6d | 6 | 0.0764 | 1.1250 |
| West Court | 7a | 7 | 0.1387 | 1.5000 |

Of all the eighty-four spaces represented in the Palace Phase 5, thirty-one were calculated to be controlling spaces, whereas the remaining fifty-three spaces (excluding the carrier position) are controlled spaces. Spatial units that display the greatest level of control over localised access to surrounding spaces are House J's western terrace (node 4d), the Tower Court (node 5s), the East Court (node 4b), the southern terrace of the Palace (node 3i), the northern terrace of Houses G and H (node 5m), the Southeast Court (node 5f), the southern substructure patio (node 4o), the second level of the Tower (node 6L), the West Court (node 7a), and the west-central room of the Toilets (node 7b). Of all the controlling spaces represented at Palace Phase 5, the Tower Court (node 5s) is the only space that contains sculpture appearing at Palace Phase 5; the remaining artworks occupy controlled spaces (see Table 5.27). The only recognisable theme located in a controlling space at Palace Phase 5 is Dynastic Historical text, appearing on the Tablet of the 96 Glyphs (GR 73; Figure 4.60), whereas the themes confined to controlled spatial context are Royal Portraiture Solo (GR 77) appearing on the exterior of House I and Cosmological ? (GR 76) assigned to the interior of House I (see Chapter 4, Section 4.3.5, The Buildings of the Palace Phase 5, House I Sculptured Figure (GR 77) and Serpent Door: Description). The increased level of control that the Tower Court exhibits at Palace Phase 5 relates specifically to the additions of House I and the Sweat Baths, as gaining access to these localities first required entering the Tower Court.

Table 5.27. *Palenque Palace Phase 5 sculpture ranked according to control values, U Pakal K'inich Janab' Pakal II, K'inich Kan B'alam III, and K'inich K'uk' B'alam II (circa* A.D. *742 to 799)*

| Node | Control value | Rank | Sculpture GR No. | Theme |
|---|---|---|---|---|
| 5s | 4.2833 | 1 | **70, 73** (Tablet 96 Glyphs), 74, 75 | **?(Stucco fret), TDynHist,** CapDis |
| 4b | 3.9999 | 2 | 58, 59, 60, 61 (Hieroglyphic Stairs) | CapDis, THist |
| 7b | 1.5000 | 3 | 69 | CosC |
| 7a | 1.5000 | 3 | 63, 64, 65, 67, 68 | RitDan?, GPWA, CosC, TCapSac |
| 6L | 1.5000 | 3 | 72 | CosC |
| 2e | 1.4999 | 4 | 80 (Bench) | TCosCU |
| 6f | 1.2916 | 5 | 3 (Bench?), 4, 5, 6 (Del Río Throne), 11, 13 | RAud, Acc, CosTU, THist, TDynHist |
| 5o | 1.0000 | 6 | 79 | CosU |
| 2a | 0.9999 | X | 42, 43, 44, 45, 46, 47 | THist, APWA, APSolo |
| 5e | 0.9761 | X | 29, 38, 40, 41, 62 | GPSolo, CosT, TDed? |
| 1b | 0.9166 | X | 48, 49, 50, 51, 52, 53 | HumSac, SacDan, Text? |
| 1a | 0.9166 | X | 55, 56, 57 (Bench) | RPSolo, RAudAccTDynHist |
| 6e | 0.9083 | X | 8 | CosCU |
| 7d | 0.8333 | X | 7 | CosU |
| 6r | 0.8333 | X | 78 | CosU |
| 6n | 0.8333 | X | 30, 31, 32, 33, 34, 35, 36, 37, 39, 40, 41 | RAud, APSolo, PSolo, CosT |
| 6m | 0.6583 | X | 71 | CapDis |
| 8a | 0.5000 | X | 66 | GPWA |
| 5c | 0.4761 | X | 24, 25, 26, 27 | RPSolo, MMot, CosU, CosCU |
| 6h | 0.3750 | X | **76,** 77 | **Cos?, RPSolo** |
| 6a | 0.2500 | X | 23, 26, 27, 28 | CosT?, MMot, CosU, CosCU |
| 6b | 0.2500 | X | 16, 17, 18, 19, 20, 21, 22, 26, 27, 28 | RAud, Text?, PSolo, RitDan, MMot, CosU |
| 5a | 0.1428 | X | 24, 25, 26, 27 | RPSolo, MMot, CosU, CosCU |
| 5b | 0.1428 | X | 24, 25, 26, 27 | RPSolo, MMot, CosU, CosCU |

*Note:* See Appendix 6 for building and location information for nodes.

Consequently, the interiors of House I and the Sweat Baths are defined as controlled spaces.

As noted in the Phase 4 discussion, due to the architectural changes implemented at Palace Phase 5, there is a shift in the order of ranking of the deeper Palace sculpture when compared to the previous architectural phase. This shift suggests that preserving the privacy of preexisting sculpture may have become less important as the Palace complex became increasingly congested. Furthermore, the absence of any significant sculptural contribution at Palace Phase 5, excluding perhaps the Tablet of the 96 Glyphs (GR 73), suggests that architectural decoration was not of primary importance to those presiding over most of the later Palace renovations; for example, as with Houses G and H, House F shows no evidence of ever being decorated. When considering the three spatial units of the Palace that display sculpture, in terms of their relative depth from the carrier position, three of the four sculptural elements appearing at Palace Phase 5 are all visible from the Tower Court (node 5s) at a depth of Level 5 from the carrier (see Table 5.28). The only Phase 5 sculpture not visible from this court space is the serpent framing the interior east entrance to House I (GR 76), which corresponds to a carrier depth of Level 6 (node 6h).

## 5.1.6. Analysis of the Palace Phase 6

The Palace as it appears at Phase 6 (see Figure 5.11) was probably the result of structural modifications implemented during the reign of Wak Kimi Janab' Pakal III. Acceding to the throne in 9.18.9.4.4 (A.D. 799), Wak Kimi Janab' Pakal III was the last ruler of Palenque (Martin and Grube 2000: 175). His main architectural contribution to the Palace Group at Phase 6 was the bulk of the secondary partitioning that still divides many of the larger structures into smaller rooms and spaces. The intensive partitioning of the Palace Group that is apparent today probably occurred late in the history of Palenque just prior to the centre's collapse (Greene Robertson 1985a: 63–64). "There seem to originally have been few partitions in the open or corridor-like vaults of the exterior as well as most of the inner ranges [of the Palace]" (Holmes 1897: 159). The only two structures added to the Palace during this period were House L and the southwest building at plaza level, both

Table 5.28. *Palenque Palace Phase 5 sculpture ranked according to depth from the carrier, U Pakal K'inich Janab' Pakal II, K'inich Kan B'alam III, and K'inich K'uk' B'alam II (circa* A.D. *742 to 799)*

| Node | Carrier R | Sculpture GR No. | Theme |
|---|---|---|---|
| c | 0 | 1, 2, 82, 83, 84, 85 | CosC, RPSoloCosCU, RPSol |
| 1a | 1 | 55, 56, 57 (Bench) | RPSolo, RAudAccTDynHist |
| 1b | 1 | 48, 49, 50, 51, 52, 53 | HumSac, SacDan, Text? |
| 2a | 2 | 42, 43, 44, 45, 46, 47 | THist, APWA, APSolo |
| 2e | 2 | 80 (Bench) | TCosCU |
| 4b | 4 | 58, 59, 60, 61 (Hieroglyphic Stairs) | CapDis, THist |
| 5a | 5 | 24, 25, 26, 27 | RPSolo, MMot, CosU, CosCU |
| 5b | 5 | 24, 25, 26, 27 | RPSolo, MMot, CosU, CosCU |
| 5c | 5 | 24, 25, 26, 27 | RPSolo, MMot, CosU, CosCU |
| 5e | 5 | 29, 38, 40, 41, 62 | GPSolo, CosT, TDed? |
| 5o | 5 | 79 | CosU |
| 5s | 5 | **70, 73** (Tablet 96 Glyphs), 74, 75 | **?(Stucco fret), TDynHist,** CapDis |
| 6a | 6 | 23, 26, 27, 28 | CosT?, MMot, CosU, CosCU |
| 6b | 6 | 16, 17, 18, 19, 20, 21, 22, 26, 27, 28 | RAud, Text?, PSolo, RitDan, MMot, CosU |
| 6e | 6 | 8 | CosCU |
| 6f | 6 | 3 (Bench?), 4, 5, 6 (Del Río Throne), 11, 13 | RAud, Acc, CosTU, THist, TDynHist |
| 6h | 6 | **76, 77** | **Cos?, RPSolo** |
| 6L | 6 | 72 | CosC |
| 6m | 6 | 71 | CapDis |
| 6n | 6 | 30, 31, 32, 33, 34, 35, 36, 37, 39, 40, 41 | RAud, APSolo, PSolo, CosT |
| 6r | 6 | 78 | CosU |
| 7a | 7 | 63, 64, 65, 67, 68 | RitDan?, GPWA, CosC, TCapSac |
| 7b | 7 | 69 | CosC |
| 7d | 7 | 7 | CosU |
| 8a | 8 | 66 | GPWA |

*Note:* See Appendix 4 for building and location information for nodes.

of which show no evidence of ever being decorated. In my view, the addition of House L was an attempt to free space within the Palace's interior to accommodate a growing population of Palace inhabitants; that attempt required building precariously over the top of the Southern Subterranean Structures. The presence of rubble fill in the southeastern corridor of the Subterraneos, directly beneath House L, suggests that both structures may have been at risk of collapsing (see Chapter 4, Section 4.3.1, The Buildings of the Palace Phase 1, The Northern and Southern Subterranean Galleries and Passages).

Seventeen buildings comprise Palace Phase 6 at Palenque, compared to the fifteen that collectively represent Phase 5 of the Palace. Due to Wak Kimi Janab' Pakal III's changes to the Palace, there was a considerable increase in the number of spaces that comprise the building group's interior. The corresponding access diagram for Palace Phase 6 comprises one hundred thirty-five nodes and one hundred seventy-two edges that connect them. This is an overall increase of fifty-one spatial units (including the carrier position) when compared to the previous architectural phase of the Palace. As at Phase 5, the effect of the overall increase in number connections (edges) between spatial units (nodes) at Phase 6 have the effect of further increasing the level of integration expressed by the Palace plan.

At Palace Phase 6, there are ten levels of depth starting from the carrier position; this is an overall increase of two levels when compared to the two previous architectural phases (see Figure 5.12). At Phase 6 of the Palace, seven spaces are represented at Levels 1, 2, and 8; fourteen spaces at Level 3; twenty-seven spaces at Level 4; twenty-four spaces at Level 5; twenty-eight spaces at Level 6; sixteen spaces at Level 7; three spaces at Level 9; and one space at Level 10. Of all the spaces that collectively comprise the Palace at Phase 6, thirty-seven were defined as transitional as it was determined that their principal function was to provide access between other spatial units. As at Palace Phases 2, 3, 4, and 5, four courtyards were also represented at Palace Phase 6, referred to in the following discussion as the East Court (node 3b), West Court (node 4g), Southeast Court (node 5e), and Southwest Court (node 3d).

When comparing the averaged integration values for the Palace houses and other structures at Phase 6

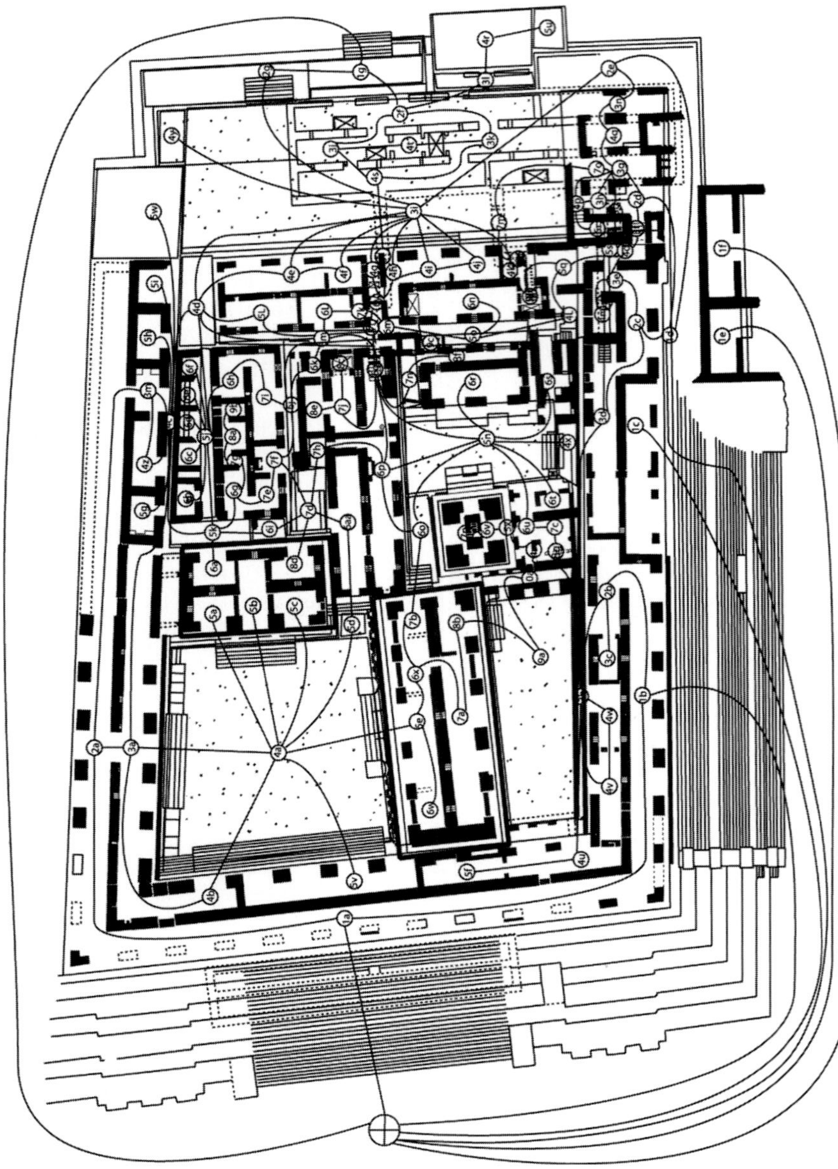

**Figure 5.11.** Palenque Palace Phase 6, Combined Convex Map and Access Diagram. Palenque Palace drawn by Greene Robertson (1985b: fig. 1).

with those of the previous architectural phase, there is a significant change in their order of ranking (see Table 5.29). Due to heavy partitioning in the Palace interior, as well as the addition of House L and the structure at plaza level at the final stage of the Palace Group's evolution, House A replaces House E as the most integrated structure of the complex, and House J replaces House K as the third most integrated structure. At Palace Phase 6 we also see a substantial increase in the level of integration that House J possesses when compared to all other architectural phases of the Palace. Furthermore, Phase 6 of the Palace Houses K, F, and C have become considerably more segregated with respect to the rest of the Palace complex, whereas the Toilets, House B, the south substructures, and the Tower remain (on average) some of the least integrated Palace buildings. At Phase 6 of the Palace, the accessibility ranking of House L falls between that of House B and that of the southern substructure, indicating the relatively restricted nature of the building.

When compared to the previous architectural phase, the courtyards at Palace Phase 6 display a shift in their relative ordering when ranked according to corresponding RA and control value calculations (see Table 5.30). In terms of their respective depths from the carrier, however, there is no change in the ordering of the court spaces when compared to Palace Phase 5. As indicated, at Phases 2, 3, 4, and 5, the Palace complex comprises four quadrants marked by the presence and proximity of the Palace courtyards, each of which displays varying levels of depth from the perspective of both the carrier position (i.e., the visitor entering the complex), as well as all other spaces in the system (i.e., those spaces occupied by Palace inhabitants). Liendo Stuardo notes that, from the perspective of the carrier, primary access to the deeper quarters of the Palace complex (i.e., the Tower Court, the Southeastern Court, and the West Court) required first passing through the East Court, which was "a public space [utilised] for official performances" (Liendo Stuardo 2003: 197), a view that is supported by my own analysis of the Palenque Palace Group. At Palace Phase 6, the shallowest courtyard from the perspective of the visitor is the East Court, followed by the Tower Court, Southeast Court, and West Court, respectively. As at Palace Phase 5, the West Court is the most restricted of the Palace court spaces from the perspective of both visitors and Palace inhabitants, as this courtyard is the most distant from the carrier position whilst displaying the lowest integration value of all the court spaces represented (see Table 5.30). At Palace Phase 6, the Tower Court replaces the Southeast Court as the most integrated (i.e., accessible) court from the perspective of persons occupying all other areas of the Palace complex. At Phase 6 of the Palace Group, the East Court displays the highest level of control, followed by the Tower Court, Southeast Court, and the West Court.

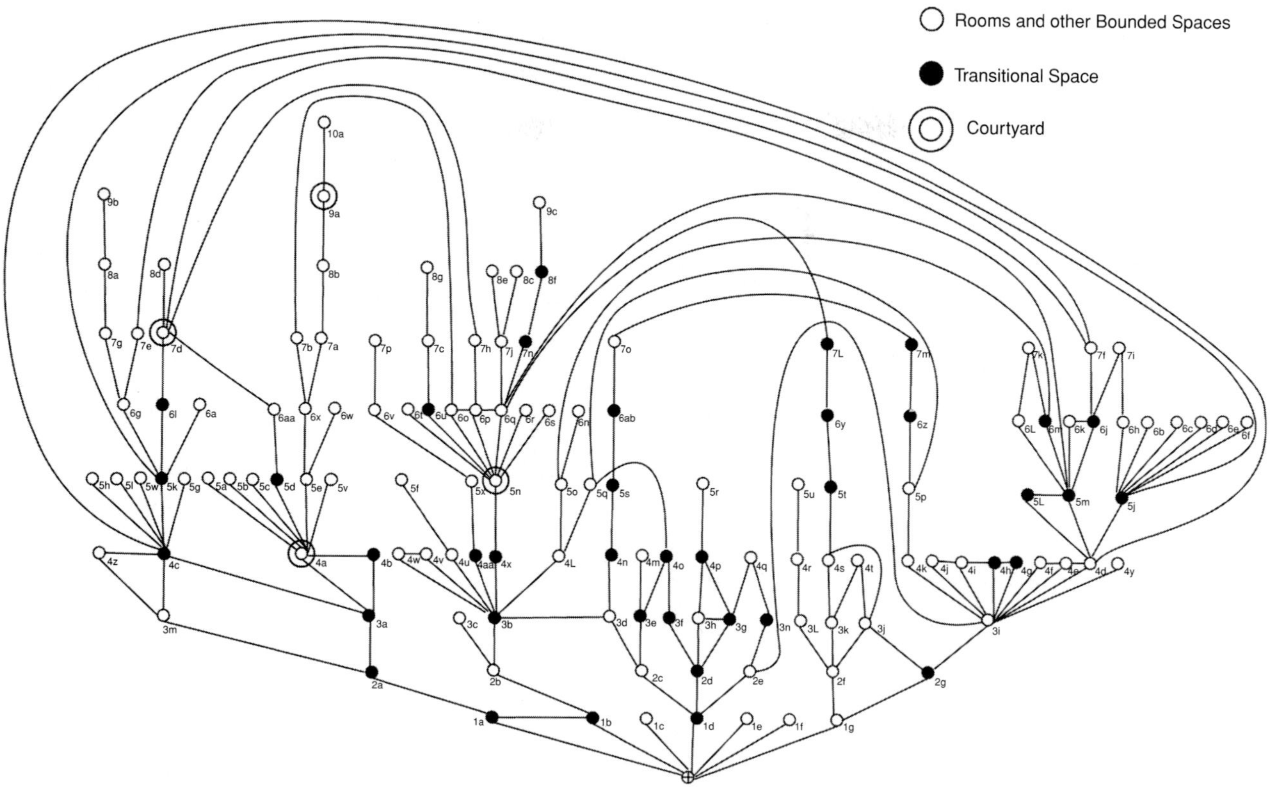


**Figure 5.12.** Palenque Palace Phase 6, "Justified" Access Diagram.

According to Liendo Stuardo (2003: 197–198), variations in the relative depths and spatial configurations of the four quadrants at Palace Phase 6 suggest that the southeastern quarter served a residential function. This view is supported by the discovery of a midden in the southeastern corner of the complex containing household materials such as jars, serving vessels, needles, obsidian blades, jade beads, as well as macro botanical residues. The southeastern quarter of the Palace at Phase 6 encompasses remnants of the Southeast Court (node 7d), House J (nodes 3m, 4z, 5g, 5h, and 5i), House G (nodes 5L and 6L), as well as the southeastern (node 6a) and southwestern (node 8d) sanctuaries of House B, House F (nodes 6g, 6h, 7e, 7f, 7g, 7i, 8a, and 9b), and the eastern gallery of House E (nodes 6aa and 7j).

As stated earlier in text, no Palace sculpture or major buildings have been attributed to Wak Kimi Janab' Pakal III's reign at Palace Phase 6. The focus of Wak Kimi Janab' Pakal III's work in the Palace was the partitioning of existing Palace architecture. The absence of any

Table 5.29. *Average RA values for Palace Phase 6 houses and other structures*

| Structure | Average RA values | Rank |
|---|---|---|
| House A | 0.0586 | 1 |
| House G | 0.0613 | 2 |
| House J | 0.0621 | 3 |
| House H | 0.0652 | 4 |
| House E | 0.0684 | 5 |
| House D | 0.0701 | 6 |
| Structure Plaza Level | 0.0744 | 7 |
| House K | 0.0746 | 8 |
| House A-D | 0.0748 | 9 |
| House I | 0.0765 | 10 |
| House F | 0.0788 | 11 |
| Toilet | 0.0799 | 12 |
| House B | 0.0811 | 13 |
| House L | 0.0818 | 14 |
| South Substructure | 0.0823 | 15 |
| House C | 0.0857 | 16 |
| Tower | 0.1026 | 17 |

*Note:* Ranked from most integrated to least integrated.

Table 5.30. *Palace Phase 6 Court carrier depths, RA values, and control values*

| Palace courts | Node | Carrier rank | RA value | Control value |
|---|---|---|---|---|
| East Court | 4a | 4 | 0.0658 | 5.5833 |
| Tower Court | 5n | 5 | 0.0543 | 4.7500 |
| Southeast Court | 7d | 7 | 0.0682 | 2.8333 |
| West Court | 9a | 9 | 0.1212 | 1.5000 |

Table 5.31. *Palenque Palace Phase 6 sculpture ranked according to RA values, Wak Kimi Janab' Pakal III* (A.D. 799–?)

| Node | RA value | Rank | Sculpture (GR No.) | Theme |
|---|---|---|---|---|
| 10a | 0.1361 | 1 | 66 | GPWA |
| 9a | 0.1212 | 2 | 63, 64, 65, 67, 68 | RitDan?, GPWA, CosC, TCapSac |
| 8b | 0.1065 | 3 | 30 | RAud |
| 6ab | 0.1030 | 4 | 79 | CosU |
| 6v | 0.1026 | 5 | 72 | CosC |
| 7a | 0.0919 | 6 | 31, 32, 33, 34, 35, 36, 37, 39, 40, 41 | APSolo, PSolo, CosT |
| 7c | 0.0834 | 7 | 69 | CosC |
| 8d | 0.0834 | 7 | 16, 17, 18, 19, 20, 21, 22, 26, 27, 28 | RAud, Text?, PSolo, RitDan, MMot, CosU |
| 5a | 0.0815 | 8 | 24, 25, 26, 27 | RPSolo, MMot, CosU, CosCU |
| 5b | 0.0815 | 8 | 24, 25, 26, 27 | RPSolo, MMot, CosU, CosCU |
| 5c | 0.0815 | 8 | 24, 25, 26, 27 | RPSolo, MMot, CosU, CosCU |
| 6z | 0.0791 | 9 | 78 | CosU |
| 3c | 0.0782 | 10 | **54** | **?(Mural)** |
| 6a | 0.0777 | 11 | 23, 28, 26, 27, 28 | CosT?, MMot, CosU, CosCU |
| 6aa | 0.0775 | 12 | 8, **15** | CosCU, **?(Geometric painting)** |
| 5e | 0.0740 | 13 | 29, 38, 40, 41, 62 | GPSolo, RAud, CosT, TDed? |
| 2f | 0.0734 | 14 | 80 (Bench) | TCosCU |
| 6r | 0.0692 | 15 | 76, 77 | Cos?, RPSolo |
| 7d | 0.0682 | 16 | 28 | CosU |
| 7L | 0.0664 | 17 | 7 | CosU |
| 4a | 0.0658 | 18 | 58, 59, 60, 61 (Hieroglyphic Stairs) | CapDis, THist |
| 6o | 0.0645 | 19 | 71 | CapDis |
| 1b | 0.0615 | 20 | 48, 49, 50, 51, 52, 53 | HumSac, SacDan, Text? |
| 1a | 0.0608 | 21 | 55, 56, 57 (Bench) | RPSolo, RAudAccTDynHist |
| 2a | 0.0603 | 22 | 42, 43, 44, 45, 46, 47 | THist, APWA, APSolo |
| 4x | 0.0596 | 23 | 70 | ?(Stucco fret) |
| 6p | 0.0580 | 25 | 3 (Bench), 4, 5, 6 (Del Río Throne), **10**, 11, 13, **14** | **Text?, CosU,** RAud, Acc, CosTU, THist, TDynHist |
| 5n | 0.0543 | 26 | 73 (Tablet 96 Glyphs), 74, 75 | TDynHist, CapDis |

*Notes:* Ranked from least integrated to most integrated. See Appendix 5 for building and location information for nodes.

significant architectural and sculptural contribution suggests that Wak Kimi Janab' Pakal III's desire to expand and embellish the Palace was surpassed by the need to maximise all available space in the existing Palace complex. It remains possible, however, that some of the painted artwork present within the Palace Group was commissioned some time during Wak Kimi Janab' Pakal III's reign (see Tables 5.31, 5.32, and 5.33). At Palace Phase 6, I have attributed remnants of a mural (GR 54) appearing in the southeastern room of House D, in addition to the fish painting (GR 14) located in the northeastern room of House E, to the reign of Wak Kimi Janab' Pakal III, as both appear to postdate the intensive partitioning of the Palace Group interior. I also propose that the eroded hieroglyphic text (GR 10) located on the eastern wall in the northeastern room of House E, in addition to the geometric painting (GR 15) appearing on the same wall of the same room, corresponds to Phase 6 of the Palace. Displaying no discernible themes, remnants of the House D mural as well as the geometric painting of House E hold intermediate positions in the RA ranking of artworks at Palace Phase 6, whereas the eroded text and the fish painting hold more integrated positions indicating their higher accessibility (see Table 5.31).

Of all the one hundred thirty-five spaces that represent Phase 6 of the Palace, forty-four were calculated as controlling spaces (excluding the carrier space), whereas the remaining ninety spaces are controlled spaces. Some of the spatial units that display the greatest level of control over localised access to surrounding spaces include the eastern and northeastern corridors of House F (nodes 5j and 5k), the East Court (node 4a), the western corridor of House J (node 4c), the southern terrace of the Palace (node 3i), the Tower Court (node 5n), remnants of the Southeast Court (node 7d), the southeastern and southwestern rooms of House E (nodes 7j and 6q), House K's south-central corridor (node 3e), House H's northern terrace (node 5o), the eastern corridor of House L (node 4p), the second level of the Tower (node 6v), the west-central room of the Toilets (node 7c), House F's central room (node 8a), the southern corridor of House I (node 8f), and the West Court (node 9a). Of all the

Table 5.32. *Palenque Palace Phase 6 sculpture ranked according to control values, Wak Kimi Janab' Pakal III* (A.D. *799–?*)

| Node | Control value | Rank | Sculpture (GR No.) | Theme |
|---|---|---|---|---|
| 4a | 5.5833 | 1 | 58, 59, 60, 61 (Hieroglyphic Stairs) | CapDis, THist |
| 5n | 4.7500 | 2 | 73 (Tablet 96 Glyphs), 74, 75 | TDynHist, CapDis |
| 7d | 2.8333 | 4 | 28 | CosU |
| 7c | 1.5000 | 5 | 69 | CosC |
| 6v | 1.5000 | 5 | 72 | CosC |
| 9a | 1.5000 | 5 | 63, 64, 65, 67, 68 | RitDan?, GPWA, CosC, TCapSac |
| 5e | 1.4583 | 6 | 29, 38, 40, 41, 62 | GPSolo, RAud, CosT, TDed? |
| 2f | 1.4166 | 7 | 80 (Bench) | TCosCU |
| 6p | 1.1249 | 9 | 3 (Bench), 4, 5, 6 (Del Río Throne) **10**, 11, 13, **14** | **Text?, CosU,** RAud, Acc, CosTU, THist, TDynHist |
| 6ab | 1.0000 | 10 | 79 | CosU |
| 8b | 1.0000 | 10 | 30 | RAud |
| 2a | 0.9166 | X | 42, 43, 44, 45, 46, 47 | THist, APWA, APSolo |
| 6o | 0.8750 | X | 71 | CapDis |
| 7a | 0.8333 | X | 31, 32, 33, 34, 35, 36, 37, 38, 39, 40, 41 | APSolo, PSolo, CosT |
| 6z | 0.8333 | X | 78 | CosU |
| 1a | 0.8095 | X | 55, 56, 57 (Bench) | RPSolo, RAudAccTDynHist |
| 1b | 0.8095 | X | 48, 49, 50, 51, 52, 53 | HumSac, SacDan, Text? |
| 6aa | 0.7000 | X | 8, **15** | CosCU, **?(Geometric painting)** |
| 7L | 0.6666 | X | 7 | CosU |
| 10a | 0.5000 | X | 66 | GPWA |
| 3c | 0.3333 | X | **54** | **?(Mural)** |
| 4x | 0.2500 | X | 70 | ?(Stucco fret) |
| 6a | 0.2000 | X | 23, 28, 26, 27, 28 | CosT?, MMot, CosU, CosCU |
| 8d | 0.2000 | X | 16, 17, 18, 19, 20, 21, 22, 26, 27, 28 | RAud, Text?, PSolo, RitDan, MMot, CosU |
| 6r | 0.1250 | X | 76, 77 | Cos?, RPSolo |
| 5a | 0.1250 | X | 24, 25, 26, 27 | RPSolo, MMot, CosU, CosCU |
| 5b | 0.1250 | X | 24, 25, 26, 27 | RPSolo, MMot, CosU, CosCU |
| 5c | 0.1250 | X | 24, 25, 26, 27 | RPSolo, MMot, CosU, CosCU |

*Note:* See Appendix 6 for building and location information for nodes.

controlling spaces represented at Palace Phase 6, the northeastern room of House E (node 6aa) is the only space that displays Phase 6 artwork, more specifically the eroded hieroglyphic text (GR 10) and the fish painting (GR 14), whereas the remains of the two other works (i.e., the mural and the geometric painting) occupy controlled spaces (see Table 5.32). At Palace Phase 6, the only discernible image located in a controlling space is the fish image that has Underworld associations.

As noted at the beginning of Chapter 4, the Palace at Palenque was not planned and constructed as a single architectural conception but is instead the end result of a gradual accumulation of buildings and other architectural features over time (Maudslay 1889–1902b: 11; Holmes 1897: 169; Martin and Grube 2000: 164). This slow but sure accumulation also holds true for all of the monumental sculpture and the other artwork that appeared in combination with the architecture of the Palace, which represents a gradual accretion of artistic accompaniments. The shift away from more ornate monumental construction to the blocking and segmenting of existing Palace architecture would have been due to a range of influencing factors; among these factors may have been: (1) declining access to the necessary raw materials, such as limestone and wood used in building and stucco production; (2) the loss and/or absence of the artistic and/or engineering skills required to undertake larger construction programs; (3) dwindling control over the necessary labour force to implement more ambitious construction projects; and (3) increasing pressure to protect and preserve the privacy of an ever-growing population of Palace inhabitants from outsiders.

The Palace Phase 6 would not have been configured solely to maintain the privacy of inhabitants with respect to the encroachment of visitors, but also to more effectively segregate inhabitants from other inhabitants occupying different areas of the Palace complex. While functioning chiefly as barriers limiting or controlling access between spatial units, the addition of transverse walls would have also prevented inhabitants from being more

Table 5.33. *Palenque Palace Phase 6 sculpture ranked according to depth from the carrier, Wak Kimi Janab' Pakal III* (A.D. *799–?)*

| Node | Carrier R | Sculpture (GR No.) | Theme |
|---|---|---|---|
| c | 0 | 1, 2, 82, 83, 84, 85 | CosC, RPSolo, CosCU, RPSol |
| 1a | 1 | 55, 56, 57 (Bench) | RPSolo, RAudAcc, TDynHist |
| 1b | 1 | 48, 49, 50, 51, 52, 53 | HumSac, SacDan, Text? |
| 2a | 2 | 42, 43, 44, 45, 46, 47 | THist, APWA, APSolo |
| 2f | 2 | 80 (Bench) | TCosCU |
| 3c | 3 | **54** | **?(Mural)** |
| 4a | 4 | 58, 59, 60, 61 (Hieroglyphic Stairs) | CapDis, THist |
| 4x | 4 | 70 | ?(Stucco fret) |
| 5a | 5 | 24, 25, 26, 27 | RPSolo, MMot, CosU, CosCU |
| 5b | 5 | 24, 25, 26, 27 | RPSolo, MMot, CosU, CosCU |
| 5c | 5 | 24, 25, 26, 27 | RPSolo, MMot, CosU, CosCU |
| 5e | 5 | 29, 38, 40, 41, 62 | GPSolo, CosT, TDed? |
| 5n | 5 | 73 (Tablet 96 Glyphs), 74, 75 | TDynHist, CapDis |
| 6a | 6 | 23, 28, 26, 27, 28 | CosT?, MMot, CosU, CosCU |
| 6aa | 6 | 8, **15** | CosCU, **?(Geometric painting)** |
| 6ab | 6 | 79 | CosU |
| 6o | 6 | 71 | CapDis |
| 6p | 6 | 3 (Bench), 4, 5, 6 (Del Río Throne) **10**, 11, 13, **14** | **Text?, CosU,** RAud, Acc, CosTU, THist, TDynHist |
| 6r | 6 | 76, 77 | Cos?, RPSolo |
| 6v | 6 | 72 | CosC |
| 6z | 6 | 78 | CosU |
| 7a | 7 | 31, 32, 33, 34, 35, 36, 37, 39, 40, 41 | APSolo, PSolo, CosT |
| 7c | 7 | 69 | CosC |
| 7d | 7 | 28 | CosU |
| 7L | 7 | 7 | CosU |
| 8b | 8 | 30 | RAud |
| 8d | 8 | 16, 17, 18, 19, 20, 21, 22, 26, 27, 28 | RAud, Text?, PSolo, RitDan, MMot, CosU |
| 9a | 9 | 63, 64, 65, 67, 68 | RitDan?, GPWA, CosC, TCapSac |
| 10a | 10 | 66 | GPWA |

*Note:* See Appendix 4 for building and location information for nodes.

readily observed. Throughout the Palace many other masonry walls were added, including those closing several of the spaces between the eastern and western piers of House C to coincide with the interior partitioning of the structure. The combined effect of these walls and partitions was to break up the east and west galleries of House C into a number of smaller rooms, changing the function of the space. Some of the other spaces segmented at Palace Phase 6 were the eastern corridor of House D and the southern corridor of House A-D. Some of the most intensive partitioning at Phase 6 occurred in the southeastern quadrant of the Palace, which included extensions to the eastern sides of Houses E (nodes 7h and 8e) and F (nodes 6c, 6d, 6e, and 6f). With the construction of House E's eastern extension, the southeastern entrance of the structure was closed off and a bench was placed in front of it. Furthermore, Houses G and H were joined by two half-vaults and additional transverse walls which, combined with a crude bench, prevented direct access between the two buildings and into the interior of the Palace (see Figure 4.52).

## 5.2. Public and private sculpture of the Palenque Palace Group

Applying various methods in access analysis, a correlation was drawn between the levels of accessibility calculated for spaces of the Palenque Palace Group and the relative privacy of sculptural themes and other artwork assigned to those spaces. The processes undertaken were, first, to discern the thematic variation in a range of sculpture and other artistic media, then to determine how these variations related to differences in corresponding carrier depths, and RA and control value calculations. As previously stated, it follows that fixed sculptural media assigned to distinct types of spaces (differentiated by progressive enclosure, channeled movement, and changes in elevation) may, like the spaces themselves, be sorted from most public to most private, and the scenes, activities, and symbolism displayed can be characterised similarly (see Chapter 2, Section 2.4, Investigative Procedures, for more information). As the Palenque Palace Group, however, represents more than 150 years of

Table 5.34. *Counts of sculptures and other artistic features assigned to different phases of the Palace at Palenque*

| Ruler | Palace construction phase | Counts of sculptures and other decorative features | Relative percentage |
|---|---|---|---|
| K'inich Janab' Pakal I | 1 | 22 | 24% |
| | 2 | 26 | 28% |
| K'inich K'an Joy Chitam II | 3 | 27 | 30% |
| K'inich Akul Mo'Nab' III | 4 | 6 | 7% |
| U Pakal K'inich Janab' Pakal II<br>K'inich Kan B'alam III<br>K'inich K'uk' B'alam II | 5 | 4 | 4% |
| Wak Kimi Janab' Pakal III | 6 | 6 | 7% |

*Note:* Sculptural features, as designated by Greene Robertson (1985b).

architectural development (see Martin and Grube 2000: 164), it was understood that any attempt to indiscriminately interpret the entire inventory of monuments and decoration in terms of its distribution on the Palace's "privacy gradient" would have been inappropriate due to the different temporal periods to which many of the artworks belong. The strategy used to overcome this interpretive problem was to order the art and architecture of the Palace to phases and the reigns of kings. This method provided a means of confining the periods investigated, whilst ensuring that any correlation made between sculptural theme and spatial context was both appropriate and meaningful. My main objective in this study has been to test whether the thematic content of Classic Maya sculpture and other media would vary depending on the type of audience that could access it. In support of this proposal, it is acknowledged that "some vehicles of information inherent in Maya built forms (e.g. painted, carved, and inscribed murals; lintels and plaques inside temples and palaces) would have been accessible only to elites" (Webster 1998: 35).

Access analysis of sculpture associated with the earliest phases of the Palace Group (see Section 5.1.1, Analysis of the Palace Phase 1, Section 5.1.2, Analysis of the Palace Phase 2, and Section 5.1.3, Analysis of the Palace Phase 3, this chapter) does indicate a correlation between the thematic variation present in Classic Maya art and differences in the accessibility of that art. Furthermore, shifts in the distribution of sculptural themes identified at Palace Phase 3 (from the inner to the outer areas of the Palace complex) imply a change in the type of audience that the ruler sought to address. It should be noted, however, that although the Palace Group investigation detected some correspondence between spatial context and variations in the sculptural theme at the earlier phases, analysis of the remaining three stages of the Palace Group (see Section 5.1.4, Analysis of the Palace Phase 4, Section 5.1.5, Analysis of the Palace Phase 5, and Section 5.1.6, Analysis of the Palace Phase 6, this chapter) suggest that the importance of sculptural embellishment in the Palace complex may have gradually diminished over time. The dwindling importance of Palace sculpture and other decoration was not only signaled by the relatively minor artistic contributions at Phases 4, 5, and 6 (see Table 5.34), but is also marked by the changing accessibility of earlier artwork with the structural modifications undertaken by later kings. More specifically, there appears to have been little or no attempt on the part of Palenque's later rulers to conserve the degree of accessibility or relative privacy of existing sculpture commissioned at the earliest phases of the Palace when carrying out their respective construction programs.

The majority of the surviving sculptures and other artworks of the Palace at Palenque correspond temporally to the first three phases of the building groups' construction, more specifically to the reigns of K'inich Janab' Pakal I (see Chapter 4, Section 4.3.1, The Buildings of the Palace Phases 1, and Section 4.3.2, The Buildings of the Palace Phase 2) and K'inich K'an Joy Chitam II (see Chapter 4, Section 4.3.3, The Buildings of the Palace Phase 3). In contrast, at the later stages of the Palace's development, during the reigns of K'inich Akul Mo'Nab' III (see Chapter 4, Section 4.3.4, The Buildings of the Palace Phase 4), U Pakal K'inich Janab' Pakal II, K'inich Kan B'alam III, K'inich K'uk' B'alam II (see Chapter 4, Section 4.3.5, The Buildings of the Palace Phase 5), and Wak Kimi Janab' Pakal III (see Chapter 6, Section 4.3.6, The Buildings of the Palace Phase 6), relatively little sculpture was added to the Palace complex (see Table 5.34). On considering all decorative features of the Palace, as listed and described by Greene Robertson (1985b), approximately 82% were found to correlate temporally to the first three phases of the building groups' construction. The bulk of the Palace Group artwork, totaling some 52%, appears during the reign of K'inich Janab' Pakal I, followed by K'inich K'an Joy Chitam II's contribution of 30%. The last three phases of the development of the Palace saw the collective contribution of the remaining 18% of decorative works.

Most of the Palace Group sculpture and other artwork are concentrated in the northeastern and northwestern quadrants of the Palace, consigned to the interior and exterior of Houses A, D, A-D, and C as well as to the northern exterior of House B. To a lesser extent, sculpture and other decorative artwork are also located in the complex's southwestern quarter, where it is associated with the interior and exterior of the Tower and House I, as well as with the eastern half of House E. At all six phases of construction, the absence of decoration is most evident in the southeastern quadrant of the Palace, where, together with its congested configuration, the implication is that this area probably remained relatively private from the perspective of those visitors permitted to enter the building group. The justification for this reasoning is that "decoration . . . is a symbolic medium and to be effective it requires an audience" (Grahame 1997: 141). The more open and decorated northeastern, northwestern, and southwestern quadrants of the Palace Group were configured to signal, among other things, wealth, ancestry, and ascendancy of the king to inhabitants and outsiders permitted to enter the complex. In contrast, the understated and progressively claustral character of the southeastern quarter (see Section 5.1.5, Analysis of the Palace Phase 5, and Section 5.1.6, Analysis of the Palace Phase 6, this chapter) indicates that it was not intended for general access, but rather it functioned more exclusively as a private domain where Palace inhabitants, such as the king and, perhaps, members of Palenque's nobility, withdrew to retire and/or to avoid public scrutiny. "During Classic times, the evidence seems to indicate an association of residential . . . [as well as] other functions within palace structures" (Liendo Stuardo 2003: 194; see also Martin and Grube 2000: 164). At Palace Phase 6, the southeastern quadrant of the Palace encompasses remnants of the Southeast Court; Houses J, G, and F; in addition to the southeastern and southwestern sanctuaries of House B and the eastern rooms of House E.

One of the objectives of the architectural amendments undertaken in the Palace was to reconfigure the complex in terms of its accessibility and control (see Section 5.1.3, Analysis of the Palace Phase 3, this chapter). With all the changes that occurred at the later stages of the Palace Group's evolution, however, there appears to have been little attempt to preserve the privacy of sculptural themes and other artwork introduced at the earlier architectural phases (see Section 5.1.4, Analysis of the Palace Phase 4, Section 5.1.5, Analysis of the Palace Phase 5, and Section 5.1.6, Analysis of the Palace Phase 6, this chapter). To draw on a specific example, because of the changes implemented at Palace Phase 4, we see a notable shift in the relative depths of the deeper sculpture when compared to the previous phase. More specifically, several of the subterranean vaults that mark the function of the underground passages as symbolic conduits to the Underworld (Schele 1986: 122) appear at a higher level of integration relative to other Palace sculpture. From the perspective of other spaces in the system, the implication is that the once highly restricted subterranean vaults were becoming increasingly accessible relative to other Palace artworks, an indication perhaps that maintaining the exclusivity of the sculptural features of the Palace was becoming less important (see Section 5.1.4, Analysis of the Palace Phase 4, this chapter, for more information).

It is reasonable to argue that, if efforts were made to control access to certain artwork appearing in the Palace, one would expect to see thematic variation in artwork subject to differences in spatial context. The fact that we see relatively limited variability in the distribution of earlier sculptural themes, when ordered according to the privacy gradients of later Palace phases, implies little or no concern by the later kings to preserve/control the accessibility of artworks commissioned by earlier kings. This apparent disregard for the sculptural programs of earlier architectural phases is especially evident with the construction of the Palace Toilets at Palace Phase 4, which overlies the sculpture at the southern end of the *Casita* Wall, which was installed at Phase 3 of the Palace (see Chapter 4, Section 4.3.3, The Buildings of the Palace Phase 3, House D: Description and Sculpture).

One of the few sculptural themes that frequently appears in the more public spaces of the Palace over several phases is "Capture." More specifically, at Palace Phase 2 during the reign of K'inich Janab' Pakal I we see the appearance of sculpture displaying captive texts and imagery in the Early East Court and the West Court of the Palace (GR 58 [*alfardas* only], 59, 60, 61, and 68; Figures 4.27, 4.28 4.29, and 4.46). The maintenance of the East Court as a consistently shallow space relative to the other Palace Courtyards signals the importance of this semi-public space at the later stages of the Palace's development. At Palace Phase 3, with the closing of the northern end of the Palace with Houses D and A-D, additional captive sculpture appears in the East Court (GR 58; Figures 4.46). Following this renovation, at Palace Phase 4, further captive art was commissioned by K'inich Akul Mo'Nab' III (GR 74 and 75; Figures 4.55 and 4.56) and placed in the Tower Court, which is the deeper of the two court spaces that display capture imagery. The captive sculptures found in the East Court, although more pronounced in numbers, are comparatively crude in appearance when compared to those located in the Tower Court. The most elaborate captive monuments of the Palace are referred to as the Tablet of the Scribe (GR 75; Figure 4.56) and the Tablet of the Orator (GR 74; Figure 4.55). Given their position

Table 5.35. *Average RA values for Palace Phases 1–6*

| Palace phase | Mean RA value |
|---|---|
| 1 | 0.1842 |
| 2 | 0.1599 |
| 3 | 0.1315 |
| 4 | 0.1165 |
| 5 | 0.1062 |
| 6 | 0.0742 |

close to House E and the skill in their manufacture, it is probable that the captive monuments appearing in the Tower Court were valued more than those appearing in the more accessible East Court of the Palace.

It is clear that the presence of monuments displaying or recording capture themes were intended for public display as indicated by their location in the Palace courtyards rather than in the more restricted interior of buildings. The positioning of captive texts and iconography in semi-public context over several phases of the construction suggests concern for maintaining the accessibility of this type of sculptural theme. For persons permitted to enter the shallower areas of the Palace complex, the display of captives was designed to instill a sense of dread among enemies of Palenque and confidence among allies.

Shifts in the relative accessibility of Palace Group sculpture over time are due to the gradual increase in the overall level of integration the Palace displays at each successive phase. Whereas the overall spatial configuration of the Palace at Palenque can be described as relatively distributed at all six phases of its construction (see Analysis of Palace Phases 1–6, this Chapter), the average integration values calculated for each of the Palace phases indicate that the spatial system of the Palace becomes more integrated at each successive stage of its development. (Note: As outlined previously [see Chapter 2, Section 2.4.5, Measuring Control Values and RA Analysis], the higher the RA value, the more segregated the corresponding space or system of spaces is said to be; conversely, the lower resulting value, the more integrated the spatial unit or system; see Table 5.35). The most dramatic change to the configuration of the Palace occurs at Palace Phase 3; from this point on we see a significant broadening of corresponding access maps with no overall increase in depth. In fact, at Palace Phase 3 there is at first a decrease in the number of permeation levels before the number gradually climbs back up again with structural modifications occurring at the final three phases of the Palace's construction.

Until Palace Phase 3, the numbers of spatial units that comprise the Palace at each level correlate closely, giving the impression of a deeper, more restricted spatial system when compared to the later phases of the Palace. Following Palace Phase 2, the number of spaces holding intermediate positions on the justified access maps far outnumber the total number of permeation levels in each chart, an indication of the increasingly integrated nature of the spaces in the system, and correspondingly the various sculpture appearing in those spaces (see Tables 5.36 and 5.37). The gradual widening of the Palace Group access diagrams at the later phases of construction is the result of the increased number of connections (edges) providing access between the spatial units (nodes) that keep spaces relatively shallow in relation to the other spaces in the systems.

My analysis of the architectural sequencing of the Palace at Palenque indicates a shift in the architectural plan of the building group from something that resembled an acropolis-type arrangement (see Chapter 4, Section 4.3.1, The Buildings of the Palace Phase 1, and

Table 5.36. *Node count at each level of depth, Palace Phases 1–6*

| Depth level | Phase 1 | Phase 2 | Phase 3 | Phase 4 | Phase 5 | Phase 6 |
|---|---|---|---|---|---|---|
| 1 | 2 | 2 | 5 | 4 | 4 | 7 |
| 2 | 5 | 5 | 7 | 5 | 5 | 7 |
| 3 | 5 | 5 | 9 | 9 | 9 | 14 |
| 4 | 5 | 5 | 17 | 17 | 17 | 27 |
| 5 | 7 | 7 | 12 | 21 | 20 | 24 |
| 6 | 6 | 9 | 6 | 10 | 19 | 28 |
| 7 | 7 | 9 | – | 5 | 6 | 16 |
| 8 | 5 | 11 | – | 1 | 3 | 7 |
| 9 | 8 | 2 | – | – | – | 3 |
| 10 | 2 | 1 | – | – | – | 1 |
| **Total** | **52** | **56** | **56** | **72** | **83** | **134** |

Table 5.37. *Counts of nodes and edges, Palace Phases 1–6*

| Palace phases | Total nodes | Total edges |
|---|---|---|
| 1 | 53 | 70 |
| 2 | 57 | 77 |
| 3 | 57 | 83 |
| 4 | 73 | 103 |
| 5 | 84 | 115 |
| 6 | 135 | 172 |

Section 4.3.2, The Buildings of the Palace Phase 2), as described by Andrews (1975: 67) and Cohodas (1973: 215–216), to one that conforms to a Palace-type configuration (see Chapter 4, Section 4.3.3, The Buildings of the Palace Phase 3, Section 4.3.4, The Buildings of the Palace Phase 4, Section 4.3.5, The Buildings of the Palace Phase 5, and Section 4.3.6, The Buildings of the Palace Phase 6). Buildings and platforms that collectively comprise acropolis-type configurations are intentionally ordered as a gradual ascent, starting at the entrance of the group and terminating with a principal structure, or structures, at its furthermost point (Cohodas 1973: 215). This type of configuration accords with the appearance of the Palace Group at Phases 1 and 2 under K'inich Janab' Pakal I, when the Palace consisted of two principal levels: the First Palace Terrace, occupied by the then-exposed Northern Galleries and Subterraneos; and all other buildings, which occupied the level of the Second Palace Terrace (see Figure 4.1). The combined effect of these platforms was to provide a progressive change in elevation from the perspective of someone entering the complex from the plaza fronting the earlier northern Palace stairs. The objective of this type of building configuration was to allow rituals undertaken by elites in the upper reaches of the complex to remain visible to the lower classes occupying the more accessible plazas below (Andrews 1975: 67). In contrast, palaces are generally characterised by a series of closely arranged private courtyards (e.g., Palace Phase 4), where lines of sight are limited and buildings are less hierarchically organised.

As stated earlier, access analysis of the sculpture appearing at the earliest phases of the Palace Group (see Section 5.1.1, Analysis of the Palace Phase 1, Section 5.1.2, Analysis of the Palace Phase 2, and Section 5.1.3, Analysis of the Palace Phase 3, this chapter) has revealed a correlation between the thematic variation present in Classic Maya art and differences in the accessibility of that art. Surviving sculpture and other artworks assigned to Phases 1 and 2 represent the remainder of K'inich Janab' Pakal I's artistic vision for the Palace at Palenque. When considered collectively, in terms of their distribution on the Phase 2 "privacy gradient," differences do become apparent in the thematic content of monuments, with changes in spatial context. More specifically, during the reign of K'inich Janab' Pakal I, the shallower, semi-public reception area of the Palace (i.e., the Early East Court) displays what appears to be war propaganda imagery (GR 58 [*alfardas* only] and 59; Figures 4.46 and 4.28). More specifically, sculpture displays capture themes and corresponding texts (GR 61; Figures 4.27 and 4.29) carved on the staircase and foundation of Houses A and C. At a deeper level, access is provided to House B, which portrays standing royal portraits (GR 24 and 25) and a Mat Motif (GR 26), marking House B as a possible *popol nah*, or "council house." In addition, at this level of permeation, access is also provided to the east gallery of House C, which portrays pantheon-style portraiture (GR 29; Figures 4.23, 4.24, 4.25, and 4.26). At a still deeper level, the House E throne room is accessible, displaying Accession and Royal Audience themes such as K'inich Janab' Pakal I's coronation scene and throne (GR 5 and 80; Figures 4.6 and 4.13). At the deepest levels of the Palace, the subterranean passages become accessible where mainly Cosmological Underworld themes are represented (GR 7, 78, and 79; Figures 4.7 and 4.8).

In apparent contrast to K'inich Janab' Pakal I's sculptural program, with the closing of the northern end of the Palace with House A-D by later ruler K'inich K'an Joy Chitam II at Palace Phase 3, we see a shift in agenda as reflected in the placement of additional furniture and sculpture at higher levels of accessibility. The addition of portrait *alfardas,* masks, and "Cosmic Monsters" on the northern monumental stairway (GR 82, 83, and 84; Figure 4.40), combined with a throne, coronation scene, and dynastic–historical documentation positioned in the northern corridor of House A-D (GR 57; Figure 4.39), implies increased concern for the addressing of the civic population in some way (see Section 5.1.3, Analysis of the Palace Phase 3, this chapter).

Positioned centrally in the northern corridor of House A-D, the Palace Tablet (GR 57) is located at the first level of permeation at Palace Phase 3, whereas the Oval Tablet (GR 5) holds a depth of Level 4. At the previous architectural phase, during the reign of K'inich Janab' Pakal I, the Oval Tablet of House E was positioned at a carrier depth of Level 7. Although the Palace Tablet and the Oval Tablet correspond to different phases of the Palace's construction, the similarities between the two monuments are obvious, with each displaying Royal Audience Accession imagery. It appears that the objective of K'inich K'an Joy Chitam II in placing Royal Audience Accession sculpture at an increased level of accessibility was to validate his authority among the broader civic population at Palenque. The significance of the Palace Tablet's location in the northern gallery of House A-D is that it marks a point of demarcation between what was principally the king's domain (i.e., the Palace) and what was the domain of the civic population (i.e., areas outside the Palace complex). The presence of a throne in the northern corridor of House A-D beneath the Palace Tablet suggests that the location was probably used by K'inich K'an Joy Chitam II to address crowds congregated in the plaza below. The space also may have been used as a vantage point from which the ruler could observe ritual proceedings associated with the ballcourt located just northeast of the Palace complex. Documenting K'inich K'an Joy Chitam II's accession, the imagery on the monument also portrays K'inich Janab' Pakal I offering the "drum-major" headdress (principal crown of rulership) to K'inich K'an Joy Chitam II. The fact that we see K'inich K'an Joy Chitam II receiving the crown from his father, K'inich Janab' Pakal I (who was deceased at the time of K'inich K'an Joy Chitam II's accession), rather than from his predecessor K'inich Kan B'alam II, signals the importance of the father's endorsement to his claim as king (see Chapter 4, Section 4.3.3, House A-D: Description and Sculpture, for more information).

Although sculpture and other artwork assigned to the Palace Group may point to activities once associated with its structures, sculptural variations appearing on buildings of different periods may also signal shifts in the significance of the subject matter represented. Although boundaries may be marked symbolically by decoration, when attempting to discern meaning in decorative features, care must be taken to confine the inquiry to appropriate social, cultural, and historical contexts because it remains possible that the relevancy and meaning of architectural decoration will change over time (Pearson and Richards 1994b: 27–28). The "early buildings of the Palace complex . . . (Houses B and C) were decorated with huge masks . . . [while the] later buildings of the Palace were adorned with figures seated on thrones" (Greene Robertson 1985a: 63), perhaps indicating a shift in the sculptural program of the Palace from one of god veneration to the worship of rulers. The placement of the north stair *alfardas* and masks at such a high level of accessibility at Palace Phase 3 was undertaken to substantiate K'inich K'an Joy Chitam II's authority and power in the eyes of the wider populous (see Section 5.1.3, Analysis of the Palace Phase 3, this chapter). This proposal, I submit, was also the purpose of the scenes portraying K'inich K'an Joy Chitam II participating in sacrificial rites connected with the Underworld that appear on the exterior piers of House D.

The sculptured piers of the Palace functioned as Palenque's version of public stelae (Schele 1986: 118), communicating the cosmological and hierarchical status quo in Maya society at Palenque. In Classic Maya architecture, piers were often decorated with iconography and, to a lesser extent, passages of text. Some of the best examples of pier sculpture are found on the exterior of the Palace houses at Palenque, where rulers are portrayed in formal garb in the company of subordinates. Not unlike the Bonampak Murals, which are "divided into horizontal registers assigning persons and gods to spaces according to their rank in the world of the Maya" (Greene Robertson 1985b: 33), sculpture appearing on the exterior of buildings at Palenque were also configured to signal social hierarchies. For example, although the central protagonist of each House A pier is a ruler, the most elevated character on each monument is a portrait of a God K portrayed as the head of staff. Below the God K representations are the portraits of the kings standing authoritatively, beneath which lesser nobles are depicted in subordinate poses seated at the feet of the rulers.

As stated earlier in this investigation, points to consider when attempting to establish relative status of individuals depicted together are: (1) the degree of embellishment in costumes; (2) the degree of embellishment in associated paraphernalia; (3) the number of costume elements; (4) furniture (e.g., the presence of thrones or benches); (5) figure orientation (whether individuals are depicted frontally or in profile); (6) elevation (e.g., whether individuals are portrayed crouching, seated, standing, or elevated); and (7) posture (whether persons portrayed are depicted in assertive or submissive pose) (see Chapter 3, Section 2.4.6, Reductionist–Comparative and Thematic–Analytical Analysis of Maya Art.)

6

# Access analysis of Maya art and architecture: Summary and conclusions

Access analysis of Maya building groups, together with thematic inquiry of monumental art, has provided a means of further assessing the role that imagery played in signaling how space was used and socially demarcated in Classic Maya city centres. At the beginning of this inquiry, I argued that, because of limitations in the breadth of literacy among the Classic Maya (A.D. 250–900; Marcus 1992b: 230), an alternative means would have been required to transmit doctrine to the wider population. One of the methods that the Maya elite would have used to communicate to the broader masses in city centres was with the strategic placement of sculpture and other iconographic media, in effect using monumental art as a form of "signposting" and as a means of signaling the function and hierarchical divisions in ritual and administrative spaces in city centres.

The objective of my investigation was to determine whether a spatial analysis of sculpture within a Classic Maya centre would influence current proposals that define elite-civic demarcation and area function, sustained by ethnohistoric, artefactual, and epigraphic assessment. As previously stated, I saw value in a study establishing whether imagery, as displayed on monuments and architecture, could further contribute to our understanding of social order and control within important Maya sites. Results of my analysis of sculpture and other artwork associated with the earlier phases of the Palace Group, as well as the Cross Group temples at Palenque, do suggest a correspondence between the thematic variations present in Classic Maya art and differences in the accessibility of that art. Furthermore, differences identified in the text and imagery of monuments suggest disparities in the types of activities associated with those spaces as well as the classes of audiences that could access them.

In this investigation, differences also have been identified in the way rulers chose to portray themselves in sculpture and other artwork, depending on whether representations were positioned in public (common) view or restricted to more private (elite) display. The implication of this finding is that Maya rulers used art to promote different perceptions of themselves, depending on whether the art was positioned in areas accessible to other elite individuals or to the general population. General questions that I addressed in this investigation were: (1) What was it that motivated the Maya elite to position sculpture where they did? (2) Is there specific iconography marking specific spaces? and (3) Acknowledging the multidimensionality of meaning communicated by Maya sculpture, are there embedded within compositions clues as to how certain space was used and socially demarcated?

## 6.1. Summary of research objectives

By using methods in access analysis, monumental art assigned to private/elite spaces was compared and contrasted with sculpture and other decorative media positioned in public/civic spaces. In this investigation, access diagrams were used to establish hierarchical contexts for the comparison of iconographic variation in monumental art resulting from differences in spatial positioning. More specifically, monuments and other iconographic media investigated included those that could be positioned in areas classified as most elite and most common as well as those that could be positioned along the privacy gradient in between (Pearson and Richards 1994a: 40). A premise underpinning my investigation was the notion that public space and private space (as differentiated by variations in accessibility) corresponded to notions of common space and elite space. Due to the stratified nature of Classic Maya society and the subsequent segregation of the populous, it is likely the monumental art contained within common and elite categories of space would have been accessed by corresponding common and elite audiences. Following this line of reasoning, the potential for using access diagrams to discern the role that iconography played in signaling hierarchical divisions in Maya building groups was explored. Access diagrams were justified from the perspective of the carrier position or, more specifically, from the viewpoint of the individual standing externally to the spatial system being investigated.

My objective in using justified access maps was to determine the hierarchical distribution of internal and external bounded spaces from the perspective of the outsider (as articulated by walls and doorways, bounding structures, and natural barriers, such as terracing and abrupt changes in elevation) and, correspondingly, all surviving artwork associated with those spaces. Access diagrams derived from the building groups investigated were also used in the calculation of indices referred to as the Relative Asymmetry (RA) Index and the Control Values Index, which assisted in discerning the relative depths and accessibility of spaces and associated sculpture from the viewpoint of visitors as well as inhabitants (i.e., persons who would have been permitted to enter and/or occupy the spatial systems being investigated). Whereas justified access maps of building groups were used to establish the accessibility of artworks from the perspective of the outsider, control values and RA analysis were used in establishing the relative accessibility of spaces and associated artworks from the perspective of all other spaces, which collectively represent the domain of the inhabitant (i.e., the entire spatial system under study; see Chapter 2, Section 2.4.5, Measuring Control Values and RA Analysis, for more information). By using these approaches, I was able to determine hierarchies of space in Maya architecture that functioned as a contextual framework for the comparative analysis of thematic variation in associated sculpture and other pictographic media.

Given the contextual variation necessary for this type of investigation, a range of spatial contexts containing iconographic media was required. This study, focused as it was on the relationship between imagery and spatial context, required groups of buildings in addition to single structures as units of investigation, as single structures alone did not provide the variety and complexity of form necessary for a study of social and hierarchical partitioning in a Maya city centre. Subsequently, architectural forms singled out for this investigation were those structures found to comply in form with Andrews's (1975) building group archetypes (i.e., Temple Groups, Palace Groups, Quadrangle Groups, and Acropolis Groups). Examples of Andrews's building group archetypes include the Palenque Cross Group, the Palenque Palace, the Uxmal Nunnery Quadrangle, and the Copán Acropolis. Given the consistency with which these structural forms are represented at different Maya centres, value was seen in their use for the purposes of my investigation. Andrews's archetypes provided broad spatial complexes suitable for my study while permitting the formulation of generalised arguments concerning the behaviour associated with art and architecture among the Classic Maya. Originally, my goal in this investigation was to investigate examples of all four of Andrews's structural types. Space in this study, however, permitted the inclusion of only two of the building group types just mentioned. Subsequently, this study focused exclusively on two of Andrews's (1975) structural types, namely the Cross (Temple) Group and the Palace Group from Palenque.

Measures were taken in this study to provide temporal control over the Palace Group component of this investigation, as the Palenque Palace represents a long period of architectural development (Martin and Grube 2000: 164). More specifically, monumental art and architecture of the Palace complex were sorted to phases and the reigns of kings. This sorting was undertaken as a way of confining the periods of time investigated, whilst assuring any correlations made among sculpture, architecture, and space were in context. Furthermore, sorting the Palace architecture into phases provided a method of establishing inferences as to the intention of different Maya rulers in the commissioning and placement of particular sculptural media within specific spatial/architectural contexts. As noted previously, "many of the architectural patterns and principals . . . documented at [Maya sites] . . . can shed light on the issue of the ideological and political strategies put to work by different dynasties and rulers in their respective dynastic building programs (Fash 1998: 224–225). General questions that I addressed using examples from Palenque were: (1) What motivated the positioning of informative sculpture and other artwork at various levels of common and elite accessibility? and (2) What are the differences between the information being transmitted by artwork in different spatial contexts?

It is acknowledged that any argument presented concerning the spatial distribution of sculpture and other artwork relies on the presence of monuments and other artistic media preserved in situ. For this reason, primary data used in my study were obtained from surviving monumental art adhered to architectural features, such as sculpture as well as some surviving mural art. In my inquiry, focus was directed at examining indoor and outdoor sculpture, the sculpture of entrances, sculpture and its depth within structures, and the sculpture of stairways and elevated architecture. The main sources used in this investigation were published and unpublished architectural plans of the Palenque Palace and Cross Group (e.g., Blom 1923; Andrews 1974, 1975; Greene Robertson 1985a, 1985b, 1991), including photographs and line drawings of artwork appearing on friezes, murals, piers, jambs, panels, roofcombs, panels, stairways, lintels, altars, thrones, and benches (see Maudslay 1889–1902a; Seler 1976; Schele and Mathews 1979; Greene Robertson 1985a, 1985b, 1991). The themes of the various sculptural media investigated were determined through compositional inquiry of monuments (Clarkson 1979; see also Parmington 2003). Inscriptions and elements such as the figural position of individuals portrayed, perspective, furniture, and costume were considered collectively to determine the behaviour portrayed and to propose activities associated with their spatial contexts (see Chapter 2, Section 2.4.7, Some Categories for Thematic Analytical Investigation).

## 6.2. Discussion and conclusions

It has been argued that "Maya buildings [functioned as] . . . symbolic devices for marking social status" (Fash 1998: 239). One way that status was expressed was in terms of varying degrees of elevation and accessibility within architectonic space. Maya elite architecture is identified by its restrictiveness. The elevated nature of many Maya building groups, combined with their small interiors, suggests that access was reserved primarily for the elite in Maya society (Cohodas 1973: 215). "Monumental stairways . . . [did] many things beyond allowing people to move up and down a building or platform. Iconographic and textual data show that stairways . . . display tribute, booty, and captives. By sitting at various levels the Maya establish or confirm relative social ranking" (Houston 1998: 522). In contrast to elite architecture, characteristics of public architectural assemblages are their spaciousness, comparatively low elevation, and increased levels of accessibility. For example, wide entrances promote entry into plazas and participation in the activities (ritual or otherwise) once undertaken within those spaces (Cohodas 1973: 216).

In my investigation, Maya sculpture and other artwork were ranked according to locations on the access diagrams and were interpreted in terms of their public/common and private/elite accessibility. Examples of some of the sculptural types assigned to public and progressively private spatial contexts in Maya centres include stelae, piers, and altars (most often associated with open plazas); lintels and jambs (typically found at the entrances to structures); and panels, murals, and benches (most often found in the interior of buildings or building groups). On analysing all surviving sculpture associated with the Palace and Cross Group at Palenque, twenty-two generic themes were defined and found to correlate to a variety of spatial contexts; these themes were as follows: (1) Accession; (2) Ancestor Portraiture Solo; (3) Ancestor Portraiture with Attendants (4) Captive Display; (5) Text, Captive Display; (6) Cosmological Underworld; (7) Cosmological Terrestrial; (8) Cosmological Celestial; (9) God Portraiture Solo; (10) God Portraiture with Attendants; (11) Human Sacrifice; (12) Auto Sacrifice; (13) Mat Motif; (14) Portraiture Solo; (15) Sacrificial Dance; (16) Text, Cosmological; (17) Text, Historical; (18) Text, Mythological; (19) Text, Dedication; (20) Royal Audience; (21) Ritual Dance; and (22) Royal Portraiture Solo (see Chapter 2, Section 2.4.7, Some Categories for Thematic Analytical Investigation).

The presence of different sculptural themes at varying levels of accessibility in the early phases of the Palace (see Chapter 4, Section 4.3.1, The Buildings of the Palace Phase 1, Section 4.3.2, The Buildings of the Palace Phase 2, and Section 4.3.3, The Buildings of the Palace Phase 3) suggests that different imagery was directed at different audiences at Palenque. More specifically, during the reign of K'inich Janab' Pakal I, the shallow reception area of the Palace (i.e., the Early East Court) displays war propaganda themes, such as Captive Display (GR 58 [*alfardas* only], 59; Figures 4.46 and 4.28). At a deeper level, royal portraits (GR 24 and 25) and a "mat" pattern (GR 26) decorating the roof of the structure mark House B as a *popol nah* ("council house"). In addition, at this level of depth, pantheon style portraiture (GR 29; Figures 4.23, 4.24, 4.25, and 4.26) is represented located in the east gallery of House C. At a deeper level, Accession and Royal Audience (GR 5 and 80; Figures 4.13 and 4.6) themes are portrayed, followed by sculpture housed in the subterranean passages depicting Cosmological Underworld themes (GR 7, 78, and 79; see Figures 4.7 and 4.8). When considering the sculpture of the early phases of the Palace Group (Palace Phases 1, 2, and 3), the three carved vaults (GR 7, 78, and 79) found in the south subterranean passages represent the deepest sculpture of the Palace from the perspective of the outsider. Located in transitional spaces, the East, West, and South Subterranean Passage vaults are also the least "integrated" (i.e., the most segregated from the perspective of Palace inhabitants) of all the surviving sculpture represented. As noted by Schele, the presence of these sculptures in the underground passages leading to the Subterraneos marked the corridors as entrances into the Underworld (Schele 1986: 122).

In apparent contrast to K'inich Janab' Pakal I, with the closing of the northern end of the Palace with House

A-D by his son K'inich K'an Joy Chitam II, there was a shift in agenda as reflected in the placement of additional furniture and sculpture at low levels of accessibility. More specifically, high concentrations of sculpture in the shallower regions of the Palace at Phase 3 imply a change in the audience that K'inich K'an Joy Chitam II was seeking to address with his sculptural program. Most notably, the positioning of a throne in the north corridor of House A-D, combined with the Palace Tablet (GR 57; Figure 4.39), implies an increased concern for the addressing of the civic population. K'inich K'an Joy Chitam II's objective in placing Royal Audience Accession imagery at a higher level of accessibility suggests an increased concern to legitimise his authority among the broader population at Palenque. As pointed out in Chapter 5, the significance of the Palace Tablet's location is that it marks a point of demarcation between what was, first and foremost, the ruler's domain (i.e., the Palace) and what was the domain of the civic population (i.e., areas outside the Palace complex). The installation of northern monumental stairs at Palace Phase 3 (while providing an impressive approach for persons permitted to ascend them) were built primarily to function as boundaries to more effectively segregate Palace occupants from the general population at Palenque. Furthermore, the presence of a throne just beneath the Palace Tablet suggests that the northern gallery of House A-D was probably used by K'inich K'an Joy Chitam II to address or observe people outside the Palace complex (see Chapter 4, Section 4.3.3, The Buildings of the Palace Phase 3).

Differences have been identified in how rulers chose to represent themselves in monumental art at Palenque. More specifically, public imagery (positioned in spaces where access is less controlled) tends to portray rulers authoritatively, whereas more private portraiture (appearing in more restricted spatial contexts) depicts rulers in a more understated manner, in some instances even as subordinates. For example, all three of the Cross Group tablets portray K'inich Kan B'alam II as a subordinate to the gods and the cosmological imagery represented at the centre of the monuments. The private character of the Cross Group sanctuaries, while signaled by the deep and elevated nature of the structures, is also implied by the understated appearance of the costumes worn by K'inich Kan B'alam II in all of the Cross Group tablets. In contrast, surviving sculpture appearing on the entablature and exterior piers of the Temple of the Cross and the Temple of the Sun indicates that elaborately dressed individuals once adorned the more public exterior of the buildings. Similar examples appear on the external piers of the Palace Group Houses A and D, where kings are portrayed authoritatively in the company of subordinate individuals (see Chapter 4, Section 4.3.3, The Buildings of the Palace Phase 3, House A-D: Description and Sculpture). In accordance with this finding, public monuments tend to portray rulers as they would have appeared before large crowds or for important public occasions, whereas simpler dress is typically found in spaces where access to the monument's viewing would have been restricted (Taylor 1983: 13–15).

A correlation was found between permeation gradient and shifts in the values calculated with increasing depth for the Cross Group. More specifically, as one moves progressively deeper into the Cross Group, spatial units become less integrated (i.e., more segregated) from the rest of the spatial system. Spaces displaying the greatest level of control over access to other spaces in the system are the rooms fronting the sanctuaries in each of the temples. These spaces were calculated to be controlling, because access to each of the temple's inner sanctuaries, as well as the rooms flanking these sacred spaces, is provided exclusively from these rooms. The central plaza of the Cross Group is also a controlling space, as access to all three of the Cross Group temples requires passing through this space. In addition to occupying controlled spaces, the tablet of each of the inner sanctums is positioned in the least integrated spaces of the building group.

The most controlling spaces in access diagrams tend to be the most distributed of the system, marking their relatively high accessibility from the perspective of other spaces in local proximity. Throughout the evolution of the Palace complex (see Chapter 4, Section 4.2, Evolutionary Sequence of the Palace), spaces displaying the greatest level of control over access to other spaces include House J's western terrace, the upper southeast platform of the Palace, and all four of the Palace courtyards. The restricted, confined, and undecorated appearance of House J marked the private character of the building. The Palace courtyards were consistently some of the most controlling spaces of the Palace Group over all six phases of development, functioning as semi-public and semi-private domains as well as spaces that could be used to regulate access to more restricted areas of the complex. Some of the other spatial units that held significant levels of control at various stages of the Palace's development included the northern substructure's southern terrace, the northern substructure's northern terrace, the southern substructure corridors, the subterranean passages, and the first and second levels of the Tower. In contrast, some of the controlled spaces included the eastern and western galleries of Houses C and E, the northern and southern rooms of House B, as well as the Palace Toilets and Sweat Baths (see Chapter 5, Section 5.1, Access Analysis of Palace Phases 1–6).

Over the development of the Palace Group, sculptural themes assigned to controlling spaces included Capture Display, as well as Cosmological Underworld. Themes appearing in controlled spatial contexts included God Portraiture Solo, Ancestor Portraiture Solo, Ancestor Portraiture with Attendant, Human Sacrifice, Sacrificial Dance, God Portraiture with Attendants, Royal Audience, and Royal Audience Accession. (Note: Monuments displaying Royal Audience Accession themes as well as Dynastic Historical texts were identified in both controlling and controlled spatial contexts.)

Although some correspondence between spatial context and thematic variations in monumental art has been detected, analysis of the last three phases of the Palace (i.e., Phases 4, 5, and 6) suggests that architectural embellishment became less important over time. This disparity is signaled not only by the absence of sculpture and other artwork at the later phases of the Palace, but also by the shifting accessibility of earlier artwork with the architectural modifications undertaken by subsequent kings. The largest proportion of decorative works, totaling some 52%, appeared during the reign of K'inich Janab' Pakal I; this was followed by K'inich K'an Joy Chitam II's contribution of 30%. Over the remaining three phases of the Palace, K'inich Akul Mo'Nab' III, U Pakal K'inich Janab' Pakal II, K'inich Kan B'alam III, K'inich K'uk' B'alam II, and Wak Kimi Janab' Pakal III jointly added the remaining 18% (see Chapter 5, Section 5.2, Public and Private Sculpture of the Palenque Palace Group).

My analysis of the architectural sequencing of the Palace at Palenque identified a major shift in the configuration of the building group after Phase 2, from what was an acropolis-type arrangement (see Chapter 4, Section 4.3.1, The Buildings of the Palace Phase 1, and Section 4.3.2, The Buildings of the Palace Phase 2) to one that conforms more precisely to a Palace-type configuration (see Chapter 4, Section 4.3.3, The Buildings of the Palace Phase 3, Section 4.3.4, The Buildings of the Palace Phase 4, Section 4.3.5, The Buildings of the Palace Phase 5, and Section 4.3.6, The Buildings of the Palace Phase 6) as described by Andrews (1975) and Cohodas (1973: 215). In accordance with Andrews's (1975: 67) definition of an Acropolis Group archetype, the combined effect of the Palace platforms at Phases 1 and 2 provided a progressive change in elevation from the perspective of someone entering via the Northern Galleries, which I have previously argued functioned as a form of gateway structure controlling access into the Palace Group interior. Such shifts in structural form, from one building group configuration to another, have been noted for Group A-5 at Uaxactun. Like the Palenque Palace, Group A-5 was the result of gradual architectural modification that occurred over the reigns of successive kings. Eight phases of construction of Group A-5 have been documented, illustrating how a building complex can change dramatically over time, in this case from what was once a triadic Temple Group design to a series of complex multichambered galleries (see Miller 1999: 30).

The effect of the overall increase in number of access routes between spatial units over successive phases of the Palace was to increase the level of integration expressed by the plan of the building group. Presently, it remains unclear why the Palace shifted from what was once a deeper, less integrated plan at Phases 1 and 2 to a gradually more integrated design at the remaining architectural phases. The shift away from monumental construction to the blocking and segmenting of existing Palace architecture would have been due to a range of factors, some of which may have been: (1) declining access to the necessary raw materials, such as limestone and wood used in building and stucco production; (2) the loss and/or absence of the artistic and/or engineering skills required to undertake larger construction programs; (3) dwindling control over the necessary labour force to implement more ambitious construction projects; and (4) increasing pressure to protect and preserve the privacy of an ever-growing population of Palace inhabitants from outsiders. Maya palaces often changed in their spatial configuration "since they grew organically to suit the needs of an expanding elite in residence" (Cohodas 1973: 216).

The increase in the number of route options between spaces, although enhancing accessibility and reducing effective control in some of the systems, may have also been used as a strategy of avoidance. More specifically, in spatial systems where control over movement does not need to be enforced (e.g., in the home, where rules of movement are acknowledged and abided by), alternate routes through a network of spaces provides a means of moving unobserved between points. Therefore, rather than increasing levels of accessibility, integrated spatial systems may also have been used to enhance privacy in the sense that certain spaces do not have to be encroached upon when accessing others. In contrast, limiting the number of path options between spaces may be used in instances where the regulation of movement needs to be enforced, as spaces ordered consecutively with no alternative routes between points provide no means of avoidance. Such a spatial configuration is apparent in the Cross Group and the earlier phases of the Palace, where there are fewer route options between spaces in the shallower regions of the Palace complex (see Chapter 4, Section 4.3.1, The Buildings of the Palace Phase 1, and Section 4.3.2, The Buildings of the Palace Phase 2).

One of the objectives of the many architectural amendments undertaken in the Palace was to reconfigure the complex in terms of accessibility and control. It seems, however, that the Palace Group at Palenque was configured not only to enhance or preserve the privacy of inhabitants with respect to the encroachment of visitors, but to also more effectively segregate inhabitants from other inhabitants. "Privacy... [operates] in two dimensions: one that relate[s] to the public world of strangers to the private world of inhabitants, and a second which relate[s] the inhabitants to one another" (Grahame 1997: 143–144). Changes in the levels of integration and control displayed by the Palace houses at each phase of construction suggest that activities associated with these buildings may have become increasingly hidden from, or exposed to, the observances of others occupying the different areas of the Palace complex. Due to the structural amendments at Palace Phases 5 and 6, the Southeast Court and the Tower Court of the Palace became substantially smaller, indicating that the activities and social encounters occurring within these spaces were becoming more intimate. It has been proposed that "architectural settings... reflect the modes of ritual communication that occurred in those spaces" (Moore 1996: 789). Among these modes of communication are "distinct spatial ranges... [which] reflect general limits to modes of interpersonal communication" (Moore 1996: 791; see also Hall 1959, 1966). These modes of communication have been referred to as "intimate" (occurring at zero distance between participants), "personal" (0.45–1.20 meters), "social" (1.20–3.60 meters), and "public" (3.60–7.50+ meters; see Chapter 2, Section 2.1, Public versus Private Space, Public versus Private Art).

At Phases 2, 3, 4, and 5, the Palace complex comprises four quadrants marked by the presence of the four Palace courtyards. The more open and decorated northeastern, northwestern, and southwestern quadrants of the Palace Group were configured to signal, among other things, wealth, ancestry, and ascendancy of the king to inhabitants and outsiders permitted to enter the complex. In contrast, the closed and understated character of the southeastern quarter (see Chapter 5, Section 5.1.5, Analysis of the Palace Phase 5, and Section 5.1.6, Analysis of the Palace Phase 6) suggests that it was not intended for general admittance, but rather it functioned as a private domain where Palace inhabitants withdrew to either retire or avoid being observed. The large size and shallow positioning of the East Court (as well as the relatively low level of integration displayed by the court space) is consistent with the view that the northeast quadrant of the Palace functioned as an interface between inhabitants and visitors permitted to enter the Palace complex. In support of this proposal, House C (the building that defines the east boundary of the northeast quadrant) is described as the *yet nah* ("companion house") on the structure's Hieroglyphic Staircase.

## 6.3. Art and access analysis: Advantages and disadvantages of methodological approach

The main problem that I encountered when employing access analysis to interpret the hierarchical ordering of thematic art in Maya building groups was the inability to avoid some level subjectivity in the investigation; specifically, when representing units of architectonic space in access diagrams and when allocating the generalised themes appearing in monumental artwork. In access analysis, choices have to be made concerning the degree of resolution that one wishes to represent in access diagrams. These choices vary depending on the individual who is undertaking the investigation. As Hillier and Hanson (1984: 98) have acknowledged, "one... has to decide when changes in the shapes of buildings or boundaries are allowed to make a difference to convex spaces," which will ultimately affect the quantity and distribution of nodes and edges in access diagrams. In relation to the designation of themes represented in sculpture and other artwork, these too may vary depending on the interpretation of the investigator.

Any interpretation of architecture and corresponding sculpture derived from analysis remains dependent on the accuracy of architectural plans used to construct access diagrams, as well as the themes assigned to artworks. Furthermore, in cases where the architectural features investigated do not correspond to a single phase of construction, accurate temporal sequencing in architecture and corresponding embellishments is essential. As stipulated earlier in this inquiry, the Palace sequence offered in this study, in addition to results drawn from the analysis of corresponding sculptural features, can only be as accurate as the supporting evidence. For example, the absence of looted or obliterated sculpture may potentially affect arguments presented in this investigation. Like the few construction sequences previously proposed (see Andrews 1974; Greene Robertson 1985b: fig. 5, 118; Schele 1986: 118–119; Miller 1999: 35–38; Martin and Grube 2000: 164), the one that is presented here is based on limited archaeological, epigraphic, and stylistic evidence. Consequently, this investigation should be viewed as a practical demonstration of an analytical technique rather than a presentation of definitive results. My main objective in this study was to demonstrate, by example, how a sequence in Maya architecture, when identified, may be combined with methods in access analysis to discern intent in the sculptural programs implemented by different rulers over different periods of time. As stated previously, the power of this procedure lies in its capacity to be repeated given the availability of new information. For example, in light of any new data (drawn preferably from further archaeological, epigraphic and iconographic investigation of the Cross Group Palace at Palenque), some refinement of the architectural and sculptural sequencing presented in this study would be expected, after which amended access maps could be reanalysed and more definitive conclusions drawn.

# Appendixes 1–7

Appendix 1. *The Palenque Cross Group Dates by Peter Mathews, reprinted with permission.*

| Monument: Text reference | Date of monument | Date of passage | Calendar round | Julian date | Event | Name |
|---|---|---|---|---|---|---|
| T.Cross, Tab. Cross: A16-B16 | 9.12.18. 5.19 ? | 12.19.13. 3. 0 | 1 Ajaw 18 Sutz' | 13 Dec 3121 BC | 819d | – |
| T.Cross, Tab. Cross: A1-A13 | 9.12.18. 5.19 ? | 12.19.13. 4. 0 | 8 Ajaw 18 Sek | 2 Jan 3120 BC | Birth | "Lady Beastie" |
| T.Cross, Tab. Cross: implied | 9.12.18. 5.19 ? | 12.19.13. 4. 0 | 8 Ajaw 18 Sek | 2 Jan 3120 BC | Birth | [*"Lady Beastie"] |
| T.Cross, Tab. Cross: D3-C5 | 9.12.18. 5.19 ? | 13. 0. 0. 0. 0 | 4 Ajaw 8 Kumk'u | 8 Sep 3114 BC | Period-ending | – |
| T.Cross, Tab. Cross: implied | 9.12.18. 5.19 ? | 13. 0. 0. 0. 0 | 4 Ajaw 8 Kumk'u | 8 Sep 3114 BC | Creation | GI "the father" |
| T.Sun, Tab. Sun: implied | 9.12.18. 5.19 ? | 13. 0. 0. 0. 0 | 4 Ajaw 8 Kumk'u | 8 Sep 3114 BC | Creation | – |
| T.Sun, Tab. Sun: O2-N3 | 9.12.18. 5.19 ? | 13. 0. 0. 0. 0 | 4 Ajaw 8 Kumk'u | 8 Sep 3114 BC | Creation | – |
| T.Cross, Tab. Cross: implied | 9.12.18. 5.19 ? | 13. 0. 1. 9. 0 | 11 Ik' 18 Mol | 1 Mar 3112 BC | "Deer-hoof" | ["Lady Beastie"] |
| T.Cross, Tab. Cross: C9-D9 | 9.12.18. 5.19 ? | 13. 0. 1. 9. 2 | 13 Ik' 20 Mol | 3 Mar 3112 BC | Dedication | [GI "the father"] |
| T.Sun, Tab. Sun: B15-A16 | 9.12.18. 5.19 ? | 1.18. 4. 7. 1 | 1 Imix 19 Pax | 2 Feb 2360 BC | 819d | – |
| TFC, Tab. Fol. Cross: B13-A14 | 9.12.18. 5.19 ? | 1.18. 4. 7. 1 | 1 Imix 19 Pax | 2 Feb 2360 BC | 819d | – |
| T.Cross, Tab. Cross: E1-F1 | 9.12.18. 5.19 ? | 1.18. 5. 3. 2 | 9 Ik' 15 Keh | 10 Nov 2360 BC | Birth | GI |
| T.Cross, *Alfardas*: A1-B1 | 9.12.19.14.12 | 1.18. 5. 3. 2 | 9 Ik' 15 Keh | 10 Nov 2360 BC | Birth | GI |
| T.Sun, Tab. Sun: A1-B12 | 9.12.18. 5.19 ? | 1.18. 5. 3. 6 | 13 Kimi 19 Keh | 14 Nov 2360 BC | Birth | GIII |
| T.Sun, *Alfardas*: A1-B1 | 9.13. 0. 0. 0 ? | 1.18. 5. 3. 6 | 13 Kimi 19 Keh | 14 Nov 2360 BC | Birth | GIII |
| TFC, Tab. Fol. Cross: A1-A12 | 9.12.18. 5.19 ? | 1.18. 5. 4. 0 | 1 Ajaw 13 Mak | 28 Nov 2360 BC | Birth | GII |
| TFC, Tab. Fol. Cross: D14-C15 | 9.12.18. 5.19 ? | 1.18. 5. 4. 0 | 1 Ajaw 13 Mak | 28 Nov 2360 BC | Birth | GII |
| TFC, *Alfardas*: A1-B1 | 9.12.19.14.12 ? | 1.18. 5. 4. 0 | 1 Ajaw 13 Mak | 28 Nov 2360 BC | Birth | GII |
| TFC, Tab. Fol. Cross: C7b-D8 | 9.12.18. 5.19 ? | 2. 0. 0. 0. 0 | 2 Ajaw 3 wayeb' | 9 Mar 2325 BC | Period-ending | "Lady Beastie" |
| T.Cross, Tab. Cross: E9-F9 | 9.12.18. 5.19 ? | 2. 1. 0.14. 2 | 9 Ik' 0 Yax | 2 Sep 2305 BC | Accession | "Lady Beastie" |
| T.Cross, Tab. Cross: F12 | 9.12.18. 5.19 ? | 2. 1. 0.14. 2 | 9 Ik' 0 Ya | 2 Sep 2305 BC | [*Accession] | [*"Lady Beastie"] |
| T.Cross, Tab. Cross: implied | 9.12.18. 5.19 ? | 5. 7.11. 8. 4 | 1 K'an 2 Kumk'u | 20 Mar 993 BC | Birth | U Kokan Chan |
| T.Cross, Tab. Cross: Q2-P3 | 9.12.18. 5.19 ? | 5. 8.17.15.17 | 11 Kab'an 0 Pop | 6 Apr 967 BC | Accession | U Kokan Chan |
| T.Cross, Tab. Cross: P4-Q4 | 9.12.18. 5.19 ? | 8.18. 0.13. 6 | 5 Kimi 14 K'ayab' ? | 30 Mar 397 | Birth | K'uk' B'alam I |
| T.Cross, Tab. Cross: P10-Q10 | 9.12.18. 5.19 ? | 8.19. 6. 8. 8 | 11 Lamat 6 Xul | 8 Aug 422 | Birth | "Casper II" |
| T.Cross, Tab. Cross: implied | 9.12.18. 5.19 ? | 8.19. 6. 8. 8 | 11 Lamat 6 Xul | 8 Aug 422 | Birth | "Casper II" |
| T.Cross, Tab. Cross: Q8-P9 | 9.12.18. 5.19 ? | 8.19.15. 3. 4 | 1 K'an 2 K'ayab' ? | 10 Mar 431 | Accession | [*K'uk' B'alam I] |
| T.Cross, Tab. Cross: P14-Q14 | 9.12.18. 5.19 ? | 8.19.19.11.17 | 2 Kab'an 10 Xul | 9 Aug 435 | Accession | "Casper II" |
| T.Cross, Tab. Cross: Q17-R2 | 9.12.18. 5.19 ? | 9. 0. 0. 0. 0 | 8 Ajaw 13 Keh | 10 Dec 435 | Period-ending | [*"Casper II"] |
| T.Cross, Tab. Cross: implied | 9.12.18. 5.19 ? | 9. 1. 4. 5. 0 | 12 Ajaw 13 Sak | 14 Nov 459 | Birth | B'utz'aj Sak Chik |
| T.Cross, Tab. Cross: R10-S10 | 9.12.18. 5.19 ? | 9. 1.10. 0. 0 | 5 Ajaw 3 Sek | 5 Jul 465 | Birth | Akul Mo' Nab' I |
| T.Cross, Tab. Cross: R7-S7 | 9.12.18. 5.19 ? | 9. 2.12. 6.18 | 3 Etz'nab' 11 Xul | 28 Jul 487 | Accession | [*B'utz'aj Sak Chik] |
| T.Cross, Tab. Cross: implied | 9.12.18. 5.19 ? | 9. 2.15. 3. 8 | 12 Lamat 6 Wo | 3 May 490 | Birth | K'an Joy Chitam I |
| T.Sun, Tab. Sun: P2-Q2 | 9.12.18. 5.19 ? | 9. 3. 1.15. 0 | 12 Ajaw 8 Keh | 19 Nov 496 | Heir-designation | K'an Joy Chitam I |
| T.Cross, Tab. Cross: S12-R13 | 9.12.18. 5.19 ? | 9. 3. 6. 7.17 | 5 Kab'an 0 Sutz' | 3 Jun 501 | Accession | [*Akul Mo' Nab' I] |
| T.Cross, Tab. Cross: implied | 9.12.18. 5.19 ? | 9. 4. 9. 0. 4 | 7 K'an 17 Mol | 3 Sep 523 | Birth | [*Akul Mo' Nab' II] |
| T.Cross, Tab. Cross: U10-T11 | 9.12.18. 5.19 ? | 9. 4. 9. 0. 4 | 7 K'an 17 Mol | 3 Sep 523 | Birth | Akul Mo' Nab' II |
| T.Cross, Tab. Cross: implied | 9.12.18. 5.19 ? | 9. 4.10. 1. 5 | 11 Chikchan 13 Ch'en | 18 Sep 524 | Birth | Kan B'alam I |
| T.Cross, Tab. Cross: T14-U14 | 9.12.18. 5.19 ? | 9. 4.10. 1. 5 | 11 Chikchan 13 Ch'en | 18 Sep 524 | Birth | Kan B'alam I |
| T.Cross, Tab. Cross: implied | 9.12.18. 5.19 ? | 9. 4.10. 1. 5 | 11 Chikchan 13 Ch'en | 18 Sep 524 | Birth | Kan B'alam I |
| T.Cross, Tab. Cross: R17-S17 | 9.12.18. 5.19 ? | 9. 4.14.10. 4 | 5 K'an 12 K'ayab' | 23 Feb 529 | Accession | [*K'an Joy Chitam I] |
| T.Cross, Tab. Cross: U5-T6 | 9.12.18. 5.19 ? | 9. 6.11. 5. 1 | 1 Imix 4 Sip | 2 May 565 | Accession | Akul Mo' Nab' II |

| | | | | | | | | |
|---|---|---|---|---|---|---|---|---|
| T.Cross, Tab. Cross: implied | 9.12.18. 5.19 ? | 9. 6.18. 5.12 | 10 Eb' | 0 Wo | 6 Apr | 572 | Accession | Kan B'alam I |
| T.Sun, Tab. Sun: P12-Q12 | 9.12.18. 5.19 ? | 9.10. 2. 6. 6 | 2 Kimi | 19 Sutz' | 20 May | 635 | Birth | K'inich Kan B'alam I |
| TFC, Tab. Fol. Cross: implied | 9.12.18. 5.19 ? | 9.10. 2. 6. 6 | 2 Kimi | 19 Sutz' | 20 May | 635 | Birth | [*K'inich Kan B'alam II] |
| T.Cross, Incensario 2: implied | 9.13. 6. 1.11 ?? | 9.10. 5. 8. 4 | 2 K'an | 2 Xul ??? | 11 Jun | 638 | Birth ? | ? |
| T.Cross, Tab. Cross: G-H | 9.12.18. 5.19 ? | 9.10. 8. 9. 3 | 9 Ak'b'al | 6 Xul | 14 Jun | 641 | Heir-designation | K'inich Kan B'alam II |
| T.Sun, Tab. Sun: P6-Q6 | 9.12.18. 5.19 ? | 9.10. 8. 9. 3 | 9 Ak'b'al | 6 Xul | 14 Jun | 641 | Heir-designation | K'inich Kan B'alam I |
| T.Sun, Tab. Sun: implied | 9.12.18. 5.19 ? | 9.10. 8. 9. 3 | 9 Ak'b'al | 6 Xul | 14 Jun | 641 | Heir-designation | K'inich Kan B'alam I |
| T.Sun, Tab. Sun: E1-F1 | 9.12.18. 5.19 ? | 9.10. 8. 9. 3 | 9 Ak'b'al | 6 Xul | 14 Jun | 641 | Heir-designation | K'inich Kan B'alam I |
| TFC, Tab. Fol. Cross: implied | 9.12.18. 5.19 ? | 9.10. 8. 9. 3 | 9 Ak'b'al | 6 Xul | 14 Jun | 641 | Heir-designation | [*K'inich Kan B'alam II] |
| T.Sun, Tab. Sun: implied | 9.12.18. 5.19 ? | 9.10. 8. 9. 8 | 1 Lamat | 11 Xul | 19 Jun | 641 | Heir-designation | K'inich Kan B'alam I |
| T.Cross, Tab. Cross: K9 | 9.12.18. 5.19 ? | 9.10.10. 0. 0 | 13 Ajaw | 18 K'ank'in | 3 Dec | 642 | ? | – |
| T.Sun, Tab. Sun: G2-H2 | 9.12.18. 5.19 ? | 9.10.10. 0. 0 | 13 Ajaw | 18 K'ank'in | 3 Dec | 642 | Period-ending | K'inich Kan B'alam I |
| T.Sun, Tab. Sun: Q14b-Q15 | 9.12.18. 5.19 ? | 9.10.10. 0. 0 | 13 Ajaw | 18 K'ank'in | 3 Dec | 642 | Period-ending | K'inich Kan B'alam I |
| T.Cross, Incensario 1: A1-? | 9.12.10. 0. 0 | 9.10.15. 6. 8 | 4 Lamat | 16 Pop | 14 Mar | 648 | [*Birth] | [*B'atz' Chan Mat] ? |
| T.Cross, Incensario 1: implied | 9.12.10. 0. 0 | 9.10.15. 6. 8 | 4 Lamat | 16 Pop ? | 14 Mar | 648 | Birth | [*B'atz' Chan Mat] ? |
| T.Cross, Incensario 1: D1-D2 | 9.12.10. 0. 0 | 9.11. 5. 0. 0 | 5 Ajaw | 3 Sak ? | 15 Sep | 657 | Period-ending | B'atz' [Chan Mat] |
| T.Cross, Incensario 1: C6-D6 | 9.12.10. 0. 0 | 9.11. 6.16.17 | 13 Kab'an | 10 Ch'en ? | 13 Aug | 659 | ? | |
| T.Cross, Incensario 1: D10-D11 | 9.12.10. 0. 0 | 9.11. 7. 0. 0 | 10 Ajaw | 13 Yax ? | 5 Sep | 659* | Period-ending ? | B'atz' [Chan Mat] |
| T.Cross, Incensario 1: E1-F1 | 9.12.10. 0. 0 | 9.11. 9.14.19 | 2 Kawak | 17 Xul ? | 20 Jun | 662 | Office-taking | B'atz' [Chan Mat] |
| T.Cross, Incensario 1: F3-F4 | 9.12.10. 0. 0 | 9.11.10. 0. 0 | 11 Ajaw | 18 Ch'en ? | 20 Aug | 662 | Period-ending | B'atz' [Chan Mat] |
| T.Cross, Incensario 1: F7-E8 | 9.12.10. 0. 0 | 9.11.13. 0. 0 | 12 Ajaw | 3 Ch'en ? | 4 Aug | 665 | ? | – |
| T.Cross, Incensario 1: E11-F11 | 9.12.10. 0. 0 | 9.11.15.10. 7 | 3 Manik' | 0 wayeb' ? | 17 Feb | 668 | ? | B'atz' [Chan Mat] |
| T.Cross, Incensario 1: G2-G3 | 9.12.10. 0. 0 | 9.12. 0. 0. 0 | 10 Ajaw | 8 Yaxk'in | 28 Jun | 672 | Period-ending | – |
| T.Cross, Incensario 1: H3 | 9.12.10. 0. 0 | 9.12. 0. 6.18 | 5 Etz'nab' | 6 K'ank'in ? | 13 Nov | 672 | Death | Lady Tz'ak Ajaw |
| T.Cross, Incensario 1: H6-G7 | 9.12.10. 0. 0 | 9.12. 8. 9.18 | 7 Etz'nab' | 6 Muwan ? | 1 Dec | 680 | Death | B'atz' [Chan Mat] |
| T.Cross, Incensario 1: I4 | 9.12.10. 0. 0 | 9.12. 8. 9.18 | 7 Etz'nab' | 6 Muwan ? | 1 Dec | 680 | [Death] | [*B'atz' Chan Mat] |
| T.Cross, Incensario 1: H9 | 9.12.10. 0. 0 | 9.12. 8.10. 0 | 9 Ajaw | 8 Muwan ? | 3 Dec | 680 | Period-ending | – |
| T.Cross, Incensario 1: I6-I7 | 9.12.10. 0. 0 | 9.12.10. 0. 0 | 9 Ajaw | 18 Sutz' | 7 May | 682 | Dedic. of Incens. | |
| T.Cross, Tab. Cross: L1-L2 | 9.12.18. 5.19 ? | 9.12.11.12.10 | 8 Ok | 3 K'ayab' | 7 Jan | 684 | Accession | K'inich Kan B'alam II |
| T.Sun, Tab. Sun: L1-M1 | 9.12.18. 5.19 ? | 9.12.11.12.10 | 8 Ok | 3 K'ayab' | 7 Jan | 684 | Accession | K'inich Kan B'alam I |
| TFC, Tab. Fol. Cross: N5-O5 | 9.12.18. 5.19 ? | 9.12.11.12.10 | 8 Ok | 3 K'ayab' | 7 Jan | 684 | Accession | K'inich Kan B'alam II |
| TFC, Tab. Fol. Cross: E1-E2 | 9.12.18. 5.19 ? | 9.12.11.12.10 | 8 Ok | 3 K'ayab' | 7 Jan | 684 | Accession | K'inich Kan B'alam II |
| T.Sun, Sanct. Fr. R. Pan.:A1-B1 | 9.12.19.14.12 ? | 9.12.15. 7.11 | 10 Chuwen | 4 Sak | 9 Sep | 687 | Conquest | K'inich Kan B'alam I |
| T.Sun, Tab. Sun: N4-O4 | 9.12.18. 5.19 ? | 9.12.18. 5.16 | 2 Kib' | 14 Mol | 20 Jul | 690 | T. Sun Dedication | K'inich Kan B'alam I |
| T.Cross, Tab. Cross: implied | 9.12.18. 5.19 ? | 9.12.18. 5.16 | 2 Kib' | 14 Mol | 20 Jul | 690 | T.Cross Dedication | [*K'inich Kan B'alam II] |
| TFC, Tab. Fol. Cross: L1-M1 | 9.12.18. 5.19 ? | 9.12.18. 5.16 | 2 Kib' | 14 Mol | 20 Jul | 690 | TFC Dedication | [*K'inich Kan B'alam II] |
| TFC, Tab. Fol. Cross: N16 | 9.12.18. 5.19 ? | 9.12.18. 5.16 | 2 Kib' | 14 Mol | 20 Jul | 690 | TFC Dedication | K'inich Kan B'alam II |
| TFC, Tab. Fol. Cross: implied | 9.12.18. 5.19 ? | 9.12.18. 5.16 | 2 Kib' | 14 Mol | 20 Jul | 690 | TFC Dedication | K'inich Kan B'alam II |
| T.Sun, Tab. Sun: O7-N8 | 9.12.18. 5.19 ? | 9.12.18. 5.17 | 3 Kab'an | 15 Mol | 21 Jul | 690 | T. Sun Dedication | K'inich Kan B'alam I |
| TFC, Tab. Fol. Cross: M5-L6 | 9.12.18. 5.19 ? | 9.12.18. 5.17 | 3 Kab'an | 15 Mol | 21 Jul | 690 | TFC Dedication | K'inich Kan B'alam II |
| T.Cross, Tab. Cross: implied | 9.12.18. 5.19 ? | 9.12.18. 5.19 | 5 Kawak | 17 Mol ? | 23 Jul | 690 | T.Cross Dedication | K'inich Kan B'alam II |
| T.Sun, Tab. Sun: implied | 9.12.18. 5.19 ? | 9.12.18. 5.19 | 5 Kawak | 17 Mol ? | 23 Jul | 690 | T. Sun Dedication | K'inich Kan B'alam I |
| TFC, Tab. Fol. Cross: implied | 9.12.18. 5.19 ? | 9.12.18. 5.19 | 5 Kawak | 17 Mol | 23 Jul | 690 | TFC Dedication | K'inich Kan B'alam II |

*(continued)*

Appendix 1 *(continued)*

| Monument: Text reference | Date of monument | Date of passage | Calendar round | | Julian date | Event | Name |
|---|---|---|---|---|---|---|---|
| T.Sun, Piers A,D: D3? | 9.12.19.14.12 ? | 9.12.18. 7. 1 | 1 Imix | 19 Ch'en | 14 Aug 690 | 819d | – |
| T.Sun, Sanct. Jambs: D6b-C7a? | 9.12.19.14.12 ? | 9.12.18. 7. 1 | 1 Imix | 19 Ch'en | 14 Aug 690 | 819d | – |
| TFC, Sanct. Jambs: D6b-C7a | 9.12.19.14.12 ? | 9.12.18. 7. 1 | 1 Imix | 19 Ch'en | 14 Aug 690 | 819d | – |
| T.Cross, *Alfardas*:E2-F2 | 9.12.19.14.12 | 9.12.19.14.12 | 5 Eb' | 5 K'ayab' | 7 Jan 692 | Sanct. Dedication | K'inich Kan B'alam II |
| T.Cross, Sanct. Jambs: Bp4-Ap5 | 9.12.19.14.12 | 9.12.19.14.12 | 5 Eb' | 5 K'ayab' | 7 Jan 692 | Sanct. Dedication | K'inich Kan B'alam II |
| T.Sun, *Alfardas*:E2-F2? | 9.13. 0. 0. 0 ? | 9.12.19.14.12 | 5 Eb' | 5 Kayab' ? | 7 Jan 692 | Sanct. Dedication | K'inich Kan B'alam I |
| T.Sun, Piers A,D: A1-C2? | 9.12.19.14.12 ? | 9.12.19.14.12 | 5 Eb' | 5 K'ayab' | 7 Jan 692 | Sanct. Dedication | K'inich Kan B'alam I |
| T.Sun, Sanct. Jambs: C1-D4? | 9.12.19.14.12 ? | 9.12.19.14.12 | 5 Eb' | 5 K'ayab' | 7 Jan 692 | Sanct. Dedication | K'inich Kan B'alam I |
| TFC, *Alfardas*:E2-F2 | 9.12.19.14.12 ? | 9.12.19.14.12 | 5 Eb' | 5 Kayab' | 7 Jan 692 | Sanct. Dedication | K'inich Kan B'alam II |
| TFC, Sanct. Jambs: C1-D4 | 9.12.19.14.12 ? | 9.12.19.14.12 | 5 Eb' | 5 K'ayab' | 7 Jan 692 | Sanct. Dedication | K'inich Kan B'alam II |
| Death's Head Monument: B1-C1 | 9.13. 0. 0. 0 | 9.12.19.14.12 | 5 Eb' | 5 Kayab' | 7 Jan 692 | Sanct. Dedication | K'inich Kan B'alam II |
| TFC, Incensario: implied | 9.13. 0. 4.12 ? | 9.12.19.14.18 ? | 11 Etz'nab' | 11 K'ayab' | 13 Jan 692 | ? | ? |
| T.Sun, *Alfardas*: H2-I2 | 9.13. 0. 0. 0 ? | 9.13. 0. 0. 0 | 8 Ajaw | 8 Wo | 15 Mar 692 | Period-ending | K'inich Kan B'alam I |
| T.Cross, Stela 1: A1 | 9.13. 0. 0. 0 | 9.13. 0. 0. 0 | 8 Ajaw | 8 Wo ?? | 15 Mar 692 | – | [*K'inich Kan B'alam II] |
| TFC, Tab. Fol. Cross: N15-O15 | 9.12.18. 5.19 ? | 9.13. 0. 0. 0 | 8 Ajaw | 8 Wo | 15 Mar 692 | Period-ending | K'inich Kan B'alam II |
| TFC, Sanct. Jambs: C11b-C12 | 9.12.19.14.12 ? | 9.13. 0. 0. 0 | 8 Ajaw | 8 Wo | 15 Mar 692 | Period-ending | K'inich Kan B'alam II |
| Death's Head Monument: G1-F2 | 9.13. 0. 0. 0 | 9.13. 0. 0. 0 | 8 Ajaw | 8 Wo | 15 Mar 692 | Sanct. dedic. | K'inich Kan B'alam II |
| TFC, Incensario: Ap4-Ap5 | 9.13. 0. 4.12 ? | 9.13. 0. 4.12 ? | 9 Eb' | 0 Yaxk'in | 15 Jun 692 | Dedic. of Incens. | Lady ? Hun Winiknal |
| T.Cross, Incensario 2: A4-B4 | 9.13. 6. 1.11 ?? | 9.13. 2. 4.13 | 2 B'en | 11 Xul ?? | 6 Jun 694 | Death | ? |
| T.Cross, Incensario 2: implied | 9.13. 6. 1.11 ?? | 9.13. 6. 1.11 | 2 Chuwen | 9 Wo ??? | 15 Mar 698 | ? | ? |

Appendix 2. *The Palenque Palace Dates by Peter Mathews, reprinted with permission*

| Monument: Text reference | Date of monument | Date of passage | Calender round | Julian date | Event | Ruler |
|---|---|---|---|---|---|---|
| Hse. A-D, Pal. Tab.: P17-O18 | 9.14. 8.14.15 | 9. 9. 2. 4. 8 | 5 Lamat 1 Mol | 26 Jul 615 | Accession | K'inich Janab' Pakal I |
| Hse. A-D, Pal. Tab.: C1-D1 | 9.14. 8.14.15 | 9.10.10.11. 2 | 1 Ik' 15 Yaxk'in | 13 Jul 643 | 819d | – |
| Hse. A-D, Pal. Tab.: A1-A18 | 9.14. 8.14.15 | 9.10.11.17. 0 | 11 Ajaw 8 Mak | 2 Nov 644 | Birth | K'inich K'an Joy Chitam II |
| Hse. A-D, Pal. Tab.: F6-E7 | 9.14. 8.14.15 | 9.10.18.17.19 | 2 Kawak 12 Keh | 16 Oct 651 | 'Deer-hoof' | K'inich K'an Joy Chitam |
| Hse. A-D, Pal. Tab.: E17-F18 | 9.14. 8.14.15 | 9.11. 0. 0. 0 | 12 Ajaw 8 Keh | 11 Oct 652 | Period-ending | K'inich Janab' Pakal I |
| Hse. A-D, Gl. Tab. 1: A1-B1? | ?? | 9.11. 0. 0. 0 | 12 Ajaw 8 Keh | 11 Oct 652 | *Period-ending | *K'inich Janab' Pakal I |
| Hse. A-D, Gl. Tab. 1: A5 | ?? | 9.11. 1.10.12 | 12 Eb' 10 Sutz' | 6 May 654 | ? | ? |
| Hse. A-D, Pal. Tab.: G10-G11 | 9.14. 8.14.15 | 9.11.13. 0. 0 | 12 Ajaw 3 Ch'en | 4 Aug 665 | 'Deer-hoof' | K'inich K'an Joy Chitam II |
| Hse. A-D, Pal. Tab.: J9-I10 | 9.14. 8.14.15 | 9.12.11. 5.18 | 6 Etz'nab 11 Yax | 28 Aug 683 | Death | K'inich Janab' Pakal I |
| Hse. A-D, Pal. Tab.: K6-L6 | 9.14. 8.14.15 | 9.12.11.12.10 | 8 Ok 3 K'ayab' | 7 Jan 684 | Accession | K'inich Kan B'alam II |
| Hse. A-D, Gl. Tab. 2: pHq3-pGq4 | ?? | 9.12.19.14.12 | 5 Eb' 5 K'ayab' | 7 Jan 692 | Dedication | *K'inich Kan B'alam II |
| Hse. A-D, Pal. Tab.: N9-N10 | 9.14. 8.14.15 | 9.13.10. 0. 0 | 7 Ajaw 3 Kumk'u | 22 Jan 702 | Period-ending | K'inich Kan B'alam II |
| Hse. A-D, Pal. Tab.: M8-N8 | 9.14. 8.14.15 | 9.13.10. 1. 5 | 6 Chikchan 3 Pop | 16 Feb 702 | Death | K'inich Kan B'alam II |
| Hse. A-D, Pal. Tab.: M15-N18? | 9.14. 8.14.15 | 9.13.10. 6. 8 | 5 Lamat 6 Xul | 30 May 702 | Accession | K'inich K'an Joy Chitam II |
| Hse. A-D, Pal. Tab.: R4-Q5,R8 | 9.14. 8.14.15 | 9.13.10. 6. 8 | 5 Lamat 6 Xul | 30 May 702 | Accession | K'inich K'an Joy Chitam II |
| Hse. A-D, Pal. Tab.: Q9-Q13 | 9.14. 8.14.15 | 9.14. 8.14.15 | 9 Men 3 Yax | 10 Aug 720 | Hse. A-D Dedic. | K'inich K'an Joy Chitam II |
| Hse. E, Painted Insc.: pL2-pN1 | 9.14.10. 4. 2 ? | 2. 0. 0.10. 2 | 9 Ik' 0 Sak | 27 Sep 2325 BC | *Accession | "Lady Beastie" |
| Hse. E, Thr. Seat: pE1-pF1 | 9.13.10. 6. 8 ? | 9. 9. 2. 4. 8 | 5 Lamat 1 Mol | 26 Jul 615 | Accession | K'inich Janab' Pakal I |
| Hse. E, Thr. Seat: pI1-pJ1 | 9.13.10. 6. 8 ? | 9.12.11.12.10 | 8 Ok 3 K'ayab' | 7 Jan 684 | Accession | K'inich Kan B'alam II |
| Hse. E, Thr. Seat: pN2 | 9.13.10. 6. 8 ? | 9.13. 7. 2. 2 | 9 Ik 15 Wo ?? | 21 Mar 699 ? | ? | ? |
| Hse. E, Painted Insc.: pX2-pZ1 | 9.14.10. 4. 2 ? | 9.14.10. 4. 2 | 9 Ik' 5 K'ayab' | 30 Dec 721 | Accession | K'inich Akul Mo' Nab' III |
| Hse. C, Hier. St.: A1-B4 | ?? | 9. 8. 9.13. 0 | 8 Ajaw 13 Pop | 23 Mar 603 | Birth | K'inich Janab' Pakal I |
| Hse. C, Hier. St.: B5aii-B5bii | ?? | 9. 9. 2. 4. 8 | 5 Lamat 1 Mol | 26 Jul 615 | Accession | K'inich Janab' Pakal I |
| Hse. C, Eaves: K1-K2 | 9.11. 9. 5.19 | 9.11. 9. 5.19 | 4 Kawak 2 Pax | 22 Dec 661 | House Dedication | [K'inich Janab' Pakal I] |
| Hse. C, West Found.: G1 | 9.11.10.16. 7 | 9.11.10.16. 7 | 13 Manik' 20 Yaxk'in | 13 Jul 663 | Sacrifice | Santa Elena lord |
| Subterr., Thr. 1: M-N | 9.11. 0. 0. 0 | 9.11. 0. 0. 0 | 12 Ajaw 8 Keh | 11 Oct 652 | Period-ending | [K'inich Janab' Pakal I] |
| Subterr., Tableritos: K1b-K2 | 9.11. 1.12. 8 | 9.11. 0. 0. 0 | 12 Ajaw 8 Keh | 11 Oct 652 | Period-ending | [K'inich Janab' Pakal I] |
| Subterr., Tableritos: E1-F1 | 9.11. 1.12. 8 | 9.11. 1.12. 6 | 7 Kimi 4 Xul | 9 Jun 654 | House Dedication | K'inich Janab' Pakal I |
| Subterr., Tableritos: A1-B2 | 9.11. 1.12. 8 | 9.11. 1.12. 8 | 9 Lamat 6 Xul | 11 Jun 654 | ? | K'inich Janab' Pakal I |
| Twr. Crt., Tab. 96 Gl.: B6-A7 | 9.17.13. 0. 7 | 9.11. 2. 1.11 | 9 Chuwen 9 Mak | 1 Nov 654 | Hse. E Dedication | K'inich Janab' Pakal I |
| Twr. Crt., Tab. 96 Gl.: D4-C5 | 9.17.13. 0. 7 | 9.13.10. 6. 8 | 5 Lamat 6 Xul | 30 May 702 | Accession | K'inich K'an Joy Chitam II |
| Twr. Crt., Tab. 96 Gl.: F2-E3 | 9.17.13. 0. 7 | 9.14.10. 4. 2 | 9 Ik' 5 K'ayab' | 30 Dec 721 | Accession | K'inich Akul Mo' Nab' III |
| Twr. Crt., Tab. 96 Gl.: H1-G2 | 9.17.13. 0. 7 | 9.16.13. 0. 7 | 9 Manik' 15 Wo | 4 Mar 764 | Accession | K'inich K'uk' B'alam II |
| Twr. Crt., Tab. 96 Gl.: K2-K3 | 9.17.13. 0. 7 | 9.17.13. 0. 0 | 13 Ajaw 13 Muwan | 13 Nov 783 | Period-ending | K'inich K'uk' B'alam II |
| Twr. Crt., Tab. 96 Gl.: H7-G8 | 9.17.13. 0. 7 | 9.17.13. 0. 7 | 7 Manik' 0 Pax | 20 Nov 783 | Anniv. of Access. | K'inich K'uk' B'alam II |
| Twr. Crt., Tab. 96 Gl.: implied | 9.17.13. 0. 7 | 9.17.13. 0. 7 | 7 Manik' 0 Pax | 20 Nov 783 | Anniv. of Access. | K'inich K'uk' B'alam II |
| Hse. A, West *Alfardas*: B1-B2 | 9.11. 9.10.13 | 9.11. 9.10.12 | 6 Eb' 10 Wo | 25 Mar 662 | Bloodletting | [K'inich Janab' Pakal I] |
| Hse. A, West *Alfardas*:A1-A2 | 9.11. 9.10.13 | 9.11. 9.10.13 | 7 B'en 11 Wo | 26 Mar 662 | Bloodletting | [K'inich Janab' Pakal I] |
| Hse. A, Pier A: C3-D3 | ?? | 9.11.15.11.11 | 1 Chuwen 19 Pop | 12 Mar 668 819d | – | |

Appendix 3. *Palenque Palace sculpture Phases 1–6*

| | | | | | Ruler/Reign | | | | | | | |
|---|---|---|---|---|---|---|---|---|---|---|---|---|
| Structure, location | Sculptural media | GR No. | Theme | Date | KJP1 | KKaB2 | KKJC2 | KAMN3 | UpKJP2 | KKaB3 | KKuB2 | JP3 |
| ***Phase 1 Sculpture*** | | | | | | | | | | | | |
| Southern Subterranean Structure | Hieroglyphic Bench | 80 | TCosCU | 9.11.0.0.0 (AD 652) | X | | | | | | | |
| Southern Subterranean Structure | Bench | 81 | – | – | X | | | | | | | |
| North Subterranean Building | Stucco and painting | 2 | – | – | X | | | | | | | |
| East Subterranean Passage | Tableritos | – | TDed | 9.11.0.0.0 (AD 652) | X | | | | | | | |
| South Subterranean Passage | Tableritos | – | TDed | 9 11.1.12.6 (AD 654) | X | | | | | | | |
| West Subterranean Passage | Tableritos | – | TDed | 9.11.1.12.8 (AD 654) | X | | | | | | | |
| House E, East Subterranean Passage | Vaults (2) | 7 | CosCU | – | X | | | | | | | |
| House H, South Subterranean Passage | Vaults (2) | 79 | CosU | – | X | | | | | | | |
| House K, west Subterranean Passage | Vaults (2) | 78 | CosU | – | X | | | | | | | |
| House E, northwestern room, northern end | Serpent Door | 3 | CosU | – | X | | | | | | | |
| House E, northwestern room, east wall | Figures by Oval Palace Tablet | 4 | Cos? | – | X | | | | | | | |
| House E, northwestern room, east wall | Oval Palace Tablet | 5 | RAudAcc | – | X | | | | | | | |
| House E, northeastern room, northern end | Bicephalic Room | 8 | CosCU | – | X | | | | | | | |
| House E, northwestern entrance, south jamb | Hieroglyphs | 11 | Text? | – | X | | | | | | | |
| House E, western external walls | Motifs on wall (4 walls of motifs) | 13 | Cos? | – | X | | | | | | | |
| House B, southwestern room, east jamb | Legs | 21 | PSolo | – | X | | | | | | | |
| House B, southwestern room, west jamb | Legs | 22 | Dance | – | X | | | | | | | |
| House B, north side | Pier B | 24 | RPSolo | – | X | | | | | | | |
| House B, north side | Pier C | 25 | RPSolo | – | X | | | | | | | |
| House B, east roof | Serpent scene and woven border | 26 | MMot, CosU | – | X | | | | | | | |
| House B, west roof | Mask and serpent scene | 27 | MMot, CosUC | – | X | | | | | | | |
| House B, south roof | Mask | 28 | CosU | – | X | | | | | | | |
| ***Phase 2 Sculpture*** | | | | – | | | | | | | | |
| House C, west side of pier | Pier C | 32 | APSolo,Seat | – | X | | | | | | | |
| House C, south side of pier | Pier C | 33 | PSolo,Stand | – | X | | | | | | | |
| House C, west side of pier | Pier D | 34 | APSolo,Seat | – | X | | | | | | | |

| | | | | | | | |
|---|---|---|---|---|---|---|---|
| House C, west side of pier | Pier E | 35 | APSolo,Seat | – | X | | |
| House C, west side of pier | Pier F | 36 | APSolo,Seat | – | X | | |
| House C, south side of pier | Pier F | 37 | PSolo,Stand | – | X | | |
| House C, eastern room, west wall | Masks | 29 | GPSolo | – | X | | |
| House C, southwestern room, east wall | Sculptured wall | 30 | RAud | – | X | | |
| House C, south surface of pier | Pier B painting | 31 | – | – | X | | |
| House C, east roof | Masks | 38 | CosT | – | X | | |
| House C, west roof | Masks | 39 | CosT | – | X | | |
| House C, south roof | Masks | 40 | CosT | – | X | | |
| House C, north roof | Masks | 41 | CosT | – | X | | |
| House C foundation, east side | Figures and glyph blocks | 59 | CapDis | – | X | | |
| House C, east side | *Alfardas* | 60 | CapDis | – | X | | |
| House C, east façade, east court | Hieroglyphic stairs | 61 | THist | 9.11. 6.16.11 (AD 659) | X | | |
| House C, northeastern corner | Eaves Hieroglyphs | 62 | TDed? | (9.11. 9. 5.19) (AD 661) | X | | |
| East Court, east side | *Alfarda*, left | 58 | CapDis | 9.11.9.10.12 (AD 662) | X | | |
| East Court, east side | *Alfarda*, right | 58 | CapDis | 9.11. 9.10.13 (AD 662) | X | | |
| House C, west foundation | Texts (7) | 68 | TCapSac | 9.11.10.16. 7 (AD 663) | X | | |
| House A, east side of pier | Pier A | 42 | THist? | 9.11.17.15.10 ? (AD 670?) | X | X | |
| House A, east side of pier | Pier B | 43 | APWAStand | – | X | X | |
| House A, east side of pier | Pier C | 44 | APWAStand | – | X | X | |
| House A, east side of pier | Pier D | 45 | APWAStand | – | X | X | |
| House A, east side of pier | Pier E | 46 | APWAStand | – | X | X | |
| House A, east medial wall | Medallions | 47 | APSolo | – | X | X | X |
| ***Phase 3 Sculpture*** | | | | | | | |
| House E, northwestern room, east wall | Del Río Throne | 6 | TDynHist | 9.13.10. 6. 8 (AD 702) | | | X |
| House D, west side of pier | Pier B | 48 | HumSac | – | | | X |
| House D, west side of pier | Pier C | 49 | SacDan | – | | | X |
| House D, west side of pier | Pier D | 50 | SacDan | – | | | X |
| House D, west side of pier | Pier E | 51 | Sac? | – | | | X |
| House D, west side of pier | Pier F | 52 | HumSac | – | | | X |
| House D, west side of pier | Pier G | 53 | Text? | – | | | X |
| House A-D, north side of pier | Pier G | 55 | RPSolo | – | | | X |
| House A-D, north side of pier | Pier J | 56 | RPSolo | – | | | X |
| Southern substructure platform | South altar | 1 | CosC | – | | | X |
| West Court, central, west basement wall | *Casita* wall | 63 | GPWA | – | | | X |

(*continued*)

Appendix 3 *(continued)*

| Structure, location | Sculptural media | GR No. | Theme | Date | Ruler/Reign | | | | | | | |
|---|---|---|---|---|---|---|---|---|---|---|---|---|
| | | | | | KJP1 | KKaB2 | KKJC2 | KAMN3 | UpKJP2 | KKaB3 | KKuB2 | JP3 |
| West Court, northern end, west basement wall | Stucco head | 64 | GPWA | – | | | X | | | | | |
| West Court, southern end, west basement wall | God K in Venus symbol | 65 | Cos C | – | | | X | | | | | |
| Tower court, west basement wall | *Casita* with two heads | 66 | GPWA | – | | | X | | | | | |
| Tower court, west basement wall | Peccary | 69 | CosC | – | | | X | | | | | |
| House A-D, northern corridor, south wall | Palace Tablet | 57 | RAudAcc,TDyn Hist | 9.14. 8.15.18 (AD 720) | | | X | | | | | |
| East Court, east side | Peripheral Figures | 58c | CapDis | – | | | X | | | | | |
| North substructure exterior, east end | Panel 1 | 82 | RPSolo, CosCU | – | | | X | | | | | |
| North substructure exterior, east end | Panel 2 | 83 | RPSolo, CosCU | – | | | X | | | | | |
| North substructure exterior, east end | Panel 4 | 84 | RPSolo, CosCU | – | | | X | | | | | |
| Lower northern palace stair, east end | *Alfarda* figure | 85 | RPSoloStand | – | | | X | | | | | |
| House B, southwestern room, north wall | Incised and sculptured figure | 16 | RAud | – | | | X | | | | | |
| House B, southwestern room, north wall | Sculptured Ik figures | 17 | RAud | – | | | X | | | | | |
| House B, southwestern room, north wall | Hieroglyphic inscription | 18 | Text? | – | | | X | | | | | |
| House B, southwestern room, north wall | *Cauac* throne | 19 | RAud | – | | | X | | | | | |
| House B, southwestern room, east wall | Wall painting | 20 | – | – | | | X | | | | | |
| House B, southeastern room, north wall | Sculptured fret | 23 | CosT? | – | | | X | | | | | |
| ***Phase 4 Sculpture*** | | | | | | | | | | | | |
| House E, vault spring | Mural | – | – | 9.14.10. 4. 2 (721) | | | X | X | | | | |
| Tower base, east side | Figures | 71 | CapDis | – | | | | X | | | | |
| Tower interior, south side | Venus painting | 72 | CosC | – | | | | X | | | | |
| Tower, lower south side, west end | Tablet of the Scribe | 74 | CapDis | 9.14.13. 5. 0 – 9.14.14. 5.4? (AD 725?) | | | | X | | | | |
| Tower, lower south side, east end | Tablet of the Orator | 75 | CapDis | 9.14.13. 5. 0 – 9.14.14. 5.4? (AD 725?) | | | | X | | | | |
| West Court, southern end, north side of pillars | Figures on pillars (3) | 67 | Dance? | – | | | | X | | | | |

| | | | | | | | | | |
|---|---|---|---|---|---|---|---|---|---|
| ***Phase 5 Sculpture*** | | | | | | | | | |
| Tower court, southwest corner | Stucco fret | 70 | – | – | X | X | X | X | |
| House I, east wall interior | Serpent door | 76 | Cos? | – | X | X | X | X | |
| House I, north side, west end | Figure | 77 | RPSolo | – | X | X | X | X | |
| Tower stair, central, south side | Tablet of 96 Hieroglyphs | 73 | TDynHist | 9.17.13. 0. 7 (AD 783) | | | | X | |
| ***Phase 6 Sculpture*** | | | | | | | | | |
| House E, northwestern room, east wall | Wall painting | 9 | – | – | | | | X | X |
| House E, northwestern room, east wall | Hieroglyphs | 10 | Text? | – | | | | X | X |
| House E, northwestern room, east wall | Eaves motif painting | 12 | – | – | | | | X | X |
| House E, northwestern room, east wall | Fish painting | 14 | CosU | – | | | | X | X |
| House E, northeastern room, south wall | Geometric wall painting | 15 | – | – | | | | X | X |
| House D, southeastern room, west wall | Wall painting | 54 | – | – | | | | X | X |

*Themes*
APSolo = Ancestor Portraiture Solo
APWA = Ancestor Portraiture with Attendants
Acc = Accession
CapDis = Captive Display
TCapDis = Text, Captive Display
CosU = Cosmological Underworld
CosT = Cosmological Terrestrial
CosC = Cosmological Celestial
GPSolo = God Portrait Solo
GPWA = God Portraiture with Attendants
HumSac = Human Sacrifice
MMot = Mat Motif
PSolo = Portraiture Solo
SacDan = Sacrificial Dance
TCos = Text, Cosmological
THist = Text, Historical
TMyth = Text, Mythological
TDed = Text, Dedication
RAud = Royal Audience
RitDan = Ritual Dance
RPSolo = Royal Portrait Solo
Stand = Standing
Seat = Seated

*Ruler*
KJP1 = K'inich Janab' Pakal I
KKaB2 = K'inich Kan B'alam II
KKJC2 = K'inich K'an Joy Chitam II
KAMN3 = K'inich Akul Mo'Nab' III
UpKJP2 = U Pakal K'inich Janab' Pakal II
KKaB3 = K'inich Kan B'alam III
KKuB2 = K'inich K'uk' B'alam II
JP3 = Wak Kimi Janab' Pakal III

Appendix 4. *Sculpture ranked according to carrier depths, Palenque Palace Phases 1–6*

| Palenque Palace Phase 1 sculpture ranked according to carrier depths | | | | | | |
|---|---|---|---|---|---|---|
| Description of spatial unit | Cell | Carrier rank | Sculpture GR No. | RA value | Control value | Theme |
| *Carrier* | c | **0** | – | 0.1863 | 0.5833 | – |
| North Substructure, north terrace lower | 1a | **1** | – | 0.1667 | 2.0000 | – |
| Southern Substructure, south terrace lower | 1b | **1** | – | 0.2059 | 1.7500 | – |
| North Substructure, east terrace lower | 2a | **2** | – | 0.1508 | 0.4500 | – |
| North Substructure, north corridor | 2b | **2** | – | 0.1772 | 0.7500 | – |
| North Substructure, west terrace lower | 2c | **2** | – | 0.1508 | 0.4500 | – |
| Southern Substructure, southeast terrace lower | 2d | **2** | – | 0.2443 | 1.4999 | – |
| North Substructure, south corridor | 3a | **3** | – | 0.1606 | 0.7000 | – |
| North Substructure, south terrace lower | 3b | **3** | – | 0.1320 | 2.2000 | – |
| South Substructure, east-central room | 3c | **3** | (Bench) | 0.2293 | 1.0833 | – |
| South Substructure, east-central corridor | 3d | **3** | – | 0.2293 | 1.0833 | – |
| South Substructure Patio, north corridor | 3e | **3** | – | 0.2609 | 0.7500 | – |
| House B, north terrace upper | 4a | **4** | – | 0.1192 | 1.5333 | – |
| Terrace, lower west | 4b | **4** | – | 0.1546 | 0.5333 | – |
| South Substructure, northeast corridor | 4c | **4** | – | 0.2172 | 1.1666 | – |
| South Substructure, central room | 4d | **4** | (Bench) | 0.2647 | 0.6666 | – |
| South Substructure, patio | 4e | **4** | – | 0.2971 | 1.5000 | – |
| House B, east terrace upper | 5a | **5** | – | 0.1576 | 0.2000 | – |
| House B, north room | 5b | **5** | **24, 25, 26, 27** | 0.1425 | 0.7000 | **RPSolo; MMot; CosU; CosCU** |
| House E, access north stair | 5c | **5** | – | 0.1184 | 0.9000 | – |
| Terrace, upper northwest | 5d | **5** | – | 0.1244 | 1.3166 | – |
| House K, west room | 5e | **5** | – | 0.1425 | 0.9000 | – |
| East Subterranean Passage, south corridor | 5f | **5** | – | 0.2006 | 0.8333 | – |
| South Substructure Patio, southwest corner | 5g | **5** | – | 0.3363 | 0.5000 | – |

| | | | | | | |
|---|---|---|---|---|---|---|
| House B, south room | 6a | **6** | **21, 22, 26, 27, 28** | 0.1501 | 0.7500 | **PSolo; RitDan; MMot; CosU** |
| House E, east room | 6b | **6** | **8** | 0.1282 | 0.9166 | **CosCU** |
| House E, west room | 6c | **6** | **3, 4, 5, 13, 11, 80** (Bench) | 0.1244 | 1.1666 | **RAud; Acc; CosTU; THist; TCosCU** |
| Southwest Court, later tower court | 6d | **6** | – | 0.1267 | 0.8666 | – |
| House K, east room | 6e | **6** | (West Subterranean Stairs) | 0.1395 | 1.6999 | – |
| East Subterranean Passage, dogs leg | 6f | **6** | – | 0.1833 | 1.0000 | – |
| Southeast Court, early | 7a | **7** | – | 0.1388 | 1.0595 | – |
| House E, southwestern room | 7b | **7** | (East Subterranean Stairs) | 0.1440 | 0.7500 | – |
| East Subterranean Passage, north corridor | 7c | **7** | 7 | 0.1637 | 1.0000 | **CosCU** |
| House H, substructure | 7d | **7** | (South Subterranean Stairs) | 0.1584 | 0.8333 | – |
| South Subterranean Passage, north corridor | 7e | **7** | – | 0.1727 | 0.7000 | – |
| Terrace, upper southwest | 7f | **7** | – | 0.1448 | 1.3666 | – |
| Terrace, upper southeast | 7g | **7** | – | 0.1267 | 2.4166 | – |
| House J, west access | 8a | **8** | – | 0.1667 | 5.0833 | – |
| South Subterranean Passage, dogs leg | 8b | **8** | 78 | 0.1916 | 1.0000 | CosU |
| West Subterranean Passage, dogs leg | 8c | **8** | – | 0.2059 | 1.0000 | – |
| Terrace, upper southeast corner | 8d | **8** | – | 0.1652 | 0.1666 | – |
| Terrace, upper southwest corner | 8e | **8** | – | 0.1817 | 0.3333 | – |
| House J, northwest room | 9a | **9** | (Bench) | 0.2051 | 0.1428 | – |
| House J west, north-central room | 9b | **9** | – | 0.2036 | 0.4761 | – |
| House J west, central room | 9c | **9** | (Bench) | 0.2029 | 1.6428 | – |
| House J west, south-central room | 9d | **9** | – | 0.2051 | 0.1428 | – |
| House J west, south room | 9e | **9** | – | 0.2051 | 0.1428 | – |
| House J, south patio | 9f | **9** | – | 0.2051 | 0.1428 | – |
| West Subterranean Passage, south corridor | 9g | **9** | 79 | 0.2330 | 1.0000 | CosU |
| South Subterranean Passage, south corridor | 9h | **9** | – | 0.2217 | 1.0000 | – |
| House J, east room | 10a | **10** | – | 0.2300 | 0.3333 | – |
| South Subterranean Structure, northwest room | 10b | **10** | 81 (Bench) | 0.2564 | 1.0000 | – |

Appendix 4 *(continued)*

| Palenque Palace Phase 2 sculpture ranked according to carrier depths | | | | | | |
|---|---|---|---|---|---|---|
| Description of spatial unit | Cell | Carrier rank | Sculpture GR No. | RA value | Control value | Theme |
| *Carrier* | C | **0** | – | 0.1682 | 0.5833 | – |
| North Substructure, north terrace lower | 1a | **1** | – | 0.1494 | 2.0000 | – |
| Southern Substructure, south terrace lower | 1b | **1** | – | 0.1870 | 1.7500 | – |
| North Substructure, east terrace lower | 2a | **2** | – | 0.1338 | 0.4500 | – |
| North Substructure, north corridor | 2b | **2** | – | 0.1584 | 0.7500 | – |
| North Substructure, west terrace lower | 2c | **2** | – | 0.1338 | 0.4500 | – |
| South Substructure, south corridor | 2d | **2** | – | 0.2045 | 1.4999 | – |
| Southern Substructure, southeast terrace lower | 2e | **2** | – | 0.2227 | 0.3333 | – |
| North Substructure, south corridor | 3a | **3** | – | 0.1416 | 0.7000 | – |
| North Substructure, south terrace lower | 3b | **3** | – | 0.1156 | 2.2000 | – |
| South Substructure, east-central room | 3c | **3** | (Bench) | 0.2013 | 1.0833 | – |
| South Substructure, east-central corridor | 3d | **3** | – | 0.2013 | 1.0833 | – |
| South Substructure Patio, north corridor | 3e | **3** | – | 0.2357 | 0.7500 | – |
| East Court, House A west foundation (early) | 4a | **4** | **58** (*alfardas* only), **59**, **60**, **61**(H.S.) | 0.1078 | 1.5333 | **CapDis; THist** |
| Terrace, lower west | 4b | **4** | – | 0.1318 | 0.5333 | – |
| South Substructure, central room | 4c | **4** | (Bench) | 0.2357 | 0.6666 | – |
| South Substructure, northeast corridor | 4d | **4** | – | 0.1896 | 1.1666 | – |
| South Substructure, patio | 4e | **4** | – | 0.2701 | 1.5000 | – |
| House A, west room | 5a | **5** | – | 0.1299 | 0.9500 | – |
| House B, north room | 5b | **5** | 24, 25, 26, 27 | 0.1286 | 1.0333 | RPSolo; MMot; CosU; CosCU |
| East Court, southwest corner | 5c | **5** | – | 0.1260 | 0.7833 | – |
| House C, east room | 5d | **5** | **29, 38, 40, 41, 62** | 0.1149 | 0.6761 | **GPSolo; CosT; TDed?** |
| House K, west room | 5e | **5** | – | 0.1338 | 0.9500 | – |
| East Subterranean Passage, south corridor | 5f | **5** | – | 0.1734 | 0.8333 | – |
| South Substructure Patio, southwest corner | 5g | **5** | – | 0.3071 | 0.5000 | – |
| House A & J, east patio | 6a | **6** | – | 0.1403 | 1.4583 | – |
| House B, south room | 6b | **6** | 21, 22, 26, 27, 28 | 0.1312 | 0.5833 | PSolo; RitDan; MMot; CosU; CosCU |
| House C, west room | 6c | **6** | **30, 31, 32, 33, 34, 35, 36, 37, 39, 40, 41** | 0.1279 | 0.7261 | **RAud; APSolo; PSolo; CosT** |

| | | | | | | |
|---|---|---|---|---|---|---|
| House C, west foundation | 6d | **6** | **68** | 0.1273 | 1.0090 | **TCapSac** |
| House E, east room | 6e | **6** | 8 | 0.1156 | 0.7499 | CosCU |
| Southwest Court, later tower court | 6f | **6** | – | 0.1071 | 2.1166 | – |
| House K, east room | 6g | **6** | (West Subterranean Stairs) | 0.1247 | 1.7261 | – |
| East Subterranean Passage, dogs leg | 6h | **6** | – | 0.1597 | 0.8333 | – |
| House A, east room | 6i | **6** | **42, 43, 44, 45, 46, 47** | 0.1539 | 0.5833 | **THist; APWA; APSolo** |
| House J, west access | 7a | **7** | – | 0.1312 | 5.3333 | – |
| Southeast Court, early | 7b | **7** | – | 0.1162 | 1.0416 | – |
| House E, west room | 7c | **7** | 3, 4, 5, 13, 11, 80 (Bench) | 0.1149 | 0.7261 | RAud; Acc; CosTU; THist; TCosCU |
| House H, substructure | 7d | **7** | (South Subterranean Stairs) | 0.1383 | 0.6428 | – |
| South Subterranean Passage, north corridor | 7e | **7** | – | 0.1552 | 0.7000 | – |
| Terrace, upper southeast | 7f | **7** | – | 0.1071 | 2.3333 | – |
| East Subterranean Passage, north corridor | 7g | **7** | 7 | 0.1396 | 1.8333 | CosU |
| Terrace, upper southwest | 7h | **7** | – | 0.1253 | 0.3666 | – |
| House J, east room | 7i | **7** | – | 0.1734 | 0.5833 | – |
| House J, northwest room | 8a | **8** | (Bench) | 0.1675 | 0.1250 | – |
| House J west, north-central room | 8b | **8** | – | 0.1662 | 0.4583 | – |
| House J west, central room | 8c | **8** | (Bench) | 0.1656 | 1.1250 | – |
| House J west, south-central room | 8d | **8** | – | 0.1675 | 0.1250 | – |
| House J west, south room | 8e | **8** | – | 0.1675 | 0.1250 | – |
| House J, south patio | 8f | **8** | – | 0.1675 | 0.1250 | – |
| Terrace, upper southeast corner | 8g | **8** | – | 0.1409 | 0.1666 | – |
| South Subterranean Passage, dogs leg | 8h | **8** | 78 | 0.1695 | 1.0000 | CosU |
| West Subterranean Passage, dogs leg | 8i | **8** | – | 0.1864 | 1.0000 | – |
| House E, southwestern room | 8j | **8** | (East Subterranean Stairs) | 0.1227 | 0.8333 | – |
| Terrace, upper southwest corner | 8k | **8** | – | 0.1617 | 0.3333 | – |
| South Subterranean Passage, south corridor | 9a | **9** | – | 0.2000 | 1.0000 | – |
| West Subterranean Passage, south corridor | 9b | **9** | 79 | 0.2169 | 1.0000 | CosU |
| South Subterranean Structure, northwest room | 10a | **10** | 81 (Bench) | 0.2286 | 1.0000 | – |

Appendix 4 *(continued)*

Palenque Palace Phase 3 sculpture ranked according to carrier depths

| Description of spatial unit | Cell | Carrier rank | Sculpture GR No. | RA value | Control value | Theme |
|---|---|---|---|---|---|---|
| *Carrier* | C | **0** | **1,2, 82, 83, 84, 85** | 0.1000 | 1.4499 | **CosC; RPSoloCosCU; RPSolo** |
| House A-D, north corridor | 1a | **1** | **55, 56, 57** (Throne?) | 0.0994 | 1.0666 | **RPSolo; RAudAccTDynHist** |
| House D, west corridor | 1b | **1** | **48, 49, 50, 51, 52, 53** | 0.1117 | 0.7000 | **HumSac; SacDan; Text?** |
| House D & K, patio | 1c | **1** | – | 0.0942 | 1.1833 | – |
| House K, west room | 1d | **1** | – | 0.1084 | 0.6000 | – |
| Southern Substructure, south terrace | 1e | **1** | – | 0.1208 | 1.4500 | – |
| House A-D, south corridor | 2a | **2** | – | 0.0864 | 1.0595 | – |
| House A, east corridor | 2b | **2** | 42, 43, 44, 45, 46, 47 | 0.1149 | 0.8333 | THist; APWA; APSolo |
| House D & K, external path | 2c | **2** | – | 0.0805 | 1.2428 | – |
| House D, east corridor | 2d | **2** | – | 0.0968 | 0.8999 | – |
| House K, east room | 2e | **2** | (West Subterranean Stair) | 0.0955 | 1.4499 | – |
| South Substructure, south corridor | 2f | **2** | **80** (Bench moved) | 0.1429 | 1.4999 | **TCosCU** |
| Southern Substructure, southeast terrace lower | 2g | **2** | – | 0.1565 | 0.3333 | – |
| House A, west corridor | 3a | **3** | – | 0.1013 | 0.8011 | – |
| East Court, House A west foundation | 3b | **3** | **58** (remaining Captive Panels), 59, 60, 61(H.S.) | 0.0961 | 3.6166 | **CapDis**; THist |
| House J west, central room | 3c | **3** | (Bench) | 0.1214 | 0.9583 | – |
| Southwest Court, later tower court | 3d | **3** | – | 0.0851 | 2.3333 | – |
| South Subterranean Passage, north corridor | 3e | **3** | – | 0.1253 | 0.7000 | – |
| South Substructure, east-central room | 3f | **3** | (Bench) | 0.1662 | 1.0833 | – |
| South Substructure, east-central corridor | 3g | **3** | – | 0.1662 | 1.0833 | – |
| Terrace, upper southwest | 3h | **3** | – | 0.0929 | 1.5333 | – |
| South Substructure Patio, north corridor | 3i | **3** | – | 0.1760 | 0.7500 | – |
| House B, northeast room | 4a | **4** | 24, 25, 26, 27 | 0.1318 | 0.1428 | RPSolo; MMot; CosU; CosCU |
| House B, central room | 4b | **4** | 24, 25, 26, 27 | 0.1318 | 0.1428 | RPSolo; MMot; CosU; CosCU |
| House B, northwest room | 4c | **4** | 24, 25, 26, 27 | 0.1240 | 0.4761 | RPSolo; MMot; CosU; CosCU |
| East Court, southwest corner | 4d | **4** | – | 0.1149 | 0.9761 | – |
| House C, east room | 4e | **4** | 29, 38, 40, 41, 62 | 0.1052 | 0.6190 | GPSolo; CosT; TDed? |

| | | | | | | |
|---|---|---|---|---|---|---|
| House C, west room | 4f | **4** | 30, 31, 32, 33, 34, 35, 36, 37, 39, 40, 41 | 0.1149 | 0.9761 | RAud; APSolo; PSolo; CosT |
| West Court, early | 4g | **4** | **63**, **64**, **65**, **66**, 68, **69** | 0.1201 | 0.4761 | **GPWA; CosC;** TCapSac |
| House J, west access path | 4h | **4** | – | 0.1058 | 5.2833 | – |
| House E, west room | 4i | **4** | 3, 4, 5, **6** (DR Throne), 13, 11 | 0.1026 | 0.7261 | RAud; Acc; CosTU; THist; **TDynHist** |
| House J west, north-central room | 4j | **4** | – | 0.1318 | 0.4583 | – |
| House H, substructure | 4k | **4** | (South Subterranean Stair) | 0.1156 | 0.6428 | – |
| Terrace, upper southeast | 4L | **4** | – | 0.0896 | 2.1761 | – |
| South Substructure, northeast corridor | 4m | **4** | – | 0.1766 | 1.1666 | – |
| South Substructure, central room | 4n | **4** | (Bench) | 0.2006 | 0.6666 | – |
| South Substructure, patio | 4o | **4** | – | 0.2104 | 1.5000 | – |
| West Subterranean Passage, dogs leg | 4p | **4** | – | 0.1558 | 1.0000 | – |
| South Terrace, southwest corner | 4q | **4** | – | 0.1286 | 0.2500 | – |
| House J, northwest room | 5a | **5** | (Bench) | 0.1416 | 0.1250 | – |
| House J west, south-central room | 5b | **5** | – | 0.1416 | 0.1250 | – |
| House J west, south room | 5c | **5** | – | 0.1416 | 0.1250 | – |
| House J, south patio | 5d | **5** | – | 0.1416 | 0.1250 | – |
| Southeast Court, early | 5e | **5** | – | 0.1032 | 2.4916 | – |
| Terrace, upper southeast corner | 5f | **5** | – | 0.1247 | 0.1666 | – |
| House E, east room | 5g | **5** | 8 | 0.1078 | 1.0333 | CosCU |
| House E, southwestern room | 5h | **5** | (East Subterranean Stair) | 0.1123 | 0.9999 | – |
| East Subterranean Passage, south corridor | 5i | **5** | – | 0.1695 | 0.8333 | – |
| South Subterranean Passage, dogs leg | 5j | **5** | 78 | 0.1461 | 1.0000 | CosU |
| West Subterranean Passage, south corridor | 5k | **5** | 79 | 0.1864 | 1.0000 | CosU |
| South Substructure Patio, southwest corner | 5L | **5** | – | 0.2468 | 0.5000 | – |
| House B, southeast room | 6a | **6** | **23**, 26, 27, 28 | 0.1390 | 0.2000 | **CosT?**; MMot; CosU; CosCU |
| House B, southwest room | 6b | **6** | **16**, **17**, **18**, **19**, **20**, 21, 22, 26, 27, 28 | 0.1390 | 0.2000 | **RAud**; **Text?**; PSolo; RitDan; MMot; CosU |
| East Subterranean Passage, north corridor | 6c | **6** | 7 | 0.1331 | 0.8333 | CosU |
| East Subterranean Passage, dogs leg | 6d | **6** | – | 0.1532 | 1.0000 | – |
| South Subterranean Passage, south corridor | 6e | **6** | – | 0.1766 | 1.0000 | – |
| South Subterranean Structure, northwest room | 6f | **6** | 81 (Bench) | 0.1974 | 1.0000 | – |

Appendix 4 *(continued)*

Palenque Palace Phase 4 sculpture ranked according to carrier depths

| Description of spatial unit | Cell | Carrier rank | Sculpture GR No. | RA value | Control value | Theme |
|---|---|---|---|---|---|---|
| *Carrier* | C | **0** | 1, 2, 82, 83, 84, 85 | 0.0959 | 1.4999 | CosC; RPSoloCosCU; RPSolo |
| House A-D, north corridor | 1a | **1** | 55, 56, 57 (Bench) | 0.0982 | 0.9166 | RPSolo; RAudAccTDynHist |
| House D, west corridor | 1b | **1** | 48, 49, 50, 51, 52, 53 | 0.0986 | 0.9166 | HumSac; SacDan; Text? |
| House K, west room | 1c | **1** | – | 0.0939 | 0.5000 | – |
| Southern Substructure, south terrace | 1d | **1** | – | 0.0994 | 1.0000 | – |
| House A, east corridor | 2a | **2** | 42, 43, 44, 45, 46, 47 | 0.1041 | 0.9166 | THist; APWA; APSolo |
| House D, east corridor | 2b | **2** | – | 0.0943 | 0.9999 | – |
| House K, east room | 2c | **2** | (West Subterranean Stair) | 0.0814 | 1.2777 | – |
| South Substructure, south corridor | 2d | **2** | 80 (Bench) | 0.1193 | 1.4999 | TCosCU |
| South Terrace stair | 2e | **2** | – | 0.0931 | 0.4444 | – |
| House A, west corridor | 3a | **3** | – | 0.0998 | 1.1011 | – |
| House A-D, southwest corridor | 3b | **3** | – | 0.1029 | 0.4999 | – |
| House D & K, external path | 3c | **3** | – | 0.0822 | 2.0333 | – |
| South Terrace | 3d | **3** | – | 0.0759 | 3.9999 | – |
| South Subterranean Passage, north corridor | 3e | **3** | – | 0.1076 | 0.7500 | – |
| South Substructure, east-central room | 3f | **3** | (Bench) | 0.1428 | 1.0833 | – |
| South Substructure, east-central corridor | 3g | **3** | – | 0.1428 | 1.0833 | – |
| South Substructure Patio, north corridor | 3h | **3** | – | 0.1475 | 0.7500 | – |
| House J west, central room | 3i | **3** | (Bench) | 0.1103 | 0.9583 | – |
| House A-D, southeast corridor | 4a | **4** | – | 0.1170 | 0.3928 | – |
| East Court, House A west foundation | 4b | **4** | 58, 59, 60, 61 (H.S.) | 0.1013 | 3.9166 | CapDis; THist |
| House J west, north-central room | 4c | **4** | – | 0.1182 | 0.4583 | – |
| House J, west access path | 4d | **4** | – | 0.0974 | 5.2499 | – |
| Southeast Court, path access | 4e | **4** | – | 0.0818 | 0.0861 | – |
| House G, south room | 4f | **4** | – | 0.0876 | 0.5277 | – |
| House G & H, central corridor | 4g | **4** | – | 0.0779 | 1.4206 | – |
| House H, south room | 4h | **4** | – | 0.0884 | 0.8611 | – |
| House H, west patio | 4i | **4** | – | 0.0767 | 0.6706 | – |
| West Subterranean Passage, dogs leg | 4j | **4** | – | 0.1354 | 1.0000 | – |
| South Substructure, northeast corridor | 4k | **4** | – | 0.1549 | 1.1666 | – |
| South Substructure, central room | 4L | **4** | (Bench) | 0.1682 | 0.6666 | – |
| South Substructure, patio | 4m | **4** | – | 0.1745 | 1.5000 | – |
| Tower Court | 4n | **4** | **74, 75** | 0.0794 | 1.3928 | **CapDis** |
| South Terrace, southeast corner | 4o | **4** | – | 0.1025 | 0.1111 | – |
| South Terrace, southwest corner | 4p | **4** | – | 0.1025 | 0.1111 | – |
| Tower, crossway access | 4q | **4** | – | 0.1072 | 0.6666 | – |

| | | | | | | |
|---|---|---|---|---|---|---|
| House B, northeast room | 5a | **5** | 24, 25, 26, 27 | 0.1279 | 0.1428 | RPSolo; MMot; CosU; CosCU |
| House B, central room | 5b | **5** | 24, 25, 26, 27 | 0.1279 | 0.1428 | RPSolo; MMot; CosU; CosCU |
| House B, northwest room | 5c | **5** | 24, 25, 26, 27 | 0.1174 | 0.4761 | RPSolo; MMot; CosU; CosCU |
| East Court, southwest corner | 5d | **5** | – | 0.1021 | 0.8928 | – |
| House C, east room | 5e | **5** | 29, 38, 40, 41, 62 | 0.1013 | 0.9761 | GPSolo; CosT; TDed? |
| Tower Court, northeast corner | 5f | **5** | **71** | 0.0908 | 0.7833 | **CapDis** |
| House E, west room | 5g | **5** | 3 (Bench?), 4, 5, 6 (DRThrone), 11, 13 | 0.0872 | 1.1166 | RAud; Acc; CosTU; THist; TDynHist |
| Tower, first level | 5h | **5** | – | 0.1330 | 1.0000 | – |
| Toilets, east corridor | 5i | **5** | – | 0.1064 | 0.7000 | – |
| House J, northwest room | 5j | **5** | (Bench) | 0.1240 | 0.1250 | – |
| House J west, south-central room | 5k | **5** | – | 0.1240 | 0.1250 | – |
| House J west, south room | 5L | **5** | – | 0.1240 | 0.1250 | – |
| House J, south patio | 5m | **5** | – | 0.1240 | 0.1250 | – |
| Southeast Court | 5n | **5** | – | 0.0841 | 2.8750 | – |
| House G, north room | 5o | **5** | – | 0.0822 | 0.7261 | – |
| House G & H, north terrace | 5p | **5** | – | 0.0736 | 1.7833 | – |
| House H, north room | 5q | **5** | – | 0.0900 | 0.6428 | – |
| House H, northwest room | 5r | **5** | (Bench) (South Subterranean Stair) | 0.1037 | 1.0833 | – |
| East Subterranean Passage, south corridor | 5s | **5** | – | 0.1475 | 0.8333 | – |
| West Subterranean Passage, south corridor | 5t | **5** | 79 | 0.1569 | 1.0000 | CosU |
| South Substructure Patio, southwest corner | 5u | **5** | – | 0.2011 | 0.5000 | – |
| House B, southeast room | 6a | **6** | 23, 26, 27, 28 | 0.1111 | 0.1666 | CosT?; MMot; CosU; CosCU |
| House B, southwest room | 6b | **6** | 16, 17, 18, 19, 20, 21, 22, 26, 27, 28 | 0.1111 | 0.1666 | RAud; Text?; PSolo; RitDan; MMot; CosU |
| House E, east room | 6c | **6** | 8 | 0.0829 | 0.8928 | CosCU |
| House E, southwestern room | 6d | **6** | (East Subterranean Stair) | 0.0919 | 0.8928 | – |
| East Subterranean Passage, dogs leg | 6e | **6** | – | 0.1311 | 1.0000 | – |
| South Subterranean Passage, dogs leg | 6f | **6** | 78 | 0.1303 | 0.8333 | CosU |
| South Substructure, northwest room | 6g | **6** | 81 (Bench) | 0.1729 | 1.0000 | – |
| Tower, second level | 6h | **6** | **72** | 0.1600 | 1.5000 | **CosC** |
| House C, west room | 6i | **6** | 30, 31, 32, 33, 34, 35, 36, 37, 39, 40, 41 | 0.1268 | 0.8333 | RAud; APSolo; PSolo; CosT |
| Toilet, west-central room | 6j | **6** | – | 0.1338 | 1.5000 | – |
| East Subterranean Passage, north corridor | 7a | **7** | 7 | 0.1146 | 0.8333 | CosU |
| South Subterranean Passage, south corridor | 7b | **7** | – | 0.1612 | 1.0000 | – |
| Tower, third level | 7c | **7** | (Bench) | 0.1878 | 0.5000 | – |
| West Court | 7d | **7** | 63, 64, 65, 66, **67**, 68,69 | 0.1557 | 1.5000 | **RitDan?**; GPWA; CosC; TCapSac |
| Toilet, north room | 7e | **7** | (Toilets) | 0.1608 | 0.5000 | – |
| West Court, south room | 8a | **8** | – | 0.1835 | 0.5000 | – |

Appendix 4 *(continued)*

Palenque Palace Phase 5 sculpture ranked according to carrier depths

| Description of spatial unit | Cell | Carrier rank | Sculpture GR No. | RA value | Control value | Theme |
|---|---|---|---|---|---|---|
| *Carrier* | c | **0** | 1, 2, 82, 83, 84, 85 | 0.0899 | 1.4999 | CosC; RPSoloCosCU; RPSolo |
| House A-D, north corridor | 1a | **1** | 55, 56, 57 (Bench) | 0.0923 | 0.9166 | RPSolo; RAudAccTDynHist |
| House D, west corridor | 1b | **1** | 48, 49, 50, 51, 52, 53 | 0.0911 | 0.9166 | HumSac; SacDan; Text? |
| House K, west room | 1c | **1** | – | 0.0926 | 0.5000 | – |
| Southern Substructure, south terrace | 1d | **1** | – | 0.0917 | 1.0000 | – |
| House A, east corridor | 2a | **2** | 42, 43, 44, 45, 46, 47 | 0.0946 | 0.9999 | THist; APWA; APSolo |
| House D, east corridor | 2b | **2** | – | 0.0840 | 0.9999 | – |
| House K, east room | 2c | **2** | (West Subterranean Stair) | 0.0714 | 1.2777 | – |
| South Terrace stair | 2d | **2** | – | 0.0811 | 0.4444 | – |
| South Substructure, south corridor | 2e | **2** | 80 (Bench) | 0.1096 | 1.4999 | TCosCU |
| House A, west corridor | 3a | **3** | – | 0.0832 | 0.9761 | – |
| House J west, central room | 3b | **3** | (Bench) | 0.0976 | 0.9761 | – |
| House A-D, southwest corridor | 3c | **3** | – | 0.0908 | 0.4999 | – |
| House D & K, external path | 3d | **3** | – | 0.0708 | 2.3333 | – |
| South Subterranean Passage, north corridor | 3e | **3** | – | 0.0940 | 0.7500 | – |
| South Substructure, east-central room | 3f | **3** | (Bench) | 0.1255 | 1.0833 | – |
| South Substructure, east-central corridor | 3g | **3** | – | 0.1255 | 1.0833 | – |
| South Substructure Patio, north corridor | 3h | **3** | – | 0.1328 | 0.7500 | – |
| South Terrace | 3i | **3** | – | 0.0661 | 3.9499 | – |
| House A-D, southeast corridor | 4a | **4** | – | 0.0979 | 0.4761 | – |
| East Court, House A west foundation | 4b | **4** | 58, 59, 60, 61 (H.S.) | 0.0861 | 3.9999 | CapDis; THist |
| House J west, north-central room | 4c | **4** | – | 0.1014 | 0.4761 | – |
| House J, west access path | 4d | **4** | – | 0.0788 | 4.9166 | – |
| Southeast Court, east side | 4e | **4** | – | 0.0702 | 1.8373 | – |
| Tower Court, southwest corner stair | 4f | **4** | – | 0.0802 | 0.2916 | – |
| House G, south room | 4g | **4** | – | 0.0767 | 0.4777 | – |
| House G & H, central corridor | 4h | **4** | – | 0.0726 | 1.3194 | – |
| House H, south room | 4i | **4** | – | 0.0767 | 0.8944 | – |
| House H, west patio | 4j | **4** | – | 0.0673 | 0.6527 | – |
| West Subterranean Passage, dogs leg | 4k | **4** | – | 0.1137 | 1.0000 | – |
| South Terrace, southeast corner | 4L | **4** | – | 0.0885 | 0.1111 | – |
| South Substructure, northeast corridor | 4m | **4** | – | 0.1331 | 1.1666 | – |
| House B, northeast room | 4n | **4** | (Bench) | 0.1490 | 0.6666 | – |
| South Substructure, patio | 4o | **4** | – | 0.1563 | 1.5000 | – |
| South Terrace, southwest corner | 4p | **4** | – | 0.0885 | 0.1111 | – |
| Tower, crossway access | 4q | **4** | – | 0.0934 | 0.6666 | – |
| House B, northeast room | 5a | **5** | 24, 25, 26, 27 | 0.1108 | 0.1428 | RPSolo; MMot; CosU; CosCU |
| House B, northwest room | 5b | **5** | 24, 25, 26, 27 | 0.1108 | 0.1428 | RPSolo; MMot; CosU; CosCU |
| House B, northwest room | 5c | **5** | 24, 25, 26, 27 | 0.1020 | 0.4761 | RPSolo; MMot; CosU; CosCU |
| East Court, southwest corner | 5d | **5** | – | 0.0911 | 0.8928 | – |
| House C, east room | 5e | **5** | 29, 38, 40, 41, 62 | 0.0926 | 0.9761 | GPSolo; CosT; TDed? |
| Southeast Court, north side | 5f | **5** | – | 0.0814 | 2.4166 | – |

| | | | | | | |
|---|---|---|---|---|---|---|
| House J, northwest room | 5g | **5** | (Bench) | 0.1029 | 0.1428 | – |
| House J west, south-central room | 5h | **5** | – | 0.1029 | 0.1428 | – |
| House J west, south room | 5i | **5** | – | 0.1029 | 0.1428 | – |
| House J, south patio | 5j | **5** | – | 0.1029 | 0.1428 | – |
| House G, northeast room | 5k | **5** | – | 0.0761 | 0.2916 | – |
| House G, northwest room | 5L | **5** | – | 0.0837 | 0.3250 | – |
| House G & H, north terrace | 5m | **5** | – | 0.0652 | 2.7833 | – |
| House H, northwest room | 5n | **5** | (Bench) (South Subterranean Stair) | 0.0911 | 1.2500 | – |
| West Subterranean Passage, south corridor | 5o | **5** | 79 | 0.1364 | 1.0000 | CosU |
| East Subterranean Passage, south corridor | 5p | **5** | – | 0.1255 | 0.8333 | – |
| South Substructure Patio, southwest corner | 5q | **5** | – | 0.1804 | 0.5000 | – |
| House F, east room | 5r | **5** | – | 0.0926 | 0.6666 | – |
| Tower Court | 5s | **5** | **70, 73** (Throne Tablet), 74, 75 | 0.0802 | 4.2833 | **?(Stucco fret); TDyn Hist**; CapDis |
| Tower, first level | 5t | **5** | – | 0.1181 | 1.0000 | – |
| House B, southeast room | 6a | **6** | 23, 26, 27, 28 | 0.1049 | 0.2500 | CosT?; MMot; CosU; CosCU |
| House B, southwest room | 6b | **6** | 16, 17, 18, 19, 20, 21, 22, 26, 27, 28 | 0.1049 | 0.2500 | RAud; Text?; PSolo; RitDan; MMot; CosU |
| House F, west room | 6c | **6** | – | 0.0917 | 0.7500 | – |
| Southeast Court, west side | 6d | **6** | – | 0.0764 | 1.1250 | – |
| House E, east room | 6e | **6** | 8 | 0.0726 | 0.9083 | CosCU |
| House I, west room | 6f | **6** | 3 (Bench?), 4, 5, 6 (DR Throne), 11, 13 | 0.0802 | 1.2916 | RAud; Acc; CosTU; THist; TDynHist |
| House I, east corridor | 6g | **6** | – | 0.0964 | 1.3250 | – |
| House I, north room | 6h | **6** | **76, 77** | 0.1049 | 0.3750 | **Cos?; RPSolo** |
| House I, west corridor | 6i | **6** | – | 0.1058 | 0.1250 | – |
| Toilets, southwest room | 6j | **6** | – | 0.1058 | 0.1250 | – |
| Toilets, east corridor | 6k | **6** | – | 0.1046 | 0.6250 | – |
| Tower, second level | 6L | **6** | 72 | 0.1411 | 1.5000 | CosC |
| Tower Court, northeast corner | 6m | **6** | 71 | 0.0864 | 0.6583 | CapDis |
| House C, west room | 6n | **6** | 30, 31, 32, 33, 34, 35, 36, 37, 39, 40, 41 | 0.1152 | 0.8333 | RAud; APSolo; PSolo; CosT |
| House E, southwestern room | 6o | **6** | (East Subterranean Stair) | 0.0752 | 0.8250 | – |
| House H, north room | 6p | **6** | (Bench) | 0.0861 | 0.4583 | – |
| East Subterranean Passage, dogs leg | 6q | **6** | – | 0.1111 | 1.0000 | – |
| South Subterranean Passage, dogs leg | 6r | **6** | 78 | 0.1128 | 0.8333 | CosU |
| South Substructure, northwest room | 6s | **6** | 81 (Bench) | 0.1484 | 1.0000 | – |
| West Court | 7a | **7** | 63, 64, 65, 67, 68 | 0.1387 | 1.5000 | RitDan?; GPWA; CosC; TCapSac |
| Toilet, west-central room | 7b | **7** | 69 | 0.1278 | 1.5000 | CosC |
| South Subterranean Passage, south corridor | 7c | **7** | – | 0.1349 | 1.0000 | – |
| East Subterranean Passage, north corridor | 7d | **7** | 7 | 0.0937 | 0.8333 | CosU |
| House I, south corridor | 7e | **7** | – | 0.1196 | 1.2500 | – |
| Tower, third level | 7f | **7** | (Bench) | 0.1651 | 0.5000 | – |
| West Court, south room | 8a | **8** | 66 | 0.1628 | 0.5000 | GPWA |
| Toilet, north room | 8b | **8** | (Toilets) | 0.1516 | 0.5000 | – |
| House I, rear | 8c | **8** | (Sweat Bath) | 0.1440 | 0.5000 | – |

Appendix 4 *(continued)*

Palenque Palace Phase 6 sculpture ranked according to carrier depths

| Description of spatial unit | Cell | Carrier rank | Sculpture (GR No.) | RA value | Control value | Theme |
|---|---|---|---|---|---|---|
| *Carrier* | c | **0** | 1, 2, 82, 83, 84, 85 | 0.0595 | 4.2499 | CosC; RPSoloCosCU; RPSolo |
| House A-D, north corridor | 1a | **1** | 55, 56, 57 (Bench) | 0.0608 | 0.8095 | RPSolo; RAudAccTDynHist |
| House D, west corridor | 1b | **1** | 48, 49, 50, 51, 52, 53 | 0.0615 | 0.8095 | HumSac; SacDan; Text? |
| House K, northwest room | 1c | **1** | – | 0.0744 | 0.1428 | – |
| House K, external path access | 1d | **1** | – | 0.0593 | 1.2261 | – |
| Southwest Corner Plaza Level, north room | 1e | **1** | – | 0.0744 | 0.1428 | – |
| Southwest Corner Plaza Level, south room | 1f | **1** | – | 0.0744 | 0.1428 | – |
| Southern Substructure, south terrace | 1g | **1** | – | 0.0616 | 0.7261 | – |
| House A, east corridor | 2a | **2** | 42, 43, 44, 45, 46, 47 | 0.0603 | 0.9166 | THist; APWA; APSolo |
| House D, southeast room | 2b | **2** | – | 0.0634 | 1.4583 | – |
| House K, southwest room | 2c | **2** | – | 0.0631 | 0.9166 | – |
| House L, north-central corridor | 2d | **2** | – | 0.0700 | 1.3333 | – |
| House L, south terrace | 2e | **2** | – | 0.0557 | 0.8409 | – |
| South Substructure, south corridor | 2f | **2** | 80 (Bench) | 0.0734 | 1.4166 | TCosCU |
| South Terrace stair | 2g | **2** | – | 0.0586 | 0.6742 | – |
| House A, west corridor | 3a | **3** | – | 0.0570 | 1.0694 | – |
| House D & K, external path | 3b | **3** | – | 0.0593 | 3.4999 | – |
| House D east, south-central room | 3c | **3** | **54** | 0.0782 | 0.3333 | **?(Mural)** |
| House K, east room | 3d | **3** | (West Subterranean Stair) | 0.0664 | 0.9583 | – |
| House K, south-central corridor | 3e | **3** | – | 0.0770 | 1.6667 | – |
| House L, northwest corridor | 3f | **3** | – | 0.0796 | 0.5833 | – |
| House L, south-central corridor | 3g | **3** | – | 0.0841 | 1.4167 | – |
| House L, central room | 3h | **3** | – | 0.0841 | 0.8333 | – |
| South Terrace | 3i | **3** | – | 0.0484 | 4.8333 | – |
| South Substructure, east-central room | 3j | **3** | (Bench) | 0.0700 | 1.4167 | – |
| South Substructure, east-central corridor | 3k | **3** | – | 0.0837 | 1.0833 | – |
| South Substructure Patio, north corridor | 3L | **3** | – | 0.0884 | 0.7500 | – |
| House J west, central room | 3m | **3** | (Bench) | 0.0605 | 0.9444 | – |
| House L, south corridor | 3n | **3** | – | 0.0698 | 0.8333 | – |
| East Court, House A west foundation | 4a | **4** | 58, 59, 60, 61 (H.S.) | 0.0658 | 5.5833 | CapDis; THist |
| House A-D, southeast corridor | 4b | **4** | – | 0.0690 | 0.3750 | – |
| House J west, access corridor | 4c | **4** | – | 0.0505 | 5.4500 | – |
| House F, south patio | 4d | **4** | – | 0.0447 | 1.3032 | – |
| House G, southeast room | 4e | **4** | – | 0.0542 | 0.7576 | – |
| House G, southwest room | 4f | **4** | – | 0.0626 | 0.4242 | – |

| | | | | | | |
|---|---|---|---|---|---|---|
| House G & H, east-central corridor | 4g | **4** | (Bench) | 0.0626 | 0.4242 | – |
| House G & H, west-central corridor | 4h | **4** | – | 0.0625 | 0.9242 | – |
| House H, southeast room | 4i | **4** | – | 0.0625 | 0.9242 | – |
| House H, south-central room | 4j | **4** | – | 0.0626 | 0.4242 | – |
| House H, southeast-central room | 4k | **4** | – | 0.0585 | 0.4242 | – |
| House H, northeast room | 4L | **4** | – | 0.0630 | 0.7917 | – |
| House K, southeast room | 4m | **4** | – | 0.0902 | 0.3333 | – |
| South Subterranean Passage, north corridor | 4n | **4** | – | 0.0784 | 0.8333 | – |
| House K, south corridor | 4o | **4** | – | 0.0769 | 1.1667 | – |
| House L, east corridor | 4p | **4** | – | 0.0984 | 1.5833 | – |
| House L, south-central room | 4q | **4** | – | 0.0819 | 0.7500 | – |
| South Substructure, patio | 4r | **4** | – | 0.1032 | 1.5000 | – |
| South Substructure, northeast corridor | 4s | **4** | – | 0.0790 | 0.8333 | – |
| South Substructure, central room | 4t | **4** | (Bench) | 0.0855 | 0.5833 | – |
| House A-D, southwest room | 4u | **4** | – | 0.0737 | 1.1250 | – |
| House D, northeast room | 4v | **4** | – | 0.0738 | 0.6250 | – |
| House D east, north-central room | 4w | **4** | – | 0.0738 | 0.6250 | – |
| Tower Court, patio stair | 4x | **4** | – | 0.0596 | 0.2500 | – |
| South Terrace, southeast corner | 4y | **4** | – | 0.0630 | 0.0909 | – |
| House J west, north-central room | 4z | **4** | – | 0.0634 | 0.4444 | – |
| Tower, crossway access | 4aa | **4** | – | 0.0734 | 0.6250 | – |
| House B, northeast room | 5a | **5** | 24, 25, 26, 27 | 0.0815 | 0.1250 | RPSolo; MMot; CosU; CosCU |
| House B, central room | 5b | **5** | 24, 25, 26, 27 | 0.0815 | 0.1250 | RPSolo; MMot; CosU; CosCU |
| House B, northwest room | 5c | **5** | 24, 25, 26, 27 | 0.0815 | 0.1250 | RPSolo; MMot; CosU; CosCU |
| East Court, southwest corner | 5d | **5** | – | 0.0759 | 0.6250 | – |
| House C east, north-central room | 5e | **5** | 29, 38, 40, 41, 62 | 0.0740 | 1.4583 | GPSolo; RAud; CosT; TDed? |
| House A-D south, west-central room | 5f | **5** | – | 0.0889 | 0.5000 | – |
| House J, northwest room | 5g | **5** | (Bench) | 0.0662 | 0.1111 | – |
| House J east, south-central room | 5h | **5** | – | 0.0662 | 0.1111 | – |
| House J east, south room | 5i | **5** | – | 0.0662 | 0.1111 | – |
| House F, east corridor | 5j | **5** | – | 0.0577 | 7.3667 | – |
| House F, northeast corridor | 5k | **5** | – | 0.0628 | 2.2778 | – |
| House G, northeast room | 5L | **5** | – | 0.0545 | 0.3095 | – |
| House G, path access | 5m | **5** | – | 0.0479 | 2.4167 | – |
| Tower Court | 5n | **5** | 70, 73 (Throne Tablet), 74, 75 | 0.0543 | 4.7500 | ?(Stucco fret); TDyn Hist; CapDis |
| House H, north terrace | 5o | **5** | – | 0.0632 | 1.6667 | – |
| House H, northwest room | 5p | **5** | (Bench) (South Subterranean Stair) | 0.0652 | 1.3333 | – |
| House H, southwest room | 5q | **5** | – | 0.0663 | 1.0000 | – |
| House L, northeast corridor | 5r | **5** | – | 0.1133 | 0.3333 | – |

(*continued*)

Appendix 4 *(continued)*

Palenque Palace Phase 6 sculpture ranked according to carrier depths

| Description of spatial unit | Cell | Carrier rank | Sculpture GR No. | RA value | Control value | Theme |
|---|---|---|---|---|---|---|
| West Subterranean Passage, dogs leg | 5s | **5** | – | 0.0926 | 1.0000 | – |
| East Subterranean Passage, south corridor | 5t | **5** | – | 0.0817 | 0.8333 | – |
| South Substructure Patio, southwest corner | 5u | **5** | – | 0.1183 | 0.5000 | – |
| House A-D, south-central room | 5v | **5** | – | 0.0818 | 0.1250 | – |
| House J, south patio | 5w | **5** | – | 0.0663 | 0.1111 | – |
| Tower, first level | 5x | **5** | – | 0.0879 | 1.0000 | – |
| House B, southeast room | 6a | **6** | 23, 28, 26, 27, 28 | 0.0777 | 0.2000 | CosT?; MMot; CosU; CosCU |
| House F, northeast corner room | 6b | **6** | – | 0.0724 | 0.1250 | – |
| House F, northeast central room | 6c | **6** | – | 0.0724 | 0.1250 | – |
| House E, northeast room | 6aa | **6** | 8, **15** | 0.0775 | 0.7000 | CosCU |
| West Subterranean Passage, south corridor | 6ab | **6** | 79 | 0.1030 | 1.0000 | CosU |
| House F, east central room | 6d | **6** | (Bench) | 0.0724 | 0.1250 | – |
| House F, southeast central room | 6e | **6** | (Bench) | 0.0724 | 0.1250 | – |
| House F, southeast corner room | 6f | **6** | – | 0.0724 | 0.1250 | – |
| House F, north-central room | 6g | **6** | – | 0.0747 | 1.2000 | – |
| House F, south room | 6h | **6** | – | 0.0700 | 0.6250 | – |
| House F, northwest corridor | 6i | **6** | – | 0.0699 | 0.4000 | – |
| House E & F, central corridor | 6j | **6** | – | 0.0604 | 1.4761 | – |
| House E, southeast corner room | 6k | **6** | – | 0.0614 | 0.3928 | – |
| House G, north-central room | 6L | **6** | – | 0.0627 | 0.6428 | – |
| House G & H, north-central corridor | 6m | **6** | – | 0.0580 | 0.9761 | – |
| House H, north room | 6n | **6** | (Bench) | 0.0784 | 0.3333 | – |
| Tower Court, northeast corner | 6o | **6** | 71 | 0.0645 | 0.8750 | CapDis |
| House E, northwestern room | 6p | **6** | 3 (Bench), 4, 5, 6 (DR Throne) **10**, 11, 13, **14** | 0.0580 | 1.1249 | **Text?**; **CosU**;RAud; Acc; CosTU; THist; TDynHist |
| House E, southwestern room | 6q | **6** | (East Subterranean Stair) | 0.0524 | 1.8511 | – |
| House I, north room | 6r | **6** | 76, 77 | 0.0692 | 0.1250 | Cos?; RPSolo |
| House I, west corridor | 6s | **6** | – | 0.0692 | 0.1250 | – |
| Toilets, southwest room | 6t | **6** | – | 0.0692 | 0.1250 | – |
| Toilets, east corridor | 6u | **6** | – | 0.0688 | 0.6250 | – |
| Tower, second level | 6v | **6** | 72 | 0.1026 | 1.5000 | CosC |
| House C, northeast room | 6w | **6** | – | 0.0888 | 0.3333 | – |
| House C east, south-central room | 6x | **6** | 29, 38, 40, 41, 62 | 0.0777 | 1.3333 | GPSolo; RAud; CosT; TDed? |
| South Subterranean Passage, dogs leg | 6z | **6** | 78 | 0.0791 | 0.8333 | CosU |
| House C, northwest room | 7a | 7 | 31, 32, 33, 34, 35, 36, 37, 39, 40, 41 | 0.0919 | 0.8333 | APSolo; PSolo; CosT |

| | | | | | | |
|---|---|---|---|---|---|---|
| House C, southeast room | 7b | **7** | – | 0.0753 | 0.6667 | – |
| Toilet, west-central room | 7c | **7** | 69 | 0.0834 | 1.5000 | CosC |
| Southeast Court | 7d | **7** | 28 | 0.0682 | 2.8333 | CosU |
| House F, northwest room | 7e | **7** | – | 0.0809 | 0.6667 | – |
| House F, west-central room | 7f | **7** | (Bench) | 0.0706 | 0.9500 | – |
| House F, north-central room | 7g | **7** | – | 0.0902 | 0.8333 | – |
| House E east extension, north room | 7h | **7** | – | 0.0688 | 0.7500 | – |
| House F, southeast room | 7i | **7** | – | 0.0732 | 0.7500 | – |
| House E, southeast room | 7j | **7** | – | 0.0667 | 2.1666 | – |
| House G, northeast room | 7k | **7** | – | 0.0725 | 0.8333 | – |
| East Subterranean Passage, north corridor | 7L | **7** | 7 | 0.0664 | 0.6666 | CosU |
| South Subterranean Passage, south corridor | 7m | **7** | – | 0.0921 | 1.0000 | – |
| House I, east corridor | 7n | **7** | – | 0.0665 | 0.6666 | – |
| South Substructure, northwest room | 7o | **7** | 81 (Bench) | 0.1025 | 1.0000 | – |
| Tower, third level | 7p | **7** | (Bench) | 0.1175 | 0.5000 | – |
| House F, central room | 8a | **8** | – | 0.1049 | 1.5000 | – |
| House C, southwest room | 8b | **8** | 31, 32, 33, 34, 35, 36, 37, 38, 39, 40, 41 | 0.1065 | 1.0000 | APSolo; PSolo; CosT |
| House E, southeast room | 8c | **8** | (Bench) | 0.0815 | 0.3333 | – |
| House B, southwest room | 8d | **8** | 16, 17, 18, 19, 20, 21, 22, 26, 27, 28, (Bench?) | 0.0834 | 0.2000 | RAud; Text?; PSolo; RitDan; MMot; CosU |
| House E east extension, south room | 8e | **8** | – | 0.0816 | 0.3333 | – |
| House I, south corridor | 8f | **8** | – | 0.0814 | 1.5000 | – |
| West court | 9a | **9** | 63, 64, 65, 67, 68 | 0.1212 | 1.5000 | RitDan?; GPWA; CosC; TCapSac |
| House F, south-central room | 9b | **9** | – | 0.1199 | 0.5000 | – |
| House I, rear | 9c | **9** | (Sweat Bath) | 0.0964 | 0.5000 | – |
| West Court, south room | 10a | **10** | 66 | 0.1361 | 0.5000 | GPWA |

Appendix 5. *Sculpture ranked according to relative asymmetry (RA) values, Palenque Palace Phases 1–6*

| Palenque Palace Phase 1 sculpture ranked according to relative asymmetry (RA) values (least integrated to most integrated) | | | | | | |
|---|---|---|---|---|---|---|
| Description of spatial unit | Cell | Carrier rank | Sculpture GR No. | RA value | Control value | Theme |
| South Substructure Patio, southwest corner | 5g | 5 | – | **0.3363** | 0.5000 | – |
| South Substructure, Patio | 4e | 4 | – | **0.2971** | 1.5000 | – |
| South Substructure, central room | 4d | 4 | (Bench) | **0.2647** | 0.6666 | – |
| South Substructure Patio, north corridor | 3e | 3 | – | **0.2609** | 0.7500 | – |
| South Subterranean Structure, northwest room | 10b | 10 | 81 (Bench) | **0.2564** | 1.0000 | – |
| Southern Substructure, southeast terrace lower | 2d | 2 | – | **0.2443** | 1.4999 | – |
| West Subterranean Passage, south corridor | 9g | 9 | **79** | **0.2330** | 1.0000 | **CosU** |
| House J, east room | 10a | 10 | – | **0.2300** | 0.3333 | – |
| South Substructure, east-central room | 3c | 3 | (Bench) | **0.2293** | 1.0833 | – |
| South Substructure, east-central corridor | 3d | 3 | – | **0.2293** | 1.0833 | – |
| South Subterranean Passage, south corridor | 9h | 9 | – | **0.2217** | 1.0000 | – |
| South Substructure, northeast corridor | 4c | 4 | – | **0.2172** | 1.1666 | – |
| Southern Substructure, south terrace lower | 1b | 1 | – | **0.2059** | 1.7500 | – |
| West Subterranean Passage, dogs leg | 8c | 8 | – | **0.2059** | 1.0000 | – |
| House J, northwest room | 9a | 9 | (Bench) | **0.2051** | 0.1428 | – |
| House J west, south-central room | 9d | 9 | – | **0.2051** | 0.1428 | – |
| House J west, south room | 9e | 9 | – | **0.2051** | 0.1428 | – |
| House J, south patio | 9f | 9 | – | **0.2051** | 0.1428 | – |
| House J west, north-central room | 9b | 9 | – | **0.2036** | 0.4761 | – |
| House J west, central room | 9c | 9 | (Bench) | **0.2029** | 1.6428 | – |
| East Subterranean Passage, south corridor | 5f | 5 | – | **0.2006** | 0.8333 | – |
| South Subterranean Passage, dogs leg | 8b | 8 | **78** | **0.1916** | 1.0000 | **CosU** |
| *Carrier* | c | 0 | – | **0.1863** | 0.5833 | – |
| East Subterranean Passage, dogs leg | 6f | 6 | – | **0.1833** | 1.0000 | – |

| | | | | | | |
|---|---|---|---|---|---|---|
| Terrace, upper southwest corner | 8e | 8 | – | **0.1817** | 0.3333 | – |
| North Substructure, north corridor | 2b | 2 | – | **0.1772** | 0.7500 | – |
| South Subterranean Passage, north corridor | 7e | 7 | – | **0.1727** | 0.7000 | – |
| North Substructure, north terrace lower | 1a | 1 | – | **0.1667** | 2.0000 | – |
| House J, west access | 8a | 8 | – | **0.1667** | 5.0833 | – |
| Terrace, upper southeast corner | 8d | 8 | – | **0.1652** | 0.1666 | – |
| East Subterranean Passage, north corridor | 7c | 7 | **7** | **0.1637** | 1.0000 | **CosCU** |
| North Substructure, south corridor | 3a | 3 | – | **0.1606** | 0.7000 | – |
| House H, substructure | 7d | 7 | (South Subterranean Stairs) | **0.1584** | 0.8333 | – |
| House B, east terrace upper | 5a | 5 | – | **0.1576** | 0.2000 | – |
| Terrace, lower west | 4b | 4 | – | **0.1546** | 0.5333 | – |
| North Substructure, east terrace lower | 2a | 2 | – | **0.1508** | 0.4500 | – |
| North Substructure, west terrace lower | 2c | 2 | – | **0.1508** | 0.4500 | – |
| House B, south room | 6a | 6 | **21, 22, 26, 27, 28** | **0.1501** | 0.7500 | **PSolo; RitDan; MMot; CosU** |
| Terrace, upper southwest | 7f | 7 | – | **0.1448** | 1.3666 | – |
| House E, southwestern room | 7b | 7 | (East Subterranean Stairs) | **0.1440** | 0.7500 | – |
| House B, north room | 5b | 5 | **24, 25, 26, 27** | **0.1425** | 0.7000 | **RPSolo; MMot; CosU; CosCU** |
| House K, west room | 5e | 5 | – | **0.1425** | 0.9000 | – |
| House K, east room | 6e | 6 | (West Subterranean Stairs) | **0.1395** | 1.6999 | – |
| Southeast Court, early | 7a | 7 | – | **0.1388** | 1.0595 | – |
| North Substructure, south terrace lower | 3b | 3 | – | **0.1320** | 2.2000 | – |
| House E, east room | 6b | 6 | **8** | **0.1282** | 0.9166 | **CosCU** |
| Southwest Court, later tower court | 6d | 6 | – | **0.1267** | 0.8666 | – |
| Terrace, upper southeast | 7g | 7 | – | **0.1267** | 2.4166 | – |
| Terrace, upper northwest | 5d | 5 | – | **0.1244** | 1.3166 | – |
| House E, west room | 6c | 6 | **3, 4, 5, 13, 11, 80** (Bench) | **0.1244** | 1.1666 | **RAud; Acc; CosTU; THist; TCosCU** |
| House B, north terrace upper | 4a | 4 | – | **0.1192** | 1.5333 | – |
| House E, access north stair | 5c | 5 | – | **0.1184** | 0.9000 | – |

Appendix 5 *(continued)*

| Palenque Palace Phase 2 sculpture ranked according to relative asymmetry (RA) values (least integrated to most integrated) | | | | | | |
|---|---|---|---|---|---|---|
| Description of spatial unit | Cell | Carrier rank | Sculpture GR No. | RA value | Control value | Theme |
| South Substructure Patio, southwest corner | 5g | 5 | – | **0.3071** | 0.5000 | – |
| South Substructure Patio | 4e | 4 | – | **0.2701** | 1.5000 | – |
| South Substructure Patio, north corridor | 3e | 3 | – | **0.2357** | 0.7500 | – |
| South Substructure, central room | 4c | 4 | (Bench) | **0.2357** | 0.6666 | – |
| South Subterranean Structure, northwest room | 10a | 10 | 81 (Bench) | **0.2286** | 1.0000 | – |
| Southern Substructure, southeast terrace lower | 2e | 2 | – | **0.2227** | 0.3333 | – |
| West Subterranean Passage, south corridor | 9b | 9 | 79 | **0.2169** | 1.0000 | CosU |
| South Substructure, south corridor | 2d | 2 | – | **0.2045** | 1.4999 | – |
| South Substructure, east-central room | 3c | 3 | (Bench) | **0.2013** | 1.0833 | – |
| South Substructure, east-central corridor | 3d | 3 | – | **0.2013** | 1.0833 | – |
| South Subterranean Passage, south corridor | 9a | 9 | – | **0.2000** | 1.0000 | – |
| South Substructure, northeast corridor | 4d | 4 | – | **0.1896** | 1.1666 | – |
| Southern Substructure, south terrace lower | 1b | 1 | – | **0.1870** | 1.7500 | – |
| West Subterranean Passage, dogs leg | 8i | 8 | – | **0.1864** | 1.0000 | – |
| East Subterranean Passage, south corridor | 5f | 5 | – | **0.1734** | 0.8333 | – |
| House J, east room | 7i | 7 | – | **0.1734** | 0.5833 | – |
| South Subterranean Passage, dogs leg | 8h | 8 | 78 | **0.1695** | 1.0000 | CosU |
| *Carrier* | C | 0 | – | **0.1682** | 0.5833 | – |
| House J, northwest room | 8a | 8 | (Bench) | **0.1675** | 0.1250 | – |
| House J west, south-central room | 8d | 8 | – | **0.1675** | 0.1250 | – |
| House J west, south room | 8e | 8 | – | **0.1675** | 0.1250 | – |
| House J, south patio | 8f | 8 | – | **0.1675** | 0.1250 | – |
| House J west, north-central room | 8b | 8 | – | **0.1662** | 0.4583 | – |
| House J west, central room | 8c | 8 | (Bench) | **0.1656** | 1.1250 | – |
| Terrace, upper southwest corner | 8k | 8 | – | **0.1617** | 0.3333 | – |
| East Subterranean Passage, dogs leg | 6h | 6 | – | **0.1597** | 0.8333 | – |
| North Substructure, north corridor | 2b | 2 | – | **0.1584** | 0.7500 | – |
| South Subterranean Passage, north corridor | 7e | 7 | – | **0.1552** | 0.7000 | – |
| House A, east room | 6i | 6 | **42, 43, 44, 45, 46, 47** | **0.1539** | 0.5833 | **THist; APWA; APSolo** |

| | | | | | | |
|---|---|---|---|---|---|---|
| North Substructure, north terrace lower | 1a | 1 | – | **0.1494** | 2.0000 | – |
| North Substructure, south corridor | 3a | 3 | – | **0.1416** | 0.7000 | – |
| Terrace, upper southeast corner | 8g | 8 | – | **0.1409** | 0.1666 | – |
| House A & J, east patio | 6a | 6 | – | **0.1403** | 1.4583 | – |
| East Subterranean Passage, north corridor | 7g | 7 | 7 | **0.1396** | 1.8333 | CosU |
| House H, substructure | 7d | 7 | (South Subterranean Stairs) | **0.1383** | 0.6428 | – |
| North Substructure, east terrace lower | 2a | 2 | – | **0.1338** | 0.4500 | – |
| North Substructure, west terrace lower | 2c | 2 | – | **0.1338** | 0.4500 | – |
| House K, west room | 5e | 5 | – | **0.1338** | 0.9500 | – |
| Terrace, lower west | 4b | 4 | – | **0.1318** | 0.5333 | – |
| House B, south room | 6b | 6 | 21, 22, 26, 27, 28 | **0.1312** | 0.5833 | PSolo; RitDan; MMot; CosU; CosCU |
| House J, west access | 7a | 7 | – | **0.1312** | 5.3333 | – |
| House A, west room | 5a | 5 | – | **0.1299** | 0.9500 | – |
| House B, north room | 5b | 5 | 24, 25, 26, 27 | **0.1286** | 1.0333 | RPSolo; MMot; CosU; CosCU |
| House C, west room | 6c | 6 | **30, 31, 32, 33, 34, 35, 36, 37, 39, 40, 41** | **0.1279** | 0.7261 | **RAud; APSolo; PSolo; CosT** |
| House C, west foundation | 6d | 6 | **68** | **0.1273** | 1.0090 | **TCapSac** |
| East Court, southwest corner | 5c | 5 | – | **0.1260** | 0.7833 | – |
| Terrace, upper southwest | 7h | 7 | – | **0.1253** | 0.3666 | – |
| House K, east room | 6g | 6 | (West Subterranean Stairs) | **0.1247** | 1.7261 | – |
| House E, southwestern room | 8j | 8 | (East Subterranean Stairs) | **0.1227** | 0.8333 | – |
| Southeast Court, early | 7b | 7 | – | **0.1162** | 1.0416 | – |
| North Substructure, south terrace lower | 3b | 3 | – | **0.1156** | 2.2000 | – |
| House E, east room | 6e | 6 | 8 | **0.1156** | 0.7499 | CosCU |
| House C, east room | 5d | 5 | **29, 30, 38, 40, 41, 62** | **0.1149** | 0.6761 | **GPSolo; CosT; TDed?** |
| House E, west room | 7c | 7 | 3, 4, 5, 13, 11, 80 (Bench) | **0.1149** | 0.7261 | RAud; Acc; CosTU; THist; TCosCU |
| East Court, House A west foundation (early) | 4a | 4 | **58** (*alfardas* only), **59, 60, 61**(H.S.) | **0.1078** | 1.5333 | **CapDis; THist** |
| Southwest Court, later tower court | 6f | 6 | – | **0.1071** | 2.1166 | – |
| Terrace, upper southeast | 7f | 7 | – | **0.1071** | 2.3333 | – |

Appendix 5 *(continued)*

Palenque Palace Phase 3 sculpture ranked according to relative asymmetry (RA) values (least integrated to most integrated)

| Description of spatial unit | Cell | Carrier rank | Sculpture GR No. | RA value | Control value | Theme |
|---|---|---|---|---|---|---|
| South Substructure Patio, southwest corner | 5L | 5 | – | **0.2468** | 0.5000 | – |
| South Substructure Patio | 4o | 4 | – | **0.2104** | 1.5000 | – |
| South Substructure, central room | 4n | 4 | (Bench) | **0.2006** | 0.6666 | – |
| South Subterranean Structure, northwest room | 6f | 6 | 81 (Bench) | **0.1974** | 1.0000 | – |
| West Subterranean Passage, south corridor | 5k | 5 | 79 | **0.1864** | 1.0000 | CosU |
| South Substructure, northeast corridor | 4m | 4 | – | **0.1766** | 1.1666 | – |
| South Subterranean Passage, south corridor | 6e | 6 | – | **0.1766** | 1.0000 | – |
| South Substructure Patio, north corridor | 3i | 3 | – | **0.1760** | 0.7500 | – |
| East Subterranean Passage, south corridor | 5i | 5 | – | **0.1695** | 0.8333 | – |
| South Substructure, east-central room | 3f | 3 | (Bench) | **0.1662** | 1.0833 | – |
| South Substructure, east-central corridor | 3g | 3 | – | **0.1662** | 1.0833 | – |
| Southern Substructure, southeast terrace lower | 2g | 2 | – | **0.1565** | 0.3333 | – |
| West Subterranean Passage, dogs leg | 4p | 4 | – | **0.1558** | 1.0000 | – |
| East Subterranean Passage, dogs leg | 6d | 6 | – | **0.1532** | 1.0000 | – |
| South Subterranean Passage, dogs leg | 5j | 5 | 78 | **0.1461** | 1.0000 | CosU |
| South Substructure, south corridor | 2f | 2 | **80** (Bench moved) | **0.1429** | 1.4999 | **TCosCU** |
| House J, northwest room | 5a | 5 | (Bench) | **0.1416** | 0.1250 | – |
| House J west, south-central room | 5b | 5 | – | **0.1416** | 0.1250 | – |
| House J west, south room | 5c | 5 | – | **0.1416** | 0.1250 | – |
| House J, south patio | 5d | 5 | – | **0.1416** | 0.1250 | – |
| House B, southeast room | 6a | 6 | **23**, 26, 27, 28 | **0.1390** | 0.2000 | **CosT?**; MMot; CosU; CosCU |
| House B, southwest room | 6b | 6 | **16**, **17**, **18**, **19**, **20**, 21, 22, 26, 27, 28 | **0.1390** | 0.2000 | **RAud**; **Text?**; PSolo; RitDan; MMot; CosU |
| East Subterranean Passage, north corridor | 6c | 6 | 7 | **0.1331** | 0.8333 | CosU |
| House B, northeast room | 4a | 4 | 24, 25, 26, 27 | **0.1318** | 0.1428 | RPSolo; MMot; CosU; CosCU |
| House B, central room | 4b | 4 | 24, 25, 26, 27 | **0.1318** | 0.1428 | RPSolo; MMot; CosU; CosCU |
| House J west, north-central room | 4j | 4 | – | **0.1318** | 0.4583 | – |
| South Terrace, southwest corner | 4q | 4 | – | **0.1286** | 0.2500 | – |
| South Subterranean Passage, north corridor | 3e | 3 | – | **0.1253** | 0.7000 | – |

| | | | | | | |
|---|---|---|---|---|---|---|
| Terrace, upper southeast corner | 5f | 5 | – | **0.1247** | 0.1666 | – |
| House B, northwest room | 4c | 4 | 24, 25, 26, 27 | **0.1240** | 0.4761 | RPSolo; MMot; CosU; CosCU |
| House J west, central room | 3c | 3 | (Bench) | **0.1214** | 0.9583 | – |
| Southern Substructure, south terrace | 1e | 1 | – | **0.1208** | 1.4500 | – |
| West Court, early | 4g | 4 | **63, 64, 65, 66**, 68, **69** | **0.1201** | 0.4761 | **GPWA; CosC;** TCapSac |
| House H, substructure | 4k | 4 | (South Subterranean Stair) | **0.1156** | 0.6428 | – |
| House A, east corridor | 2b | 2 | 42, 43, 44, 45, 46, 47 | **0.1149** | 0.8333 | THist; APWA; APSolo |
| East Court, southwest corner | 4d | 4 | – | **0.1149** | 0.9761 | – |
| House C, west room | 4f | 4 | 30, 31, 32, 33, 34, 35, 36, 37, 39, 40, 41 | **0.1149** | 0.9761 | RAud; APSolo; PSolo; CosT |
| House E, southwestern room | 5h | 5 | (East Subterranean Stair) | **0.1123** | 0.9999 | – |
| House D, west corridor | 1b | 1 | **48, 49, 50, 51, 52, 53** | **0.1117** | 0.7000 | **HumSac; SacDan; Text?** |
| House K, west room | 1d | 1 | – | **0.1084** | 0.6000 | – |
| House E, east room | 5g | 5 | 8 | **0.1078** | 1.0333 | CosCU |
| House J, west access path | 4h | 4 | – | **0.1058** | 5.2833 | – |
| House C, east room | 4e | 4 | 29, 38, 40, 41, 62 | **0.1052** | 0.6190 | GPSolo; CosT; TDed? |
| Southeast Court, early | 5e | 5 | – | **0.1032** | 2.4916 | – |
| House E, west room | 4i | 4 | 3, 4, 5, **6** (DR Throne), 13, 11 | **0.1026** | 0.7261 | RAud; Acc; CosTU; THist; **TDynHist** |
| House A, west corridor | 3a | 3 | – | **0.1013** | 0.8011 | – |
| *Carrier* | C | 0 | **1,2, 82, 83, 84, 85** | **0.1000** | 1.4499 | **CosC; RPSoloCosCU; RPSolo** |
| House A-D, north corridor | 1a | 1 | **55, 56, 57** (Throne?) | **0.0994** | 1.0666 | **RPSolo; RAudAccTDynHist** |
| House D, east corridor | 2d | 2 | – | **0.0968** | 0.8999 | – |
| East Court, House A west foundation | 3b | 3 | **58** (remaining Captive Panels), 59, 60, 61(H.S.) | **0.0961** | 3.6166 | **CapDis**; THist |
| House K, east room | 2e | 2 | (West Subterranean Stair) | **0.0955** | 1.4499 | – |
| House D & K, patio | 1c | 1 | – | **0.0942** | 1.1833 | – |
| Terrace, upper southwest | 3h | 3 | – | **0.0929** | 1.5333 | – |
| Terrace, upper southeast | 4L | 4 | – | **0.0896** | 2.1761 | – |
| House A-D, south corridor | 2a | 2 | – | **0.0864** | 1.0595 | – |
| Southwest Court, later tower court | 3d | 3 | – | **0.0851** | 2.3333 | – |
| House D & K, external path | 2c | 2 | – | **0.0805** | 1.2428 | – |

Appendix 5 *(continued)*

| Palenque Palace Phase 4 sculpture ranked according to relative asymmetry (RA) values (least integrated to most integrated) | | | | | | |
|---|---|---|---|---|---|---|
| Description of spatial unit | Cell | Carrier rank | Sculpture GR No. | RA value | Control value | Theme |
| South Substructure Patio, southwest corner | 5u | 5 | – | **0.2011** | 0.5000 | – |
| Tower, third level | 7c | 7 | (Bench) | **0.1878** | 0.5000 | – |
| West Court, south room | 8a | 8 | – | **0.1835** | 0.5000 | – |
| South Substructure, patio | 4m | 4 | – | **0.1745** | 1.5000 | – |
| South Substructure, northwest room | 6g | 6 | 81 (Bench) | **0.1729** | 1.0000 | – |
| South Substructure, central room | 4L | 4 | (Bench) | **0.1682** | 0.6666 | – |
| South Subterranean Passage, south corridor | 7b | 7 | – | **0.1612** | 1.0000 | – |
| Toilet, north room | 7e | 7 | (Toilets) | **0.1608** | 0.5000 | – |
| Tower, second level | 6h | 6 | **72** | **0.1600** | 1.5000 | **CosC** |
| West Subterranean Passage, south corridor | 5t | 5 | 79 | **0.1569** | 1.0000 | CosU |
| West Court | 7d | 7 | 63, 64, 65, 66, **67**, 68,69 | **0.1557** | 1.5000 | **RitDan?**; GPWA; CosC; TCapSac |
| South Substructure, northeast corridor | 4k | 4 | – | **0.1549** | 1.1666 | – |
| East Subterranean Passage, south corridor | 5s | 5 | – | **0.1475** | 0.8333 | – |
| South Substructure Patio, north corridor | 3h | 3 | – | **0.1475** | 0.7500 | – |
| South Substructure, east-central corridor | 3g | 3 | – | **0.1428** | 1.0833 | – |
| South Substructure, east-central room | 3f | 3 | (Bench) | **0.1428** | 1.0833 | – |
| West Subterranean Passage, dogs leg | 4j | 4 | – | **0.1354** | 1.0000 | – |
| Toilet, west-central room | 6j | 6 | – | **0.1338** | 1.5000 | – |
| Tower, first level | 5h | 5 | – | **0.1330** | 1.0000 | – |
| East Subterranean Passage, dogs leg | 6e | 6 | – | **0.1311** | 1.0000 | – |
| South Subterranean Passage, dogs leg | 6f | 6 | 78 | **0.1303** | 0.8333 | CosU |
| House B, central room | 5b | 5 | 24, 25, 26, 27 | **0.1279** | 0.1428 | RPSolo; MMot; CosU; CosCU |
| House B, northeast room | 5a | 5 | 24, 25, 26, 27 | **0.1279** | 0.1428 | RPSolo; MMot; CosU; CosCU |
| House C, west room | 6i | 6 | 30, 31, 32, 33, 34, 35, 36, 37, 39, 40, 41 | **0.1268** | 0.8333 | RAud; APSolo; PSolo; CosT |
| House J, south patio | 5m | 5 | – | **0.1240** | 0.1250 | – |
| House J west, south room | 5L | 5 | – | **0.1240** | 0.1250 | – |
| House J west, south-central room | 5k | 5 | – | **0.1240** | 0.1250 | – |
| House J, northwest room | 5j | 5 | (Bench) | **0.1240** | 0.1250 | – |
| South Substructure, south corridor | 2d | 2 | 80 (Bench) | **0.1193** | 1.4999 | TCosCU |
| House J west, north-central room | 4c | 4 | – | **0.1182** | 0.4583 | – |
| House B, northwest room | 5c | 5 | 24, 25, 26, 27 | **0.1174** | 0.4761 | RPSolo; MMot; CosU; CosCU |
| House A-D, southeast corridor | 4a | 4 | – | **0.1170** | 0.3928 | – |
| East Subterranean Passage, north corridor | 7a | 7 | 7 | **0.1146** | 0.8333 | CosU |
| House B, southwest room | 6b | 6 | 16, 17, 18, 19, 20, 21, 22, 26, 27, 28 | **0.1111** | 0.1666 | RAud; Text?; PSolo; RitDan; MMot; CosU |

| | | | | | | |
|---|---|---|---|---|---|---|
| House B, southeast room | 6a | 6 | 23, 26, 27, 28 | **0.1111** | 0.1666 | CosT?; MMot; CosU; CosCU |
| House J west, central room | 3i | 3 | (Bench) | **0.1103** | 0.9583 | – |
| South Subterranean Passage, north corridor | 3e | 3 | – | **0.1076** | 0.7500 | – |
| Tower, crossway access | 4q | 4 | – | **0.1072** | 0.6666 | – |
| Toilets, east corridor | 5i | 5 | – | **0.1064** | 0.7000 | – |
| House A, east corridor | 2a | 2 | 42, 43, 44, 45, 46, 47 | **0.1041** | 0.9166 | THist; APWA; APSolo |
| House H, northwest room | 5r | 5 | (Bench) (South Subterranean Stair) | **0.1037** | 1.0833 | – |
| House A-D, southwest corridor | 3b | 3 | – | **0.1029** | 0.4999 | – |
| South Terrace, southwest corner | 4p | 4 | – | **0.1025** | 0.1111 | – |
| South Terrace, southeast corner | 4o | 4 | – | **0.1025** | 0.1111 | – |
| East Court, southwest corner | 5d | 5 | – | **0.1021** | 0.8928 | – |
| House C, east room | 5e | 5 | 29, 38, 40, 41, 62 | **0.1013** | 0.9761 | GPSolo; CosT; TDed? |
| East Court, House A west foundation | 4b | 4 | 58, 59, 60, 61 (H.S.) | **0.1013** | 3.9166 | CapDis; THist |
| House A, west corridor | 3a | 3 | – | **0.0998** | 1.1011 | – |
| Southern Substructure, south terrace | 1d | 1 | – | **0.0994** | 1.0000 | – |
| House D, west corridor | 1b | 1 | 48, 49, 50, 51, 52, 53 | **0.0986** | 0.9166 | HumSac; SacDan; Text? |
| House A-D, north corridor | 1a | 1 | 55, 56, 57 (Bench) | **0.0982** | 0.9166 | RPSolo; RAudAccTDynHist |
| House J, west access path | 4d | 4 | – | **0.0974** | 5.2499 | – |
| *Carrier* | C | 0 | 1, 2, 82, 83, 84, 85 | **0.0959** | 1.4999 | CosC; RPSoloCosCU; RPSolo |
| House D, east corridor | 2b | 2 | – | **0.0943** | 0.9999 | – |
| House K, west room | 1c | 1 | – | **0.0939** | 0.5000 | – |
| South Terrace stair | 2e | 2 | – | **0.0931** | 0.4444 | – |
| House E, southwestern room | 6d | 6 | (East Subterranean Stair) | **0.0919** | 0.8928 | – |
| Tower Court, northeast corner | 5f | 5 | **71** | **0.0908** | 0.7833 | **CapDis** |
| House H, north room | 5q | 5 | – | **0.0900** | 0.6428 | – |
| House H, south room | 4h | 4 | – | **0.0884** | 0.8611 | – |
| House G, south room | 4f | 4 | – | **0.0876** | 0.5277 | – |
| House E, west room | 5g | 5 | 3 (Bench?), 4, 5, 6 (DR Throne), 11, 13 | **0.0872** | 1.1166 | RAud; Acc; CosTU; THist; TDynHist |
| Southeast Court | 5n | 5 | – | **0.0841** | 2.8750 | – |
| House E, east room | 6c | 6 | 8 | **0.0829** | 0.8928 | CosCU |
| House G, north room | 5o | 5 | – | **0.0822** | 0.7261 | – |
| House D & K, external path | 3c | 3 | – | **0.0822** | 2.0333 | – |
| Southeast Court, path access | 4e | 4 | – | **0.0818** | 0.0861 | – |
| House K, east room | 2c | 2 | (West Subterranean Stair) | **0.0814** | 1.2777 | – |
| Tower Court | 4n | 4 | **74, 75** | **0.0794** | 1.3928 | **CapDis** |
| House G & H, central corridor | 4g | 4 | – | **0.0779** | 1.4206 | – |
| House H, west patio | 4i | 4 | – | **0.0767** | 0.6706 | – |
| South Terrace | 3d | 3 | – | **0.0759** | 3.9999 | – |
| House G & H, north terrace | 5p | 5 | – | **0.0736** | 1.7833 | – |

Palenque Palace Phase 5 sculpture ranked according to relative asymmetry (RA) values (least integrated to most integrated)

| Description of spatial unit | Cell | Carrier rank | Sculpture GR No. | RA value | Control value | Theme |
|---|---|---|---|---|---|---|
| South Substructure Patio, southwest corner | 5q | 5 | – | **0.1804** | 0.5000 | – |
| Tower, third level | 7f | 7 | (Bench) | **0.1651** | 0.5000 | – |
| West Court, south room | 8a | 8 | 66 | **0.1628** | 0.5000 | GPWA |
| South Substructure, patio | 4o | 4 | – | **0.1563** | 1.5000 | – |
| Toilet, north room | 8b | 8 | (Toilets) | **0.1516** | 0.5000 | – |
| House B, northeast room | 4n | 4 | (Bench) | **0.1490** | 0.6666 | – |
| South Substructure, northwest room | 6s | 6 | 81 (Bench) | **0.1484** | 1.0000 | – |
| House I, rear | 8c | 8 | (Sweat Baths) | **0.1440** | 0.5000 | – |
| Tower, second level | 6L | 6 | 72 | **0.1411** | 1.5000 | CosC |
| West Court | 7a | 7 | 63, 64, 65, 67, 68 | **0.1387** | 1.5000 | RitDan?; GPWA; CosC; TCapSac |
| West Subterranean Passage, south corridor | 5o | 5 | 79 | **0.1364** | 1.0000 | CosU |
| South Subterranean Passage, south corridor | 7c | 7 | – | **0.1349** | 1.0000 | – |
| South Substructure, northeast corridor | 4m | 4 | – | **0.1331** | 1.1666 | – |
| South Substructure Patio, north corridor | 3h | 3 | – | **0.1328** | 0.7500 | – |
| Toilet, west-central room | 7b | 7 | 69 | **0.1278** | 1.5000 | CosC |
| South Substructure, east-central room | 3f | 3 | (Bench) | **0.1255** | 1.0833 | – |
| South Substructure, east-central corridor | 3g | 3 | – | **0.1255** | 1.0833 | – |
| East Subterranean Passage, south corridor | 5p | 5 | – | **0.1255** | 0.8333 | – |
| House I, south corridor | 7e | 7 | – | **0.1196** | 1.2500 | – |
| Tower, first level | 5t | 5 | – | **0.1181** | 1.0000 | – |
| House C, west room | 6n | 6 | 30, 31, 32, 33, 34, 35, 36, 37, 39, 40, 41 | **0.1152** | 0.8333 | RAud; APSolo; PSolo; CosT |
| West Subterranean Passage, dogs leg | 4k | 4 | – | **0.1137** | 1.0000 | – |
| South Subterranean Passage, dogs leg | 6r | 6 | 78 | **0.1128** | 0.8333 | CosU |
| East Subterranean Passage, dogs leg | 6q | 6 | – | **0.1111** | 1.0000 | – |
| House B, northeast room | 5a | 5 | 24, 25, 26, 27 | **0.1108** | 0.1428 | RPSolo; MMot; CosU; CosCU |
| House B, northwest room | 5b | 5 | 24, 25, 26, 27 | **0.1108** | 0.1428 | RPSolo; MMot; CosU; CosCU |
| South Substructure, south corridor | 2e | 2 | 80 (Bench) | **0.1096** | 1.4999 | TCosCU |
| House I, west corridor | 6i | 6 | – | **0.1058** | 0.1250 | – |
| Toilets, southwest room | 6j | 6 | – | **0.1058** | 0.1250 | – |
| House B, southeast room | 6a | 6 | 23, 26, 27, 28 | **0.1049** | 0.2500 | CosT?; MMot; CosU; CosCU |
| House B, southwest room | 6b | 6 | 16, 17, 18, 19, 20, 21, 22, 26, 27, 28 | **0.1049** | 0.2500 | RAud; Text?; PSolo; RitDan; MMot; CosU |
| House I, north room | 6h | 6 | 76, 77 | **0.1049** | 0.3750 | **Cos?; RPSolo** |
| Toilets, east corridor | 6k | 6 | – | **0.1046** | 0.6250 | – |
| House J, northwest room | 5g | 5 | (Bench) | **0.1029** | 0.1428 | – |
| House J west, south-central room | 5h | 5 | – | **0.1029** | 0.1428 | – |
| House J west, south room | 5i | 5 | – | **0.1029** | 0.1428 | – |
| House J, south patio | 5j | 5 | – | **0.1029** | 0.1428 | – |
| House B, northwest room | 5c | 5 | 24, 25, 26, 27 | **0.1020** | 0.4761 | RPSolo; MMot; CosU; CosCU |
| House J west, north-central room | 4c | 4 | – | **0.1014** | 0.4761 | – |
| House A-D, southeast corridor | 4a | 4 | – | **0.0979** | 0.4761 | – |

| | | | | | | |
|---|---|---|---|---|---|---|
| House J west, central room | 3b | 3 | (Bench) | **0.0976** | 0.9761 | – |
| House I, east corridor | 6g | 6 | – | **0.0964** | 1.3250 | – |
| House A, east corridor | 2a | 2 | 42, 43, 44, 45, 46, 47 | **0.0946** | 0.9999 | THist; APWA; APSolo |
| South Subterranean Passage, north corridor | 3e | 3 | – | **0.0940** | 0.7500 | – |
| East Subterranean Passage, north corridor | 7d | 7 | 7 | **0.0937** | 0.8333 | CosU |
| Tower, crossway access | 4q | 4 | – | **0.0934** | 0.6666 | – |
| House K, west room | 1c | 1 | – | **0.0926** | 0.5000 | – |
| House C, east room | 5e | 5 | 29, 38, 40, 41, 62 | **0.0926** | 0.9761 | GPSolo; CosT; TDed? |
| House F, east room | 5r | 5 | – | **0.0926** | 0.6666 | – |
| House A-D, north corridor | 1a | 1 | 55, 56, 57 (Bench) | **0.0923** | 0.9166 | RPSolo; RAudAccTDynHist |
| Southern Substructure, south terrace | 1d | 1 | – | **0.0917** | 1.0000 | – |
| House F, west room | 6c | 6 | – | **0.0917** | 0.7500 | – |
| House D, west corridor | 1b | 1 | 48, 49, 50, 51, 52, 53 | **0.0911** | 0.9166 | HumSac; SacDan; Text? |
| East Court, southwest corner | 5d | 5 | – | **0.0911** | 0.8928 | – |
| House H, northwest room | 5n | 5 | (Bench) (South Subterranean Stair) | **0.0911** | 1.2500 | – |
| House A-D, southwest corridor | 3c | 3 | – | **0.0908** | 0.4999 | – |
| *Carrier* | c | 0 | 1, 2, 82, 83, 84, 85 | **0.0899** | 1.4999 | CosC; RPSoloCosCU; RPSolo |
| South Terrace, southeast corner | 4L | 4 | – | **0.0885** | 0.1111 | – |
| South Terrace, southwest corner | 4p | 4 | – | **0.0885** | 0.1111 | – |
| Tower Court, northeast corner | 6m | 6 | 71 | **0.0864** | 0.6583 | CapDis |
| East Court, House A west foundation | 4b | 4 | 58, 59, 60, 61 (H.S.) | **0.0861** | 3.9999 | CapDis; THist |
| House H, north room | 6p | 6 | (Bench) | **0.0861** | 0.4583 | – |
| House D, east corridor | 2b | 2 | – | **0.0840** | 0.9999 | – |
| House G, northwest room | 5L | 5 | – | **0.0837** | 0.3250 | – |
| House A, west corridor | 3a | 3 | – | **0.0832** | 0.9761 | – |
| Southeast Court, north side | 5f | 5 | – | **0.0814** | 2.4166 | – |
| South Terrace stair | 2d | 2 | – | **0.0811** | 0.4444 | – |
| Tower Court, southwest corner stair | 4f | 4 | – | **0.0802** | 0.2916 | – |
| Tower Court | 5s | 5 | **70, 73** (Throne Tablet), 74, 75 | **0.0802** | 4.2833 | **?(Stucco fret)**; **TDyn Hist**; CapDis |
| House I, west room | 6f | 6 | 3 (Bench?), 4, 5, 6 (DR Throne), 11, 13 | **0.0802** | 1.2916 | RAud; Acc; CosTU; THist; TDynHist |
| House J, west access path | 4d | 4 | – | **0.0788** | 4.9166 | – |
| House G, south room | 4g | 4 | – | **0.0767** | 0.4777 | – |
| House H, south room | 4i | 4 | – | **0.0767** | 0.8944 | – |
| Southeast Court, west side | 6d | 6 | – | **0.0764** | 1.1250 | – |
| House G, northeast room | 5k | 5 | – | **0.0761** | 0.2916 | – |
| House E, southwestern room | 6o | 6 | (East Subterranean Stair) | **0.0752** | 0.8250 | – |
| House G & H, central corridor | 4h | 4 | – | **0.0726** | 1.3194 | – |
| House E, east room | 6e | 6 | 8 | **0.0726** | 0.9083 | CosCU |
| House K, east room | 2c | 2 | (West Subterranean Stair) | **0.0714** | 1.2777 | – |
| House D & K, external path | 3d | 3 | – | **0.0708** | 2.3333 | – |
| Southeast Court, east side | 4e | 4 | – | **0.0702** | 1.8373 | – |
| House H, west patio | 4j | 4 | – | **0.0673** | 0.6527 | – |
| South Terrace | 3i | 3 | – | **0.0661** | 3.9499 | – |
| House G & H, north terrace | 5m | 5 | – | **0.0652** | 2.7833 | – |

Appendix 5 *(continued)*

| Palenque Palace Phase 6 sculpture ranked according to relative asymmetry (RA) values (least integrated to most integrated) | | | | | | |
|---|---|---|---|---|---|---|
| Description of spatial unit | Cell | Carrier rank | Sculpture GR No. | RA value | Control value | Theme |
| West Court, south room | 10a | 10 | 66 | **0.1361** | 0.5000 | GPWA |
| West court | 9a | 9 | 63, 64, 65, 67, 68 | **0.1212** | 1.5000 | RitDan?; GPWA; CosC; TCapSac |
| House F, south-central room | 9b | 9 | – | **0.1199** | 0.5000 | – |
| South Substructure Patio, southwest corner | 5u | 5 | – | **0.1183** | 0.5000 | – |
| Tower, third level | 7p | 7 | (Bench) | **0.1175** | 0.5000 | – |
| House L, northeast corridor | 5r | 5 | – | **0.1133** | 0.3333 | – |
| House C, southwest room | 8b | 8 | 31, 32, 33, 34, 35, 36, 37, 38, 39, 40, 41 | **0.1065** | 1.0000 | APSolo; PSolo; CosT |
| House F, central room | 8a | 8 | – | **0.1049** | 1.5000 | – |
| South Substructure, patio | 4r | 4 | – | **0.1032** | 1.5000 | – |
| West Subterranean Passage, south corridor | 6ab | 6 | 79 | **0.1030** | 1.0000 | CosU |
| Tower, second level | 6v | 6 | 72 | **0.1026** | 1.5000 | CosC |
| South Substructure, northwest room | 7o | 7 | 81 (Bench) | **0.1025** | 1.0000 | – |
| House L, east corridor | 4p | 4 | – | **0.0984** | 1.5833 | – |
| House I, rear | 9c | 9 | (Sweat Bath) | **0.0964** | 0.5000 | – |
| West Subterranean Passage, dogs leg | 5s | 5 | – | **0.0926** | 1.0000 | – |
| South Subterranean Passage, south corridor | 7m | 7 | – | **0.0921** | 1.0000 | – |
| House C, northwest room | 7a | 7 | 31, 32, 33, 34, 35, 36, 37, 39, 40, 41 | **0.0919** | 0.8333 | APSolo; PSolo; CosT |
| House K, southeast room | 4m | 4 | – | **0.0902** | 0.3333 | – |
| House F, north-central room | 7g | 7 | – | **0.0902** | 0.8333 | – |
| House A-D south, west-central room | 5f | 5 | – | **0.0889** | 0.5000 | – |
| House C, northeast room | 6w | 6 | – | **0.0888** | 0.3333 | – |
| South Substructure Patio, north corridor | 3L | 3 | – | **0.0884** | 0.7500 | – |
| Tower, first level | 5x | 5 | – | **0.0879** | 1.0000 | – |
| South Substructure, central room | 4t | 4 | (Bench) | **0.0855** | 0.5833 | – |
| House L, south-central corridor | 3g | 3 | – | **0.0841** | 1.4167 | – |
| House L, central room | 3h | 3 | – | **0.0841** | 0.8333 | – |
| South Substructure, east-central corridor | 3k | 3 | – | **0.0837** | 1.0833 | – |
| Toilet, west-central room | 7c | 7 | 69 | **0.0834** | 1.5000 | CosC |
| House B, southwest room | 8d | 8 | 16, 17, 18, 19, 20, 21, 22, 26, 27, 28 (Bench?) | **0.0834** | 0.2000 | RAud; Text?; PSolo; RitDan; MMot; CosU |
| House L, south-central room | 4q | 4 | – | **0.0819** | 0.7500 | – |
| House A-D, south-central room | 5v | 5 | – | **0.0818** | 0.1250 | – |
| East Subterranean Passage, south corridor | 5t | 5 | – | **0.0817** | 0.8333 | – |

| | | | | | | |
|---|---|---|---|---|---|---|
| House E east extension, south room | 8e | 8 | – | **0.0816** | 0.3333 | – |
| House B, northeast room | 5a | 5 | 24, 25, 26, 27 | **0.0815** | 0.1250 | RPSolo; MMot; CosU; CosCU |
| House B, central room | 5b | 5 | 24, 25, 26, 27 | **0.0815** | 0.1250 | RPSolo; MMot; CosU; CosCU |
| House B, northwest room | 5c | 5 | 24, 25, 26, 27 | **0.0815** | 0.1250 | RPSolo; MMot; CosU; CosCU |
| House E, southeast room | 8c | 8 | (Bench) | **0.0815** | 0.3333 | – |
| House I, south corridor | 8f | 8 | – | **0.0814** | 1.5000 | – |
| House F, northwest room | 7e | 7 | – | **0.0809** | 0.6667 | – |
| House L, northwest corridor | 3f | 3 | – | **0.0796** | 0.5833 | – |
| South Subterranean Passage, dogs leg | 6z | 6 | 78 | **0.0791** | 0.8333 | CosU |
| South Substructure, northeast corridor | 4s | 4 | – | **0.0790** | 0.8333 | – |
| South Subterranean Passage, north corridor | 4n | 4 | – | **0.0784** | 0.8333 | – |
| House H, north room | 6n | 6 | (Bench) | **0.0784** | 0.3333 | – |
| House D east, south-central room | 3c | 3 | **54** | **0.0782** | 0.3333 | **?(Mural)** |
| House B, southeast room | 6a | 6 | 23, 28, 26, 27, 28 | **0.0777** | 0.2000 | CosT?; MMot; CosU; CosCU |
| House C east, south-central room | 6x | 6 | 29, 38, 40, 41, 62 | **0.0777** | 1.3333 | GPSolo; RAud; CosT; TDed? |
| House E, northeast room | 6aa | 6 | 8, **15** | **0.0775** | 0.7000 | CosCU |
| House K, south-central corridor | 3e | 3 | – | **0.0770** | 1.6667 | – |
| House K, south corridor | 4o | 4 | – | **0.0769** | 1.1667 | – |
| East Court, southwest corner | 5d | 5 | – | **0.0759** | 0.6250 | – |
| House C, southeast room | 7b | 7 | – | **0.0753** | 0.6667 | – |
| House F, north-central room | 6g | 6 | – | **0.0747** | 1.2000 | – |
| House K, northwest room | 1c | 1 | – | **0.0744** | 0.1428 | – |
| Southwest Corner Plaza Level, north room | 1e | 1 | – | **0.0744** | 0.1428 | – |
| Southwest Corner Plaza Level, south room | 1f | 1 | – | **0.0744** | 0.1428 | – |
| House C east, north-central room | 5e | 5 | 29, 38, 40, 41, 62 | **0.0740** | 1.4583 | GPSolo; RAud; CosT; TDed? |
| House D, northeast room | 4v | 4 | – | **0.0738** | 0.6250 | – |
| House D east, north-central room | 4w | 4 | – | **0.0738** | 0.6250 | – |
| House A-D, southwest room | 4u | 4 | – | **0.0737** | 1.1250 | – |
| South Substructure, south corridor | 2f | 2 | 80 (Bench) | **0.0734** | 1.4166 | TCosCU |
| Tower, crossway access | 4aa | 4 | – | **0.0734** | 0.6250 | – |
| House F, southeast room | 7i | 7 | – | **0.0732** | 0.7500 | – |
| House G, northeast room | 7k | 7 | – | **0.0725** | 0.8333 | – |
| House F, northeast corner room | 6b | 6 | – | **0.0724** | 0.1250 | – |
| House F, northeast central room | 6c | 6 | – | **0.0724** | 0.1250 | – |
| House F, east-central room | 6d | 6 | (Bench) | **0.0724** | 0.1250 | – |
| House F, southeast central room | 6e | 6 | (Bench) | **0.0724** | 0.1250 | – |
| House F, southeast corner room | 6f | 6 | – | **0.0724** | 0.1250 | – |
| House F, west-central room | 7f | 7 | (Bench) | **0.0706** | 0.9500 | – |
| House L, north-central corridor | 2d | 2 | – | **0.0700** | 1.3333 | – |
| South Substructure, east-central room | 3j | 3 | (Bench) | **0.0700** | 1.4167 | – |
| House F, south room | 6h | 6 | – | **0.0700** | 0.6250 | – |
| House F, northwest corridor | 6i | 6 | – | **0.0699** | 0.4000 | – |

(*continued*)

Appendix 5 *(continued)*

Palenque Palace Phase 6 sculpture ranked according to relative asymmetry (RA) values (least integrated to most integrated)

| Description of spatial unit | Cell | Carrier rank | Sculpture GR No. | RA value | Control value | Theme |
|---|---|---|---|---|---|---|
| House L, south corridor | 3n | 3 | – | **0.0698** | 0.8333 | – |
| House I, north room | 6r | 6 | 76, 77 | **0.0692** | 0.1250 | Cos?; RPSolo |
| House I, west corridor | 6s | 6 | – | **0.0692** | 0.1250 | – |
| Toilets, southwest room | 6t | 6 | – | **0.0692** | 0.1250 | – |
| House A-D, southeast corridor | 4b | 4 | – | **0.0690** | 0.3750 | – |
| Toilets, east corridor | 6u | 6 | – | **0.0688** | 0.6250 | – |
| House E east extension, north room | 7h | 7 | – | **0.0688** | 0.7500 | – |
| Southeast Court | 7d | 7 | 28 | **0.0682** | 2.8333 | CosU |
| House E, southeast room | 7j | 7 | – | **0.0667** | 2.1666 | – |
| House I, east corridor | 7n | 7 | – | **0.0665** | 0.6666 | – |
| House K, east room | 3d | 3 | (West Subterranean Stair) | **0.0664** | 0.9583 | – |
| East Subterranean Passage, north corridor | 7L | 7 | 7 | **0.0664** | 0.6666 | CosU |
| House H, southwest room | 5q | 5 | – | **0.0663** | 1.0000 | – |
| House J, south patio | 5w | 5 | – | **0.0663** | 0.1111 | – |
| House J, northwest room | 5g | 5 | (Bench) | **0.0662** | 0.1111 | – |
| House J east, south-central room | 5h | 5 | – | **0.0662** | 0.1111 | – |
| House J east, south room | 5i | 5 | – | **0.0662** | 0.1111 | – |
| East Court, House A west foundation | 4a | 4 | 58, 59, 60, 61 (H.S.) | **0.0658** | 5.5833 | CapDis; THist |
| House H, northwest room | 5p | 5 | (Bench) (South Subterranean Stair) | **0.0652** | 1.3333 | – |
| Tower Court, northeast corner | 6o | 6 | 71 | **0.0645** | 0.8750 | CapDis |
| House D, southeast room | 2b | 2 | – | **0.0634** | 1.4583 | – |
| House J west, north-central room | 4z | 4 | – | **0.0634** | 0.4444 | – |
| House H, north terrace | 5o | 5 | – | **0.0632** | 1.6667 | – |
| House K, southwest room | 2c | 2 | – | **0.0631** | 0.9166 | – |
| House H, northeast room | 4L | 4 | – | **0.0630** | 0.7917 | – |
| South Terrace, southeast corner | 4y | 4 | – | **0.0630** | 0.0909 | – |
| House F, northeast corridor | 5k | 5 | – | **0.0628** | 2.2778 | – |
| House G, north-central room | 6L | 6 | – | **0.0627** | 0.6428 | – |
| House G, southwest room | 4f | 4 | – | **0.0626** | 0.4242 | – |
| House G & H, east-central corridor | 4g | 4 | (Bench) | **0.0626** | 0.4242 | – |
| House H, south-central room | 4j | 4 | – | **0.0626** | 0.4242 | – |

| | | | | | | |
|---|---|---|---|---|---|---|
| House G & H, west-central corridor | 4h | 4 | – | **0.0625** | 0.9242 | – |
| House H, southeast room | 4i | 4 | – | **0.0625** | 0.9242 | – |
| Southern Substructure, south terrace | 1g | 1 | – | **0.0616** | 0.7261 | – |
| House D, west corridor | 1b | 1 | 48, 49, 50, 51, 52, 53 | **0.0615** | 0.8095 | HumSac; SacDan; Text? |
| House E, southeast corner room | 6k | 6 | – | **0.0614** | 0.3928 | – |
| House A-D, north corridor | 1a | 1 | 55, 56, 57 (Bench) | **0.0608** | 0.8095 | RPSolo; RAudAccTDynHist |
| House J west, central room | 3m | 3 | (Bench) | **0.0605** | 0.9444 | – |
| House E & F, central corridor | 6j | 6 | – | **0.0604** | 1.4761 | – |
| House A, east corridor | 2a | 2 | 42, 43, 44, 45, 46, 47 | **0.0603** | 0.9166 | THist; APWA; APSolo |
| Tower Court, patio stair | 4x | 4 | 70 | **0.0596** | 0.2500 | ?(Stucco fret) |
| *Carrier* | c | 0 | 1, 2, 82, 83, 84, 85 | **0.0595** | 4.2499 | CosC; RPSoloCosCU; RPSolo |
| House K, external path access | 1d | 1 | – | **0.0593** | 1.2261 | – |
| House D & K, external path | 3b | 3 | – | **0.0593** | 3.4999 | – |
| South Terrace stair | 2g | 2 | – | **0.0586** | 0.6742 | – |
| House H, southeast-central room | 4k | 4 | – | **0.0585** | 0.4242 | – |
| House G & H, north-central corridor | 6m | 6 | – | **0.0580** | 0.9761 | – |
| House E, northwestern room | 6p | 6 | 3 (Bench), 4, 5, 6 (DR Throne) **10**, 11, 13, **14** | **0.0580** | 1.1249 | **Text?**; **CosU**;RAud; Acc; CosTU; THist; TDynHist |
| House F, east corridor | 5j | 5 | – | **0.0577** | 7.3667 | – |
| House A, west corridor | 3a | 3 | – | **0.0570** | 1.0694 | – |
| House L, south terrace | 2e | 2 | – | **0.0557** | 0.8409 | – |
| House G, northeast room | 5L | 5 | – | **0.0545** | 0.3095 | – |
| Tower Court | 5n | 5 | 70, 73 (Throne Tablet), 74, 75 | **0.0543** | 4.7500 | ?(Stucco fret); TDyn Hist; CapDis |
| House G, southeast room | 4e | 4 | – | **0.0542** | 0.7576 | – |
| House E, southwestern room | 6q | 6 | (East Subterranean Stair) | **0.0524** | 1.8511 | – |
| House J west, access corridor | 4c | 4 | – | **0.0505** | 5.4500 | – |
| South Terrace | 3i | 3 | – | **0.0484** | 4.8333 | – |
| House G, path access | 5m | 5 | – | **0.0479** | 2.4167 | – |
| House F, south patio | 4d | 4 | – | **0.0447** | 1.3032 | – |

Appendix 6. *Sculpture ranked according to control values, Palenque Palace Phases 1–6*

| Palenque Palace Phase 1 sculpture ranked according to control values (most controlling to least controlling) | | | | | | |
|---|---|---|---|---|---|---|
| Description of spatial unit | Cell | Carrier rank | Sculpture GR No. | RA value | Control value | Theme |
| House J, west access | 8a | 8 | – | 0.1667 | **5.0833** | – |
| Terrace, upper southeast | 7g | 7 | – | 0.1267 | **2.4166** | – |
| North Substructure, south terrace lower | 3b | 3 | – | 0.1320 | **2.2000** | – |
| North Substructure, north terrace lower | 1a | 1 | – | 0.1667 | **2.0000** | – |
| Southern Substructure, south terrace lower | 1b | 1 | – | 0.2059 | **1.7500** | – |
| House K, east room | 6e | 6 | (West Subterranean Stairs) | 0.1395 | **1.6999** | – |
| House J west, central room | 9c | 9 | (Bench) | 0.2029 | **1.6428** | – |
| House B, north terrace upper | 4a | 4 | – | 0.1192 | **1.5333** | – |
| South Substructure, patio | 4e | 4 | – | 0.2971 | **1.5000** | – |
| Southern Substructure, southeast terrace lower | 2d | 2 | – | 0.2443 | **1.4999** | – |
| Terrace, upper southwest | 7f | 7 | – | 0.1448 | **1.3666** | – |
| Terrace, upper northwest | 5d | 5 | – | 0.1244 | **1.3166** | – |
| South Substructure, northeast corridor | 4c | 4 | – | 0.2172 | **1.1666** | – |
| House E, west room | 6c | 6 | **3, 4, 5, 13, 11, 80** (Bench) | 0.1244 | **1.1666** | **RAud; Acc; CosTU; THist; TCosCU** |
| South Substructure, east-central room | 3c | 3 | (Bench) | 0.2293 | **1.0833** | – |
| South Substructure, east-central corridor | 3d | 3 | – | 0.2293 | **1.0833** | – |
| Southeast Court, early | 7a | 7 | – | 0.1388 | **1.0595** | – |
| East Subterranean Passage, dogs leg | 6f | 6 | – | 0.1833 | **1.0000** | – |
| East Subterranean Passage, north corridor | 7c | 7 | **7** | 0.1637 | **1.0000** | **CosCU** |
| South Subterranean Passage, dogs leg | 8b | 8 | **78** | 0.1916 | **1.0000** | **CosU** |
| West Subterranean Passage, dogs leg | 8c | 8 | – | 0.2059 | **1.0000** | – |
| West Subterranean Passage, south corridor | 9g | 9 | **79** | 0.2330 | **1.0000** | **CosU** |
| South Subterranean Passage, south corridor | 9h | 9 | – | 0.2217 | **1.0000** | – |
| South Subterranean Structure, northwest room | 10b | 10 | 81 (Bench) | 0.2564 | **1.0000** | – |

| | | | | | | |
|---|---|---|---|---|---|---|
| House E, east room | 6b | 6 | **8** | 0.1282 | **0.9166** | **CosCU** |
| House E, access north stair | 5c | 5 | – | 0.1184 | **0.9000** | – |
| House K, west room | 5e | 5 | – | 0.1425 | **0.9000** | – |
| Southwest Court, later tower court | 6d | 6 | – | 0.1267 | **0.8666** | – |
| East Subterranean Passage, south corridor | 5f | 5 | – | 0.2006 | **0.8333** | – |
| House H, substructure | 7d | 7 | (South Subterranean Stairs) | 0.1584 | **0.8333** | – |
| North Substructure, north corridor | 2b | 2 | – | 0.1772 | **0.7500** | – |
| South Substructure Patio, north corridor | 3e | 3 | – | 0.2609 | **0.7500** | – |
| House B, south room | 6a | 6 | **21, 22, 26, 27, 28** | 0.1501 | **0.7500** | **PSolo; RitDan; MMot; CosU** |
| House E, southwestern room | 7b | 7 | (East Subterranean Stairs) | 0.1440 | **0.7500** | – |
| North Substructure, south corridor | 3a | 3 | – | 0.1606 | **0.7000** | – |
| House B, north room | 5b | 5 | **24, 25, 26, 27** | 0.1425 | **0.7000** | **RPSolo; MMot; CosU; CosCU** |
| South Subterranean Passage, north corridor | 7c | 7 | – | 0.1727 | **0.7000** | – |
| South Substructure, central room | 4d | 4 | (Bench) | 0.2647 | **0.6666** | – |
| *Carrier* | c | 0 | – | 0.1863 | **0.5833** | – |
| Terrace, lower west | 4b | 4 | – | 0.1546 | **0.5333** | – |
| South Substructure Patio, southwest corner | 5g | 5 | – | 0.3363 | **0.5000** | – |
| House J west, north-central room | 9b | 9 | – | 0.2036 | **0.4761** | – |
| North Substructure, east terrace lower | 2a | 2 | – | 0.1508 | **0.4500** | – |
| North Substructure, west terrace lower | 2c | 2 | – | 0.1508 | **0.4500** | – |
| Terrace, upper southwest corner | 8e | 8 | – | 0.1817 | **0.3333** | – |
| House J, east room | 10a | 10 | – | 0.2300 | **0.3333** | – |
| House B, east terrace upper | 5a | 5 | – | 0.1576 | **0.2000** | – |
| Terrace, upper southeast corner | 8d | 8 | – | 0.1652 | **0.1666** | – |
| House J, northwest room | 9a | 9 | (Bench) | 0.2051 | **0.1428** | – |
| House J west, south-central room | 9d | 9 | – | 0.2051 | **0.1428** | – |
| House J west, south room | 9e | 9 | – | 0.2051 | **0.1428** | – |
| House J, south patio | 9f | 9 | – | 0.2051 | **0.1428** | – |

Appendix 6 *(continued)*

Palenque Palace Phase 2 sculpture ranked according to control values (most controlling to least controlling)

| Description of spatial unit | Cell | Carrier rank | Sculpture GR No. | RA value | Control value | Theme |
|---|---|---|---|---|---|---|
| House J, west access | 7a | 7 | – | 0.1312 | **5.3333** | – |
| Terrace, upper southeast | 7f | 7 | – | 0.1071 | **2.3333** | – |
| North Substructure, south terrace lower | 3b | 3 | – | 0.1156 | **2.2000** | – |
| Southwest Court, later tower court | 6f | 6 | – | 0.1071 | **2.1166** | – |
| North Substructure, north terrace lower | 1a | 1 | – | 0.1494 | **2.0000** | – |
| East Subterranean Passage, north corridor | 7g | 7 | 7 | 0.1396 | **1.8333** | CosU |
| Southern Substructure, south terrace lower | 1b | 1 | – | 0.1870 | **1.7500** | – |
| House K, east room | 6g | 6 | (West Subterranean Stairs) | 0.1247 | **1.7261** | – |
| East Court, House A. west foundation (early) | 4a | 4 | **58** (*alfardas* only), **59**, **60**, **61**(H.S.) | 0.1078 | **1.5333** | **CapDis; THist** |
| South Substructure, patio | 4e | 4 | – | 0.2701 | **1.5000** | – |
| South Substructure, south corridor | 2d | 2 | – | 0.2045 | **1.4999** | – |
| House A & J, east patio | 6a | 6 | – | 0.1403 | **1.4583** | – |
| South Substructure, northeast corridor | 4d | 4 | – | 0.1896 | **1.1666** | – |
| House J west, central room | 8c | 8 | (Bench) | 0.1656 | **1.1250** | – |
| South Substructure, east-central room | 3c | 3 | (Bench) | 0.2013 | **1.0833** | – |
| South Substructure, east-central corridor | 3d | 3 | – | 0.2013 | **1.0833** | – |
| Southeast Court, early | 7b | 7 | – | 0.1162 | **1.0416** | – |
| House B, north room | 5b | 5 | 24, 25, 26, 27 | 0.1286 | **1.0333** | RPSolo; MMot; CosU; CosCU |
| House C, west foundation | 6d | 6 | **68** | 0.1273 | **1.0090** | **TCapSac** |
| South Subterranean Passage, dogs leg | 8h | 8 | 78 | 0.1695 | **1.0000** | CosU |
| West Subterranean Passage, dogs leg | 8i | 8 | – | 0.1864 | **1.0000** | – |
| South Subterranean Passage, south corridor | 9a | 9 | – | 0.2000 | **1.0000** | – |
| West Subterranean Passage, south corridor | 9b | 9 | 79 | 0.2169 | **1.0000** | CosU |
| South Subterranean Structure, northwest room | 10a | 10 | 81 (Bench) | 0.2286 | **1.0000** | – |
| House A, west room | 5a | 5 | – | 0.1299 | **0.9500** | – |
| House K, west room | 5e | 5 | – | 0.1338 | **0.9500** | – |

| | | | | | | |
|---|---|---|---|---|---|---|
| House E, southwestern room | 8j | 8 | (East Subterranean Stairs) | 0.1227 | **0.8333** | – |
| East Subterranean Passage, south corridor | 5f | 5 | – | 0.1734 | **0.8333** | – |
| East Subterranean Passage, dogs leg | 6h | 6 | – | 0.1597 | **0.8333** | – |
| East Court, southwest corner | 5c | 5 | – | 0.1260 | **0.7833** | – |
| North Substructure, north corridor | 2b | 2 | – | 0.1584 | **0.7500** | – |
| South Substructure Patio, north corridor | 3e | 3 | – | 0.2357 | **0.7500** | – |
| House E, east room | 6e | 6 | 8 | 0.1156 | **0.7499** | CosCU |
| House C, west room | 6c | 6 | **30, 31, 32, 33, 34, 35, 36, 37, 39, 40, 41** | 0.1279 | **0.7261** | **RAud; APSolo; PSolo; CosT** |
| House E, west room | 7c | 7 | 3, 4, 5, 13, 11, 80 (Bench) | 0.1149 | **0.7261** | RAud; Acc; CosTU; THist; TCosCU |
| North Substructure, south corridor | 3a | 3 | – | 0.1416 | **0.7000** | – |
| South Subterranean Passage, north corridor | 7e | 7 | – | 0.1552 | **0.7000** | – |
| House C, east room | 5d | 5 | **29, 30, 38, 40, 41, 62** | 0.1149 | **0.6761** | **GPSolo; CosT; TDed?** |
| South Substructure, central room | 4c | 4 | (Bench) | 0.2357 | **0.6666** | – |
| House H, substructure | 7d | 7 | (South Subterranean Stairs) | 0.1383 | **0.6428** | – |
| *Carrier* | C | 0 | – | 0.1682 | **0.5833** | – |
| House B, south room | 6b | 6 | 21, 22, 26, 27, 28 | 0.1312 | **0.5833** | PSolo; RitDan; MMot; CosU; CosCU |
| House A, east room | 6i | 6 | **42, 43, 44, 45, 46, 47** | 0.1539 | **0.5833** | **THist; APWA; APSolo** |
| House J, east room | 7i | 7 | – | 0.1734 | **0.5833** | – |
| Terrace, lower west | 4b | 4 | – | 0.1318 | **0.5333** | – |
| South Substructure Patio, southwest corner | 5g | 5 | – | 0.3071 | **0.5000** | – |
| House J west, north-central room | 8b | 8 | – | 0.1662 | **0.4583** | – |
| North Substructure, east terrace lower | 2a | 2 | – | 0.1338 | **0.4500** | – |
| North Substructure, west terrace lower | 2c | 2 | – | 0.1338 | **0.4500** | – |
| Terrace, upper southwest | 7h | 7 | – | 0.1253 | **0.3666** | – |
| Southern Substructure, southeast terrace lower | 2e | 2 | – | 0.2227 | **0.3333** | – |
| Terrace, upper southwest corner | 8k | 8 | – | 0.1617 | **0.3333** | – |
| Terrace, upper southeast corner | 8g | 8 | – | 0.1409 | **0.1666** | – |
| House J, northwest room | 8a | 8 | (Bench) | 0.1675 | **0.1250** | – |
| House J west, south-central room | 8d | 8 | – | 0.1675 | **0.1250** | – |
| House J west, south room | 8e | 8 | – | 0.1675 | **0.1250** | – |
| House J, south patio | 8f | 8 | – | 0.1675 | **0.1250** | – |

Appendix 6 *(continued)*

Palenque Palace Phase 3 sculpture ranked according to control values (most controlling to least controlling)

| Description of spatial unit | Cell | Carrier rank | Sculpture GR No. | RA value | Control value | Theme |
|---|---|---|---|---|---|---|
| House J, west access path | 4h | 4 | – | 0.1058 | **5.2833** | – |
| East Court, House A. west foundation | 3b | 3 | **58** (remaining Captive Panels), 59, 60, 61(H.S.) | 0.0961 | **3.6166** | **CapDis**; THist |
| Southeast Court, early | 5e | 5 | – | 0.1032 | **2.4916** | – |
| Southwest Court, later tower court | 3d | 3 | – | 0.0851 | **2.3333** | – |
| Terrace, upper southeast | 4L | 4 | – | 0.0896 | **2.1761** | – |
| Terrace, upper southwest | 3h | 3 | – | 0.0929 | **1.5333** | – |
| South Substructure, patio | 4o | 4 | – | 0.2104 | **1.5000** | – |
| South Substructure, south corridor | 2f | 2 | **80** (Bench moved) | 0.1429 | **1.4999** | **TCosCU** |
| Southern Substructure, south terrace | 1e | 1 | – | 0.1208 | **1.4500** | – |
| *Carrier* | C | 0 | **1, 2, 82, 83, 84, 85** | 0.1000 | **1.4499** | **CosC; RPSoloCosCU; RPSolo** |
| House K, east room | 2e | 2 | (West Subterranean Stair) | 0.0955 | **1.4499** | – |
| House D & K, external path | 2c | 2 | – | 0.0805 | **1.2428** | – |
| House D & K, patio | 1c | 1 | – | 0.0942 | **1.1833** | – |
| South Substructure, northeast corridor | 4m | 4 | – | 0.1766 | **1.1666** | – |
| South Substructure, east-central room | 3f | 3 | (Bench) | 0.1662 | **1.0833** | – |
| South Substructure, east-central corridor | 3g | 3 | – | 0.1662 | **1.0833** | – |
| House A-D, north corridor | 1a | 1 | **55, 56, 57** (Throne?) | 0.0994 | **1.0666** | **RPSolo; RAudAccTDynHist** |
| House A-D, south corridor | 2a | 2 | – | 0.0864 | **1.0595** | – |
| House E, east room | 5g | 5 | 8 | 0.1078 | **1.0333** | CosCU |
| West Subterranean Passage, dogs leg | 4p | 4 | – | 0.1558 | **1.0000** | – |
| South Subterranean Passage, dogs leg | 5j | 5 | 78 | 0.1461 | **1.0000** | CosU |
| West Subterranean Passage, south corridor | 5k | 5 | 79 | 0.1864 | **1.0000** | CosU |
| East Subterranean Passage, dogs leg | 6d | 6 | – | 0.1532 | **1.0000** | – |
| South Subterranean Passage, south corridor | 6e | 6 | – | 0.1766 | **1.0000** | – |
| South Subterranean Structure, northwest room | 6f | 6 | 81 (Bench) | 0.1974 | **1.0000** | – |
| House E, southwestern room | 5h | 5 | (East Subterranean Stair) | 0.1123 | **0.9999** | – |
| East Court, southwest corner | 4d | 4 | – | 0.1149 | **0.9761** | – |

| | | | | | | |
|---|---|---|---|---|---|---|
| House C, west room | 4f | 4 | 30, 31, 32, 33, 34, 35, 36, 37, 39, 40, 41 | 0.1149 | **0.9761** | RAud; APSolo; PSolo; CosT |
| House J west, central room | 3c | 3 | (Bench) | 0.1214 | **0.9583** | – |
| House D, east corridor | 2d | 2 | – | 0.0968 | **0.8999** | – |
| House A, east corridor | 2b | 2 | 42, 43, 44, 45, 46, 47 | 0.1149 | **0.8333** | THist; APWA; APSolo |
| East Subterranean Passage, south corridor | 5i | 5 | – | 0.1695 | **0.8333** | – |
| East Subterranean Passage, north corridor | 6c | 6 | 7 | 0.1331 | **0.8333** | CosU |
| House A, west corridor | 3a | 3 | – | 0.1013 | **0.8011** | – |
| South Substructure Patio, north corridor | 3i | 3 | – | 0.1760 | **0.7500** | – |
| House E, west room | 4i | 4 | 3, 4, 5, **6** (DR Throne), 13, 11 | 0.1026 | **0.7261** | RAud; Acc; CosTU; THist; **TDynHist** |
| House D, west corridor | 1b | 1 | **48, 49, 50, 51, 52, 53** | 0.1117 | **0.7000** | **HumSac; SacDan; Text?** |
| South Subterranean Passage, north corridor | 3e | 3 | – | 0.1253 | **0.7000** | – |
| South Substructure, central room | 4n | 4 | (Bench) | 0.2006 | **0.6666** | – |
| House H, substructure | 4k | 4 | (South Subterranean Stair) | 0.1156 | **0.6428** | – |
| House C, east room | 4e | 4 | 29, 38, 40, 41, 62 | 0.1052 | **0.6190** | GPSolo; CosT; TDed? |
| House K, west room | 1d | 1 | – | 0.1084 | **0.6000** | – |
| South Substructure Patio, southwest corner | 5L | 5 | – | 0.2468 | **0.5000** | – |
| House B, northwest room | 4c | 4 | 24, 25, 26, 27 | 0.1240 | **0.4761** | RPSolo; MMot; CosU; CosCU |
| West Court, early | 4g | 4 | **63, 64, 65, 66**, 68, **69** | 0.1201 | **0.4761** | **GPWA; CosC;** TCapSac |
| House J west, north-central room | 4j | 4 | – | 0.1318 | **0.4583** | – |
| Southern Substructure, southeast terrace lower | 2g | 2 | – | 0.1565 | **0.3333** | – |
| South Terrace, southwest corner | 4q | 4 | – | 0.1286 | **0.2500** | – |
| House B, southeast room | 6a | 6 | **23**, 26, 27, 28 | 0.1390 | **0.2000** | **CosT?**; MMot; CosU; CosCU |
| House B, southwest room | 6b | 6 | **16, 17, 18, 19, 20**, 21, 22, 26, 27, 28 | 0.1390 | **0.2000** | **RAud; Text?**; PSolo; RitDan; MMot; CosU |
| Terrace, upper southeast corner | 5f | 5 | – | 0.1247 | **0.1666** | – |
| House B, northeast room | 4a | 4 | 24, 25, 26, 27 | 0.1318 | **0.1428** | RPSolo; MMot; CosU; CosCU |
| House B, central room | 4b | 4 | 24, 25, 26, 27 | 0.1318 | **0.1428** | RPSolo; MMot; CosU; CosCU |
| House J, northwest room | 5a | 5 | (Bench) | 0.1416 | **0.1250** | – |
| House J west, south-central room | 5b | 5 | – | 0.1416 | **0.1250** | – |
| House J west, south room | 5c | 5 | – | 0.1416 | **0.1250** | – |
| House J, south patio | 5d | 5 | – | 0.1416 | **0.1250** | – |

Appendix 6 *(continued)*

| Palenque Palace Phase 4 sculpture ranked according to control values (most controlling to least controlling) | | | | | | |
|---|---|---|---|---|---|---|
| Description of spatial unit | Cell | Carrier rank | Sculpture GR No. | RA value | Control value | Theme |
| House J, west access path | 4d | 4 | – | 0.0974 | **5.2499** | – |
| South Terrace | 3d | 3 | – | 0.0759 | **3.9999** | – |
| East Court, House A, west foundation | 4b | 4 | 58, 59, 60, 61 (H.S.) | 0.1013 | **3.9166** | CapDis; THist |
| Southeast Court | 5n | 5 | – | 0.0841 | **2.8750** | – |
| House D & K, external path | 3c | 3 | – | 0.0822 | **2.0333** | – |
| House G & H, north terrace | 5p | 5 | – | 0.0736 | **1.7833** | – |
| Toilet, west-central room | 6j | 6 | – | 0.1338 | **1.5000** | – |
| West Court | 7d | 7 | 63, 64, 65, 66, **67**, 68,69 | 0.1557 | **1.5000** | **RitDan?**; GPWA; CosC; TCapSac |
| Tower, second level | 6h | 6 | **72** | 0.1600 | **1.5000** | **CosC** |
| South Substructure, patio | 4m | 4 | – | 0.1745 | **1.5000** | – |
| *Carrier* | C | 0 | 1, 2, 82, 83, 84, 85 | 0.0959 | **1.4999** | CosC; RPSoloCosCU; RPSolo |
| South Substructure, south corridor | 2d | 2 | 80 (Bench) | 0.1193 | **1.4999** | TCosCU |
| House G & H, central corridor | 4g | 4 | – | 0.0779 | **1.4206** | – |
| Tower Court | 4n | 4 | **74, 75** | 0.0794 | **1.3928** | **CapDis** |
| House K, east room | 2c | 2 | (West Subterranean Stair) | 0.0814 | **1.2777** | – |
| South Substructure, northeast corridor | 4k | 4 | – | 0.1549 | **1.1666** | – |
| House E, west room | 5g | 5 | 3 (Bench?), 4, 5, 6 (DR Throne), 11, 13 | 0.0872 | **1.1166** | RAud; Acc; CosTU; THist; TDynHist |
| House A, west corridor | 3a | 3 | – | 0.0998 | **1.1011** | – |
| House H, northwest room | 5r | 5 | (Bench) (South Subterranean Stair) | 0.1037 | **1.0833** | – |
| South Substructure, east-central room | 3f | 3 | (Bench) | 0.1428 | **1.0833** | – |
| South Substructure, east central corridor | 3g | 3 | – | 0.1428 | **1.0833** | – |
| Southern Substructure, south terrace | 1d | 1 | – | 0.0994 | **1.0000** | – |
| East Subterranean Passage, dogs leg | 6e | 6 | – | 0.1311 | **1.0000** | – |
| Tower, first level | 5h | 5 | – | 0.1330 | **1.0000** | – |
| West Subterranean Passage, dogs leg | 4j | 4 | – | 0.1354 | **1.0000** | – |
| West Subterranean Passage, south corridor | 5t | 5 | 79 | 0.1569 | **1.0000** | CosU |
| South Subterranean Passage, south corridor | 7b | 7 | – | 0.1612 | **1.0000** | – |
| South Substructure, northwest room | 6g | 6 | 81 (Bench) | 0.1729 | **1.0000** | – |
| House D, east corridor | 2b | 2 | – | 0.0943 | **0.9999** | – |
| House C, east room | 5e | 5 | 29, 38, 40, 41, 62 | 0.1013 | **0.9761** | GPSolo; CosT; TDed? |
| House J west, central room | 3i | 3 | (Bench) | 0.1103 | **0.9583** | – |
| House A-D, north corridor | 1a | 1 | 55, 56, 57 (Bench) | 0.0982 | **0.9166** | RPSolo; RAudAccTDynHist |
| House D, west corridor | 1b | 1 | 48, 49, 50, 51, 52, 53 | 0.0986 | **0.9166** | HumSac; SacDan; Text? |
| House A, east corridor | 2a | 2 | 42, 43, 44, 45, 46, 47 | 0.1041 | **0.9166** | THist; APWA; APSolo |

| | | | | | | |
|---|---|---|---|---|---|---|
| House E, east room | 6c | 6 | 8 | 0.0829 | **0.8928** | CosCU |
| House E, southwestern room | 6d | 6 | (East Subterranean Stair) | 0.0919 | **0.8928** | – |
| East Court, southwest corner | 5d | 5 | – | 0.1021 | **0.8928** | – |
| House H, south room | 4h | 4 | – | 0.0884 | **0.8611** | – |
| East Subterranean Passage, north corridor | 7a | 7 | 7 | 0.1146 | **0.8333** | CosU |
| House C, west room | 6i | 6 | 30, 31, 32, 33, 34, 35, 36, 37, 39, 40, 41 | 0.1268 | **0.8333** | RAud; APSolo; PSolo; CosT |
| South Subterranean Passage, dogs leg | 6f | 6 | 78 | 0.1303 | **0.8333** | CosU |
| East Subterranean Passage, south corridor | 5s | 5 | – | 0.1475 | **0.8333** | – |
| Tower Court, northeast corner | 5f | 5 | **71** | 0.0908 | **0.7833** | **CapDis** |
| South Subterranean Passage, north corridor | 3e | 3 | – | 0.1076 | **0.7500** | – |
| South Substructure Patio, north corridor | 3h | 3 | – | 0.1475 | **0.7500** | – |
| House G, north room | 5o | 5 | – | 0.0822 | **0.7261** | – |
| Toilets, east corridor | 5i | 5 | – | 0.1064 | **0.7000** | – |
| House H, west patio | 4i | 4 | – | 0.0767 | **0.6706** | – |
| Tower, crossway access | 4q | 4 | – | 0.1072 | **0.6666** | – |
| South Substructure, central room | 4L | 4 | (Bench) | 0.1682 | **0.6666** | – |
| House H, north room | 5q | 5 | – | 0.0900 | **0.6428** | – |
| House G, south room | 4f | 4 | – | 0.0876 | **0.5277** | – |
| House K, west room | 1c | 1 | – | 0.0939 | **0.5000** | – |
| Toilet, north room | 7e | 7 | (Toilets) | 0.1608 | **0.5000** | – |
| West Court, south room | 8a | 8 | – | 0.1835 | **0.5000** | – |
| Tower, third level | 7c | 7 | (Bench) | 0.1878 | **0.5000** | – |
| South Substructure Patio, southwest corner | 5u | 5 | – | 0.2011 | **0.5000** | – |
| House A-D, southwest corridor | 3b | 3 | – | 0.1029 | **0.4999** | – |
| House B, northwest room | 5c | 5 | 24, 25, 26, 27 | 0.1174 | **0.4761** | RPSolo; MMot; CosU; CosCU |
| House J west, north-central room | 4c | 4 | – | 0.1182 | **0.4583** | – |
| South Terrace stair | 2e | 2 | – | 0.0931 | **0.4444** | – |
| House A-D, southeast corridor | 4a | 4 | – | 0.1170 | **0.3928** | – |
| House B, southeast room | 6a | 6 | 23, 26, 27, 28 | 0.1111 | **0.1666** | CosT?; MMot; CosU; CosCU |
| House B, southwest room | 6b | 6 | 16, 17, 18, 19, 20, 21, 22, 26, 27, 28 | 0.1111 | **0.1666** | RAud; Text?; PSolo; RitDan; MMot; CosU |
| House B, northeast room | 5a | 5 | 24, 25, 26, 27 | 0.1279 | **0.1428** | RPSolo; MMot; CosU; CosCU |
| House B, central room | 5b | 5 | 24, 25, 26, 27 | 0.1279 | **0.1428** | RPSolo; MMot; CosU; CosCU |
| House J, northwest room | 5j | 5 | (Bench) | 0.1240 | **0.1250** | – |
| House J west, south-central room | 5k | 5 | – | 0.1240 | **0.1250** | – |
| House J west, south room | 5L | 5 | – | 0.1240 | **0.1250** | – |
| House J, south patio | 5m | 5 | – | 0.1240 | **0.1250** | – |
| South Terrace, southeast corner | 4o | 4 | – | 0.1025 | **0.1111** | – |
| South Terrace, southwest corner | 4p | 4 | – | 0.1025 | **0.1111** | – |
| Southeast Court, path access | 4e | 4 | – | 0.0818 | **0.0861** | – |

Appendix 6 *(continued)*

Palenque Palace Phase 5 sculpture ranked according to control values (most controlling to least controlling)

| Description of spatial unit | Cell | Carrier rank | Sculpture GR No. | RA value | Control value | Theme |
|---|---|---|---|---|---|---|
| House J, west access path | 4d | 4 | – | 0.0788 | **4.9166** | – |
| Tower Court | 5s | 5 | **70, 73** (Throne Tablet), 74, 75 | 0.0802 | **4.2833** | **?(Stucco fret)**; **TDyn Hist**; CapDis |
| East Court, House A west foundation | 4b | 4 | 58, 59, 60, 61 (H.S.) | 0.0861 | **3.9999** | CapDis; THist |
| South Terrace | 3i | 3 | – | 0.0661 | **3.9499** | – |
| House G & H, north terrace | 5m | 5 | – | 0.0652 | **2.7833** | – |
| Southeast Court, north side | 5f | 5 | – | 0.0814 | **2.4166** | – |
| House D & K, external path | 3d | 3 | – | 0.0708 | **2.3333** | – |
| Southeast Court, east side | 4e | 4 | – | 0.0702 | **1.8373** | – |
| South Substructure, patio | 4o | 4 | – | 0.1563 | **1.5000** | – |
| Tower, second level | 6L | 6 | 72 | 0.1411 | **1.5000** | CosC |
| West Court | 7a | 7 | 63, 64, 65, 67, 68 | 0.1387 | **1.5000** | RitDan?; GPWA; CosC; TCapSac |
| Toilet, west-central room | 7b | 7 | 69 | 0.1278 | **1.5000** | CosC |
| *Carrier* | c | 0 | 1, 2, 82, 83, 84, 85 | 0.0899 | **1.4999** | CosC; RPSoloCosCU; RPSolo |
| South Substructure, south corridor | 2e | 2 | 80 (Bench) | 0.1096 | **1.4999** | TCosCU |
| House I, east corridor | 6g | 6 | – | 0.0964 | **1.3250** | – |
| House G & H, central corridor | 4h | 4 | – | 0.0726 | **1.3194** | – |
| House I, west room | 6f | 6 | 3 (Bench?), 4, 5, 6 (DR Throne), 11, 13 | 0.0802 | **1.2916** | RAud; Acc; CosTU; THist; TDynHist |
| House K, east room | 2c | 2 | (West Subterranean Stair) | 0.0714 | **1.2777** | – |
| House H, northwest room | 5n | 5 | (Bench) (South Subterranean Stair) | 0.0911 | **1.2500** | – |
| House I, south corridor | 7e | 7 | – | 0.1196 | **1.2500** | – |
| South Substructure, northeast corridor | 4m | 4 | – | 0.1331 | **1.1666** | – |
| Southeast Court, west side | 6d | 6 | – | 0.0764 | **1.1250** | – |
| South Substructure, east-central room | 3f | 3 | (Bench) | 0.1255 | **1.0833** | – |
| South Substructure, east-central corridor | 3g | 3 | – | 0.1255 | **1.0833** | – |
| Southern Substructure, south terrace | 1d | 1 | – | 0.0917 | **1.0000** | – |
| West Subterranean Passage, dogs leg | 4k | 4 | – | 0.1137 | **1.0000** | – |
| West Subterranean Passage, south corridor | 5o | 5 | 79 | 0.1364 | **1.0000** | CosU |
| Tower, first level | 5t | 5 | – | 0.1181 | **1.0000** | – |
| East Subterranean Passage, dogs leg | 6q | 6 | – | 0.1111 | **1.0000** | – |
| South Substructure, northwest room | 6s | 6 | 81 (Bench) | 0.1484 | **1.0000** | – |
| South Subterranean Passage, south corridor | 7c | 7 | – | 0.1349 | **1.0000** | – |
| House A, east corridor | 2a | 2 | 42, 43, 44, 45, 46, 47 | 0.0946 | **0.9999** | THist; APWA; APSolo |
| House D, east corridor | 2b | 2 | – | 0.0840 | **0.9999** | – |
| House A, west corridor | 3a | 3 | – | 0.0832 | **0.9761** | – |
| House J west, central room | 3b | 3 | (Bench) | 0.0976 | **0.9761** | – |
| House C, east room | 5e | 5 | 29, 38, 40, 41, 62 | 0.0926 | **0.9761** | GPSolo; CosT; TDed? |
| House A-D, north corridor | 1a | 1 | 55, 56, 57 (Bench) | 0.0923 | **0.9166** | RPSolo; RAudAccTDynHist |
| House D, west corridor | 1b | 1 | 48, 49, 50, 51, 52, 53 | 0.0911 | **0.9166** | HumSac; SacDan; Text? |
| House E, east room | 6e | 6 | 8 | 0.0726 | **0.9083** | CosCU |

| | | | | | | |
|---|---|---|---|---|---|---|
| House H, south room | 4i | 4 | – | 0.0767 | **0.8944** | – |
| East Court, southwest corner | 5d | 5 | – | 0.0911 | **0.8928** | – |
| East Subterranean Passage, south corridor | 5p | 5 | – | 0.1255 | **0.8333** | – |
| House C, west room | 6n | 6 | 30, 31, 32, 33, 34, 35, 36, 37, 39, 40, 41 | 0.1152 | **0.8333** | RAud; APSolo; PSolo; CosT |
| South Subterranean Passage, dogs leg | 6r | 6 | 78 | 0.1128 | **0.8333** | CosU |
| East Subterranean Passage, north corridor | 7d | 7 | 7 | 0.0937 | **0.8333** | CosU |
| House E, southwestern room | 6o | 6 | (East Subterranean Stair) | 0.0752 | **0.8250** | – |
| South Subterranean Passage, north corridor | 3e | 3 | – | 0.0940 | **0.7500** | – |
| South Substructure Patio, north corridor | 3h | 3 | – | 0.1328 | **0.7500** | – |
| House F, west room | 6c | 6 | – | 0.0917 | **0.7500** | – |
| House B, northeast room | 4n | 4 | (Bench) | 0.1490 | **0.6666** | – |
| Tower, crossway access | 4q | 4 | – | 0.0934 | **0.6666** | – |
| House F, east room | 5r | 5 | – | 0.0926 | **0.6666** | – |
| Tower Court, northeast corner | 6m | 6 | 71 | 0.0864 | **0.6583** | CapDis |
| House H, west patio | 4j | 4 | – | 0.0673 | **0.6527** | – |
| Toilets, east corridor | 6k | 6 | – | 0.1046 | **0.6250** | – |
| House K, west room | 1c | 1 | – | 0.0926 | **0.5000** | – |
| South Substructure Patio, southwest corner | 5q | 5 | – | 0.1804 | **0.5000** | – |
| Tower, third level | 7f | 7 | (Bench) | 0.1651 | **0.5000** | – |
| West Court, south room | 8a | 8 | 66 | 0.1628 | **0.5000** | GPWA |
| Toilet, north room | 8b | 8 | (Toilets) | 0.1516 | **0.5000** | – |
| House I, rear | 8c | 8 | (Sweat Bath) | 0.1440 | **0.5000** | – |
| House A-D, southwest corridor | 3c | 3 | – | 0.0908 | **0.4999** | – |
| House G, south room | 4g | 4 | – | 0.0767 | **0.4777** | – |
| House A-D, southeast corridor | 4a | 4 | – | 0.0979 | **0.4761** | – |
| House J west, north-central room | 4c | 4 | – | 0.1014 | **0.4761** | – |
| House B, northwest room | 5c | 5 | 24, 25, 26, 27 | 0.1020 | **0.4761** | RPSolo; MMot; CosU; CosCU |
| House H, north room | 6p | 6 | (Bench) | 0.0861 | **0.4583** | – |
| South Terrace stair | 2d | 2 | – | 0.0811 | **0.4444** | – |
| House I, north room | 6h | 6 | **76, 77** | 0.1049 | **0.3750** | **Cos?; RPSolo** |
| House G, northwest room | 5L | 5 | – | 0.0837 | **0.3250** | – |
| Tower Court, southwest corner stair | 4f | 4 | – | 0.0802 | **0.2916** | – |
| House G, northeast room | 5k | 5 | – | 0.0761 | **0.2916** | – |
| House B, southeast room | 6a | 6 | 23, 26, 27, 28 | 0.1049 | **0.2500** | CosT?; MMot; CosU; CosCU |
| House B, southwest room | 6b | 6 | 16, 17, 18, 19, 20, 21, 22, 26, 27, 28 | 0.1049 | **0.2500** | RAud; Text?; PSolo; RitDan; MMot; CosU |
| House B, northeast room | 5a | 5 | 24, 25, 26, 27 | 0.1108 | **0.1428** | RPSolo; MMot; CosU; CosCU |
| House B, northwest room | 5b | 5 | 24, 25, 26, 27 | 0.1108 | **0.1428** | RPSolo; MMot; CosU; CosCU |
| House J, northwest room | 5g | 5 | (Bench) | 0.1029 | **0.1428** | – |
| House J west, south-central room | 5h | 5 | – | 0.1029 | **0.1428** | – |
| House J west, south room | 5i | 5 | – | 0.1029 | **0.1428** | – |
| House J, south patio | 5j | 5 | – | 0.1029 | **0.1428** | – |
| House I, west corridor | 6i | 6 | – | 0.1058 | **0.1250** | – |
| Toilets, southwest room | 6j | 6 | – | 0.1058 | **0.1250** | – |
| South Terrace, southeast corner | 4L | 4 | – | 0.0885 | **0.1111** | – |
| South Terrace, southwest corner | 4p | 4 | – | 0.0885 | **0.1111** | – |

Appendix 6 *(continued)*

Palenque Palace Phase 6 sculpture ranked according to control values (most controlling to least controlling)

| Description of spatial unit | Cell | Carrier rank | Sculpture GR No. | RA value | Control value | Theme |
|---|---|---|---|---|---|---|
| House F, east corridor | 5j | 5 | – | 0.0577 | **7.3667** | – |
| East Court, House A west foundation | 4a | 4 | 58, 59, 60, 61 (H.S.) | 0.0658 | **5.5833** | CapDis; THist |
| House J west, access corridor | 4c | 4 | – | 0.0505 | **5.4500** | – |
| South Terrace | 3i | 3 | – | 0.0484 | **4.8333** | – |
| Tower Court | 5n | 5 | **70, 73** (Throne Tablet), 74, 75 | 0.0543 | **4.7500** | **?(Stucco fret)**; **TDyn Hist**; CapDis |
| *Carrier* | c | 0 | 1, 2, 82, 83, 84, 85 | 0.0595 | **4.2499** | CosC; RPSoloCosCU; RPSolo |
| House D & K, external path | 3b | 3 | – | 0.0593 | **3.4999** | – |
| Southeast Court | 7d | 7 | 28 | 0.0682 | **2.8333** | CosU |
| House G, path access | 5m | 5 | – | 0.0479 | **2.4167** | – |
| House F, northeast corridor | 5k | 5 | – | 0.0628 | **2.2778** | – |
| House E, southeast room | 7j | 7 | – | 0.0667 | **2.1666** | – |
| House E, southwestern room | 6q | 6 | (East Subterranean Stair) | 0.0524 | **1.8511** | – |
| House K, south-central corridor | 3e | 3 | – | 0.0770 | **1.6667** | – |
| House H, north terrace | 5o | 5 | – | 0.0632 | **1.6667** | – |
| House L, east corridor | 4p | 4 | – | 0.0984 | **1.5833** | – |
| South Substructure, patio | 4r | 4 | – | 0.1032 | **1.5000** | – |
| Tower, second level | 6v | 6 | 72 | 0.1026 | **1.5000** | CosC |
| Toilet, west-central room | 7c | 7 | 69 | 0.0834 | **1.5000** | CosC |
| House F, central room | 8a | 8 | – | 0.1049 | **1.5000** | – |
| House I, south corridor | 8f | 8 | – | 0.0814 | **1.5000** | – |
| West court | 9a | 9 | 63, 64, 65, 67, 68 | 0.1212 | **1.5000** | RitDan?; GPWA; CosC; TCapSac |
| House E & F, central corridor | 6j | 6 | – | 0.0604 | **1.4761** | – |
| House C east, north-central room | 5e | 5 | 29, 38, 40, 41, 62 | 0.0740 | **1.4583** | GPSolo; RAud; CosT; TDed? |
| House D, southeast room | 2b | 2 | – | 0.0634 | **1.4583** | – |
| House L, south-central corridor | 3g | 3 | – | 0.0841 | **1.4167** | – |
| South Substructure, east-central room | 3j | 3 | (Bench) | 0.0700 | **1.4167** | – |
| South Substructure, south corridor | 2f | 2 | 80 (Bench) | 0.0734 | **1.4166** | TCosCU |
| House H, northwest room | 5p | 5 | (Bench) (South Subterranean Stair) | 0.0652 | **1.3333** | – |
| House C east, south-central room | 6x | 6 | 29, 38, 40, 41, 62 | 0.0777 | **1.3333** | GPSolo; RAud; CosT; TDed? |
| House L, north-central corridor | 2d | 2 | – | 0.0700 | **1.3333** | – |
| House F, south patio | 4d | 4 | – | 0.0447 | **1.3032** | – |
| House K, external path access | 1d | 1 | – | 0.0593 | **1.2261** | – |
| House F, north-central room | 6g | 6 | – | 0.0747 | **1.2000** | – |
| House K, south corridor | 4o | 4 | – | 0.0769 | **1.1667** | – |
| House A-D, southwest room | 4u | 4 | – | 0.0737 | **1.1250** | – |

| | | | | | | |
|---|---|---|---|---|---|---|
| House E, northwestern room | 6p | 6 | 3 (Bench), 4, 5, 6 (DR Throne) **10**, 11, 13, **14** | 0.0580 | **1.1249** | **Text?**; **CosU**;RAud; Acc; CosTU; THist; TDynHist |
| South Substructure, east-central corridor | 3k | 3 | – | 0.0837 | **1.0833** | – |
| House A, west corridor | 3a | 3 | – | 0.0570 | **1.0694** | – |
| West Subterranean Passage, dogs leg | 5s | 5 | – | 0.0926 | **1.0000** | – |
| Tower, first level | 5x | 5 | – | 0.0879 | **1.0000** | – |
| West Subterranean Passage, south corridor | 6ab | 6 | 79 | 0.1030 | **1.0000** | CosU |
| South Subterranean Passage, south corridor | 7m | 7 | – | 0.0921 | **1.0000** | – |
| South Substructure, northwest room | 7o | 7 | 81 (Bench) | 0.1025 | **1.0000** | – |
| House C, southwest room | 8b | 8 | 31, 32, 33, 34, 35, 36, 37, 38, 39, 40, 41 | 0.1065 | **1.0000** | APSolo; PSolo; CosT |
| House H, southwest room | 5q | 5 | – | 0.0663 | **1.0000** | – |
| House G & H, north-central corridor | 6m | 6 | – | 0.0580 | **0.9761** | – |
| House K, east room | 3d | 3 | (West Subterranean Stair) | 0.0664 | **0.9583** | – |
| House F, west-central room | 7f | 7 | (Bench) | 0.0706 | **0.9500** | – |
| House J west, central room | 3m | 3 | (Bench) | 0.0605 | **0.9444** | – |
| House G & H, west-central corridor | 4h | 4 | – | 0.0625 | **0.9242** | – |
| House H, southeast room | 4i | 4 | – | 0.0625 | **0.9242** | – |
| House A, east corridor | 2a | 2 | 42, 43, 44, 45, 46, 47 | 0.0603 | **0.9166** | THist; APWA; APSolo |
| House K, southwest room | 2c | 2 | – | 0.0631 | **0.9166** | – |
| Tower Court, northeast corner | 6o | 6 | 71 | 0.0645 | **0.8750** | CapDis |
| House L, south terrace | 2e | 2 | – | 0.0557 | **0.8409** | – |
| House C, northwest room | 7a | 7 | 31, 32, 33, 34, 35, 36, 37, 39, 40, 41 | 0.0919 | **0.8333** | APSolo; PSolo; CosT |
| House L, central room | 3h | 3 | – | 0.0841 | **0.8333** | – |
| House L, south corridor | 3n | 3 | – | 0.0698 | **0.8333** | – |
| South Subterranean Passage, north corridor | 4n | 4 | – | 0.0784 | **0.8333** | – |
| South Substructure, northeast corridor | 4s | 4 | – | 0.0790 | **0.8333** | – |
| East Subterranean Passage, south corridor | 5t | 5 | – | 0.0817 | **0.8333** | – |
| South Subterranean Passage, dogs leg | 6z | 6 | 78 | 0.0791 | **0.8333** | CosU |
| House F, north-central room | 7g | 7 | – | 0.0902 | **0.8333** | – |
| House G, northeast room | 7k | 7 | – | 0.0725 | **0.8333** | – |
| House A-D, north corridor | 1a | 1 | 55, 56, 57 (Bench) | 0.0608 | **0.8095** | RPSolo; RAudAccTDynHist |
| House D, west corridor | 1b | 1 | 48, 49, 50, 51, 52, 53 | 0.0615 | **0.8095** | HumSac; SacDan; Text? |
| House H, northeast room | 4L | 4 | – | 0.0630 | **0.7917** | – |
| House G, southeast room | 4e | 4 | – | 0.0542 | **0.7576** | – |
| South Substructure Patio, north corridor | 3L | 3 | – | 0.0884 | **0.7500** | – |
| House L, south-central room | 4q | 4 | – | 0.0819 | **0.7500** | – |
| House E east extension, north room | 7h | 7 | – | 0.0688 | **0.7500** | – |
| House F, southeast room | 7i | 7 | – | 0.0732 | **0.7500** | – |

*(continued)*

Appendix 6 *(continued)*

| Palenque Palace Phase 6 sculpture ranked according to control values (most controlling to least controlling) | | | | | | |
|---|---|---|---|---|---|---|
| Description of spatial unit | Cell | Carrier rank | Sculpture GR No. | RA value | Control value | Theme |
| Southern Substructure, south terrace | 1g | 1 | – | 0.0616 | **0.7261** | – |
| House E, northeast room | 6aa | 6 | 8, **15** | 0.0775 | **0.7000** | CosCU |
| South Terrace stair | 2g | 2 | – | 0.0586 | **0.6742** | – |
| House C, southeast room | 7b | 7 | – | 0.0753 | **0.6667** | – |
| House F, northwest room | 7e | 7 | – | 0.0809 | **0.6667** | – |
| East Subterranean Passage, north corridor | 7L | 7 | 7 | 0.0664 | **0.6666** | CosU |
| House I, east corridor | 7n | 7 | – | 0.0665 | **0.6666** | – |
| House G, north-central room | 6L | 6 | – | 0.0627 | **0.6428** | – |
| House D, northeast room | 4v | 4 | – | 0.0738 | **0.6250** | – |
| House D east, north-central room | 4w | 4 | – | 0.0738 | **0.6250** | – |
| Tower, crossway access | 4aa | 4 | – | 0.0734 | **0.6250** | – |
| East Court, southwest corner | 5d | 5 | – | 0.0759 | **0.6250** | – |
| House F, south room | 6h | 6 | – | 0.0700 | **0.6250** | – |
| Toilets, east corridor | 6u | 6 | – | 0.0688 | **0.6250** | – |
| South Substructure, central room | 4t | 4 | (Bench) | 0.0855 | **0.5833** | – |
| House L, northwest corridor | 3f | 3 | – | 0.0796 | **0.5833** | – |
| House A-D south, west-central room | 5f | 5 | – | 0.0889 | **0.5000** | – |
| South Substructure Patio, southwest corner | 5u | 5 | – | 0.1183 | **0.5000** | – |
| Tower, third level | 7p | 7 | (Bench) | 0.1175 | **0.5000** | – |
| House F, south-central room | 9b | 9 | – | 0.1199 | **0.5000** | – |
| House I, rear | 9c | 9 | (Sweat Bath) | 0.0964 | **0.5000** | – |
| West Court, south room | 10a | 10 | 66 | 0.1361 | **0.5000** | GPWA |
| House J west, north-central room | 4z | 4 | – | 0.0634 | **0.4444** | – |
| House G, southwest room | 4f | 4 | – | 0.0626 | **0.4242** | – |
| House G & H, east-central corridor | 4g | 4 | (Bench) | 0.0626 | **0.4242** | – |
| House H, south-central room | 4j | 4 | – | 0.0626 | **0.4242** | – |
| House H, southeast-central room | 4k | 4 | – | 0.0585 | **0.4242** | – |
| House F, northwest corridor | 6i | 6 | – | 0.0699 | **0.4000** | – |
| House E, southeast corner room | 6k | 6 | – | 0.0614 | **0.3928** | – |

| | | | | | | |
|---|---|---|---|---|---|---|
| House A-D, southeast corridor | 4b | 4 | – | 0.0690 | **0.3750** | – |
| House K, southeast room | 4m | 4 | – | 0.0902 | **0.3333** | – |
| House L, northeast corridor | 5r | 5 | – | 0.1133 | **0.3333** | – |
| House D east, south-central room | 3c | 3 | **54** | 0.0782 | **0.3333** | **?(Mural)** |
| House H, north room | 6n | 6 | (Bench) | 0.0784 | **0.3333** | – |
| House C, northeast room | 6w | 6 | – | 0.0888 | **0.3333** | – |
| House E, southeast room | 8c | 8 | (Bench) | 0.0815 | **0.3333** | – |
| House E east extension, south room | 8e | 8 | – | 0.0816 | **0.3333** | – |
| House G, northeast room | 5L | 5 | – | 0.0545 | **0.3095** | – |
| Tower Court, patio stair | 4x | 4 | – | 0.0596 | **0.2500** | – |
| House B, southeast room | 6a | 6 | 23, 28, 26, 27, 28 | 0.0777 | **0.2000** | CosT?; MMot; CosU; CosCU |
| House B, southwest room | 8d | 8 | 16, 17, 18, 19, 20, 21, 22, 26, 27, 28 (Bench?) | 0.0834 | **0.2000** | RAud; Text?; PSolo; RitDan; MMot; CosU |
| House K, northwest room | 1c | 1 | – | 0.0744 | **0.1428** | – |
| Southwest Corner Plaza Level, north room | 1e | 1 | – | 0.0744 | **0.1428** | – |
| Southwest Corner Plaza Level, south room | 1f | 1 | – | 0.0744 | **0.1428** | – |
| House B, northeast room | 5a | 5 | 24, 25, 26, 27 | 0.0815 | **0.1250** | RPSolo; MMot; CosU; CosCU |
| House B, central room | 5b | 5 | 24, 25, 26, 27 | 0.0815 | **0.1250** | RPSolo; MMot; CosU; CosCU |
| House B, northwest room | 5c | 5 | 24, 25, 26, 27 | 0.0815 | **0.1250** | RPSolo; MMot; CosU; CosCU |
| House A-D, south-central room | 5v | 5 | – | 0.0818 | **0.1250** | – |
| House F, northeast corner room | 6b | 6 | – | 0.0724 | **0.1250** | – |
| House F, northeast central room | 6c | 6 | – | 0.0724 | **0.1250** | – |
| House F, east central room | 6d | 6 | (Bench) | 0.0724 | **0.1250** | – |
| House F, southeast central room | 6e | 6 | (Bench) | 0.0724 | **0.1250** | – |
| House F, southeast corner room | 6f | 6 | – | 0.0724 | **0.1250** | – |
| House I, north room | 6r | 6 | 76, 77 | 0.0692 | **0.1250** | Cos?; RPSolo |
| House I, west corridor | 6s | 6 | – | 0.0692 | **0.1250** | – |
| Toilets, southwest room | 6t | 6 | – | 0.0692 | **0.1250** | – |
| House J, northwest room | 5g | 5 | (Bench) | 0.0662 | **0.1111** | – |
| House J east, south-central room | 5h | 5 | – | 0.0662 | **0.1111** | – |
| House J east, south room | 5i | 5 | – | 0.0662 | **0.1111** | – |
| House J, south patio | 5w | 5 | – | 0.0663 | **0.1111** | – |
| South Terrace, southeast corner | 4y | 4 | – | 0.0630 | **0.0909** | – |

Appendix 7. *Path Matrix of the Palenque Cross Group and Palace Phases 1–6*

Palenque Cross Group Path Matrix

| | c | 1a | 2a | 2b | 2c | 3a | 3b | 3c | 4a | 4b | 4c | 4d | 4e | 4f | 4g | 4h | 4i | 5a | 5b | 5c | 5d | 5e |
|---|---|---|---|---|---|---|---|---|---|---|---|---|---|---|---|---|---|---|---|---|---|---|
| **c** | x | 1 | 2 | 2 | 2 | 3 | 3 | 3 | 4 | 4 | 4 | 4 | 4 | 4 | 4 | 4 | 4 | 5 | 5 | 5 | 5 | 5 |
| **1a** | 1 | x | 1 | 1 | 1 | 2 | 2 | 2 | 3 | 3 | 3 | 3 | 3 | 3 | 3 | 3 | 3 | 4 | 4 | 4 | 4 | 4 |
| **2a** | 2 | 1 | x | 2 | 2 | 1 | 3 | 3 | 2 | 2 | 2 | 4 | 4 | 4 | 4 | 4 | 4 | 3 | 3 | 5 | 5 | 5 |
| **2b** | 2 | 1 | 2 | x | 2 | 3 | 1 | 3 | 4 | 4 | 4 | 2 | 2 | 2 | 4 | 4 | 4 | 5 | 5 | 3 | 5 | 5 |
| **2c** | 2 | 1 | 2 | 2 | x | 3 | 3 | 1 | 4 | 4 | 4 | 4 | 4 | 4 | 2 | 2 | 2 | 5 | 5 | 5 | 3 | 3 |
| **3a** | 3 | 2 | 1 | 3 | 3 | x | 4 | 4 | 1 | 1 | 1 | 5 | 5 | 5 | 5 | 5 | 5 | 2 | 2 | 6 | 6 | 6 |
| **3b** | 3 | 2 | 3 | 1 | 3 | 4 | x | 4 | 5 | 5 | 5 | 1 | 1 | 1 | 5 | 5 | 5 | 6 | 6 | 2 | 6 | 6 |
| **3c** | 3 | 2 | 3 | 3 | 1 | 4 | 4 | x | 5 | 5 | 5 | 5 | 5 | 5 | 1 | 1 | 1 | 1 | 1 | 6 | 2 | 2 |
| **4a** | 4 | 3 | 2 | 4 | 4 | 1 | 5 | 5 | x | 1 | 2 | 6 | 6 | 6 | 6 | 6 | 6 | 2 | 3 | 7 | 7 | 7 |
| **4b** | 4 | 3 | 2 | 4 | 4 | 1 | 5 | 5 | 1 | x | 2 | 6 | 6 | 6 | 6 | 6 | 6 | 1 | 3 | 7 | 7 | 7 |
| **4c** | 4 | 3 | 2 | 4 | 4 | 1 | 5 | 5 | 2 | 2 | x | 6 | 6 | 6 | 6 | 6 | 6 | 3 | 1 | 7 | 7 | 7 |
| **4d** | 4 | 3 | 4 | 2 | 4 | 5 | 1 | 5 | 6 | 6 | 6 | x | 2 | 2 | 6 | 6 | 6 | 7 | 7 | 3 | 7 | 7 |
| **4e** | 4 | 3 | 4 | 2 | 4 | 5 | 1 | 5 | 6 | 6 | 6 | 2 | x | 2 | 6 | 6 | 6 | 7 | 7 | 1 | 7 | 7 |
| **4f** | 4 | 3 | 4 | 2 | 4 | 5 | 1 | 5 | 6 | 6 | 6 | 2 | 2 | x | 6 | 6 | 6 | 7 | 7 | 3 | 7 | 7 |
| **4g** | 4 | 3 | 4 | 4 | 2 | 5 | 5 | 1 | 6 | 6 | 6 | 6 | 6 | 6 | x | 2 | 2 | 7 | 7 | 7 | 1 | 3 |
| **4h** | 4 | 3 | 4 | 4 | 2 | 5 | 5 | 1 | 6 | 6 | 6 | 6 | 6 | 6 | 2 | x | 2 | 7 | 7 | 7 | 3 | 1 |
| **4i** | 4 | 3 | 4 | 4 | 2 | 5 | 5 | 1 | 6 | 6 | 6 | 6 | 6 | 6 | 2 | 2 | x | 7 | 7 | 7 | 3 | 3 |
| **5a** | 5 | 4 | 3 | 5 | 5 | 2 | 6 | 1 | 2 | 1 | 3 | 7 | 7 | 7 | 7 | 7 | 7 | x | 4 | 8 | 8 | 8 |
| **5b** | 5 | 4 | 3 | 5 | 5 | 2 | 6 | 1 | 3 | 3 | 1 | 7 | 7 | 7 | 7 | 7 | 7 | 4 | x | 8 | 8 | 8 |
| **5c** | 5 | 4 | 5 | 3 | 5 | 6 | 2 | 6 | 7 | 7 | 7 | 3 | 1 | 3 | 7 | 7 | 7 | 8 | 8 | x | 8 | 8 |
| **5d** | 5 | 4 | 5 | 5 | 3 | 6 | 6 | 2 | 7 | 7 | 7 | 7 | 7 | 7 | 1 | 3 | 3 | 8 | 8 | 8 | x | 4 |
| **5e** | 5 | 4 | 5 | 5 | 3 | 6 | 6 | 2 | 7 | 7 | 7 | 7 | 7 | 7 | 3 | 1 | 3 | 8 | 8 | 8 | 4 | x |

Palenque Palace Path Matrix Phase 1

| | C | 1a | 1b | 2a | 2b | 2c | 2d | 2e | 3a | 3b | 3c | 3d | 3e | 4a | 4b | 4c | 4d | 4e | 5a | 5b | 5c | 5d | 5e | 5f | 5g | 6a | 6b | 6c | 6d | 6e | 6f | 7a | 7b | 7c | 7d | 7e | 7f | 7g | 8a | 8b | 8c | 8d | 8e | 9a | 9b | 9c | 9d | 9e | 9f | 9g | 9h | 10a | 10b |
|---|---|---|---|---|---|---|---|---|---|---|---|---|---|---|---|---|---|---|---|---|---|---|---|---|---|---|---|---|---|---|---|---|---|---|---|---|---|---|---|---|---|---|---|---|---|---|---|---|---|---|---|---|---|
| **C** | x | 1 | 1 | 2 | 2 | 2 | 2 | 2 | 3 | 3 | 3 | 3 | 3 | 4 | 4 | 4 | 4 | 4 | 5 | 5 | 5 | 5 | 5 | 5 | 5 | 6 | 6 | 6 | 6 | 6 | 6 | 7 | 7 | 7 | 7 | 7 | 7 | 7 | 8 | 8 | 8 | 8 | 8 | 9 | 9 | 9 | 9 | 9 | 9 | 9 | 9 | 10 | 10 |
| **1a** | 1 | x | 2 | 1 | 1 | 1 | 3 | 3 | 2 | 2 | 4 | 4 | 4 | 3 | 3 | 5 | 5 | 5 | 4 | 4 | 4 | 4 | 4 | 6 | 6 | 5 | 5 | 5 | 5 | 5 | 7 | 6 | 6 | 7 | 6 | 6 | 6 | 6 | 7 | 7 | 7 | 7 | 7 | 8 | 8 | 8 | 8 | 8 | 8 | 8 | 8 | 9 | 9 |
| **1b** | 1 | 2 | x | 3 | 3 | 3 | 1 | 1 | 4 | 4 | 2 | 2 | 2 | 5 | 5 | 3 | 3 | 3 | 6 | 6 | 6 | 6 | 6 | 4 | 4 | 7 | 7 | 7 | 7 | 7 | 5 | 8 | 7 | 6 | 8 | 8 | 8 | 8 | 9 | 9 | 9 | 9 | 9 | 10 | 10 | 10 | 10 | 10 | 10 | 10 | 10 | 11 | 11 |
| **2a** | 2 | 1 | 3 | x | 2 | 2 | 4 | 4 | 2 | 1 | 5 | 5 | 5 | 2 | 2 | 6 | 6 | 6 | 3 | 3 | 3 | 3 | 3 | 7 | 7 | 4 | 4 | 4 | 4 | 4 | 7 | 5 | 5 | 6 | 5 | 5 | 5 | 5 | 6 | 6 | 6 | 6 | 6 | 7 | 7 | 7 | 7 | 7 | 7 | 7 | 7 | 8 | 8 |
| **2b** | 2 | 1 | 3 | 2 | x | 2 | 4 | 4 | 1 | 2 | 5 | 5 | 5 | 3 | 3 | 6 | 6 | 6 | 4 | 4 | 4 | 4 | 4 | 7 | 7 | 5 | 5 | 5 | 5 | 5 | 8 | 6 | 6 | 7 | 6 | 6 | 6 | 6 | 7 | 7 | 7 | 7 | 7 | 8 | 8 | 8 | 8 | 8 | 8 | 8 | 8 | 9 | 9 |
| **2c** | 2 | 1 | 3 | 2 | 2 | x | 4 | 4 | 2 | 1 | 5 | 5 | 5 | 2 | 2 | 6 | 6 | 6 | 3 | 3 | 3 | 3 | 3 | 7 | 7 | 4 | 4 | 4 | 4 | 4 | 7 | 5 | 5 | 6 | 5 | 5 | 5 | 5 | 6 | 6 | 6 | 6 | 6 | 7 | 7 | 7 | 7 | 7 | 7 | 7 | 7 | 8 | 8 |
| **2d** | 2 | 3 | 1 | 4 | 4 | 4 | x | 2 | 5 | 5 | 3 | 3 | 3 | 6 | 6 | 4 | 4 | 4 | 7 | 7 | 7 | 7 | 7 | 5 | 5 | 8 | 8 | 8 | 8 | 8 | 6 | 9 | 8 | 7 | 9 | 9 | 9 | 9 | 10 | 10 | 10 | 10 | 10 | 11 | 11 | 11 | 11 | 11 | 11 | 11 | 11 | 12 | 12 |
| **2e** | 2 | 3 | 1 | 4 | 4 | 4 | 2 | x | 5 | 5 | 1 | 1 | 1 | 6 | 6 | 2 | 2 | 2 | 7 | 7 | 7 | 7 | 7 | 3 | 3 | 8 | 8 | 7 | 8 | 8 | 4 | 9 | 6 | 5 | 9 | 9 | 9 | 8 | 10 | 10 | 10 | 9 | 10 | 11 | 11 | 11 | 11 | 11 | 11 | 11 | 11 | 12 | 12 |
| **3a** | 3 | 2 | 4 | 2 | 1 | 2 | 5 | 5 | x | 1 | 6 | 6 | 6 | 2 | 2 | 7 | 7 | 7 | 3 | 3 | 3 | 3 | 3 | 8 | 8 | 4 | 4 | 4 | 4 | 4 | 7 | 5 | 5 | 6 | 5 | 5 | 5 | 5 | 6 | 6 | 6 | 6 | 6 | 7 | 7 | 7 | 7 | 7 | 7 | 7 | 7 | 8 | 8 |
| **3b** | 3 | 2 | 4 | 1 | 2 | 1 | 5 | 5 | 1 | x | 6 | 6 | 6 | 1 | 1 | 7 | 7 | 7 | 2 | 2 | 2 | 2 | 2 | 7 | 8 | 3 | 3 | 3 | 3 | 3 | 6 | 4 | 4 | 5 | 4 | 4 | 4 | 4 | 5 | 5 | 5 | 5 | 5 | 6 | 6 | 6 | 6 | 6 | 6 | 6 | 6 | 7 | 7 |
| **3c** | 3 | 4 | 2 | 5 | 5 | 5 | 3 | 1 | 6 | 6 | x | 2 | 2 | 7 | 7 | 1 | 1 | 3 | 8 | 8 | 7 | 8 | 8 | 2 | 4 | 9 | 7 | 6 | 8 | 9 | 3 | 8 | 5 | 4 | 9 | 10 | 8 | 7 | 9 | 10 | 11 | 8 | 9 | 10 | 10 | 10 | 10 | 10 | 10 | 12 | 11 | 11 | 12 |
| **3d** | 3 | 4 | 2 | 5 | 5 | 5 | 3 | 1 | 6 | 6 | 2 | x | 2 | 7 | 7 | 1 | 1 | 3 | 8 | 8 | 7 | 8 | 8 | 2 | 4 | 9 | 7 | 6 | 8 | 9 | 3 | 8 | 5 | 4 | 9 | 10 | 8 | 7 | 9 | 10 | 11 | 8 | 9 | 10 | 10 | 10 | 10 | 10 | 10 | 12 | 11 | 11 | 12 |
| **3e** | 3 | 4 | 2 | 5 | 5 | 5 | 3 | 1 | 6 | 6 | 2 | 2 | x | 7 | 7 | 3 | 3 | 1 | 8 | 8 | 8 | 8 | 8 | 4 | 2 | 9 | 9 | 8 | 9 | 9 | 5 | 10 | 7 | 6 | 10 | 10 | 10 | 9 | 11 | 11 | 11 | 10 | 11 | 12 | 12 | 12 | 12 | 12 | 12 | 12 | 12 | 13 | 13 |
| **4a** | 4 | 3 | 5 | 2 | 3 | 2 | 6 | 6 | 2 | 1 | 7 | 7 | 7 | x | 2 | 7 | 8 | 8 | 1 | 1 | 1 | 1 | 2 | 6 | 9 | 2 | 2 | 2 | 2 | 2 | 5 | 3 | 3 | 4 | 3 | 3 | 3 | 3 | 4 | 4 | 4 | 4 | 4 | 5 | 5 | 5 | 5 | 5 | 5 | 5 | 5 | 6 | 6 |
| **4b** | 4 | 3 | 5 | 2 | 3 | 2 | 6 | 6 | 2 | 1 | 7 | 7 | 7 | 2 | x | 8 | 8 | 8 | 3 | 3 | 3 | 2 | 1 | 8 | 9 | 4 | 4 | 4 | 3 | 2 | 7 | 5 | 5 | 6 | 4 | 3 | 3 | 5 | 6 | 5 | 4 | 6 | 4 | 7 | 7 | 7 | 7 | 7 | 7 | 5 | 6 | 8 | 6 |
| **4c** | 4 | 5 | 3 | 6 | 6 | 6 | 4 | 2 | 7 | 7 | 1 | 1 | 3 | 7 | 8 | x | 2 | 4 | 8 | 8 | 6 | 7 | 8 | 1 | 5 | 9 | 7 | 5 | 7 | 8 | 2 | 7 | 4 | 3 | 8 | 9 | 7 | 6 | 8 | 9 | 10 | 7 | 8 | 9 | 9 | 9 | 9 | 9 | 9 | 11 | 10 | 10 | 12 |
| **4d** | 4 | 5 | 3 | 6 | 6 | 6 | 4 | 2 | 7 | 7 | 1 | 1 | 3 | 8 | 8 | 2 | x | 4 | 9 | 9 | 8 | 9 | 9 | 3 | 5 | 10 | 8 | 7 | 9 | 10 | 4 | 9 | 6 | 5 | 10 | 11 | 9 | 8 | 10 | 11 | 12 | 9 | 10 | 11 | 11 | 11 | 11 | 11 | 11 | 13 | 12 | 12 | 13 |
| **4e** | 4 | 5 | 3 | 6 | 6 | 6 | 4 | 2 | 7 | 7 | 3 | 3 | 1 | 8 | 8 | 4 | 4 | x | 9 | 9 | 9 | 9 | 9 | 5 | 1 | 10 | 10 | 9 | 10 | 10 | 6 | 11 | 8 | 7 | 11 | 11 | 11 | 10 | 12 | 11 | 12 | 11 | 12 | 13 | 13 | 13 | 13 | 13 | 13 | 13 | 13 | 14 | 14 |
| **5a** | 5 | 4 | 6 | 3 | 4 | 3 | 7 | 7 | 3 | 2 | 8 | 8 | 8 | 1 | 3 | 8 | 9 | 9 | x | 2 | 2 | 2 | 3 | 7 | 10 | 3 | 3 | 3 | 3 | 3 | 6 | 4 | 4 | 5 | 4 | 4 | 4 | 4 | 5 | 5 | 5 | 5 | 5 | 6 | 6 | 6 | 6 | 6 | 6 | 6 | 6 | 7 | 7 |
| **5b** | 5 | 4 | 6 | 3 | 4 | 3 | 7 | 7 | 3 | 2 | 8 | 8 | 8 | 1 | 3 | 8 | 9 | 9 | 2 | x | 2 | 2 | 3 | 7 | 10 | 1 | 3 | 3 | 3 | 3 | 6 | 2 | 4 | 5 | 4 | 4 | 4 | 4 | 3 | 5 | 5 | 5 | 5 | 4 | 4 | 4 | 4 | 4 | 4 | 6 | 6 | 5 | 7 |
| **5c** | 5 | 4 | 6 | 3 | 4 | 3 | 7 | 7 | 3 | 2 | 7 | 7 | 8 | 1 | 3 | 6 | 8 | 9 | 2 | 2 | x | 1 | 2 | 5 | 10 | 3 | 1 | 1 | 2 | 2 | 4 | 2 | 2 | 3 | 3 | 3 | 3 | 2 | 3 | 4 | 4 | 3 | 4 | 4 | 4 | 4 | 4 | 4 | 4 | 5 | 5 | 5 | 6 |
| **5d** | 5 | 4 | 6 | 3 | 4 | 3 | 7 | 7 | 3 | 2 | 8 | 8 | 8 | 1 | 2 | 7 | 9 | 9 | 2 | 2 | 1 | x | 1 | 6 | 10 | 3 | 3 | 3 | 1 | 1 | 5 | 3 | 3 | 4 | 2 | 2 | 2 | 2 | 4 | 3 | 3 | 3 | 3 | 5 | 5 | 5 | 5 | 5 | 5 | 4 | 4 | 6 | 5 |
| **5e** | 5 | 4 | 6 | 3 | 4 | 3 | 7 | 7 | 3 | 2 | 8 | 8 | 8 | 2 | 1 | 8 | 9 | 9 | 3 | 3 | 2 | 1 | x | 7 | 10 | 4 | 3 | 3 | 2 | 1 | 6 | 4 | 4 | 5 | 3 | 2 | 2 | 3 | 5 | 4 | 3 | 4 | 3 | 6 | 6 | 6 | 6 | 6 | 6 | 4 | 5 | 7 | 5 |
| **5f** | 5 | 6 | 4 | 7 | 7 | 7 | 5 | 3 | 8 | 7 | 2 | 2 | 4 | 6 | 8 | 1 | 3 | 5 | 7 | 7 | 5 | 6 | 7 | x | 6 | 7 | 5 | 4 | 6 | 7 | 1 | 6 | 3 | 2 | 7 | 8 | 6 | 5 | 7 | 8 | 9 | 6 | 7 | 8 | 8 | 8 | 8 | 8 | 8 | 10 | 9 | 9 | 10 |
| **5g** | 5 | 6 | 4 | 7 | 7 | 7 | 5 | 3 | 8 | 8 | 4 | 4 | 2 | 9 | 9 | 5 | 5 | 1 | 10 | 10 | 10 | 10 | 10 | 6 | x | 11 | 11 | 10 | 11 | 11 | 7 | 12 | 9 | 8 | 12 | 12 | 12 | 11 | 13 | 13 | 13 | 12 | 13 | 14 | 14 | 14 | 14 | 14 | 14 | 14 | 14 | 15 | 15 |
| **6a** | 6 | 5 | 7 | 4 | 5 | 4 | 8 | 8 | 4 | 3 | 9 | 9 | 9 | 2 | 4 | 9 | 10 | 10 | 3 | 1 | 3 | 3 | 4 | 7 | 11 | x | 2 | 3 | 3 | 4 | 6 | 1 | 4 | 5 | 3 | 5 | 3 | 2 | 2 | 5 | 6 | 3 | 4 | 3 | 3 | 3 | 3 | 3 | 3 | 7 | 6 | 4 | 7 |
| **6b** | 6 | 5 | 7 | 4 | 5 | 4 | 8 | 8 | 4 | 3 | 9 | 7 | 9 | 2 | 4 | 7 | 8 | 10 | 3 | 3 | 1 | 3 | 3 | 5 | 11 | 2 | x | 1 | 2 | 3 | 4 | 1 | 2 | 3 | 3 | 4 | 2 | 1 | 2 | 4 | 5 | 2 | 3 | 3 | 3 | 3 | 3 | 3 | 3 | 6 | 5 | 4 | 6 |
| **6c** | 6 | 5 | 7 | 4 | 5 | 4 | 8 | 7 | 4 | 3 | 6 | 6 | 8 | 2 | 4 | 5 | 7 | 9 | 3 | 3 | 1 | 3 | 3 | 4 | 10 | 3 | 1 | x | 2 | 3 | 3 | 2 | 1 | 2 | 3 | 4 | 2 | 1 | 3 | 4 | 5 | 2 | 3 | 4 | 4 | 4 | 4 | 4 | 4 | 6 | 5 | 5 | 6 |
| **6d** | 6 | 5 | 7 | 4 | 5 | 4 | 8 | 8 | 4 | 3 | 8 | 8 | 9 | 2 | 3 | 7 | 9 | 10 | 3 | 3 | 2 | 1 | 2 | 6 | 11 | 3 | 2 | 2 | x | 1 | 5 | 2 | 3 | 4 | 1 | 2 | 2 | 1 | 3 | 2 | 3 | 2 | 3 | 4 | 4 | 4 | 4 | 4 | 4 | 4 | 3 | 5 | 4 |
| **6e** | 6 | 5 | 7 | 4 | 5 | 4 | 8 | 8 | 4 | 3 | 9 | 9 | 9 | 2 | 2 | 8 | 10 | 10 | 3 | 3 | 2 | 1 | 1 | 7 | 11 | 4 | 3 | 3 | 1 | x | 6 | 3 | 4 | 5 | 2 | 1 | 1 | 2 | 4 | 3 | 2 | 3 | 2 | 5 | 5 | 5 | 5 | 5 | 5 | 3 | 4 | 6 | 4 |
| **6f** | 6 | 7 | 5 | 7 | 8 | 7 | 6 | 4 | 7 | 6 | 3 | 3 | 5 | 5 | 7 | 2 | 4 | 6 | 6 | 6 | 4 | 5 | 6 | 1 | 7 | 6 | 4 | 3 | 5 | 6 | x | 5 | 2 | 1 | 6 | 7 | 5 | 4 | 6 | 7 | 8 | 5 | 6 | 7 | 7 | 7 | 7 | 7 | 7 | 9 | 8 | 8 | 9 |
| **7a** | 7 | 6 | 8 | 5 | 6 | 5 | 9 | 9 | 5 | 4 | 8 | 8 | 10 | 3 | 5 | 7 | 9 | 11 | 4 | 2 | 2 | 3 | 4 | 6 | 12 | 1 | 1 | 2 | 2 | 3 | 5 | x | 3 | 4 | 3 | 4 | 2 | 1 | 1 | 4 | 5 | 2 | 3 | 2 | 2 | 2 | 2 | 2 | 2 | 6 | 5 | 3 | 6 |
| **7b** | 7 | 6 | 7 | 5 | 6 | 5 | 8 | 6 | 5 | 4 | 5 | 5 | 7 | 3 | 5 | 4 | 6 | 8 | 4 | 4 | 2 | 3 | 4 | 3 | 9 | 4 | 2 | 1 | 3 | 4 | 2 | 3 | x | 1 | 4 | 5 | 3 | 2 | 4 | 5 | 6 | 3 | 4 | 5 | 5 | 5 | 5 | 5 | 5 | 7 | 6 | 6 | 7 |
| **7c** | 7 | 7 | 6 | 6 | 7 | 6 | 7 | 5 | 6 | 5 | 4 | 4 | 6 | 4 | 6 | 3 | 5 | 7 | 5 | 5 | 3 | 4 | 5 | 2 | 8 | 5 | 3 | 2 | 4 | 5 | 1 | 4 | 1 | x | 5 | 6 | 4 | 3 | 5 | 6 | 7 | 4 | 5 | 6 | 6 | 6 | 6 | 6 | 6 | 8 | 7 | 7 | 8 |
| **7d** | 7 | 6 | 8 | 5 | 6 | 5 | 9 | 9 | 5 | 4 | 9 | 9 | 10 | 3 | 4 | 8 | 10 | 11 | 4 | 4 | 3 | 2 | 3 | 7 | 12 | 3 | 3 | 3 | 1 | 2 | 6 | 3 | 4 | 5 | x | 3 | 3 | 2 | 4 | 1 | 4 | 3 | 4 | 5 | 5 | 5 | 5 | 5 | 5 | 4 | 2 | 6 | 3 |
| **7e** | 7 | 6 | 8 | 5 | 6 | 5 | 9 | 9 | 5 | 4 | 10 | 10 | 10 | 3 | 3 | 9 | 11 | 11 | 4 | 4 | 3 | 2 | 2 | 8 | 12 | 5 | 4 | 4 | 3 | 1 | 7 | 4 | 5 | 6 | 3 | x | 2 | 3 | 5 | 4 | 1 | 4 | 3 | 6 | 6 | 6 | 6 | 6 | 6 | 2 | 4 | 7 | 3 |
| **7f** | 7 | 6 | 8 | 5 | 6 | 5 | 9 | 9 | 5 | 4 | 10 | 8 | 10 | 3 | 3 | 7 | 9 | 11 | 4 | 4 | 3 | 2 | 2 | 6 | 12 | 3 | 2 | 2 | 2 | 1 | 5 | 2 | 3 | 4 | 3 | 2 | x | 1 | 3 | 4 | 3 | 2 | 1 | 4 | 4 | 4 | 4 | 4 | 4 | 4 | 5 | 5 | 5 |
| **7g** | 7 | 6 | 8 | 5 | 6 | 5 | 9 | 8 | 5 | 4 | 7 | 7 | 9 | 3 | 5 | 6 | 8 | 10 | 4 | 4 | 2 | 2 | 3 | 5 | 11 | 2 | 1 | 1 | 1 | 2 | 4 | 1 | 2 | 3 | 2 | 3 | 1 | x | 2 | 3 | 4 | 1 | 2 | 3 | 3 | 3 | 3 | 3 | 3 | 5 | 4 | 4 | 5 |
| **8a** | 8 | 7 | 9 | 6 | 7 | 6 | 10 | 10 | 6 | 5 | 9 | 9 | 11 | 4 | 6 | 8 | 10 | 12 | 5 | 3 | 3 | 4 | 5 | 7 | 13 | 2 | 2 | 3 | 3 | 4 | 6 | 1 | 4 | 5 | 4 | 5 | 3 | 2 | x | 5 | 6 | 3 | 4 | 1 | 1 | 1 | 1 | 1 | 1 | 7 | 6 | 2 | 7 |
| **8b** | 8 | 7 | 9 | 6 | 7 | 6 | 10 | 10 | 6 | 5 | 10 | 11 | 11 | 4 | 5 | 9 | 11 | 11 | 5 | 5 | 4 | 3 | 4 | 8 | 13 | 5 | 4 | 4 | 2 | 3 | 7 | 4 | 5 | 6 | 1 | 4 | 4 | 3 | 5 | x | 4 | 4 | 5 | 6 | 6 | 6 | 6 | 6 | 6 | 3 | 1 | 7 | 2 |
| **8c** | 8 | 7 | 9 | 6 | 7 | 6 | 10 | 10 | 6 | 5 | 11 | 11 | 11 | 4 | 4 | 10 | 12 | 12 | 5 | 5 | 4 | 3 | 3 | 9 | 13 | 6 | 5 | 5 | 3 | 2 | 8 | 5 | 6 | 7 | 4 | 1 | 3 | 4 | 6 | 4 | x | 5 | 4 | 7 | 7 | 7 | 7 | 7 | 7 | 1 | 3 | 8 | 2 |
| **8d** | 8 | 7 | 9 | 6 | 7 | 6 | 10 | 9 | 7 | 5 | 8 | 8 | 10 | 4 | 6 | 7 | 9 | 11 | 5 | 5 | 3 | 3 | 4 | 6 | 12 | 3 | 2 | 2 | 2 | 3 | 5 | 2 | 3 | 4 | 3 | 4 | 2 | 1 | 3 | 4 | 5 | x | 3 | 4 | 4 | 4 | 4 | 4 | 4 | 6 | 5 | 5 | 6 |
| **8e** | 8 | 7 | 9 | 6 | 7 | 6 | 10 | 10 | 6 | 5 | 9 | 9 | 11 | 4 | 4 | 8 | 10 | 12 | 5 | 5 | 4 | 3 | 3 | 7 | 13 | 4 | 3 | 3 | 3 | 2 | 6 | 3 | 4 | 5 | 4 | 3 | 1 | 2 | 4 | 5 | 4 | 3 | x | 5 | 5 | 5 | 5 | 5 | 5 | 5 | 6 | 6 | 6 |
| **9a** | 9 | 8 | 10 | 7 | 8 | 7 | 11 | 11 | 7 | 6 | 10 | 10 | 12 | 5 | 7 | 9 | 11 | 13 | 6 | 4 | 4 | 5 | 6 | 8 | 14 | 3 | 3 | 4 | 4 | 5 | 7 | 2 | 5 | 6 | 5 | 6 | 4 | 3 | 1 | 6 | 7 | 4 | 5 | x | 2 | 2 | 2 | 2 | 2 | 8 | 7 | 3 | 8 |
| **9b** | 9 | 8 | 10 | 7 | 8 | 7 | 11 | 11 | 7 | 6 | 10 | 10 | 12 | 5 | 7 | 9 | 11 | 13 | 6 | 4 | 4 | 5 | 6 | 8 | 14 | 3 | 3 | 4 | 4 | 5 | 7 | 2 | 5 | 6 | 5 | 6 | 4 | 3 | 1 | 6 | 7 | 4 | 5 | 2 | x | 1 | 2 | 2 | 2 | 8 | 7 | 2 | 8 |
| **9c** | 9 | 8 | 10 | 7 | 8 | 7 | 11 | 11 | 7 | 6 | 10 | 10 | 12 | 5 | 7 | 9 | 11 | 13 | 6 | 4 | 4 | 5 | 6 | 8 | 14 | 3 | 3 | 4 | 4 | 5 | 7 | 2 | 5 | 6 | 5 | 6 | 4 | 3 | 1 | 6 | 7 | 4 | 5 | 2 | 1 | x | 2 | 2 | 2 | 8 | 7 | 1 | 8 |
| **9d** | 9 | 8 | 10 | 7 | 8 | 7 | 11 | 11 | 7 | 6 | 10 | 10 | 12 | 5 | 7 | 9 | 11 | 13 | 6 | 4 | 4 | 5 | 6 | 8 | 14 | 3 | 3 | 4 | 4 | 5 | 7 | 2 | 5 | 6 | 5 | 6 | 4 | 3 | 1 | 6 | 7 | 4 | 5 | 2 | 2 | 2 | x | 2 | 2 | 8 | 7 | 3 | 8 |
| **9e** | 9 | 8 | 10 | 7 | 8 | 7 | 11 | 11 | 7 | 6 | 10 | 10 | 12 | 5 | 7 | 9 | 11 | 13 | 6 | 4 | 4 | 5 | 6 | 8 | 14 | 3 | 3 | 4 | 4 | 5 | 7 | 2 | 5 | 6 | 5 | 6 | 4 | 3 | 1 | 6 | 7 | 4 | 5 | 2 | 2 | 2 | 2 | x | 2 | 8 | 7 | 3 | 8 |
| **9f** | 9 | 8 | 10 | 7 | 8 | 7 | 11 | 11 | 7 | 6 | 10 | 10 | 12 | 5 | 7 | 9 | 11 | 13 | 6 | 4 | 4 | 5 | 6 | 8 | 14 | 3 | 3 | 4 | 4 | 5 | 7 | 2 | 5 | 6 | 5 | 6 | 4 | 3 | 1 | 6 | 7 | 4 | 5 | 2 | 2 | 2 | 2 | 2 | x | 8 | 7 | 3 | 8 |
| **9g** | 9 | 8 | 10 | 7 | 8 | 7 | 11 | 11 | 7 | 6 | 12 | 12 | 12 | 5 | 5 | 11 | 13 | 13 | 6 | 6 | 5 | 4 | 4 | 10 | 14 | 7 | 6 | 6 | 4 | 3 | 9 | 6 | 7 | 8 | 4 | 2 | 4 | 5 | 7 | 3 | 1 | 6 | 5 | 8 | 8 | 8 | 8 | 8 | 8 | x | 2 | 1 | 1 |
| **9h** | 9 | 8 | 10 | 7 | 8 | 7 | 11 | 11 | 7 | 6 | 11 | 11 | 12 | 5 | 6 | 10 | 12 | 13 | 6 | 6 | 5 | 4 | 5 | 9 | 14 | 6 | 5 | 5 | 3 | 4 | 8 | 5 | 6 | 7 | 2 | 4 | 5 | 4 | 6 | 1 | 3 | 5 | 6 | 7 | 7 | 7 | 7 | 7 | 7 | 2 | x | 1 | 1 |
| **10a** | 10 | 9 | 11 | 8 | 9 | 8 | 12 | 12 | 8 | 7 | 11 | 11 | 13 | 6 | 8 | 10 | 12 | 14 | 7 | 5 | 5 | 6 | 7 | 9 | 15 | 4 | 4 | 5 | 5 | 6 | 8 | 3 | 6 | 7 | 6 | 7 | 5 | 4 | 2 | 7 | 8 | 5 | 6 | 3 | 2 | 1 | 3 | 3 | 3 | 1 | 1 | x | 9 |
| **10b** | 10 | 9 | 11 | 8 | 9 | 8 | 12 | 12 | 8 | 7 | 12 | 12 | 13 | 6 | 6 | 12 | 13 | 14 | 7 | 7 | 6 | 5 | 5 | 10 | 15 | 7 | 6 | 6 | 4 | 4 | 9 | 6 | 7 | 8 | 3 | 3 | 5 | 5 | 7 | 2 | 2 | 6 | 6 | 8 | 8 | 8 | 8 | 8 | 8 | 1 | 1 | 9 | x |

Palenque Palace Path Matrix Phase 2

| | c | 1a | 1b | 2a | 2b | 2c | 2d | 2e | 3a | 3b | 3c | 3d | 3e | 4a | 4b | 4c | 4d | 4e | 5a | 5b | 5c | 5d | 5e | 5f | 5g | 6a | 6b | 6c | 6d | 6e | 6f |
|---|---|---|---|---|---|---|---|---|---|---|---|---|---|---|---|---|---|---|---|---|---|---|---|---|---|---|---|---|---|---|---|
| **c** | x | 1 | 1 | 2 | 2 | 2 | 2 | 2 | 3 | 3 | 3 | 3 | 3 | 4 | 4 | 4 | 4 | 4 | 5 | 5 | 5 | 5 | 5 | 5 | 5 | 6 | 6 | 6 | 6 | 6 | 6 |
| **1a** | 1 | x | 2 | 1 | 1 | 1 | 3 | 3 | 2 | 2 | 4 | 4 | 4 | 3 | 3 | 5 | 5 | 5 | 4 | 4 | 4 | 4 | 4 | 6 | 6 | 5 | 5 | 5 | 5 | 5 | 5 |
| **1b** | 1 | 2 | x | 3 | 3 | 3 | 1 | 1 | 4 | 4 | 2 | 2 | 2 | 5 | 5 | 3 | 3 | 3 | 6 | 6 | 6 | 6 | 6 | 4 | 4 | 7 | 7 | 7 | 7 | 7 | 7 |
| **2a** | 2 | 1 | 3 | x | 2 | 2 | 4 | 4 | 2 | 1 | 5 | 5 | 5 | 2 | 2 | 6 | 6 | 6 | 3 | 3 | 3 | 3 | 3 | 7 | 7 | 4 | 4 | 4 | 4 | 4 | 4 |
| **2b** | 2 | 1 | 3 | 2 | x | 2 | 4 | 4 | 1 | 2 | 5 | 5 | 5 | 3 | 3 | 6 | 6 | 6 | 4 | 4 | 4 | 4 | 4 | 7 | 7 | 5 | 5 | 5 | 5 | 5 | 5 |
| **2c** | 2 | 1 | 3 | 2 | 2 | x | 4 | 4 | 2 | 1 | 5 | 5 | 5 | 2 | 2 | 6 | 6 | 6 | 3 | 3 | 3 | 3 | 3 | 7 | 7 | 4 | 4 | 4 | 4 | 4 | 4 |
| **2d** | 2 | 3 | 1 | 4 | 4 | 4 | x | 2 | 5 | 5 | 1 | 1 | 1 | 6 | 6 | 2 | 2 | 2 | 7 | 7 | 7 | 7 | 7 | 3 | 3 | 8 | 8 | 8 | 8 | 8 | 8 |
| **2e** | 2 | 3 | 1 | 4 | 4 | 4 | 2 | x | 5 | 5 | 3 | 3 | 3 | 6 | 6 | 4 | 4 | 4 | 7 | 7 | 7 | 7 | 7 | 5 | 5 | 8 | 8 | 8 | 8 | 8 | 8 |
| **3a** | 3 | 2 | 4 | 2 | 1 | 2 | 5 | 5 | x | 1 | 6 | 6 | 6 | 2 | 2 | 7 | 7 | 7 | 3 | 3 | 3 | 3 | 3 | 8 | 8 | 4 | 4 | 4 | 4 | 4 | 4 |
| **3b** | 3 | 2 | 4 | 1 | 2 | 1 | 5 | 5 | 1 | x | 6 | 6 | 6 | 1 | 1 | 7 | 7 | 7 | 2 | 2 | 2 | 2 | 2 | 8 | 8 | 3 | 3 | 3 | 3 | 3 | 3 |
| **3c** | 3 | 4 | 2 | 5 | 5 | 5 | 1 | 3 | 6 | 6 | x | 2 | 2 | 7 | 7 | 1 | 1 | 3 | 8 | 8 | 8 | 8 | 8 | 2 | 4 | 9 | 8 | 8 | 8 | 7 | 7 |
| **3d** | 3 | 4 | 2 | 5 | 5 | 5 | 1 | 3 | 6 | 6 | 2 | x | 2 | 7 | 7 | 1 | 1 | 3 | 8 | 8 | 8 | 8 | 8 | 2 | 4 | 9 | 8 | 8 | 8 | 7 | 7 |
| **3e** | 3 | 4 | 2 | 5 | 5 | 5 | 1 | 3 | 6 | 6 | 2 | 2 | x | 7 | 7 | 3 | 3 | 1 | 8 | 8 | 8 | 8 | 8 | 4 | 2 | 9 | 9 | 9 | 9 | 9 | 9 |
| **4a** | 4 | 3 | 5 | 2 | 3 | 2 | 6 | 6 | 2 | 1 | 7 | 7 | 7 | x | 2 | 8 | 8 | 8 | 1 | 1 | 1 | 1 | 3 | 7 | 9 | 2 | 2 | 2 | 3 | 2 | 2 |
| **4b** | 4 | 3 | 5 | 2 | 3 | 2 | 6 | 6 | 2 | 1 | 7 | 7 | 7 | 2 | x | 8 | 8 | 8 | 3 | 3 | 3 | 3 | 1 | 8 | 9 | 4 | 4 | 3 | 2 | 4 | 3 |
| **4c** | 4 | 5 | 3 | 6 | 6 | 6 | 2 | 4 | 7 | 7 | 1 | 1 | 3 | 8 | 8 | x | 2 | 4 | 9 | 9 | 9 | 9 | 9 | 3 | 5 | 10 | 9 | 9 | 9 | 8 | 8 |
| **4d** | 4 | 5 | 3 | 6 | 6 | 6 | 2 | 4 | 7 | 7 | 1 | 1 | 3 | 8 | 8 | 2 | x | 4 | 9 | 8 | 7 | 7 | 8 | 1 | 5 | 8 | 7 | 7 | 7 | 6 | 6 |
| **4e** | 4 | 5 | 3 | 6 | 6 | 6 | 2 | 4 | 7 | 7 | 3 | 3 | 1 | 8 | 8 | 4 | 4 | x | 9 | 9 | 9 | 9 | 9 | 5 | 1 | 10 | 10 | 10 | 10 | 10 | 10 |
| **5a** | 5 | 4 | 6 | 3 | 4 | 3 | 7 | 7 | 3 | 2 | 8 | 8 | 8 | 1 | 3 | 9 | 9 | 9 | x | 2 | 2 | 2 | 4 | 8 | 10 | 1 | 3 | 3 | 4 | 3 | 3 |
| **5b** | 5 | 4 | 6 | 3 | 4 | 3 | 7 | 7 | 3 | 2 | 8 | 8 | 8 | 1 | 3 | 9 | 8 | 9 | 2 | x | 1 | 2 | 4 | 7 | 10 | 3 | 1 | 3 | 4 | 2 | 3 |
| **5c** | 5 | 4 | 6 | 3 | 4 | 3 | 7 | 7 | 3 | 2 | 8 | 8 | 8 | 1 | 3 | 9 | 7 | 9 | 2 | 1 | x | 2 | 4 | 6 | 10 | 3 | 2 | 3 | 4 | 1 | 3 |
| **5d** | 5 | 4 | 6 | 3 | 4 | 3 | 7 | 7 | 3 | 2 | 8 | 8 | 8 | 1 | 3 | 9 | 7 | 9 | 2 | 2 | 2 | x | 3 | 6 | 10 | 3 | 3 | 1 | 2 | 3 | 1 |
| **5e** | 5 | 4 | 6 | 3 | 4 | 3 | 7 | 7 | 3 | 2 | 8 | 8 | 8 | 3 | 1 | 9 | 8 | 9 | 4 | 4 | 4 | 3 | x | 7 | 10 | 5 | 5 | 2 | 1 | 4 | 2 |
| **5f** | 5 | 6 | 4 | 7 | 7 | 7 | 3 | 5 | 8 | 8 | 2 | 2 | 4 | 7 | 8 | 3 | 1 | 5 | 8 | 7 | 6 | 6 | 7 | x | 6 | 7 | 6 | 6 | 6 | 5 | 5 |
| **5g** | 5 | 6 | 4 | 7 | 7 | 7 | 3 | 5 | 8 | 8 | 4 | 4 | 2 | 9 | 9 | 5 | 5 | 1 | 10 | 10 | 10 | 10 | 10 | 6 | x | 11 | 11 | 11 | 11 | 11 | 11 |
| **6a** | 6 | 5 | 7 | 4 | 5 | 4 | 8 | 8 | 4 | 3 | 9 | 9 | 9 | 2 | 4 | 10 | 8 | 10 | 1 | 3 | 3 | 3 | 5 | 7 | 11 | x | 3 | 5 | 5 | 3 | 4 |
| **6b** | 6 | 5 | 7 | 4 | 5 | 4 | 8 | 8 | 4 | 3 | 8 | 8 | 9 | 2 | 4 | 9 | 7 | 10 | 3 | 1 | 2 | 3 | 5 | 6 | 11 | 3 | x | 4 | 4 | 2 | 3 |
| **6c** | 6 | 5 | 7 | 4 | 5 | 4 | 8 | 8 | 4 | 3 | 8 | 8 | 9 | 2 | 3 | 9 | 7 | 10 | 3 | 3 | 3 | 1 | 2 | 6 | 11 | 5 | 4 | x | 1 | 3 | 1 |
| **6d** | 6 | 5 | 7 | 4 | 5 | 4 | 8 | 8 | 4 | 3 | 8 | 8 | 9 | 3 | 2 | 9 | 7 | 10 | 4 | 4 | 4 | 2 | 1 | 6 | 11 | 5 | 4 | 1 | x | 3 | 1 |
| **6e** | 6 | 5 | 7 | 4 | 5 | 4 | 8 | 8 | 4 | 3 | 7 | 7 | 9 | 2 | 4 | 8 | 6 | 10 | 3 | 2 | 1 | 3 | 4 | 5 | 11 | 3 | 2 | 3 | 3 | x | 2 |
| **6f** | 6 | 5 | 7 | 4 | 5 | 4 | 8 | 8 | 4 | 3 | 7 | 7 | 9 | 2 | 3 | 8 | 6 | 10 | 3 | 3 | 3 | 1 | 2 | 5 | 11 | 4 | 3 | 1 | 1 | 2 | x |
| **6g** | 6 | 5 | 7 | 4 | 5 | 4 | 8 | 8 | 4 | 3 | 8 | 8 | 9 | 3 | 2 | 9 | 7 | 10 | 4 | 4 | 4 | 2 | 1 | 6 | 11 | 5 | 4 | 2 | 1 | 3 | 1 |
| **6h** | 6 | 7 | 5 | 8 | 8 | 8 | 4 | 6 | 8 | 7 | 3 | 3 | 5 | 6 | 7 | 4 | 2 | 6 | 7 | 6 | 5 | 5 | 6 | 1 | 7 | 6 | 5 | 5 | 5 | 4 | 4 |
| **6i** | 6 | 5 | 7 | 4 | 5 | 4 | 8 | 8 | 4 | 3 | 9 | 9 | 9 | 2 | 4 | 10 | 9 | 10 | 1 | 3 | 3 | 3 | 5 | 8 | 11 | 1 | 4 | 4 | 5 | 4 | 4 |
| **7a** | 7 | 6 | 8 | 5 | 6 | 5 | 9 | 9 | 5 | 4 | 8 | 8 | 10 | 3 | 5 | 9 | 7 | 11 | 2 | 3 | 3 | 4 | 5 | 6 | 12 | 1 | 2 | 4 | 4 | 2 | 3 |
| **7b** | 7 | 6 | 8 | 5 | 6 | 5 | 8 | 9 | 5 | 4 | 7 | 7 | 9 | 3 | 5 | 8 | 6 | 10 | 3 | 2 | 2 | 3 | 4 | 5 | 12 | 2 | 1 | 3 | 3 | 1 | 2 |
| **7c** | 7 | 6 | 8 | 5 | 6 | 5 | 7 | 9 | 5 | 4 | 6 | 6 | 8 | 3 | 4 | 7 | 5 | 9 | 4 | 3 | 2 | 2 | 3 | 1 | 10 | 4 | 3 | 2 | 2 | 1 | 1 |
| **7d** | 7 | 6 | 8 | 5 | 6 | 5 | 9 | 9 | 5 | 4 | 8 | 8 | 10 | 3 | 4 | 9 | 7 | 11 | 4 | 4 | 4 | 2 | 3 | 6 | 12 | 5 | 4 | 2 | 2 | 3 | 1 |
| **7e** | 7 | 6 | 8 | 5 | 6 | 5 | 9 | 9 | 5 | 4 | 9 | 9 | 10 | 4 | 3 | 10 | 8 | 11 | 5 | 5 | 5 | 3 | 2 | 7 | 12 | 6 | 5 | 3 | 2 | 4 | 2 |
| **7f** | 7 | 6 | 8 | 5 | 6 | 5 | 9 | 9 | 5 | 4 | 6 | 6 | 8 | 3 | 4 | 7 | 5 | 9 | 4 | 3 | 2 | 2 | 3 | 4 | 11 | 3 | 2 | 2 | 2 | 1 | 1 |
| **7g** | 7 | 8 | 6 | 7 | 8 | 7 | 5 | 7 | 7 | 6 | 4 | 4 | 6 | 5 | 6 | 5 | 3 | 7 | 6 | 5 | 4 | 4 | 5 | 2 | 8 | 5 | 4 | 4 | 4 | 3 | 3 |
| **7h** | 7 | 6 | 8 | 5 | 6 | 5 | 9 | 9 | 5 | 4 | 7 | 7 | 9 | 4 | 3 | 8 | 6 | 10 | 5 | 4 | 3 | 3 | 2 | 5 | 11 | 4 | 3 | 3 | 2 | 2 | 2 |
| **7i** | 7 | 6 | 8 | 5 | 6 | 5 | 9 | 9 | 5 | 4 | 10 | 10 | 10 | 3 | 5 | 11 | 9 | 11 | 2 | 4 | 4 | 4 | 6 | 8 | 12 | 1 | 4 | 5 | 6 | 4 | 5 |
| **8a** | 8 | 7 | 9 | 6 | 7 | 6 | 10 | 10 | 6 | 5 | 9 | 9 | 11 | 4 | 6 | 10 | 8 | 12 | 3 | 4 | 5 | 5 | 6 | 7 | 13 | 2 | 3 | 5 | 5 | 3 | 4 |
| **8b** | 8 | 7 | 9 | 6 | 7 | 6 | 10 | 10 | 6 | 5 | 9 | 9 | 11 | 4 | 6 | 10 | 8 | 12 | 3 | 4 | 5 | 5 | 6 | 7 | 13 | 2 | 3 | 5 | 5 | 3 | 4 |
| **8c** | 8 | 7 | 9 | 6 | 7 | 6 | 10 | 10 | 6 | 5 | 9 | 9 | 11 | 4 | 6 | 10 | 8 | 12 | 3 | 4 | 5 | 5 | 6 | 7 | 13 | 2 | 3 | 5 | 5 | 3 | 4 |
| **8d** | 8 | 7 | 9 | 6 | 7 | 6 | 10 | 10 | 6 | 5 | 9 | 9 | 11 | 4 | 6 | 10 | 8 | 12 | 3 | 4 | 5 | 5 | 6 | 7 | 13 | 2 | 3 | 5 | 5 | 3 | 4 |
| **8e** | 8 | 7 | 9 | 6 | 7 | 6 | 10 | 10 | 6 | 5 | 9 | 9 | 11 | 4 | 6 | 10 | 8 | 12 | 3 | 4 | 5 | 5 | 6 | 7 | 13 | 2 | 3 | 5 | 5 | 3 | 4 |
| **8f** | 8 | 7 | 9 | 6 | 7 | 6 | 10 | 10 | 6 | 5 | 9 | 9 | 11 | 4 | 6 | 10 | 8 | 12 | 3 | 4 | 5 | 5 | 6 | 7 | 13 | 2 | 3 | 5 | 5 | 3 | 4 |
| **8g** | 8 | 7 | 9 | 6 | 7 | 6 | 8 | 10 | 6 | 5 | 7 | 7 | 9 | 4 | 5 | 8 | 6 | 10 | 5 | 4 | 3 | 3 | 4 | 5 | 11 | 4 | 3 | 3 | 3 | 2 | 2 |
| **8h** | 8 | 7 | 9 | 6 | 7 | 6 | 10 | 10 | 6 | 5 | 9 | 9 | 11 | 4 | 5 | 10 | 8 | 12 | 5 | 5 | 5 | 3 | 4 | 7 | 13 | 6 | 5 | 3 | 3 | 4 | 2 |
| **8i** | 8 | 7 | 9 | 6 | 7 | 6 | 10 | 10 | 6 | 5 | 10 | 10 | 11 | 5 | 5 | 11 | 9 | 12 | 6 | 6 | 6 | 4 | 3 | 8 | 13 | 7 | 6 | 4 | 3 | 5 | 3 |
| **8j** | 8 | 7 | 7 | 6 | 7 | 6 | 6 | 8 | 6 | 5 | 5 | 5 | 7 | 4 | 4 | 6 | 4 | 8 | 5 | 4 | 3 | 3 | 4 | 3 | 9 | 4 | 3 | 3 | 3 | 2 | 2 |
| **8k** | 8 | 7 | 9 | 6 | 7 | 6 | 9 | 10 | 6 | 5 | 8 | 8 | 10 | 5 | 4 | 9 | 7 | 11 | 6 | 5 | 4 | 4 | 3 | 6 | 12 | 5 | 4 | 4 | 3 | 5 | 3 |
| **9a** | 9 | 8 | 10 | 7 | 8 | 7 | 11 | 11 | 7 | 6 | 10 | 10 | 12 | 5 | 6 | 11 | 9 | 13 | 6 | 6 | 6 | 4 | 5 | 8 | 14 | 7 | 6 | 4 | 4 | 5 | 3 |
| **9b** | 9 | 8 | 10 | 7 | 8 | 7 | 11 | 11 | 7 | 6 | 11 | 11 | 12 | 6 | 5 | 12 | 10 | 13 | 7 | 7 | 7 | 5 | 4 | 9 | 14 | 8 | 7 | 5 | 4 | 6 | 4 |
| **10a** | 10 | 9 | 11 | 8 | 9 | 8 | 12 | 12 | 8 | 7 | 11 | 11 | 13 | 6 | 6 | 12 | 10 | 14 | 7 | 7 | 7 | 5 | 5 | 9 | 15 | 8 | 7 | 5 | 5 | 6 | 4 |

| | 6g | 6h | 6i | 7a | 7b | 7c | 7d | 7e | 7f | 7g | 7h | 7i | 8a | 8b | 8c | 8d | 8e | 8f | 8g | 8h | 8i | 8j | 8k | 9a | 9b | 10a |
|---|---|---|---|---|---|---|---|---|---|---|---|---|---|---|---|---|---|---|---|---|---|---|---|---|---|---|
| **c** | 6 | 6 | 6 | 7 | 7 | 7 | 7 | 7 | 7 | 7 | 7 | 7 | 8 | 8 | 8 | 8 | 8 | 8 | 8 | 8 | 8 | 8 | 8 | 9 | 9 | 10 |
| **1a** | 5 | 7 | 5 | 6 | 6 | 6 | 6 | 6 | 6 | 8 | 6 | 6 | 7 | 7 | 7 | 7 | 7 | 7 | 7 | 7 | 7 | 7 | 7 | 8 | 8 | 9 |
| **1b** | 7 | 5 | 7 | 8 | 8 | 8 | 8 | 8 | 8 | 6 | 8 | 8 | 9 | 9 | 9 | 9 | 9 | 9 | 9 | 9 | 9 | 7 | 9 | 10 | 10 | 11 |
| **2a** | 4 | 8 | 4 | 5 | 5 | 5 | 5 | 5 | 5 | 7 | 5 | 5 | 6 | 6 | 6 | 6 | 6 | 6 | 6 | 6 | 6 | 6 | 6 | 7 | 7 | 8 |
| **2b** | 5 | 8 | 5 | 6 | 6 | 6 | 6 | 6 | 6 | 8 | 6 | 6 | 7 | 7 | 7 | 7 | 7 | 7 | 7 | 7 | 7 | 7 | 7 | 8 | 8 | 9 |
| **2c** | 4 | 8 | 4 | 5 | 5 | 5 | 5 | 5 | 5 | 7 | 5 | 5 | 6 | 6 | 6 | 6 | 6 | 6 | 6 | 6 | 6 | 6 | 6 | 7 | 7 | 8 |
| **2d** | 8 | 4 | 8 | 9 | 8 | 7 | 9 | 9 | 9 | 5 | 9 | 9 | 10 | 10 | 10 | 10 | 10 | 10 | 8 | 10 | 10 | 6 | 9 | 11 | 11 | 12 |
| **2e** | 8 | 6 | 8 | 9 | 9 | 9 | 9 | 9 | 9 | 7 | 9 | 9 | 10 | 10 | 10 | 10 | 10 | 10 | 10 | 10 | 10 | 8 | 10 | 11 | 11 | 12 |
| **3a** | 4 | 8 | 4 | 5 | 5 | 5 | 5 | 5 | 5 | 7 | 5 | 5 | 6 | 6 | 6 | 6 | 6 | 6 | 6 | 6 | 6 | 6 | 6 | 7 | 7 | 8 |
| **3b** | 3 | 7 | 3 | 4 | 4 | 4 | 4 | 4 | 4 | 6 | 4 | 4 | 5 | 5 | 5 | 5 | 5 | 5 | 5 | 5 | 5 | 5 | 5 | 6 | 6 | 7 |
| **3c** | 8 | 3 | 9 | 8 | 7 | 6 | 8 | 9 | 6 | 4 | 7 | 10 | 9 | 9 | 9 | 9 | 9 | 9 | 7 | 9 | 10 | 5 | 8 | 10 | 11 | 11 |
| **3d** | 8 | 3 | 9 | 8 | 7 | 6 | 8 | 9 | 6 | 4 | 7 | 10 | 9 | 9 | 9 | 9 | 9 | 9 | 7 | 9 | 10 | 5 | 8 | 10 | 11 | 11 |
| **3e** | 9 | 5 | 9 | 10 | 9 | 8 | 10 | 10 | 8 | 6 | 9 | 10 | 11 | 11 | 11 | 11 | 11 | 11 | 9 | 11 | 11 | 7 | 10 | 12 | 12 | 13 |
| **4a** | 3 | 6 | 2 | 3 | 3 | 3 | 3 | 4 | 3 | 5 | 4 | 3 | 4 | 4 | 4 | 4 | 4 | 4 | 4 | 4 | 5 | 4 | 5 | 5 | 6 | 6 |
| **4b** | 2 | 7 | 4 | 5 | 5 | 4 | 4 | 3 | 4 | 6 | 3 | 5 | 6 | 6 | 6 | 6 | 6 | 6 | 5 | 5 | 4 | 5 | 4 | 6 | 5 | 6 |
| **4c** | 9 | 4 | 10 | 9 | 8 | 7 | 9 | 10 | 7 | 5 | 8 | 11 | 10 | 10 | 10 | 10 | 10 | 10 | 8 | 10 | 11 | 6 | 9 | 11 | 12 | 12 |
| **4d** | 7 | 2 | 9 | 7 | 6 | 5 | 7 | 8 | 5 | 3 | 6 | 9 | 8 | 8 | 8 | 8 | 8 | 8 | 6 | 8 | 9 | 4 | 7 | 9 | 10 | 10 |
| **4e** | 10 | 6 | 10 | 11 | 10 | 9 | 11 | 11 | 9 | 7 | 10 | 11 | 12 | 12 | 12 | 12 | 12 | 12 | 10 | 12 | 12 | 8 | 11 | 13 | 13 | 14 |
| **5a** | 4 | 7 | 1 | 2 | 3 | 4 | 4 | 5 | 4 | 6 | 5 | 2 | 3 | 3 | 3 | 3 | 3 | 3 | 5 | 5 | 6 | 5 | 6 | 6 | 7 | 7 |
| **5b** | 4 | 6 | 3 | 3 | 2 | 3 | 4 | 5 | 3 | 5 | 4 | 4 | 4 | 4 | 4 | 4 | 4 | 4 | 4 | 5 | 6 | 4 | 5 | 6 | 7 | 7 |
| **5c** | 4 | 5 | 3 | 3 | 2 | 2 | 4 | 5 | 2 | 4 | 3 | 4 | 5 | 5 | 5 | 5 | 5 | 5 | 3 | 5 | 6 | 3 | 4 | 6 | 7 | 7 |
| **5d** | 2 | 5 | 3 | 4 | 3 | 2 | 2 | 3 | 2 | 4 | 3 | 4 | 5 | 5 | 5 | 5 | 5 | 5 | 3 | 3 | 4 | 3 | 4 | 4 | 5 | 5 |
| **5e** | 1 | 6 | 5 | 5 | 4 | 3 | 3 | 2 | 3 | 5 | 2 | 6 | 6 | 6 | 6 | 6 | 6 | 6 | 4 | 4 | 3 | 4 | 3 | 5 | 4 | 5 |
| **5f** | 6 | 1 | 8 | 6 | 5 | 1 | 6 | 7 | 4 | 2 | 5 | 8 | 7 | 7 | 7 | 7 | 7 | 7 | 5 | 7 | 8 | 3 | 6 | 8 | 9 | 9 |
| **5g** | 11 | 7 | 11 | 12 | 12 | 10 | 12 | 12 | 11 | 8 | 11 | 12 | 13 | 13 | 13 | 13 | 13 | 13 | 11 | 13 | 13 | 9 | 12 | 14 | 14 | 15 |
| **6a** | 5 | 6 | 1 | 1 | 2 | 4 | 5 | 6 | 3 | 5 | 4 | 1 | 2 | 2 | 2 | 2 | 2 | 2 | 4 | 6 | 7 | 4 | 5 | 7 | 8 | 8 |
| **6b** | 4 | 5 | 4 | 2 | 1 | 3 | 4 | 5 | 2 | 4 | 3 | 4 | 3 | 3 | 3 | 3 | 3 | 3 | 3 | 5 | 6 | 3 | 4 | 6 | 7 | 7 |
| **6c** | 2 | 5 | 4 | 4 | 3 | 2 | 2 | 3 | 2 | 4 | 3 | 5 | 5 | 5 | 5 | 5 | 5 | 5 | 3 | 3 | 4 | 3 | 4 | 4 | 5 | 5 |
| **6d** | 1 | 5 | 5 | 4 | 3 | 2 | 2 | 2 | 2 | 4 | 2 | 6 | 5 | 5 | 5 | 5 | 5 | 5 | 3 | 3 | 3 | 3 | 3 | 4 | 4 | 5 |
| **6e** | 3 | 4 | 4 | 2 | 1 | 1 | 3 | 4 | 1 | 3 | 2 | 4 | 3 | 3 | 3 | 3 | 3 | 3 | 2 | 4 | 5 | 2 | 5 | 5 | 6 | 6 |
| **6f** | 1 | 4 | 4 | 3 | 2 | 1 | 1 | 2 | 1 | 3 | 2 | 5 | 4 | 4 | 4 | 4 | 4 | 4 | 2 | 2 | 3 | 2 | 3 | 3 | 4 | 4 |
| **6g** | x | 5 | 6 | 4 | 3 | 2 | 2 | 1 | 2 | 4 | 1 | 6 | 5 | 5 | 5 | 5 | 5 | 5 | 3 | 3 | 2 | 3 | 2 | 4 | 3 | 4 |
| **6h** | 5 | x | 7 | 5 | 4 | 3 | 5 | 6 | 3 | 1 | 4 | 7 | 6 | 6 | 6 | 6 | 6 | 6 | 4 | 6 | 7 | 2 | 5 | 7 | 8 | 8 |
| **6i** | 6 | 7 | x | 2 | 3 | 5 | 5 | 6 | 4 | 6 | 5 | 2 | 3 | 3 | 3 | 3 | 3 | 3 | 5 | 6 | 7 | 5 | 6 | 7 | 8 | 8 |
| **7a** | 4 | 5 | 2 | x | 1 | 3 | 4 | 5 | 2 | 4 | 3 | 2 | 1 | 1 | 1 | 1 | 1 | 1 | 3 | 5 | 6 | 3 | 4 | 6 | 7 | 7 |
| **7b** | 3 | 4 | 3 | 1 | x | 2 | 3 | 4 | 1 | 3 | 2 | 3 | 2 | 2 | 2 | 2 | 2 | 2 | 2 | 4 | 5 | 2 | 3 | 5 | 6 | 6 |
| **7c** | 2 | 3 | 5 | 3 | 2 | x | 2 | 3 | 2 | 2 | 3 | 5 | 4 | 4 | 4 | 4 | 4 | 4 | 3 | 3 | 4 | 1 | 4 | 4 | 5 | 5 |
| **7d** | 2 | 5 | 5 | 4 | 3 | 2 | x | 3 | 2 | 4 | 3 | 6 | 5 | 5 | 5 | 5 | 5 | 5 | 3 | 1 | 4 | 3 | 4 | 2 | 4 | 3 |
| **7e** | 1 | 6 | 6 | 5 | 4 | 3 | 3 | x | 3 | 5 | 2 | 7 | 6 | 6 | 6 | 6 | 6 | 6 | 4 | 4 | 1 | 4 | 3 | 4 | 2 | 3 |
| **7f** | 2 | 3 | 4 | 2 | 1 | 2 | 2 | 3 | x | 2 | 1 | 4 | 3 | 3 | 3 | 3 | 3 | 3 | 1 | 3 | 4 | 1 | 2 | 4 | 5 | 5 |
| **7g** | 4 | 1 | 6 | 4 | 3 | 2 | 4 | 5 | 2 | x | 3 | 6 | 5 | 5 | 5 | 5 | 5 | 5 | 3 | 5 | 6 | 1 | 4 | 5 | 6 | 6 |
| **7h** | 1 | 4 | 5 | 3 | 2 | 3 | 3 | 2 | 1 | 3 | x | 5 | 4 | 4 | 4 | 4 | 4 | 4 | 2 | 4 | 3 | 2 | 1 | 5 | 4 | 5 |
| **7i** | 6 | 7 | 2 | 2 | 3 | 5 | 6 | 7 | 4 | 6 | 5 | x | 3 | 2 | 1 | 3 | 3 | 3 | 5 | 7 | 8 | 5 | 6 | 8 | 9 | 9 |
| **8a** | 5 | 6 | 3 | 1 | 2 | 4 | 5 | 6 | 3 | 5 | 4 | 3 | x | 2 | 2 | 2 | 2 | 2 | 4 | 6 | 7 | 4 | 5 | 7 | 8 | 8 |
| **8b** | 5 | 6 | 3 | 1 | 2 | 4 | 5 | 6 | 3 | 5 | 4 | 2 | 2 | x | 1 | 2 | 2 | 2 | 4 | 6 | 7 | 4 | 5 | 7 | 8 | 8 |
| **8c** | 5 | 6 | 3 | 1 | 2 | 4 | 5 | 6 | 3 | 5 | 4 | 1 | 2 | 1 | x | 2 | 2 | 2 | 4 | 6 | 7 | 4 | 5 | 7 | 8 | 8 |
| **8d** | 5 | 6 | 3 | 1 | 2 | 4 | 5 | 6 | 3 | 5 | 4 | 3 | 2 | 2 | 2 | x | 2 | 2 | 4 | 6 | 7 | 4 | 5 | 7 | 8 | 8 |
| **8e** | 5 | 6 | 3 | 1 | 2 | 4 | 5 | 6 | 3 | 5 | 4 | 3 | 2 | 2 | 2 | 2 | x | 2 | 4 | 6 | 7 | 4 | 5 | 7 | 8 | 8 |
| **8f** | 5 | 6 | 3 | 1 | 2 | 4 | 5 | 6 | 3 | 5 | 4 | 3 | 2 | 2 | 2 | 2 | 2 | x | 4 | 6 | 7 | 4 | 5 | 7 | 8 | 8 |
| **8g** | 3 | 4 | 5 | 3 | 2 | 3 | 3 | 4 | 1 | 3 | 2 | 5 | 4 | 4 | 4 | 4 | 4 | 4 | x | 4 | 5 | 2 | 3 | 5 | 6 | 6 |
| **8h** | 3 | 6 | 6 | 5 | 4 | 3 | 1 | 4 | 3 | 5 | 4 | 7 | 6 | 6 | 6 | 6 | 6 | 6 | 4 | x | 4 | 4 | 5 | 1 | 3 | 2 |
| **8i** | 2 | 7 | 7 | 6 | 5 | 4 | 4 | 1 | 4 | 6 | 3 | 8 | 7 | 7 | 7 | 7 | 7 | 7 | 5 | 4 | x | 5 | 4 | 3 | 1 | 2 |
| **8j** | 3 | 2 | 5 | 3 | 2 | 1 | 3 | 4 | 1 | 1 | 2 | 5 | 4 | 4 | 4 | 4 | 4 | 4 | 2 | 4 | 5 | x | 3 | 5 | 6 | 6 |
| **8k** | 2 | 5 | 6 | 4 | 3 | 4 | 4 | 3 | 2 | 4 | 1 | 6 | 5 | 5 | 5 | 5 | 5 | 5 | 3 | 5 | 4 | 3 | x | 6 | 5 | 6 |
| **9a** | 4 | 7 | 7 | 6 | 5 | 4 | 2 | 4 | 4 | 5 | 5 | 8 | 7 | 7 | 7 | 7 | 7 | 7 | 5 | 1 | 3 | 5 | 6 | x | 2 | 1 |
| **9b** | 3 | 8 | 8 | 7 | 6 | 5 | 4 | 2 | 5 | 6 | 4 | 9 | 8 | 8 | 8 | 8 | 8 | 8 | 6 | 3 | 1 | 6 | 5 | 2 | x | 1 |
| **10a** | 4 | 8 | 8 | 7 | 6 | 5 | 3 | 3 | 5 | 6 | 5 | 9 | 8 | 8 | 8 | 8 | 8 | 8 | 6 | 2 | 2 | 6 | 6 | 1 | 1 | x |

Palenque Palace Path Matrix Phase 3

| | c | 1a | 1b | 1c | 1d | 1e | 2a | 2b | 2c | 2d | 2e | 2f | 2g | 3a | 3b | 3c | 3d | 3e | 3f | 3g | 3h | 3i | 4a | 4b | 4c | 4d | 4e | 4f | 4g | 4h | 4i | 4j | 4k | 4L | 4m | 4n | 4o | 4p | 4q | 5a | 5b | 5c | 5d | 5e | 5f | 5g | 5h | 5i | 5j | 5k | 5L | 6a | 6b | 6c | 6d | 6e | 6f |
|---|---|---|---|---|---|---|---|---|---|---|---|---|---|---|---|---|---|---|---|---|---|---|---|---|---|---|---|---|---|---|---|---|---|---|---|---|---|---|---|---|---|---|---|---|---|---|---|---|---|---|---|---|---|---|---|---|---|
| **c** | x | 1 | 1 | 1 | 1 | 1 | 2 | 2 | 2 | 2 | 2 | 2 | 2 | 3 | 3 | 3 | 3 | 3 | 3 | 3 | 3 | 3 | 4 | 4 | 4 | 4 | 4 | 4 | 4 | 4 | 4 | 4 | 4 | 4 | 4 | 4 | 4 | 4 | 4 | 5 | 5 | 5 | 5 | 5 | 5 | 5 | 5 | 5 | 5 | 5 | 5 | 6 | 6 | 6 | 6 | 6 | 6 |
| **1a** | 1 | x | 1 | 2 | 2 | 2 | 1 | 1 | 1 | 2 | 3 | 3 | 3 | 2 | 2 | 2 | 3 | 4 | 4 | 4 | 3 | 4 | | 3 | 3 | 3 | 3 | 4 | 4 | 3 | 4 | 3 | 4 | 4 | 5 | 5 | 5 | 5 | 4 | 4 | 4 | 4 | 4 | 4 | 5 | 5 | 5 | 6 | 5 | 6 | 6 | 5 | 5 | 6 | 7 | 6 | 7 |
| **1b** | 1 | 1 | x | 2 | 2 | 2 | 2 | 2 | 2 | 1 | 3 | 3 | 3 | 3 | 3 | 3 | 3 | 4 | 4 | 4 | 3 | 4 | 4 | 4 | 4 | 4 | 4 | 4 | 4 | 4 | 4 | 4 | 4 | 4 | 5 | 5 | 5 | 5 | 4 | 5 | 5 | 5 | 5 | 5 | 5 | 5 | 5 | 6 | 5 | 6 | 6 | 6 | 6 | 6 | 7 | 6 | 7 |
| **1c** | 1 | 2 | 2 | x | 1 | 2 | 2 | 3 | 1 | 1 | 1 | 3 | 3 | 3 | 3 | 4 | 2 | 2 | 4 | 4 | 2 | 4 | 4 | 4 | 4 | 4 | 3 | 3 | 3 | 4 | 3 | 5 | 3 | 3 | 5 | 5 | 5 | 3 | 3 | 5 | 5 | 5 | 5 | 4 | 4 | 4 | 4 | 6 | 4 | 4 | 6 | 5 | 5 | 5 | 6 | 5 | 5 |
| **1d** | 1 | 2 | 2 | 1 | x | 2 | 3 | 3 | 2 | 2 | 1 | 3 | 3 | 4 | 4 | 4 | 3 | 2 | 4 | 4 | 2 | 4 | 5 | 5 | 5 | 5 | 4 | 4 | 4 | 5 | 4 | 5 | 4 | 3 | 5 | 5 | 5 | 3 | 3 | 6 | 6 | 6 | 6 | 4 | 4 | 4 | 4 | 6 | 5 | 4 | 6 | 5 | 5 | 5 | 6 | 6 | 5 |
| **1e** | 1 | 2 | 2 | 2 | 2 | x | 3 | 3 | 3 | 3 | 3 | 1 | 1 | 4 | 4 | 4 | 4 | 4 | 2 | 2 | 4 | 2 | 5 | 5 | 5 | 5 | 5 | 5 | 5 | 5 | 5 | 5 | 5 | 5 | 3 | 3 | 3 | 5 | 5 | 6 | 6 | 6 | 6 | 6 | 6 | 6 | 6 | 4 | 6 | 6 | 4 | 7 | 7 | 6 | 5 | 7 | 7 |
| **2a** | 2 | 1 | 2 | 2 | 3 | 3 | x | 2 | 1 | 1 | 2 | 4 | 4 | 1 | 1 | 3 | 2 | 3 | 5 | 5 | 2 | 5 | 2 | 2 | 2 | 2 | 2 | 3 | 3 | 2 | 3 | 3 | 3 | 3 | 6 | 6 | 6 | 4 | 3 | 3 | 3 | 3 | 3 | 3 | 4 | 4 | 4 | 7 | 4 | 5 | 7 | 4 | 4 | 5 | 6 | 5 | 6 |
| **2b** | 2 | 1 | 2 | 3 | 3 | 3 | 2 | x | 3 | 3 | 4 | 4 | 4 | 1 | 2 | 1 | 4 | 5 | 5 | 5 | 4 | 5 | 3 | 3 | 3 | 3 | 3 | 4 | 5 | 2 | 5 | 2 | 5 | 5 | 6 | 6 | 6 | 6 | 5 | 3 | 3 | 3 | 3 | 3 | 6 | 4 | 6 | 7 | 6 | 7 | 7 | 4 | 4 | 7 | 7 | 7 | 8 |
| **2c** | 2 | 1 | 2 | 1 | 2 | 3 | 1 | 3 | x | 1 | 1 | 4 | 4 | 2 | 2 | 4 | 1 | 2 | 5 | 5 | 1 | 5 | 3 | 3 | 3 | 3 | 2 | 2 | 2 | 3 | 2 | 4 | 2 | 2 | 6 | 6 | 6 | 3 | 2 | 4 | 4 | 4 | 4 | 3 | 3 | 3 | 3 | 6 | 3 | 4 | 7 | 4 | 4 | 4 | 5 | 4 | 5 |
| **2d** | 2 | 2 | 1 | 1 | 2 | 3 | 1 | 3 | 1 | x | 2 | 4 | 4 | 2 | 2 | 4 | 2 | 3 | 5 | 5 | 2 | 5 | 3 | 3 | 3 | 3 | 3 | 3 | 3 | 3 | 3 | 4 | 3 | 3 | 6 | 6 | 6 | 4 | 3 | 4 | 4 | 4 | 4 | 4 | 4 | 4 | 4 | 7 | 4 | 5 | 7 | 5 | 5 | 5 | 6 | 5 | 6 |
| **2e** | 2 | 3 | 3 | 1 | 1 | 3 | 2 | 4 | 1 | 2 | x | 4 | 4 | 3 | 3 | 5 | 2 | 1 | 5 | 5 | 1 | 5 | 4 | 4 | 4 | 4 | 3 | 3 | 3 | 4 | 3 | 5 | 3 | 3 | 6 | 6 | 6 | 2 | 2 | 5 | 5 | 5 | 5 | 3 | 3 | 3 | 3 | 6 | 4 | 3 | 7 | 4 | 4 | 4 | 5 | 5 | 4 |
| **2f** | 2 | 3 | 3 | 3 | 3 | 1 | 4 | 4 | 4 | 4 | 4 | x | 2 | 5 | 5 | 5 | 5 | 5 | 1 | 1 | 5 | 1 | 6 | 6 | 6 | 6 | 6 | 6 | 6 | 6 | 6 | 6 | 6 | 6 | 2 | 2 | 2 | 6 | 6 | 7 | 7 | 7 | 7 | 7 | 7 | 7 | 6 | 3 | 7 | 7 | 3 | 8 | 8 | 5 | 4 | 8 | 8 |
| **2g** | 2 | 3 | 3 | 3 | 3 | 1 | 4 | 4 | 4 | 4 | 4 | 2 | x | 5 | 5 | 5 | 5 | 5 | 3 | 3 | 5 | 3 | 6 | 6 | 6 | 6 | 6 | 6 | 6 | 6 | 6 | 6 | 6 | 6 | 4 | 4 | 4 | 6 | 6 | 7 | 7 | 7 | 7 | 7 | 7 | 7 | 7 | 5 | 7 | 7 | 5 | 8 | 8 | 7 | 6 | 8 | 8 |
| **3a** | 3 | 2 | 3 | 3 | 4 | 4 | 1 | 1 | 2 | 2 | 3 | 5 | 5 | x | 1 | 2 | 3 | 4 | 6 | 6 | 3 | 6 | 2 | 2 | 2 | 2 | 2 | 3 | 4 | 1 | 4 | 2 | 4 | 4 | 7 | 7 | 7 | 5 | 4 | 2 | 2 | 2 | 2 | 2 | 5 | 3 | 5 | 7 | 5 | 6 | 8 | 3 | 3 | 6 | 7 | 6 | 7 |
| **3b** | 3 | 2 | 3 | 3 | 4 | 4 | 1 | 2 | 2 | 2 | 3 | 5 | 5 | 1 | x | 3 | 2 | 4 | 6 | 6 | 3 | 6 | 1 | 1 | 1 | 1 | 1 | 2 | 3 | 2 | 3 | 3 | 3 | 3 | 7 | 7 | 7 | 5 | 4 | 3 | 3 | 3 | 3 | 3 | 4 | 2 | 4 | 7 | 4 | 6 | 8 | 4 | 4 | 5 | 6 | 5 | 6 |
| **3c** | 3 | 2 | 3 | 4 | 4 | 4 | 3 | 1 | 4 | 4 | 5 | 5 | 5 | 2 | 3 | x | 4 | 6 | 6 | 6 | 4 | 6 | 4 | 4 | 4 | 4 | 4 | 5 | 5 | 1 | 4 | 1 | 5 | 3 | 7 | 7 | 7 | 7 | 5 | 2 | 2 | 2 | 2 | 2 | 4 | 3 | 4 | 7 | 6 | 8 | 8 | 3 | 3 | 5 | 6 | 7 | 8 |
| **3d** | 3 | 3 | 3 | 2 | 3 | 4 | 2 | 4 | 1 | 2 | 2 | 5 | 5 | 3 | 2 | 4 | x | 3 | 6 | 6 | 2 | 6 | 3 | 3 | 3 | 3 | 1 | 1 | 1 | 3 | 1 | 4 | 1 | 1 | 6 | 7 | 7 | 4 | 3 | 4 | 4 | 4 | 4 | 2 | 2 | 2 | 2 | 5 | 2 | 5 | 8 | 3 | 3 | 3 | 4 | 3 | 4 |
| **3e** | 3 | 4 | 4 | 2 | 2 | 4 | 3 | 5 | 2 | 3 | 1 | 5 | 5 | 4 | 4 | 6 | 3 | x | 6 | 6 | 2 | 6 | 5 | 5 | 5 | 5 | 4 | 4 | 4 | 5 | 4 | 6 | 4 | 3 | 7 | 7 | 7 | 1 | 3 | 6 | 6 | 6 | 6 | 4 | 4 | 4 | 4 | 7 | 5 | 2 | 8 | 5 | 5 | 5 | 6 | 4 | 3 |
| **3f** | 3 | 4 | 4 | 4 | 4 | 2 | 5 | 5 | 5 | 5 | 5 | 1 | 3 | 6 | 6 | 6 | 6 | 6 | x | 2 | 6 | 2 | 7 | 7 | 7 | 7 | 7 | 7 | 7 | 7 | 6 | 7 | 7 | 6 | 1 | 1 | 3 | 7 | 7 | 8 | 8 | 8 | 8 | 7 | 7 | 7 | 5 | 2 | 8 | 8 | 4 | 8 | 8 | 4 | 3 | 9 | 9 |
| **3g** | 3 | 4 | 4 | 4 | 4 | 2 | 5 | 5 | 5 | 5 | 5 | 1 | 3 | 6 | 6 | 6 | 6 | 6 | 2 | x | 6 | 2 | 7 | 7 | 7 | 7 | 7 | 7 | 7 | 7 | 6 | 7 | 7 | 6 | 1 | 1 | 3 | 7 | 7 | 8 | 8 | 8 | 8 | 7 | 7 | 7 | 5 | 2 | 8 | 8 | 4 | 8 | 8 | 4 | 3 | 9 | 9 |
| **3h** | 3 | 3 | 3 | 2 | 2 | 4 | 2 | 4 | 1 | 2 | 1 | 5 | 5 | 3 | 3 | 4 | 2 | 2 | 6 | 6 | x | 6 | 4 | 4 | 4 | 3 | 3 | 3 | 3 | 3 | 3 | 4 | 3 | 1 | 6 | 7 | 7 | 3 | 1 | 4 | 4 | 4 | 4 | 2 | 2 | 2 | 2 | 5 | 4 | 4 | 8 | 3 | 3 | 3 | 4 | 5 | 5 |
| **3i** | 3 | 4 | 4 | 4 | 4 | 2 | 5 | 5 | 5 | 5 | 5 | 1 | 3 | 6 | 6 | 6 | 6 | 6 | 2 | 2 | 6 | x | 7 | 7 | 7 | 7 | 7 | 7 | 7 | 7 | 7 | 7 | 7 | 7 | 3 | 3 | 1 | 7 | 7 | 8 | 8 | 8 | 8 | 8 | 8 | 8 | 7 | 4 | 8 | 8 | 2 | 9 | 9 | 6 | 5 | 9 | 9 |
| **4a** | 4 | 3 | 4 | 4 | 5 | 5 | 2 | 3 | 3 | 3 | 4 | 6 | 6 | 2 | 1 | 4 | 3 | 5 | 7 | 7 | 4 | 7 | x | 2 | 2 | 2 | 2 | 3 | 4 | 3 | 4 | 4 | 4 | 4 | 8 | 8 | 8 | 6 | 5 | 4 | 4 | 4 | 4 | 4 | 5 | 3 | 5 | 8 | 5 | 7 | 9 | 5 | 5 | 6 | 7 | 6 | 7 |
| **4b** | 4 | 3 | 4 | 4 | 5 | 5 | 2 | 3 | 3 | 3 | 4 | 6 | 6 | 2 | 1 | 4 | 3 | 5 | 7 | 7 | 4 | 7 | 2 | x | 2 | 2 | 2 | 3 | 4 | 3 | 4 | 4 | 4 | 4 | 8 | 8 | 8 | 6 | 5 | 4 | 4 | 4 | 4 | 4 | 5 | 3 | 5 | 8 | 5 | 7 | 9 | 5 | 5 | 6 | 7 | 6 | 7 |
| **4c** | 4 | 3 | 4 | 4 | 5 | 5 | 2 | 3 | 3 | 3 | 4 | 6 | 6 | 2 | 1 | 4 | 3 | 5 | 7 | 7 | 4 | 7 | 2 | 2 | x | 1 | 2 | 3 | 4 | 3 | 3 | 4 | 4 | 3 | 8 | 8 | 8 | 6 | 5 | 4 | 4 | 4 | 4 | 3 | 4 | 2 | 4 | 7 | 5 | 7 | 9 | 4 | 4 | 5 | 6 | 6 | 7 |
| **4d** | 4 | 3 | 4 | 4 | 5 | 5 | 2 | 3 | 3 | 3 | 4 | 6 | 6 | 2 | 1 | 4 | 3 | 5 | 7 | 7 | 3 | 7 | 2 | 2 | 1 | x | 2 | 3 | 4 | 3 | 2 | 4 | 4 | 2 | 7 | 8 | 8 | 6 | 4 | 4 | 4 | 4 | 4 | 2 | 3 | 1 | 3 | 6 | 5 | 7 | 9 | 3 | 3 | 4 | 5 | 6 | 7 |
| **4e** | 4 | 3 | 4 | 3 | 4 | 5 | 2 | 3 | 2 | 3 | 3 | 6 | 6 | 2 | 1 | 4 | 1 | 4 | 7 | 7 | 3 | 7 | 2 | 2 | 2 | 2 | x | 1 | 2 | 3 | 2 | 4 | 2 | 2 | 7 | 8 | 8 | 5 | 4 | 4 | 4 | 4 | 4 | 3 | 3 | 3 | 3 | 6 | 3 | 6 | 9 | 4 | 4 | 4 | 5 | 4 | 5 |
| **4f** | 4 | 4 | 4 | 3 | 4 | 5 | 3 | 4 | 2 | 3 | 3 | 6 | 6 | 3 | 2 | 5 | 1 | 4 | 7 | 7 | 3 | 7 | 3 | 3 | 3 | 3 | 1 | x | 1 | 4 | 2 | 5 | 2 | 2 | 7 | 8 | 8 | 5 | 4 | 5 | 5 | 5 | 5 | 3 | 3 | 3 | 3 | 6 | 3 | 6 | 9 | 4 | 4 | 4 | 5 | 4 | 5 |
| **4g** | 4 | 4 | 4 | 3 | 4 | 5 | 3 | 5 | 2 | 3 | 3 | 6 | 6 | 4 | 3 | 5 | 1 | 4 | 7 | 7 | 3 | 7 | 4 | 4 | 4 | 4 | 2 | 1 | x | 4 | 2 | 5 | 2 | 2 | 7 | 8 | 8 | 5 | 4 | 5 | 5 | 5 | 5 | 3 | 3 | 3 | 3 | 6 | 3 | 6 | 9 | 4 | 4 | 4 | 5 | 4 | 5 |
| **4h** | 4 | 3 | 4 | 4 | 5 | 5 | 2 | 2 | 3 | 3 | 4 | 6 | 6 | 1 | 2 | 1 | 3 | 5 | 7 | 7 | 3 | 7 | 3 | 3 | 3 | 3 | 3 | 4 | 4 | x | 3 | 1 | 4 | 2 | 7 | 8 | 8 | 6 | 4 | 1 | 1 | 1 | 1 | 1 | 3 | 2 | 3 | 6 | 5 | 7 | 9 | 2 | 2 | 4 | 5 | 6 | 7 |
| **4i** | 4 | 4 | 4 | 3 | 4 | 5 | 3 | 5 | 2 | 3 | 3 | 6 | 6 | 4 | 3 | 4 | 1 | 4 | 6 | 6 | 3 | 7 | 4 | 4 | 3 | 2 | 2 | 2 | 2 | 3 | x | 4 | 2 | 2 | 5 | 7 | 8 | 5 | 4 | 4 | 4 | 4 | 4 | 2 | 3 | 1 | 1 | 4 | 3 | 6 | 9 | 3 | 3 | 2 | 3 | 4 | 5 |
| **4j** | 4 | 3 | 4 | 5 | 5 | 5 | 3 | 2 | 4 | 4 | 5 | 6 | 6 | 2 | 3 | 1 | 4 | 6 | 7 | 7 | 4 | 7 | 4 | 4 | 4 | 4 | 4 | 5 | 5 | 1 | 4 | x | 5 | 3 | 8 | 8 | 8 | 7 | 5 | 2 | 2 | 2 | 2 | 2 | 4 | 3 | 4 | 7 | 6 | 8 | 9 | 3 | 3 | 5 | 6 | 7 | 8 |
| **4k** | 4 | 4 | 4 | 3 | 4 | 5 | 3 | 5 | 2 | 3 | 3 | 6 | 6 | 4 | 3 | 5 | 1 | 4 | 7 | 7 | 3 | 7 | 4 | 4 | 4 | 4 | 2 | 2 | 2 | 4 | 2 | 5 | x | 2 | 7 | 8 | 8 | 5 | 4 | 5 | 5 | 5 | 5 | 3 | 3 | 3 | 3 | 6 | 1 | 4 | 9 | 4 | 4 | 4 | 5 | 2 | 3 |
| **4L** | 4 | 4 | 4 | 3 | 3 | 5 | 3 | 5 | 2 | 3 | 3 | 6 | 6 | 4 | 3 | 3 | 1 | 3 | 6 | 6 | 1 | 7 | 4 | 4 | 3 | 2 | 2 | 2 | 2 | 2 | 2 | 3 | 2 | x | 5 | 7 | 8 | 4 | 2 | 3 | 3 | 3 | 3 | 1 | 1 | 1 | 1 | 4 | 3 | 5 | 9 | 2 | 2 | 2 | 3 | 4 | 5 |
| **4m** | 4 | 5 | 5 | 5 | 5 | 3 | 6 | 6 | 6 | 6 | 6 | 2 | 4 | 7 | 7 | 7 | 6 | 7 | 1 | 1 | 6 | 3 | 8 | 8 | 8 | 7 | 7 | 7 | 7 | 7 | 5 | 8 | 7 | 5 | x | 2 | 4 | 8 | 7 | 8 | 8 | 8 | 8 | 6 | 6 | 6 | 4 | 1 | 8 | 9 | 5 | 7 | 7 | 3 | 2 | 9 | 10 |
| **4n** | 4 | 5 | 5 | 5 | 5 | 3 | 6 | 6 | 6 | 6 | 6 | 2 | 4 | 7 | 7 | 7 | 7 | 7 | 1 | 1 | 7 | 3 | 8 | 8 | 8 | 8 | 8 | 8 | 8 | 8 | 7 | 8 | 8 | 7 | 2 | x | 4 | 8 | 8 | 9 | 9 | 9 | 9 | 8 | 8 | 8 | 6 | 3 | 9 | 9 | 5 | 9 | 9 | 5 | 4 | 10 | 10 |
| **4o** | 4 | 5 | 5 | 5 | 5 | 3 | 6 | 6 | 6 | 6 | 6 | 2 | 4 | 7 | 7 | 7 | 7 | 7 | 3 | 3 | 7 | 1 | 8 | 8 | 8 | 8 | 8 | 8 | 8 | 8 | 8 | 8 | 8 | 8 | 4 | 4 | x | 8 | 8 | 9 | 9 | 9 | 9 | 9 | 9 | 9 | 8 | 5 | 9 | 9 | 1 | 10 | 10 | 7 | 6 | 10 | 10 |
| **4p** | 4 | 5 | 5 | 3 | 3 | 5 | 4 | 6 | 3 | 4 | 2 | 6 | 6 | 5 | 5 | 7 | 4 | 1 | 7 | 7 | 3 | 7 | 6 | 6 | 6 | 6 | 5 | 5 | 5 | 6 | 5 | 7 | 5 | 4 | 8 | 8 | 8 | x | 4 | 7 | 7 | 7 | 7 | 5 | 5 | 5 | 5 | 8 | 4 | 1 | 9 | 6 | 6 | 6 | 7 | 3 | 2 |
| **4q** | 4 | 4 | 4 | 3 | 3 | 5 | 3 | 5 | 2 | 3 | 2 | 6 | 6 | 4 | 4 | 5 | 3 | 3 | 7 | 7 | 1 | 7 | 5 | 5 | 5 | 4 | 4 | 4 | 4 | 4 | 4 | 5 | 4 | 2 | 7 | 8 | 8 | 4 | x | 5 | 5 | 5 | 5 | 3 | 3 | 3 | 3 | 6 | 5 | 5 | 9 | 4 | 4 | 4 | 5 | 6 | 6 |
| **5a** | 5 | 4 | 5 | 5 | 6 | 6 | 3 | 3 | 4 | 4 | 5 | 7 | 7 | 2 | 3 | 2 | 4 | 6 | 8 | 8 | 4 | 8 | 4 | 4 | 4 | 4 | 4 | 5 | 5 | 1 | 4 | 2 | 5 | 3 | 8 | 9 | 9 | 7 | 5 | x | 2 | 2 | 2 | 2 | 4 | 3 | 4 | 7 | 6 | 8 | 10 | 3 | 3 | 5 | 6 | 7 | 8 |
| **5b** | 5 | 4 | 5 | 5 | 6 | 6 | 3 | 3 | 4 | 4 | 5 | 7 | 7 | 2 | 3 | 2 | 4 | 6 | 8 | 8 | 4 | 8 | 4 | 4 | 4 | 4 | 4 | 5 | 5 | 1 | 4 | 2 | 5 | 3 | 8 | 9 | 9 | 7 | 5 | 2 | x | 2 | 2 | 2 | 4 | 3 | 4 | 7 | 6 | 8 | 10 | 3 | 3 | 5 | 6 | 7 | 8 |
| **5c** | 5 | 4 | 5 | 5 | 6 | 6 | 3 | 3 | 4 | 4 | 5 | 7 | 7 | 2 | 3 | 2 | 4 | 6 | 8 | 8 | 4 | 8 | 4 | 4 | 4 | 4 | 4 | 5 | 5 | 1 | 4 | 2 | 5 | 3 | 8 | 9 | 9 | 7 | 5 | 2 | 2 | x | 2 | 2 | 4 | 3 | 4 | 7 | 6 | 8 | 10 | 3 | 3 | 5 | 6 | 7 | 8 |
| **5d** | 5 | 4 | 5 | 5 | 6 | 6 | 3 | 3 | 4 | 4 | 5 | 7 | 7 | 2 | 3 | 2 | 4 | 6 | 8 | 8 | 4 | 8 | 4 | 4 | 4 | 4 | 4 | 5 | 5 | 1 | 4 | 2 | 5 | 3 | 8 | 9 | 9 | 7 | 5 | 2 | 2 | 2 | x | 2 | 4 | 3 | 4 | 7 | 6 | 8 | 10 | 3 | 3 | 5 | 6 | 7 | 8 |
| **5e** | 5 | 4 | 5 | 4 | 4 | 6 | 3 | 3 | 3 | 4 | 3 | 7 | 7 | 2 | 3 | 2 | 2 | 4 | 7 | 7 | 2 | 8 | 4 | 4 | 3 | 2 | 3 | 3 | 3 | 1 | 2 | 2 | 3 | 1 | 6 | 8 | 9 | 5 | 3 | 2 | 2 | 2 | 2 | x | 2 | 1 | 2 | 5 | 4 | 6 | 10 | 1 | 1 | 3 | 4 | 5 | 6 |
| **5f** | 5 | 5 | 5 | 4 | 4 | 6 | 4 | 6 | 3 | 4 | 3 | 7 | 7 | 5 | 4 | 4 | 2 | 4 | 7 | 7 | 2 | 8 | 5 | 5 | 4 | 3 | 3 | 3 | 3 | 3 | 3 | 4 | 3 | 1 | 6 | 8 | 9 | 5 | 3 | 4 | 4 | 4 | 4 | 2 | x | 2 | 2 | 5 | 4 | 6 | 10 | 3 | 3 | 3 | 4 | 5 | 6 |
| **5g** | 5 | 5 | 5 | 4 | 4 | 6 | 4 | 4 | 3 | 4 | 3 | 7 | 7 | 3 | 2 | 3 | 2 | 4 | 7 | 7 | 2 | 8 | 3 | 3 | 2 | 1 | 3 | 3 | 3 | 2 | 1 | 3 | 3 | 1 | 6 | 8 | 9 | 5 | 3 | 3 | 3 | 3 | 3 | 1 | 2 | x | 2 | 5 | 4 | 6 | 10 | 2 | 2 | 3 | 4 | 5 | 6 |
| **5h** | 5 | 5 | 5 | 4 | 4 | 6 | 4 | 6 | 3 | 4 | 3 | 6 | 7 | 5 | 4 | 4 | 2 | 4 | 5 | 5 | 2 | 7 | 5 | 5 | 4 | 3 | 3 | 3 | 3 | 3 | 1 | 4 | 3 | 1 | 4 | 6 | 8 | 5 | 3 | 4 | 4 | 4 | 4 | 2 | 2 | 2 | x | 3 | 4 | 6 | 10 | 3 | 3 | 1 | 2 | 5 | 6 |
| **5i** | 5 | 6 | 6 | 6 | 6 | 4 | 7 | 7 | 6 | 7 | 6 | 3 | 5 | 7 | 7 | 7 | 5 | 7 | 2 | 2 | 5 | 4 | 8 | 8 | 7 | 6 | 6 | 6 | 6 | 6 | 4 | 7 | 6 | 4 | 1 | 3 | 5 | 8 | 6 | 7 | 7 | 7 | 7 | 5 | 5 | 5 | 3 | x | 7 | 9 | 6 | 6 | 6 | 2 | 1 | 8 | 9 |
| **5j** | 5 | 5 | 5 | 4 | 5 | 6 | 4 | 6 | 3 | 4 | 4 | 7 | 7 | 5 | 4 | 6 | 2 | 5 | 8 | 8 | 4 | 8 | 5 | 5 | 5 | 5 | 3 | 3 | 3 | 5 | 3 | 6 | 1 | 3 | 8 | 9 | 9 | 4 | 5 | 6 | 6 | 6 | 6 | 4 | 4 | 4 | 4 | 7 | x | 3 | 10 | 5 | 5 | 5 | 6 | 1 | 2 |
| **5k** | 5 | 6 | 6 | 4 | 4 | 6 | 5 | 7 | 4 | 5 | 3 | 7 | 7 | 6 | 6 | 8 | 5 | 2 | 8 | 8 | 4 | 8 | 7 | 7 | 7 | 7 | 6 | 6 | 6 | 7 | 6 | 8 | 4 | 5 | 9 | 9 | 9 | 1 | 5 | 8 | 8 | 8 | 8 | 6 | 6 | 6 | 6 | 9 | 3 | x | 10 | 7 | 7 | 7 | 8 | 2 | 1 |
| **5L** | 5 | 6 | 6 | 6 | 6 | 4 | 7 | 7 | 7 | 7 | 7 | 3 | 5 | 8 | 8 | 8 | 8 | 8 | 4 | 4 | 8 | 2 | 9 | 9 | 9 | 9 | 9 | 9 | 9 | 9 | 9 | 9 | 9 | 9 | 5 | 5 | 1 | 9 | 9 | 10 | 10 | 10 | 10 | 10 | 10 | 10 | 10 | 6 | 10 | 10 | x | 11 | 11 | 8 | 7 | 11 | 11 |
| **6a** | 6 | 5 | 6 | 5 | 5 | 7 | 4 | 4 | 4 | 5 | 4 | 8 | 8 | 3 | 4 | 3 | 3 | 5 | 8 | 8 | 3 | 9 | 5 | 5 | 4 | 3 | 4 | 4 | 4 | 2 | 3 | 3 | 4 | 2 | 7 | 9 | 10 | 6 | 4 | 3 | 3 | 3 | 3 | 1 | 3 | 2 | 3 | 6 | 5 | 7 | 11 | x | 2 | 4 | 5 | 6 | 7 |
| **6b** | 6 | 5 | 6 | 5 | 5 | 7 | 4 | 4 | 4 | 5 | 4 | 8 | 8 | 3 | 4 | 3 | 3 | 5 | 8 | 8 | 3 | 9 | 5 | 5 | 4 | 3 | 4 | 4 | 4 | 2 | 3 | 3 | 4 | 2 | 7 | 9 | 10 | 6 | 4 | 3 | 3 | 3 | 3 | 1 | 3 | 2 | 3 | 6 | 5 | 7 | 11 | 2 | x | 4 | 5 | 6 | 7 |
| **6c** | 6 | 6 | 6 | 5 | 5 | 6 | 5 | 7 | 4 | 5 | 4 | 5 | 7 | 6 | 5 | 5 | 3 | 5 | 4 | 4 | 3 | 6 | 6 | 6 | 5 | 4 | 4 | 4 | 4 | 4 | 2 | 5 | 4 | 2 | 3 | 5 | 7 | 6 | 4 | 5 | 5 | 5 | 5 | 3 | 3 | 3 | 1 | 2 | 5 | 7 | 8 | 4 | 4 | x | 1 | 6 | 7 |
| **6d** | 6 | 7 | 7 | 6 | 6 | 5 | 6 | 7 | 5 | 6 | 5 | 4 | 6 | 7 | 6 | 6 | 4 | 6 | 3 | 3 | 4 | 5 | 7 | 7 | 6 | 5 | 5 | 5 | 5 | 5 | 3 | 6 | 5 | 3 | 2 | 4 | 6 | 7 | 5 | 6 | 6 | 6 | 6 | 4 | 4 | 4 | 2 | 1 | 6 | 8 | 7 | 5 | 5 | 1 | x | 7 | 8 |
| **6e** | 6 | 6 | 6 | 5 | 6 | 7 | 5 | 7 | 4 | 5 | 5 | 8 | 8 | 6 | 5 | 7 | 3 | 4 | 9 | 9 | 5 | 9 | 6 | 6 | 6 | 6 | 4 | 4 | 4 | 6 | 4 | 7 | 2 | 4 | 9 | 10 | 10 | 3 | 6 | 7 | 7 | 7 | 7 | 5 | 5 | 5 | 5 | 8 | 1 | 2 | 11 | 6 | 6 | 6 | 7 | x | 1 |
| **6f** | 6 | 7 | 7 | 5 | 5 | 7 | 6 | 8 | 5 | 6 | 4 | 8 | 8 | 7 | 6 | 8 | 4 | 3 | 9 | 9 | 5 | 9 | 7 | 7 | 7 | 7 | 5 | 5 | 5 | 7 | 5 | 8 | 3 | 5 | 10 | 10 | 10 | 2 | 6 | 8 | 8 | 8 | 8 | 6 | 6 | 6 | 6 | 9 | 2 | 1 | 11 | 7 | 7 | 7 | 8 | 1 | x |

Appendix 7 *(continued)*

Palenque Palace Path Matrix Phase 4

| | c | 1a | 1b | 1c | 1d | 2a | 2b | 2c | 2d | 2e | 3a | 3b | 3c | 3d | 3e | 3f | 3g | 3h | 3i | 4a | 4b | 4c | 4d | 4e | 4f | 4g | 4h | 4i | 4j | 4k | 4L | 4m | 4n | 4o | 4p | 4q | 5a | 5b |
|---|---|---|---|---|---|---|---|---|---|---|---|---|---|---|---|---|---|---|---|---|---|---|---|---|---|---|---|---|---|---|---|---|---|---|---|---|---|---|
| **c** | x | 1 | 1 | 1 | 1 | 2 | 2 | 2 | 2 | 2 | 3 | 3 | 3 | 3 | 3 | 3 | 3 | 3 | 3 | 4 | 4 | 4 | 4 | 4 | 4 | 4 | 4 | 4 | 4 | 4 | 4 | 4 | 4 | 4 | 4 | 4 | 5 | 5 |
| **1a** | 1 | x | 1 | 2 | 2 | 1 | 2 | 3 | 3 | 3 | 2 | 3 | 3 | 4 | 4 | 4 | 4 | 4 | 2 | 3 | 3 | 3 | 3 | 5 | 5 | 5 | 5 | 4 | 5 | 5 | 5 | 5 | 4 | 5 | 5 | 4 | 4 | 4 |
| **1b** | 1 | 1 | x | 2 | 2 | 2 | 1 | 3 | 3 | 3 | 3 | 2 | 2 | 4 | 4 | 4 | 4 | 4 | 3 | 4 | 4 | 4 | 4 | 5 | 5 | 5 | 4 | 3 | 5 | 5 | 5 | 5 | 3 | 5 | 5 | 3 | 5 | 5 |
| **1c** | 1 | 2 | 2 | x | 2 | 3 | 3 | 1 | 3 | 3 | 4 | 4 | 2 | 2 | 2 | 4 | 4 | 4 | 4 | 5 | 5 | 5 | 5 | 3 | 3 | 3 | 3 | 3 | 3 | 5 | 5 | 5 | 3 | 3 | 3 | 3 | 6 | 6 |
| **1d** | 1 | 2 | 2 | 2 | x | 3 | 3 | 3 | 1 | 1 | 4 | 4 | 4 | 2 | 4 | 2 | 2 | 2 | 4 | 5 | 5 | 5 | 5 | 3 | 3 | 3 | 3 | 3 | 5 | 3 | 3 | 3 | 5 | 3 | 3 | 5 | 6 | 6 |
| **2a** | 2 | 1 | 2 | 3 | 3 | x | 3 | 4 | 4 | 4 | 1 | 4 | 4 | 5 | 5 | 5 | 5 | 5 | 1 | 2 | 2 | 2 | 2 | 4 | 5 | 5 | 6 | 5 | 6 | 6 | 6 | 6 | 5 | 6 | 6 | 5 | 3 | 3 |
| **2b** | 2 | 2 | 1 | 3 | 3 | 3 | x | 2 | 4 | 4 | 4 | 1 | 1 | 3 | 3 | 5 | 5 | 5 | 4 | 5 | 5 | 5 | 5 | 4 | 4 | 4 | 3 | 2 | 4 | 6 | 6 | 6 | 2 | 4 | 4 | 2 | 6 | 6 |
| **2c** | 2 | 3 | 3 | 1 | 3 | 4 | 2 | x | 4 | 2 | 5 | 2 | 1 | 1 | 1 | 5 | 5 | 5 | 5 | 6 | 6 | 5 | 4 | 2 | 2 | 2 | 2 | 2 | 2 | 6 | 6 | 6 | 2 | 2 | 2 | 2 | 6 | 6 |
| **2d** | 2 | 3 | 3 | 3 | 1 | 4 | 4 | 4 | x | 2 | 5 | 5 | 5 | 3 | 5 | 1 | 1 | 1 | 5 | 6 | 6 | 6 | 6 | 4 | 4 | 4 | 4 | 4 | 6 | 2 | 2 | 2 | 6 | 4 | 4 | 6 | 7 | 7 |
| **2e** | 2 | 3 | 3 | 3 | 1 | 4 | 4 | 2 | 2 | x | 5 | 4 | 3 | 1 | 3 | 3 | 3 | 3 | 5 | 6 | 6 | 5 | 4 | 2 | 2 | 2 | 2 | 2 | 4 | 4 | 4 | 4 | 4 | 2 | 2 | 4 | 7 | 7 |
| **3a** | 3 | 2 | 3 | 4 | 4 | 1 | 4 | 5 | 5 | 5 | x | 5 | 5 | 4 | 6 | 6 | 6 | 6 | 2 | 1 | 1 | 2 | 1 | 3 | 4 | 5 | 5 | 5 | 7 | 7 | 7 | 7 | 4 | 5 | 5 | 6 | 2 | 2 |
| **3b** | 3 | 3 | 2 | 4 | 4 | 4 | 1 | 2 | 5 | 4 | 5 | x | 1 | 3 | 3 | 6 | 6 | 6 | 5 | 6 | 5 | 6 | 6 | 4 | 4 | 4 | 3 | 2 | 4 | 7 | 7 | 7 | 2 | 4 | 4 | 2 | 6 | 6 |
| **3c** | 3 | 3 | 2 | 2 | 4 | 4 | 1 | 1 | 5 | 3 | 5 | 1 | x | 2 | 2 | 6 | 6 | 6 | 5 | 6 | 5 | 6 | 5 | 3 | 3 | 3 | 2 | 1 | 3 | 7 | 7 | 7 | 1 | 3 | 3 | 1 | 5 | 5 |
| **3d** | 3 | 4 | 4 | 2 | 2 | 5 | 3 | 1 | 3 | 1 | 4 | 3 | 2 | x | 2 | 4 | 4 | 4 | 4 | 6 | 5 | 4 | 3 | 1 | 1 | 1 | 1 | 1 | 3 | 5 | 5 | 5 | 3 | 1 | 1 | 3 | 6 | 6 |
| **3e** | 3 | 4 | 4 | 2 | 4 | 5 | 3 | 1 | 5 | 3 | 6 | 3 | 2 | 2 | x | 6 | 6 | 6 | 6 | 7 | 7 | 6 | 5 | 3 | 3 | 3 | 3 | 3 | 1 | 7 | 7 | 7 | 3 | 3 | 3 | 3 | 8 | 8 |
| **3f** | 3 | 4 | 4 | 4 | 2 | 5 | 5 | 5 | 1 | 3 | 6 | 6 | 6 | 4 | 6 | x | 2 | 2 | 6 | 7 | 7 | 7 | 7 | 5 | 5 | 5 | 5 | 5 | 7 | 1 | 1 | 3 | 7 | 5 | 5 | 7 | 8 | 8 |
| **3g** | 3 | 4 | 4 | 4 | 2 | 5 | 5 | 5 | 1 | 3 | 6 | 6 | 6 | 4 | 6 | 2 | x | 2 | 6 | 7 | 7 | 7 | 7 | 5 | 5 | 5 | 5 | 5 | 7 | 1 | 1 | 3 | 7 | 5 | 5 | 7 | 8 | 8 |
| **3h** | 3 | 4 | 4 | 4 | 2 | 5 | 5 | 5 | 1 | 3 | 6 | 6 | 6 | 4 | 6 | 2 | 2 | x | 6 | 7 | 7 | 7 | 7 | 5 | 5 | 5 | 5 | 5 | 7 | 3 | 3 | 1 | 7 | 5 | 5 | 7 | 8 | 8 |
| **3i** | 3 | 2 | 3 | 4 | 4 | 1 | 4 | 5 | 5 | 5 | 2 | 5 | 5 | 4 | 6 | 6 | 6 | 6 | x | 3 | 3 | 1 | 1 | 3 | 4 | 4 | 5 | 6 | 7 | 7 | 7 | 7 | 5 | 5 | 5 | 6 | 4 | 4 |
| **4a** | 4 | 3 | 4 | 5 | 5 | 2 | 5 | 6 | 6 | 6 | 1 | 6 | 6 | 6 | 7 | 7 | 7 | 7 | 3 | x | 1 | 3 | 2 | 4 | 5 | 6 | 7 | 6 | 8 | 8 | 8 | 8 | 4 | 6 | 6 | 6 | 2 | 2 |
| **4b** | 4 | 3 | 4 | 5 | 5 | 2 | 5 | 6 | 6 | 6 | 1 | 5 | 5 | 5 | 7 | 7 | 7 | 7 | 3 | 1 | x | 3 | 2 | 4 | 5 | 4 | 5 | 4 | 8 | 8 | 8 | 8 | 3 | 6 | 6 | 5 | 1 | 1 |
| **4c** | 4 | 3 | 4 | 5 | 5 | 2 | 5 | 5 | 6 | 5 | 2 | 6 | 6 | 4 | 6 | 7 | 7 | 7 | 1 | 3 | 3 | x | 1 | 3 | 4 | 4 | 5 | 5 | 7 | 8 | 8 | 8 | 5 | 5 | 5 | 7 | 4 | 4 |
| **4d** | 4 | 3 | 4 | 5 | 5 | 2 | 5 | 4 | 6 | 4 | 1 | 6 | 5 | 3 | 5 | 7 | 7 | 7 | 1 | 2 | 2 | 1 | x | 2 | 3 | 3 | 4 | 4 | 6 | 8 | 8 | 8 | 4 | 4 | 4 | 6 | 3 | 3 |
| **4e** | 4 | 5 | 5 | 3 | 3 | 4 | 4 | 2 | 4 | 2 | 3 | 4 | 3 | 1 | 3 | 5 | 5 | 5 | 3 | 4 | 4 | 3 | 2 | x | 1 | 2 | 2 | 2 | 4 | 6 | 6 | 6 | 3 | 2 | 2 | 4 | 5 | 5 |
| **4f** | 4 | 5 | 5 | 3 | 3 | 5 | 4 | 2 | 4 | 2 | 4 | 4 | 3 | 1 | 3 | 5 | 5 | 5 | 4 | 5 | 5 | 4 | 3 | 1 | x | 1 | 2 | 2 | 4 | 6 | 6 | 6 | 3 | 2 | 2 | 4 | 6 | 6 |
| **4g** | 4 | 5 | 5 | 3 | 3 | 5 | 4 | 2 | 4 | 2 | 5 | 4 | 3 | 1 | 3 | 5 | 5 | 5 | 4 | 6 | 4 | 4 | 3 | 2 | 1 | x | 1 | 2 | 4 | 6 | 6 | 6 | 2 | 2 | 2 | 4 | 5 | 5 |
| **4h** | 4 | 5 | 4 | 3 | 3 | 6 | 3 | 2 | 4 | 2 | 5 | 3 | 2 | 1 | 3 | 5 | 5 | 5 | 5 | 7 | 5 | 5 | 4 | 2 | 2 | 1 | x | 1 | 4 | 6 | 6 | 6 | 3 | 2 | 2 | 3 | 7 | 7 |
| **4i** | 4 | 4 | 3 | 3 | 3 | 5 | 2 | 2 | 4 | 2 | 5 | 2 | 1 | 1 | 3 | 5 | 5 | 5 | 6 | 6 | 4 | 5 | 4 | 2 | 2 | 2 | 1 | x | 4 | 6 | 6 | 6 | 2 | 2 | 2 | 2 | 5 | 5 |
| **4j** | 4 | 5 | 5 | 3 | 5 | 6 | 4 | 2 | 6 | 4 | 7 | 4 | 3 | 3 | 1 | 7 | 7 | 7 | 7 | 8 | 8 | 7 | 6 | 4 | 4 | 4 | 4 | 4 | x | 8 | 8 | 8 | 4 | 4 | 4 | 4 | 8 | 8 |
| **4k** | 4 | 5 | 5 | 5 | 3 | 6 | 6 | 6 | 2 | 4 | 7 | 7 | 7 | 5 | 7 | 1 | 1 | 3 | 7 | 8 | 8 | 8 | 8 | 6 | 6 | 6 | 6 | 6 | 8 | x | 2 | 4 | 6 | 6 | 6 | 8 | 9 | 9 |
| **4L** | 4 | 5 | 5 | 5 | 3 | 6 | 6 | 6 | 2 | 4 | 7 | 7 | 7 | 5 | 7 | 1 | 1 | 3 | 7 | 8 | 8 | 8 | 8 | 6 | 6 | 6 | 6 | 6 | 8 | 2 | x | 4 | 8 | 6 | 6 | 8 | 9 | 9 |
| **4m** | 4 | 5 | 5 | 5 | 3 | 6 | 6 | 6 | 2 | 4 | 7 | 7 | 7 | 5 | 7 | 3 | 3 | 1 | 7 | 8 | 8 | 8 | 8 | 6 | 6 | 6 | 6 | 6 | 8 | 4 | 4 | x | 8 | 6 | 6 | 8 | 9 | 9 |
| **4n** | 4 | 4 | 3 | 3 | 5 | 5 | 2 | 2 | 6 | 4 | 4 | 2 | 1 | 3 | 3 | 7 | 7 | 7 | 5 | 4 | 3 | 5 | 4 | 3 | 3 | 2 | 3 | 2 | 4 | 6 | 8 | 8 | x | 4 | 4 | 2 | 4 | 4 |
| **4o** | 4 | 5 | 5 | 3 | 3 | 6 | 4 | 2 | 4 | 2 | 5 | 4 | 3 | 1 | 3 | 5 | 5 | 5 | 5 | 6 | 6 | 5 | 4 | 2 | 2 | 2 | 2 | 2 | 4 | 6 | 6 | 6 | 4 | x | 2 | 4 | 7 | 7 |
| **4p** | 4 | 5 | 5 | 3 | 3 | 6 | 4 | 2 | 4 | 2 | 5 | 4 | 3 | 1 | 3 | 5 | 5 | 5 | 5 | 6 | 6 | 5 | 4 | 2 | 2 | 2 | 2 | 2 | 4 | 6 | 6 | 6 | 4 | 2 | x | 4 | 7 | 7 |
| **4q** | 4 | 4 | 3 | 3 | 5 | 5 | 2 | 2 | 6 | 4 | 6 | 2 | 1 | 3 | 3 | 7 | 7 | 7 | 6 | 6 | 5 | 7 | 6 | 4 | 4 | 4 | 3 | 2 | 4 | 8 | 8 | 8 | 2 | 4 | 4 | x | 6 | 6 |

| | | | | | | | | | | | | | | | | | | | | | | | | | | | | | | | | | | | | | | |
|---|---|---|---|---|---|---|---|---|---|---|---|---|---|---|---|---|---|---|---|---|---|---|---|---|---|---|---|---|---|---|---|---|---|---|---|---|---|---|
| **5a** | 5 | 4 | 5 | 6 | 6 | 3 | 6 | 6 | 7 | 7 | 2 | 6 | 5 | 6 | 8 | 8 | 8 | 8 | 4 | 2 | 1 | 4 | 3 | 5 | 6 | 5 | 7 | 5 | 8 | 9 | 9 | 9 | 4 | 7 | 7 | 6 | x | 2 |
| **5b** | 5 | 4 | 5 | 6 | 6 | 3 | 6 | 6 | 7 | 7 | 2 | 6 | 5 | 6 | 8 | 8 | 8 | 8 | 4 | 2 | 1 | 4 | 3 | 5 | 6 | 5 | 7 | 5 | 8 | 9 | 9 | 9 | 4 | 7 | 7 | 6 | 2 | x |
| **5c** | 5 | 4 | 5 | 6 | 6 | 3 | 6 | 6 | 7 | 6 | 2 | 6 | 5 | 5 | 7 | 8 | 8 | 8 | 4 | 2 | 1 | 4 | 3 | 4 | 5 | 4 | 6 | 4 | 8 | 9 | 9 | 9 | 4 | 6 | 6 | 6 | 2 | 2 |
| **5d** | 5 | 4 | 5 | 6 | 6 | 3 | 5 | 5 | 7 | 5 | 2 | 5 | 4 | 4 | 6 | 8 | 8 | 8 | 4 | 2 | 1 | 4 | 3 | 3 | 4 | 3 | 5 | 3 | 7 | 9 | 9 | 9 | 3 | 5 | 5 | 5 | 2 | 2 |
| **5e** | 5 | 4 | 5 | 5 | 6 | 3 | 4 | 4 | 7 | 6 | 2 | 4 | 3 | 5 | 5 | 8 | 8 | 8 | 4 | 2 | 1 | 4 | 3 | 5 | 5 | 4 | 5 | 4 | 6 | 7 | 9 | 9 | 2 | 6 | 6 | 4 | 2 | 2 |
| **5f** | 5 | 5 | 4 | 4 | 6 | 4 | 3 | 3 | 7 | 5 | 3 | 3 | 2 | 4 | 4 | 7 | 7 | 8 | 5 | 3 | 2 | 5 | 4 | 4 | 4 | 3 | 4 | 3 | 5 | 6 | 8 | 9 | 1 | 5 | 5 | 3 | 3 | 3 |
| **5g** | 5 | 5 | 4 | 4 | 6 | 5 | 3 | 3 | 7 | 5 | 4 | 3 | 2 | 4 | 4 | 6 | 6 | 8 | 4 | 4 | 3 | 4 | 3 | 4 | 4 | 3 | 4 | 3 | 5 | 5 | 7 | 8 | 1 | 5 | 5 | 3 | 4 | 4 |
| **5h** | 5 | 5 | 4 | 4 | 6 | 6 | 3 | 3 | 7 | 5 | 7 | 3 | 2 | 4 | 4 | 8 | 8 | 8 | 7 | 7 | 6 | 8 | 7 | 5 | 5 | 5 | 4 | 3 | 5 | 9 | 9 | 9 | 3 | 5 | 5 | 1 | 7 | 7 |
| **5i** | 5 | 5 | 4 | 4 | 6 | 6 | 3 | 3 | 7 | 5 | 7 | 3 | 2 | 4 | 4 | 8 | 8 | 8 | 6 | 5 | 4 | 6 | 5 | 4 | 4 | 3 | 4 | 3 | 5 | 7 | 9 | 9 | 1 | 5 | 5 | 3 | 5 | 5 |
| **5j** | 5 | 4 | 5 | 6 | 6 | 3 | 6 | 5 | 7 | 5 | 2 | 7 | 6 | 4 | 6 | 8 | 8 | 8 | 2 | 3 | 3 | 2 | 1 | 3 | 4 | 4 | 5 | 5 | 7 | 9 | 9 | 9 | 5 | 5 | 5 | 7 | 4 | 4 |
| **5k** | 5 | 4 | 5 | 6 | 6 | 3 | 6 | 5 | 7 | 5 | 2 | 7 | 6 | 4 | 6 | 8 | 8 | 8 | 2 | 3 | 3 | 2 | 1 | 3 | 4 | 4 | 5 | 5 | 7 | 9 | 9 | 9 | 5 | 5 | 5 | 7 | 4 | 4 |
| **5L** | 5 | 4 | 5 | 6 | 6 | 3 | 6 | 5 | 7 | 5 | 2 | 7 | 6 | 4 | 6 | 8 | 8 | 8 | 2 | 3 | 3 | 2 | 1 | 3 | 4 | 4 | 5 | 5 | 7 | 9 | 9 | 9 | 5 | 5 | 5 | 7 | 4 | 4 |
| **5m** | 5 | 4 | 5 | 6 | 6 | 3 | 6 | 5 | 7 | 5 | 2 | 7 | 6 | 4 | 6 | 8 | 8 | 8 | 2 | 3 | 3 | 2 | 1 | 3 | 4 | 4 | 5 | 5 | 7 | 9 | 9 | 9 | 5 | 5 | 5 | 7 | 4 | 4 |
| **5n** | 5 | 4 | 5 | 4 | 4 | 3 | 5 | 3 | 5 | 3 | 2 | 5 | 4 | 2 | 4 | 6 | 6 | 6 | 2 | 3 | 3 | 2 | 1 | 1 | 2 | 2 | 3 | 3 | 5 | 7 | 7 | 7 | 3 | 3 | 3 | 5 | 4 | 4 |
| **5o** | 5 | 5 | 5 | 4 | 4 | 4 | 4 | 3 | 5 | 3 | 3 | 4 | 3 | 2 | 4 | 6 | 6 | 6 | 3 | 4 | 4 | 3 | 2 | 1 | 2 | 1 | 2 | 2 | 5 | 6 | 7 | 9 | 2 | 3 | 3 | 4 | 5 | 5 |
| **5p** | 5 | 5 | 4 | 4 | 4 | 5 | 3 | 3 | 5 | 3 | 4 | 3 | 2 | 2 | 4 | 6 | 6 | 6 | 4 | 5 | 3 | 4 | 3 | 2 | 2 | 1 | 2 | 1 | 5 | 5 | 7 | 8 | 1 | 3 | 3 | 3 | 4 | 4 |
| **5q** | 5 | 6 | 5 | 4 | 4 | 6 | 4 | 3 | 5 | 3 | 5 | 4 | 3 | 2 | 4 | 6 | 6 | 6 | 5 | 6 | 4 | 5 | 4 | 3 | 2 | 1 | 2 | 2 | 5 | 6 | 7 | 9 | 2 | 3 | 3 | 4 | 5 | 5 |
| **5r** | 5 | 6 | 5 | 4 | 4 | 7 | 5 | 3 | 5 | 3 | 6 | 4 | 3 | 2 | 4 | 6 | 6 | 6 | 6 | 7 | 5 | 6 | 5 | 4 | 3 | 2 | 1 | 2 | 5 | 7 | 7 | 9 | 3 | 3 | 3 | 4 | 6 | 6 |
| **5s** | 5 | 6 | 6 | 6 | 4 | 7 | 7 | 7 | 3 | 5 | 8 | 7 | 6 | 6 | 8 | 2 | 2 | 4 | 8 | 8 | 7 | 8 | 7 | 6 | 6 | 5 | 6 | 5 | 9 | 1 | 3 | 5 | 5 | 7 | 7 | 7 | 8 | 8 |
| **5t** | 5 | 6 | 6 | 4 | 6 | 7 | 5 | 3 | 7 | 5 | 8 | 5 | 4 | 4 | 2 | 8 | 8 | 8 | 8 | 9 | 9 | 9 | 9 | 5 | 5 | 5 | 5 | 5 | 1 | 9 | 9 | 9 | 5 | 5 | 5 | 5 | 10 | 10 |
| **5u** | 5 | 6 | 6 | 6 | 4 | 7 | 7 | 7 | 3 | 5 | 8 | 8 | 8 | 8 | 8 | 4 | 4 | 2 | 8 | 9 | 9 | 9 | 9 | 7 | 7 | 7 | 7 | 7 | 9 | 5 | 5 | 1 | 9 | 7 | 7 | 9 | 10 | 10 |
| **6a** | 6 | 5 | 6 | 5 | 5 | 4 | 6 | 4 | 6 | 4 | 3 | 6 | 5 | 3 | 5 | 7 | 7 | 7 | 3 | 4 | 4 | 3 | 2 | 2 | 3 | 3 | 4 | 4 | 6 | 8 | 8 | 8 | 4 | 4 | 4 | 6 | 5 | 5 |
| **6b** | 6 | 5 | 6 | 5 | 5 | 4 | 6 | 4 | 6 | 4 | 3 | 6 | 5 | 3 | 5 | 7 | 7 | 7 | 3 | 4 | 4 | 3 | 2 | 2 | 3 | 3 | 4 | 4 | 6 | 8 | 8 | 8 | 4 | 4 | 4 | 6 | 5 | 5 |
| **6c** | 6 | 5 | 5 | 5 | 5 | 4 | 4 | 4 | 6 | 4 | 3 | 4 | 3 | 3 | 5 | 7 | 7 | 7 | 3 | 3 | 2 | 3 | 2 | 2 | 3 | 2 | 3 | 2 | 6 | 6 | 8 | 8 | 2 | 4 | 4 | 4 | 3 | 3 |
| **6d** | 6 | 6 | 5 | 5 | 5 | 6 | 4 | 4 | 6 | 4 | 5 | 4 | 3 | 3 | 5 | 5 | 5 | 7 | 5 | 5 | 4 | 5 | 4 | 3 | 3 | 2 | 3 | 2 | 6 | 4 | 6 | 7 | 2 | 4 | 4 | 4 | 5 | 5 |
| **6e** | 6 | 7 | 7 | 7 | 5 | 8 | 6 | 6 | 4 | 6 | 7 | 6 | 5 | 5 | 7 | 3 | 3 | 5 | 7 | 7 | 6 | 7 | 6 | 5 | 5 | 4 | 5 | 4 | 8 | 2 | 4 | 6 | 4 | 6 | 6 | 6 | 7 | 7 |
| **6f** | 6 | 7 | 6 | 5 | 5 | 8 | 5 | 4 | 6 | 4 | 7 | 5 | 4 | 3 | 5 | 7 | 7 | 7 | 7 | 7 | 6 | 8 | 7 | 5 | 4 | 3 | 2 | 3 | 6 | 8 | 8 | 8 | 4 | 4 | 4 | 5 | 7 | 7 |
| **6g** | 6 | 7 | 7 | 5 | 7 | 8 | 6 | 4 | 8 | 6 | 9 | 6 | 5 | 5 | 3 | 9 | 9 | 9 | 9 | 9 | 8 | 10 | 9 | 6 | 6 | 5 | 4 | 5 | 8 | 10 | 10 | 10 | 6 | 6 | 6 | 7 | 9 | 9 |
| **6h** | 6 | 6 | 5 | 5 | 7 | 7 | 4 | 4 | 8 | 6 | 8 | 4 | 3 | 5 | 5 | 9 | 9 | 9 | 8 | 8 | 7 | 9 | 8 | 6 | 6 | 6 | 5 | 4 | 6 | 10 | 10 | 10 | 4 | 6 | 6 | 2 | 8 | 8 |
| **6i** | 6 | 5 | 6 | 6 | 7 | 4 | 5 | 5 | 8 | 7 | 3 | 5 | 4 | 6 | 6 | 9 | 9 | 10 | 5 | 3 | 2 | 5 | 4 | 6 | 6 | 5 | 6 | 4 | 7 | 8 | 10 | 10 | 3 | 7 | 7 | 5 | 3 | 3 |
| **6j** | 6 | 6 | 5 | 5 | 7 | 7 | 4 | 4 | 8 | 6 | 6 | 4 | 3 | 5 | 5 | 9 | 9 | 9 | 8 | 6 | 5 | 7 | 6 | 5 | 5 | 4 | 5 | 5 | 6 | 8 | 10 | 10 | 2 | 6 | 6 | 4 | 6 | 6 |
| **7a** | 7 | 7 | 6 | 6 | 6 | 8 | 5 | 5 | 5 | 7 | 6 | 5 | 4 | 4 | 6 | 4 | 4 | 6 | 8 | 6 | 6 | 6 | 5 | 4 | 4 | 3 | 4 | 3 | 7 | 3 | 5 | 7 | 3 | 5 | 5 | 5 | 6 | 6 |
| **7b** | 7 | 8 | 8 | 6 | 6 | 9 | 6 | 5 | 7 | 7 | 8 | 6 | 5 | 4 | 6 | 10 | 10 | 10 | 8 | 8 | 7 | 9 | 8 | 6 | 5 | 4 | 3 | 4 | 7 | 9 | 9 | 9 | 5 | 5 | 5 | 6 | 8 | 8 |
| **7c** | 7 | 7 | 6 | 6 | 8 | 8 | 5 | 5 | 9 | 7 | 9 | 5 | 4 | 6 | 6 | 10 | 10 | 10 | 9 | 9 | 8 | 10 | 9 | 7 | 7 | 7 | 6 | 5 | 7 | 11 | 11 | 11 | 5 | 7 | 7 | 3 | 9 | 9 |
| **7d** | 7 | 6 | 7 | 7 | 8 | 5 | 6 | 6 | 9 | 8 | 4 | 6 | 5 | 7 | 7 | 11 | 11 | 11 | 6 | 4 | 3 | 6 | 5 | 7 | 7 | 6 | 7 | 6 | 8 | 9 | 11 | 11 | 4 | 8 | 8 | 6 | 4 | 4 |
| **7e** | 7 | 7 | 6 | 6 | 8 | 8 | 5 | 5 | 9 | 7 | 7 | 5 | 4 | 6 | 6 | 10 | 10 | 10 | 9 | 7 | 6 | 8 | 7 | 6 | 6 | 5 | 6 | 5 | 7 | 9 | 11 | 11 | 3 | 7 | 7 | 5 | 7 | 7 |
| **8a** | 8 | 7 | 8 | 8 | 9 | 6 | 7 | 7 | 10 | 9 | 5 | 7 | 6 | 8 | 8 | 12 | 12 | 12 | 7 | 5 | 4 | 7 | 6 | 8 | 8 | 7 | 8 | 7 | 9 | 10 | 12 | 12 | 5 | 9 | 9 | 7 | 5 | 5 |

(*continued*)

Palenque Palace Path Matrix Phase 4 (Cont.)

| | 5c | 5d | 5e | 5f | 5g | 5h | 5i | 5j | 5k | 5l | 5m | 5n | 5o | 5p | 5q | 5r | 5s | 5t | 5u | 6a | 6b | 6c | 6d | 6e | 6f | 6g | 6h | 6i | 6j | 7a | 7b | 7c | 7d | 7e | 8a |
|---|---|---|---|---|---|---|---|---|---|---|---|---|---|---|---|---|---|---|---|---|---|---|---|---|---|---|---|---|---|---|---|---|---|---|---|
| **c** | 5 | 5 | 5 | 5 | 5 | 5 | 5 | 5 | 5 | 5 | 5 | 5 | 5 | 5 | 5 | 5 | 5 | 5 | 5 | 6 | 6 | 6 | 6 | 6 | 6 | 6 | 6 | 6 | 6 | 7 | 7 | 7 | 7 | 7 | 8 |
| **1a** | 4 | 4 | 4 | 5 | 5 | 5 | 5 | 4 | 4 | 4 | 4 | 4 | 5 | 5 | 6 | 6 | 6 | 6 | 6 | 5 | 5 | 5 | 6 | 7 | 7 | 7 | 6 | 5 | 6 | 7 | 8 | 7 | 6 | 7 | 7 |
| **1b** | 5 | 5 | 5 | 4 | 4 | 4 | 4 | 5 | 5 | 5 | 5 | 5 | 5 | 4 | 5 | 5 | 6 | 6 | 6 | 6 | 6 | 5 | 5 | 7 | 6 | 7 | 5 | 6 | 5 | 6 | 8 | 6 | 7 | 6 | 8 |
| **1c** | 6 | 6 | 5 | 4 | 4 | 4 | 4 | 6 | 6 | 6 | 6 | 4 | 4 | 4 | 4 | 4 | 6 | 4 | 6 | 5 | 5 | 5 | 5 | 7 | 5 | 5 | 5 | 6 | 5 | 6 | 6 | 6 | 7 | 6 | 8 |
| **1d** | 6 | 6 | 6 | 6 | 6 | 6 | 6 | 6 | 6 | 6 | 6 | 4 | 4 | 4 | 4 | 4 | 4 | 6 | 4 | 5 | 5 | 5 | 5 | 5 | 5 | 7 | 7 | 7 | 7 | 6 | 6 | 8 | 8 | 8 | 9 |
| **2a** | 3 | 3 | 3 | 4 | 5 | 6 | 6 | 3 | 3 | 3 | 3 | 3 | 4 | 5 | 6 | 7 | 7 | 7 | 7 | 4 | 4 | 4 | 6 | 8 | 8 | 8 | 7 | 4 | 7 | 8 | 9 | 8 | 5 | 8 | 6 |
| **2b** | 6 | 5 | 4 | 3 | 3 | 3 | 3 | 6 | 6 | 6 | 6 | 5 | 4 | 3 | 4 | 5 | 7 | 5 | 7 | 6 | 6 | 4 | 4 | 6 | 5 | 6 | 4 | 5 | 4 | 5 | 6 | 5 | 6 | 5 | 7 |
| **2c** | 6 | 5 | 4 | 3 | 3 | 3 | 3 | 5 | 5 | 5 | 5 | 3 | 3 | 3 | 3 | 3 | 7 | 3 | 7 | 4 | 4 | 4 | 4 | 6 | 4 | 4 | 4 | 5 | 4 | 5 | 5 | 5 | 6 | 5 | 7 |
| **2d** | 7 | 7 | 7 | 7 | 7 | 7 | 7 | 7 | 7 | 7 | 7 | 5 | 5 | 5 | 5 | 5 | 3 | 7 | 3 | 6 | 6 | 6 | 6 | 4 | 6 | 8 | 8 | 8 | 8 | 5 | 7 | 9 | 9 | 9 | 10 |
| **2e** | 6 | 5 | 6 | 5 | 5 | 5 | 5 | 5 | 5 | 5 | 5 | 3 | 3 | 3 | 3 | 3 | 5 | 5 | 5 | 4 | 4 | 4 | 4 | 6 | 4 | 6 | 6 | 7 | 6 | 7 | 7 | 7 | 8 | 7 | 9 |
| **3a** | 2 | 2 | 2 | 3 | 4 | 7 | 7 | 2 | 2 | 2 | 2 | 2 | 3 | 4 | 5 | 6 | 8 | 8 | 8 | 3 | 3 | 3 | 5 | 7 | 7 | 9 | 8 | 3 | 6 | 6 | 8 | 9 | 4 | 7 | 5 |
| **3b** | 6 | 5 | 4 | 3 | 3 | 3 | 3 | 7 | 7 | 7 | 7 | 5 | 4 | 3 | 4 | 4 | 7 | 5 | 8 | 6 | 6 | 4 | 4 | 6 | 5 | 6 | 4 | 5 | 4 | 5 | 6 | 5 | 6 | 5 | 7 |
| **3c** | 5 | 4 | 3 | 2 | 2 | 2 | 2 | 6 | 6 | 6 | 6 | 4 | 3 | 2 | 3 | 3 | 6 | 4 | 8 | 5 | 5 | 3 | 3 | 5 | 4 | 5 | 3 | 4 | 3 | 4 | 5 | 4 | 5 | 4 | 6 |
| **3d** | 5 | 4 | 5 | 4 | 4 | 4 | 4 | 4 | 4 | 4 | 4 | 2 | 2 | 2 | 2 | 2 | 6 | 4 | 8 | 3 | 3 | 3 | 3 | 5 | 3 | 5 | 5 | 6 | 5 | 4 | 4 | 6 | 7 | 6 | 8 |
| **3e** | 7 | 6 | 5 | 4 | 4 | 4 | 4 | 6 | 6 | 6 | 6 | 4 | 4 | 4 | 4 | 4 | 8 | 2 | 8 | 5 | 5 | 5 | 5 | 7 | 5 | 3 | 5 | 6 | 5 | 6 | 6 | 6 | 7 | 6 | 8 |
| **3f** | 8 | 8 | 8 | 7 | 6 | 8 | 8 | 8 | 8 | 8 | 8 | 6 | 6 | 6 | 6 | 6 | 2 | 8 | 4 | 7 | 7 | 7 | 5 | 3 | 7 | 9 | 9 | 9 | 9 | 4 | 10 | 10 | 11 | 10 | 12 |
| **3g** | 8 | 8 | 8 | 7 | 6 | 8 | 8 | 8 | 8 | 8 | 8 | 6 | 6 | 6 | 6 | 6 | 2 | 8 | 4 | 7 | 7 | 7 | 5 | 3 | 7 | 9 | 9 | 9 | 9 | 4 | 10 | 10 | 11 | 10 | 12 |
| **3h** | 8 | 8 | 8 | 8 | 8 | 8 | 8 | 8 | 8 | 8 | 8 | 6 | 6 | 6 | 6 | 6 | 4 | 8 | 2 | 7 | 7 | 7 | 7 | 5 | 7 | 9 | 9 | 10 | 9 | 6 | 10 | 10 | 11 | 10 | 12 |
| **3i** | 4 | 4 | 4 | 5 | 4 | 7 | 6 | 2 | 2 | 2 | 2 | 2 | 3 | 4 | 5 | 6 | 8 | 8 | 8 | 3 | 3 | 3 | 5 | 7 | 7 | 9 | 8 | 5 | 8 | 8 | 8 | 9 | 6 | 9 | 7 |
| **4a** | 2 | 2 | 2 | 3 | 4 | 7 | 5 | 3 | 3 | 3 | 3 | 3 | 4 | 5 | 6 | 7 | 8 | 9 | 9 | 4 | 4 | 3 | 5 | 7 | 7 | 9 | 8 | 3 | 6 | 6 | 8 | 9 | 4 | 7 | 5 |
| **4b** | 1 | 1 | 1 | 2 | 3 | 6 | 4 | 3 | 3 | 3 | 3 | 3 | 4 | 3 | 4 | 5 | 7 | 9 | 9 | 4 | 4 | 2 | 4 | 6 | 6 | 8 | 7 | 2 | 5 | 6 | 7 | 8 | 3 | 6 | 4 |
| **4c** | 4 | 4 | 4 | 5 | 4 | 8 | 6 | 2 | 2 | 2 | 2 | 2 | 3 | 4 | 5 | 6 | 8 | 9 | 9 | 3 | 3 | 3 | 5 | 7 | 8 | 10 | 9 | 5 | 7 | 6 | 9 | 10 | 6 | 8 | 7 |
| **4d** | 3 | 3 | 3 | 4 | 3 | 7 | 5 | 1 | 1 | 1 | 1 | 1 | 2 | 3 | 4 | 5 | 7 | 9 | 9 | 2 | 2 | 2 | 4 | 6 | 7 | 9 | 8 | 4 | 6 | 5 | 8 | 9 | 5 | 7 | 6 |
| **4e** | 4 | 3 | 5 | 4 | 4 | 5 | 4 | 3 | 3 | 3 | 3 | 1 | 1 | 2 | 3 | 4 | 6 | 5 | 7 | 2 | 2 | 2 | 3 | 5 | 5 | 6 | 6 | 6 | 5 | 4 | 6 | 7 | 7 | 6 | 8 |
| **4f** | 5 | 4 | 5 | 4 | 4 | 5 | 4 | 4 | 4 | 4 | 4 | 2 | 2 | 2 | 2 | 3 | 6 | 5 | 7 | 3 | 3 | 3 | 3 | 5 | 4 | 6 | 6 | 6 | 5 | 4 | 5 | 7 | 7 | 6 | 8 |
| **4g** | 4 | 3 | 4 | 3 | 3 | 5 | 3 | 4 | 4 | 4 | 4 | 2 | 1 | 1 | 1 | 2 | 5 | 5 | 7 | 3 | 3 | 2 | 2 | 4 | 3 | 5 | 6 | 5 | 4 | 3 | 4 | 7 | 6 | 5 | 7 |
| **4h** | 6 | 5 | 5 | 4 | 4 | 4 | 4 | 5 | 5 | 5 | 5 | 3 | 2 | 2 | 2 | 1 | 6 | 5 | 7 | 4 | 4 | 3 | 3 | 5 | 2 | 4 | 5 | 6 | 5 | 4 | 3 | 6 | 7 | 6 | 8 |
| **4i** | 4 | 3 | 4 | 3 | 3 | 3 | 3 | 5 | 5 | 5 | 5 | 3 | 2 | 1 | 2 | 2 | 5 | 5 | 7 | 4 | 4 | 2 | 2 | 4 | 3 | 5 | 4 | 4 | 5 | 3 | 4 | 5 | 6 | 5 | 7 |
| **4j** | 8 | 7 | 6 | 5 | 5 | 5 | 5 | 7 | 7 | 7 | 7 | 5 | 5 | 5 | 5 | 5 | 9 | 1 | 9 | 6 | 6 | 6 | 6 | 8 | 6 | 8 | 6 | 7 | 6 | 7 | 7 | 7 | 8 | 7 | 9 |
| **4k** | 9 | 9 | 7 | 6 | 5 | 9 | 7 | 9 | 9 | 9 | 9 | 7 | 6 | 5 | 6 | 7 | 1 | 9 | 5 | 8 | 8 | 6 | 4 | 2 | 8 | 10 | 10 | 8 | 8 | 3 | 9 | 11 | 9 | 9 | 10 |
| **4L** | 9 | 9 | 9 | 8 | 7 | 9 | 9 | 9 | 9 | 9 | 9 | 7 | 7 | 7 | 7 | 7 | 3 | 9 | 5 | 8 | 8 | 8 | 6 | 4 | 8 | 10 | 10 | 10 | 10 | 5 | 9 | 11 | 11 | 11 | 12 |
| **4m** | 9 | 9 | 9 | 9 | 8 | 9 | 9 | 9 | 9 | 9 | 9 | 7 | 9 | 8 | 9 | 9 | 5 | 9 | 1 | 8 | 8 | 8 | 7 | 6 | 8 | 10 | 10 | 10 | 10 | 7 | 9 | 11 | 11 | 11 | 12 |
| **4n** | 4 | 3 | 2 | 1 | 1 | 3 | 1 | 5 | 5 | 5 | 5 | 3 | 2 | 1 | 2 | 3 | 5 | 5 | 9 | 4 | 4 | 2 | 2 | 4 | 4 | 6 | 4 | 3 | 2 | 3 | 5 | 5 | 4 | 3 | 5 |
| **4o** | 6 | 5 | 6 | 5 | 5 | 5 | 5 | 5 | 5 | 5 | 5 | 3 | 3 | 3 | 3 | 3 | 7 | 5 | 7 | 4 | 4 | 4 | 4 | 6 | 4 | 6 | 6 | 7 | 6 | 5 | 5 | 7 | 8 | 7 | 9 |
| **4p** | 6 | 5 | 6 | 5 | 5 | 5 | 5 | 5 | 5 | 5 | 5 | 3 | 3 | 3 | 3 | 3 | 7 | 5 | 7 | 4 | 4 | 4 | 4 | 6 | 4 | 6 | 6 | 7 | 6 | 5 | 5 | 7 | 8 | 7 | 9 |
| **4q** | 6 | 5 | 4 | 3 | 3 | 1 | 3 | 7 | 7 | 7 | 7 | 5 | 4 | 3 | 4 | 4 | 7 | 5 | 9 | 6 | 6 | 4 | 4 | 6 | 5 | 7 | 2 | 5 | 4 | 5 | 6 | 3 | 6 | 5 | 7 |

| | | | | | | | | | | | | | | | | | | | | | | | | | | | | | | | | | | | |
|---|---|---|---|---|---|---|---|---|---|---|---|---|---|---|---|---|---|---|---|---|---|---|---|---|---|---|---|---|---|---|---|---|---|---|---|
| **5a** | 2 | 2 | 2 | 3 | 4 | 7 | 5 | 4 | 4 | 4 | 4 | 4 | 5 | 4 | 5 | 6 | 8 | 10 | 10 | 5 | 5 | 3 | 5 | 7 | 7 | 9 | 8 | 3 | 6 | 6 | 8 | 9 | 4 | 7 | 5 |
| **5b** | 2 | 2 | 2 | 3 | 4 | 7 | 5 | 4 | 4 | 4 | 4 | 4 | 5 | 4 | 5 | 6 | 8 | 10 | 10 | 5 | 5 | 3 | 5 | 7 | 7 | 9 | 8 | 3 | 6 | 6 | 8 | 9 | 4 | 7 | 5 |
| **5c** | x | 1 | 2 | 3 | 4 | 7 | 5 | 4 | 4 | 4 | 4 | 3 | 4 | 3 | 4 | 5 | 7 | 9 | 10 | 4 | 4 | 2 | 4 | 6 | 6 | 8 | 8 | 3 | 6 | 5 | 7 | 9 | 4 | 7 | 5 |
| **5d** | 1 | x | 2 | 3 | 4 | 6 | 4 | 4 | 4 | 4 | 4 | 2 | 3 | 2 | 3 | 4 | 6 | 8 | 10 | 3 | 3 | 1 | 3 | 5 | 5 | 7 | 7 | 3 | 5 | 4 | 6 | 8 | 4 | 6 | 5 |
| **5e** | 2 | 2 | x | 1 | 2 | 5 | 3 | 4 | 4 | 4 | 4 | 4 | 4 | 3 | 4 | 5 | 6 | 7 | 10 | 5 | 5 | 3 | 3 | 5 | 6 | 8 | 6 | 1 | 4 | 4 | 8 | 7 | 2 | 5 | 3 |
| **5f** | 3 | 3 | 1 | x | 1 | 4 | 2 | 5 | 5 | 5 | 5 | 3 | 3 | 2 | 3 | 4 | 6 | 6 | 10 | 4 | 4 | 2 | 2 | 4 | 5 | 7 | 5 | 2 | 3 | 3 | 6 | 6 | 3 | 4 | 4 |
| **5g** | 4 | 4 | 2 | 1 | x | 4 | 2 | 4 | 4 | 4 | 4 | 2 | 3 | 2 | 3 | 4 | 4 | 6 | 10 | 3 | 3 | 1 | 1 | 3 | 5 | 7 | 5 | 3 | 3 | 2 | 6 | 6 | 4 | 4 | 5 |
| **5h** | 7 | 6 | 5 | 4 | 4 | x | 4 | 8 | 8 | 8 | 8 | 6 | 5 | 4 | 5 | 5 | 8 | 6 | 10 | 7 | 7 | 5 | 5 | 7 | 6 | 7 | 1 | 6 | 5 | 6 | 7 | 2 | 7 | 6 | 8 |
| **5i** | 5 | 4 | 3 | 2 | 2 | 4 | x | 6 | 6 | 6 | 6 | 4 | 3 | 2 | 3 | 4 | 6 | 6 | 10 | 5 | 5 | 3 | 3 | 5 | 5 | 7 | 5 | 4 | 1 | 4 | 6 | 6 | 5 | 2 | 6 |
| **5j** | 4 | 4 | 4 | 5 | 4 | 8 | 6 | x | 2 | 2 | 2 | 2 | 3 | 4 | 5 | 6 | 8 | 8 | 10 | 3 | 3 | 3 | 5 | 7 | 8 | 9 | 9 | 5 | 7 | 6 | 9 | 10 | 6 | 8 | 7 |
| **5k** | 4 | 4 | 4 | 5 | 4 | 8 | 6 | 2 | x | 2 | 2 | 2 | 3 | 4 | 5 | 6 | 8 | 8 | 10 | 3 | 3 | 3 | 5 | 7 | 8 | 9 | 9 | 5 | 7 | 6 | 9 | 10 | 6 | 8 | 7 |
| **5L** | 4 | 4 | 4 | 5 | 4 | 8 | 6 | 2 | 2 | x | 2 | 2 | 3 | 4 | 5 | 6 | 8 | 8 | 10 | 3 | 3 | 3 | 5 | 7 | 8 | 9 | 9 | 5 | 7 | 6 | 9 | 10 | 6 | 8 | 7 |
| **5m** | 4 | 4 | 4 | 5 | 4 | 8 | 6 | 2 | 2 | 2 | x | 2 | 3 | 4 | 5 | 6 | 8 | 8 | 10 | 3 | 3 | 3 | 5 | 7 | 8 | 9 | 9 | 5 | 7 | 6 | 9 | 10 | 6 | 8 | 7 |
| **5n** | 3 | 2 | 4 | 3 | 2 | 6 | 4 | 2 | 2 | 2 | 2 | x | 1 | 2 | 3 | 4 | 6 | 6 | 8 | 1 | 1 | 1 | 3 | 5 | 6 | 7 | 7 | 5 | 5 | 4 | 7 | 8 | 6 | 6 | 7 |
| **5o** | 4 | 3 | 4 | 3 | 3 | 5 | 3 | 3 | 3 | 3 | 3 | 1 | x | 1 | 2 | 3 | 5 | 6 | 8 | 2 | 2 | 2 | 2 | 4 | 4 | 6 | 6 | 5 | 4 | 3 | 5 | 7 | 6 | 5 | 7 |
| **5p** | 3 | 2 | 3 | 2 | 2 | 4 | 2 | 4 | 4 | 4 | 4 | 2 | 1 | x | 1 | 2 | 4 | 6 | 8 | 3 | 3 | 1 | 1 | 3 | 3 | 5 | 5 | 4 | 3 | 2 | 4 | 6 | 5 | 4 | 6 |
| **5q** | 4 | 3 | 4 | 3 | 3 | 5 | 3 | 5 | 5 | 5 | 5 | 3 | 2 | 1 | x | 1 | 5 | 5 | 8 | 4 | 4 | 2 | 2 | 4 | 2 | 4 | 6 | 5 | 4 | 3 | 3 | 7 | 6 | 5 | 7 |
| **5r** | 5 | 4 | 5 | 4 | 4 | 5 | 4 | 6 | 6 | 6 | 6 | 4 | 3 | 2 | 1 | x | 6 | 4 | 8 | 5 | 5 | 3 | 3 | 5 | 1 | 3 | 6 | 6 | 5 | 4 | 2 | 7 | 7 | 6 | 8 |
| **5s** | 7 | 6 | 6 | 6 | 4 | 8 | 6 | 8 | 8 | 8 | 8 | 6 | 5 | 4 | 5 | 6 | x | 10 | 6 | 7 | 7 | 5 | 3 | 1 | 7 | 9 | 9 | 7 | 7 | 2 | 8 | 10 | 8 | 8 | 9 |
| **5t** | 9 | 8 | 7 | 6 | 6 | 6 | 6 | 8 | 8 | 8 | 8 | 6 | 6 | 6 | 5 | 4 | 10 | x | 10 | 7 | 7 | 7 | 7 | 9 | 3 | 1 | 7 | 7 | 8 | 8 | 2 | 8 | 9 | 8 | 10 |
| **5u** | 10 | 10 | 10 | 10 | 10 | 10 | 10 | 10 | 10 | 10 | 10 | 8 | 8 | 8 | 8 | 8 | 6 | 10 | x | 9 | 9 | 9 | 9 | 7 | 9 | 11 | 11 | 11 | 11 | 8 | 10 | 12 | 12 | 12 | 13 |
| **6a** | 4 | 3 | 5 | 4 | 3 | 7 | 5 | 3 | 3 | 3 | 3 | 1 | 2 | 3 | 4 | 5 | 7 | 7 | 9 | x | 2 | 2 | 4 | 6 | 6 | 8 | 8 | 6 | 6 | 5 | 7 | 9 | 7 | 7 | 8 |
| **6b** | 4 | 3 | 5 | 4 | 3 | 7 | 5 | 3 | 3 | 3 | 3 | 1 | 2 | 3 | 4 | 5 | 7 | 7 | 9 | 2 | x | 2 | 4 | 6 | 6 | 8 | 8 | 6 | 6 | 5 | 7 | 9 | 7 | 7 | 8 |
| **6c** | 2 | 1 | 3 | 2 | 1 | 5 | 3 | 3 | 3 | 3 | 3 | 1 | 2 | 1 | 2 | 3 | 5 | 7 | 9 | 2 | 2 | x | 2 | 4 | 4 | 6 | 6 | 4 | 4 | 3 | 5 | 7 | 5 | 5 | 6 |
| **6d** | 4 | 3 | 3 | 2 | 1 | 5 | 3 | 5 | 5 | 5 | 5 | 3 | 2 | 1 | 2 | 3 | 3 | 7 | 9 | 4 | 4 | 2 | x | 2 | 4 | 6 | 6 | 4 | 4 | 1 | 5 | 7 | 5 | 5 | 6 |
| **6e** | 6 | 5 | 5 | 4 | 3 | 7 | 5 | 7 | 7 | 7 | 7 | 5 | 4 | 3 | 4 | 5 | 1 | 9 | 7 | 6 | 6 | 4 | 2 | x | 6 | 8 | 8 | 6 | 6 | 1 | 7 | 9 | 7 | 7 | 8 |
| **6f** | 6 | 5 | 6 | 5 | 5 | 6 | 5 | 8 | 8 | 8 | 8 | 6 | 4 | 3 | 2 | 1 | 7 | 3 | 9 | 6 | 6 | 4 | 4 | 6 | x | 2 | 7 | 7 | 6 | 5 | 1 | 8 | 8 | 7 | 9 |
| **6g** | 8 | 7 | 8 | 7 | 7 | 7 | 7 | 9 | 9 | 9 | 9 | 7 | 6 | 5 | 4 | 3 | 9 | 1 | 11 | 8 | 8 | 6 | 6 | 8 | 2 | x | 8 | 9 | 8 | 7 | 1 | 9 | 10 | 9 | 11 |
| **6h** | 8 | 7 | 6 | 5 | 5 | 1 | 5 | 9 | 9 | 9 | 9 | 7 | 6 | 5 | 6 | 6 | 9 | 7 | 11 | 8 | 8 | 6 | 6 | 8 | 7 | 8 | x | 7 | 6 | 7 | 8 | 1 | 8 | 7 | 9 |
| **6i** | 3 | 3 | 1 | 2 | 3 | 6 | 4 | 5 | 5 | 5 | 5 | 5 | 5 | 4 | 5 | 6 | 7 | 7 | 11 | 6 | 6 | 4 | 4 | 6 | 7 | 9 | 7 | x | 5 | 5 | 8 | 8 | 1 | 6 | 2 |
| **6j** | 6 | 5 | 4 | 3 | 3 | 5 | 1 | 7 | 7 | 7 | 7 | 5 | 4 | 3 | 4 | 5 | 7 | 8 | 11 | 6 | 6 | 4 | 4 | 6 | 6 | 8 | 6 | 5 | x | 5 | 7 | 7 | 6 | 1 | 7 |
| **7a** | 5 | 4 | 4 | 3 | 2 | 6 | 4 | 6 | 6 | 6 | 6 | 4 | 3 | 2 | 3 | 4 | 2 | 8 | 8 | 5 | 5 | 3 | 1 | 1 | 5 | 7 | 7 | 5 | 5 | x | 6 | 8 | 6 | 6 | 7 |
| **7b** | 7 | 6 | 8 | 6 | 6 | 7 | 6 | 9 | 9 | 9 | 9 | 7 | 5 | 4 | 3 | 2 | 8 | 2 | 10 | 7 | 7 | 5 | 5 | 7 | 1 | 1 | 8 | 8 | 7 | 6 | x | 9 | 10 | 8 | 11 |
| **7c** | 9 | 8 | 7 | 6 | 6 | 2 | 6 | 10 | 10 | 10 | 10 | 8 | 7 | 6 | 7 | 7 | 10 | 8 | 12 | 9 | 9 | 7 | 7 | 9 | 8 | 9 | 1 | 8 | 7 | 8 | 9 | x | 9 | 8 | 10 |
| **7d** | 4 | 4 | 2 | 3 | 4 | 7 | 5 | 6 | 6 | 6 | 6 | 6 | 6 | 5 | 6 | 7 | 8 | 9 | 12 | 7 | 7 | 5 | 5 | 7 | 8 | 10 | 8 | 1 | 6 | 6 | 10 | 9 | x | 7 | 1 |
| **7e** | 7 | 6 | 5 | 4 | 4 | 6 | 2 | 8 | 8 | 8 | 8 | 6 | 5 | 4 | 5 | 6 | 8 | 8 | 12 | 7 | 7 | 5 | 5 | 7 | 7 | 9 | 7 | 6 | 1 | 6 | 8 | 8 | 7 | x | 8 |
| **8a** | 5 | 5 | 3 | 4 | 5 | 8 | 6 | 7 | 7 | 7 | 7 | 7 | 7 | 6 | 7 | 8 | 9 | 10 | 13 | 8 | 8 | 6 | 6 | 8 | 9 | 11 | 9 | 2 | 7 | 7 | 11 | 10 | 1 | 8 | x |

Palenque Palace Path Matrix Phase 5

| | c | 1a | 1b | 1c | 1d | 2a | 2b | 2c | 2d | 2e | 3a | 3b | 3c | 3d | 3e | 3f | 3g | 3h | 3i | 4a | 4b | 4c | 4d | 4e | 4f | 4g | 4h | 4i | 4j | 4k | 4L | 4m | 4n | 4o | 4p | 4q | 5a | 5b | 5c | 5d | 5e | 5f |
|---|---|---|---|---|---|---|---|---|---|---|---|---|---|---|---|---|---|---|---|---|---|---|---|---|---|---|---|---|---|---|---|---|---|---|---|---|---|---|---|---|---|---|
| **c** | x | 1 | 1 | 1 | 1 | 2 | 2 | 2 | 2 | 2 | 3 | 3 | 3 | 3 | 3 | 3 | 3 | 3 | 3 | 4 | 4 | 4 | 4 | 4 | 4 | 4 | 4 | 4 | 4 | 4 | 4 | 4 | 4 | 4 | 4 | 4 | 5 | 5 | 5 | 5 | 5 | 5 |
| **1a** | 1 | x | 1 | 2 | 2 | 1 | 2 | 3 | 3 | 3 | 2 | 2 | 3 | 3 | 4 | 4 | 4 | 4 | 4 | 3 | 3 | 3 | 3 | 4 | 4 | 5 | 5 | 5 | 4 | 5 | 5 | 5 | 5 | 5 | 5 | 4 | 4 | 4 | 4 | 4 | 4 | 6 |
| **1b** | 1 | 1 | x | 2 | 2 | 2 | 1 | 3 | 3 | 3 | 3 | 3 | 2 | 2 | 4 | 4 | 4 | 4 | 4 | 4 | 4 | 4 | 4 | 5 | 3 | 5 | 5 | 5 | 5 | 5 | 5 | 5 | 5 | 5 | 5 | 3 | 5 | 5 | 5 | 5 | 5 | 6 |
| **1c** | 1 | 2 | 2 | x | 2 | 3 | 3 | 1 | 3 | 3 | 4 | 4 | 3 | 2 | 2 | 4 | 4 | 4 | 2 | 5 | 5 | 5 | 4 | 3 | 3 | 3 | 3 | 3 | 3 | 3 | 3 | 5 | 5 | 5 | 3 | 3 | 6 | 6 | 6 | 6 | 6 | 4 |
| **1d** | 1 | 2 | 2 | 2 | x | 3 | 3 | 3 | 1 | 1 | 4 | 4 | 4 | 4 | 4 | 2 | 2 | 2 | 2 | 5 | 5 | 5 | 4 | 5 | 3 | 3 | 3 | 3 | 5 | 3 | 3 | 3 | 3 | 3 | 3 | 5 | 6 | 6 | 6 | 6 | 6 | 4 |
| **2a** | 2 | 1 | 2 | 3 | 3 | x | 3 | 4 | 4 | 4 | 1 | 1 | 4 | 4 | 5 | 5 | 5 | 5 | 5 | 2 | 2 | 2 | 2 | 3 | 5 | 4 | 5 | 6 | 5 | 6 | 5 | 6 | 6 | 6 | 5 | 5 | 3 | 3 | 3 | 3 | 3 | 4 |
| **2b** | 2 | 2 | 1 | 3 | 3 | 3 | x | 2 | 4 | 4 | 4 | 4 | 1 | 1 | 3 | 5 | 5 | 5 | 3 | 5 | 5 | 5 | 5 | 4 | 2 | 4 | 4 | 3 | 2 | 4 | 4 | 6 | 6 | 6 | 4 | 2 | 6 | 6 | 6 | 5 | 5 | 5 |
| **2c** | 2 | 3 | 3 | 1 | 3 | 4 | 2 | x | 2 | 4 | 4 | 4 | 2 | 1 | 1 | 5 | 5 | 5 | 1 | 5 | 5 | 4 | 3 | 2 | 2 | 2 | 2 | 2 | 2 | 2 | 2 | 6 | 6 | 6 | 2 | 2 | 7 | 7 | 6 | 5 | 5 | 3 |
| **2d** | 2 | 3 | 3 | 3 | 1 | 4 | 4 | 2 | x | 2 | 4 | 4 | 4 | 3 | 3 | 3 | 3 | 3 | 1 | 5 | 5 | 4 | 3 | 2 | 4 | 2 | 2 | 2 | 2 | 4 | 2 | 4 | 4 | 4 | 2 | 4 | 7 | 7 | 6 | 5 | 6 | 3 |
| **2e** | 2 | 3 | 3 | 3 | 1 | 4 | 4 | 4 | 2 | x | 5 | 5 | 5 | 5 | 5 | 1 | 1 | 1 | 3 | 6 | 6 | 6 | 5 | 4 | 6 | 4 | 4 | 4 | 4 | 6 | 4 | 2 | 2 | 2 | 4 | 6 | 7 | 7 | 7 | 7 | 7 | 5 |
| **3a** | 3 | 2 | 3 | 4 | 4 | 1 | 4 | 4 | 4 | 5 | x | 2 | 5 | 5 | 5 | 6 | 6 | 6 | 3 | 1 | 1 | 2 | 1 | 2 | 5 | 3 | 4 | 4 | 4 | 6 | 4 | 7 | 7 | 7 | 4 | 6 | 2 | 2 | 2 | 2 | 2 | 3 |
| **3b** | 3 | 2 | 3 | 4 | 4 | 1 | 4 | 4 | 4 | 5 | 2 | x | 5 | 5 | 6 | 6 | 6 | 6 | 6 | 3 | 3 | 1 | 1 | 2 | 6 | 3 | 4 | 4 | 4 | 6 | 4 | 7 | 7 | 7 | 4 | 6 | 4 | 4 | 4 | 4 | 4 | 3 |
| **3c** | 3 | 3 | 2 | 3 | 4 | 4 | 1 | 2 | 4 | 5 | 5 | 5 | x | 1 | 3 | 6 | 6 | 6 | 3 | 6 | 6 | 6 | 5 | 4 | 2 | 4 | 4 | 4 | 4 | 4 | 4 | 7 | 7 | 7 | 4 | 2 | 7 | 7 | 6 | 5 | 5 | 5 |
| **3d** | 3 | 3 | 2 | 2 | 4 | 4 | 1 | 1 | 3 | 5 | 5 | 5 | 1 | x | 2 | 6 | 6 | 6 | 2 | 6 | 6 | 5 | 4 | 4 | 1 | 3 | 3 | 2 | 1 | 3 | 3 | 7 | 7 | 7 | 3 | 1 | 6 | 6 | 5 | 4 | 4 | 4 |
| **3e** | 3 | 4 | 4 | 2 | 4 | 5 | 3 | 1 | 3 | 5 | 5 | 6 | 3 | 2 | x | 6 | 6 | 6 | 2 | 6 | 6 | 5 | 4 | 3 | 3 | 3 | 3 | 3 | 3 | 1 | 3 | 7 | 7 | 7 | 3 | 3 | 8 | 8 | 7 | 6 | 6 | 4 |
| **3f** | 3 | 4 | 4 | 4 | 2 | 5 | 5 | 5 | 3 | 1 | 6 | 6 | 6 | 6 | 6 | x | 2 | 2 | 4 | 7 | 7 | 7 | 6 | 5 | 7 | 5 | 5 | 5 | 5 | 7 | 5 | 1 | 1 | 3 | 5 | 7 | 8 | 8 | 8 | 8 | 8 | 6 |
| **3g** | 3 | 4 | 4 | 4 | 2 | 5 | 5 | 5 | 3 | 1 | 6 | 6 | 6 | 6 | 6 | 2 | x | 2 | 4 | 7 | 7 | 7 | 6 | 5 | 7 | 5 | 5 | 5 | 5 | 7 | 5 | 1 | 1 | 3 | 5 | 7 | 8 | 8 | 8 | 8 | 8 | 6 |
| **3h** | 3 | 4 | 4 | 4 | 2 | 5 | 5 | 5 | 3 | 1 | 6 | 6 | 6 | 6 | 6 | 2 | 2 | x | 4 | 7 | 7 | 7 | 6 | 5 | 7 | 5 | 5 | 5 | 5 | 7 | 5 | 3 | 3 | 1 | 5 | 7 | 8 | 8 | 8 | 8 | 8 | 6 |
| **3i** | 3 | 4 | 4 | 2 | 2 | 5 | 3 | 1 | 1 | 3 | 3 | 6 | 3 | 2 | 2 | 4 | 4 | 4 | x | 4 | 4 | 3 | 2 | 1 | 3 | 1 | 1 | 1 | 1 | 3 | 1 | 5 | 5 | 5 | 1 | 3 | 6 | 6 | 5 | 4 | 5 | 2 |
| **4a** | 4 | 3 | 4 | 5 | 5 | 2 | 5 | 5 | 5 | 6 | 1 | 3 | 6 | 6 | 6 | 7 | 7 | 7 | 4 | x | 1 | 3 | 2 | 3 | 5 | 4 | 5 | 5 | 5 | 7 | 5 | 8 | 8 | 8 | 5 | 7 | 2 | 2 | 2 | 2 | 2 | 4 |
| **4b** | 4 | 3 | 4 | 5 | 5 | 2 | 5 | 5 | 5 | 6 | 1 | 3 | 6 | 6 | 6 | 7 | 7 | 7 | 4 | 1 | x | 3 | 2 | 3 | 4 | 4 | 4 | 5 | 4 | 7 | 5 | 8 | 8 | 8 | 5 | 6 | 1 | 1 | 1 | 1 | 1 | 4 |
| **4c** | 4 | 3 | 4 | 5 | 5 | 2 | 5 | 4 | 4 | 6 | 2 | 1 | 6 | 5 | 5 | 7 | 7 | 7 | 3 | 3 | 3 | x | 1 | 2 | 6 | 3 | 4 | 4 | 4 | 6 | 4 | 8 | 8 | 8 | 4 | 7 | 4 | 4 | 4 | 4 | 4 | 3 |
| **4d** | 4 | 3 | 4 | 4 | 4 | 2 | 5 | 3 | 3 | 5 | 1 | 1 | 5 | 4 | 4 | 6 | 6 | 6 | 2 | 2 | 2 | 1 | x | 1 | 5 | 2 | 3 | 3 | 3 | 5 | 3 | 7 | 7 | 7 | 3 | 5 | 3 | 3 | 3 | 3 | 3 | 2 |
| **4e** | 4 | 4 | 5 | 3 | 5 | 3 | 4 | 2 | 2 | 4 | 2 | 2 | 4 | 4 | 3 | 5 | 5 | 5 | 1 | 3 | 3 | 2 | 1 | x | 4 | 1 | 2 | 2 | 2 | 4 | 2 | 6 | 6 | 6 | 2 | 4 | 4 | 4 | 4 | 4 | 4 | 1 |
| **4f** | 4 | 4 | 3 | 3 | 3 | 5 | 2 | 2 | 4 | 6 | 5 | 6 | 2 | 1 | 3 | 7 | 7 | 7 | 3 | 5 | 4 | 6 | 5 | 4 | x | 4 | 4 | 3 | 2 | 4 | 4 | 8 | 8 | 8 | 4 | 2 | 5 | 5 | 5 | 4 | 3 | 5 |
| **4g** | 4 | 5 | 5 | 3 | 3 | 4 | 4 | 2 | 2 | 4 | 3 | 3 | 4 | 3 | 3 | 5 | 5 | 5 | 1 | 4 | 4 | 3 | 2 | 1 | 4 | x | 1 | 2 | 2 | 4 | 2 | 6 | 6 | 6 | 2 | 4 | 5 | 5 | 5 | 5 | 5 | 2 |
| **4h** | 4 | 5 | 5 | 3 | 3 | 5 | 4 | 2 | 2 | 4 | 4 | 4 | 4 | 3 | 3 | 5 | 5 | 5 | 1 | 5 | 4 | 4 | 3 | 2 | 4 | 1 | x | 1 | 2 | 4 | 2 | 6 | 6 | 6 | 2 | 4 | 5 | 5 | 4 | 3 | 5 | 3 |
| **4i** | 4 | 5 | 5 | 3 | 3 | 6 | 3 | 2 | 2 | 4 | 4 | 4 | 4 | 2 | 3 | 5 | 5 | 5 | 1 | 5 | 5 | 4 | 3 | 2 | 3 | 2 | 1 | x | 1 | 4 | 2 | 6 | 6 | 6 | 2 | 3 | 6 | 6 | 5 | 4 | 6 | 3 |
| **4j** | 4 | 4 | 5 | 3 | 5 | 5 | 2 | 2 | 2 | 4 | 4 | 4 | 4 | 1 | 3 | 5 | 5 | 5 | 1 | 5 | 4 | 4 | 3 | 2 | 2 | 2 | 2 | 1 | x | 4 | 2 | 6 | 6 | 6 | 2 | 2 | 5 | 5 | 4 | 3 | 5 | 3 |
| **4k** | 4 | 5 | 5 | 3 | 3 | 6 | 4 | 2 | 4 | 6 | 6 | 6 | 4 | 3 | 1 | 7 | 7 | 7 | 3 | 7 | 7 | 6 | 5 | 4 | 4 | 4 | 4 | 4 | 4 | x | 4 | 8 | 8 | 8 | 4 | 4 | 8 | 8 | 8 | 7 | 7 | 5 |
| **4L** | 4 | 5 | 5 | 3 | 3 | 5 | 4 | 2 | 2 | 4 | 4 | 4 | 4 | 3 | 3 | 5 | 5 | 5 | 1 | 5 | 5 | 4 | 3 | 2 | 4 | 2 | 2 | 2 | 2 | 4 | x | 6 | 6 | 6 | 2 | 4 | 6 | 6 | 6 | 5 | 6 | 3 |
| **4m** | 4 | 5 | 5 | 5 | 3 | 6 | 6 | 6 | 4 | 2 | 7 | 7 | 7 | 7 | 7 | 1 | 1 | 3 | 5 | 8 | 8 | 8 | 7 | 6 | 8 | 6 | 6 | 6 | 6 | 8 | 6 | x | 2 | 4 | 6 | 8 | 9 | 9 | 8 | 7 | 7 | 7 |
| **4n** | 4 | 5 | 5 | 5 | 3 | 6 | 6 | 6 | 4 | 2 | 7 | 7 | 7 | 7 | 7 | 1 | 1 | 3 | 5 | 8 | 8 | 8 | 7 | 6 | 8 | 6 | 6 | 6 | 6 | 8 | 6 | 2 | x | 4 | 6 | 8 | 9 | 9 | 9 | 9 | 9 | 7 |
| **4o** | 4 | 5 | 5 | 5 | 3 | 6 | 6 | 6 | 4 | 2 | 7 | 7 | 7 | 7 | 7 | 3 | 3 | 1 | 5 | 8 | 8 | 8 | 7 | 6 | 8 | 6 | 6 | 6 | 6 | 8 | 6 | 4 | 4 | x | 6 | 8 | 9 | 9 | 9 | 9 | 9 | 7 |
| **4p** | 4 | 5 | 5 | 3 | 3 | 5 | 4 | 2 | 2 | 4 | 4 | 4 | 4 | 3 | 3 | 5 | 5 | 5 | 1 | 5 | 5 | 4 | 3 | 2 | 4 | 2 | 2 | 2 | 2 | 4 | 2 | 6 | 6 | 6 | x | 4 | 6 | 6 | 6 | 5 | 6 | 3 |
| **4q** | 4 | 4 | 3 | 3 | 5 | 5 | 2 | 2 | 4 | 6 | 6 | 6 | 2 | 1 | 3 | 7 | 7 | 7 | 3 | 7 | 6 | 7 | 5 | 4 | 2 | 4 | 4 | 3 | 2 | 4 | 4 | 8 | 8 | 8 | 4 | x | 7 | 7 | 6 | 5 | 5 | 5 |
| **5a** | 5 | 4 | 5 | 6 | 6 | 3 | 6 | 7 | 7 | 7 | 2 | 4 | 7 | 6 | 8 | 8 | 8 | 8 | 6 | 2 | 1 | 4 | 3 | 4 | 5 | 5 | 5 | 6 | 5 | 8 | 6 | 9 | 9 | 9 | 6 | 7 | x | 2 | 2 | 2 | 2 | 5 |
| **5b** | 5 | 4 | 5 | 6 | 6 | 3 | 6 | 7 | 7 | 7 | 2 | 4 | 7 | 6 | 8 | 8 | 8 | 8 | 6 | 2 | 1 | 4 | 3 | 4 | 5 | 5 | 5 | 6 | 5 | 8 | 6 | 9 | 9 | 9 | 6 | 7 | 2 | x | 2 | 2 | 2 | 5 |
| **5c** | 5 | 4 | 5 | 6 | 6 | 3 | 6 | 6 | 6 | 7 | 2 | 4 | 6 | 5 | 7 | 8 | 8 | 8 | 5 | 2 | 1 | 4 | 3 | 4 | 5 | 5 | 4 | 5 | 4 | 8 | 6 | 8 | 9 | 9 | 6 | 6 | 2 | 2 | x | 1 | 2 | 4 |
| **5d** | 5 | 4 | 5 | 6 | 6 | 3 | 5 | 5 | 5 | 7 | 2 | 4 | 5 | 4 | 6 | 8 | 8 | 8 | 4 | 2 | 1 | 4 | 3 | 4 | 4 | 5 | 3 | 4 | 3 | 7 | 5 | 7 | 9 | 9 | 5 | 5 | 2 | 2 | 1 | x | 2 | 3 |
| **5e** | 5 | 4 | 5 | 6 | 6 | 3 | 5 | 5 | 6 | 7 | 2 | 4 | 5 | 4 | 6 | 8 | 8 | 8 | 5 | 2 | 1 | 4 | 3 | 4 | 3 | 5 | 5 | 6 | 5 | 7 | 6 | 7 | 9 | 9 | 6 | 5 | 2 | 2 | 2 | 2 | x | 5 |

| | | | | | | | | | | | | | | | | | | | | | | | | | | | | | | | | | | | | | | | | | | |
|---|---|---|---|---|---|---|---|---|---|---|---|---|---|---|---|---|---|---|---|---|---|---|---|---|---|---|---|---|---|---|---|---|---|---|---|---|---|---|---|---|---|---|
| **5f** | 5 | 6 | 6 | 4 | 4 | 4 | 5 | 3 | 3 | 5 | 3 | 3 | 5 | 4 | 4 | 6 | 6 | 6 | 2 | 4 | 4 | 3 | 2 | 1 | 5 | 2 | 3 | 3 | 3 | 5 | 3 | 7 | 7 | 7 | 3 | 5 | 5 | 5 | 4 | 3 | 5 | x |
| **5g** | 5 | 4 | 5 | 5 | 5 | 3 | 6 | 4 | 4 | 6 | 2 | 2 | 6 | 5 | 5 | 7 | 7 | 7 | 3 | 3 | 3 | 2 | 1 | 2 | 6 | 3 | 4 | 4 | 4 | 6 | 4 | 8 | 8 | 8 | 4 | 6 | 4 | 4 | 4 | 4 | 4 | 3 |
| **5h** | 5 | 4 | 5 | 5 | 5 | 3 | 6 | 4 | 4 | 6 | 2 | 2 | 6 | 5 | 5 | 7 | 7 | 7 | 3 | 3 | 3 | 2 | 1 | 2 | 6 | 3 | 4 | 4 | 4 | 6 | 4 | 8 | 8 | 8 | 4 | 6 | 4 | 4 | 4 | 4 | 4 | 3 |
| **5i** | 5 | 4 | 5 | 5 | 5 | 3 | 6 | 4 | 4 | 6 | 2 | 2 | 6 | 5 | 5 | 7 | 7 | 7 | 3 | 3 | 3 | 2 | 1 | 2 | 6 | 3 | 4 | 4 | 4 | 6 | 4 | 8 | 8 | 8 | 4 | 6 | 4 | 4 | 4 | 4 | 4 | 3 |
| **5j** | 5 | 4 | 5 | 5 | 5 | 3 | 6 | 4 | 4 | 6 | 2 | 2 | 6 | 5 | 5 | 7 | 7 | 7 | 3 | 3 | 3 | 2 | 1 | 2 | 6 | 3 | 4 | 4 | 4 | 6 | 4 | 8 | 8 | 8 | 4 | 6 | 4 | 4 | 4 | 4 | 4 | 3 |
| **5k** | 5 | 5 | 5 | 4 | 4 | 4 | 4 | 3 | 3 | 5 | 3 | 3 | 4 | 3 | 4 | 6 | 6 | 6 | 2 | 4 | 4 | 3 | 2 | 1 | 4 | 2 | 2 | 3 | 2 | 5 | 3 | 6 | 7 | 7 | 3 | 4 | 5 | 5 | 4 | 3 | 5 | 2 |
| **5L** | 5 | 6 | 5 | 4 | 4 | 7 | 4 | 3 | 3 | 5 | 5 | 5 | 4 | 3 | 4 | 6 | 6 | 6 | 2 | 5 | 4 | 5 | 4 | 3 | 4 | 2 | 1 | 2 | 2 | 5 | 3 | 6 | 7 | 7 | 3 | 4 | 5 | 5 | 4 | 3 | 5 | 3 |
| **5m** | 5 | 5 | 4 | 4 | 4 | 5 | 3 | 3 | 3 | 5 | 4 | 4 | 3 | 2 | 4 | 6 | 6 | 6 | 2 | 4 | 3 | 4 | 3 | 2 | 3 | 2 | 1 | 2 | 1 | 5 | 3 | 5 | 7 | 7 | 3 | 3 | 4 | 4 | 3 | 2 | 4 | 2 |
| **5n** | 5 | 6 | 5 | 4 | 4 | 7 | 4 | 3 | 3 | 5 | 5 | 5 | 4 | 3 | 4 | 6 | 6 | 6 | 2 | 6 | 5 | 5 | 4 | 3 | 4 | 3 | 2 | 1 | 2 | 5 | 3 | 7 | 7 | 7 | 3 | 4 | 6 | 6 | 5 | 4 | 6 | 4 |
| **5o** | 5 | 6 | 6 | 4 | 6 | 7 | 5 | 3 | 5 | 7 | 7 | 6 | 5 | 4 | 2 | 8 | 8 | 8 | 4 | 8 | 8 | 7 | 6 | 5 | 5 | 5 | 5 | 5 | 5 | 1 | 5 | 9 | 9 | 9 | 5 | 5 | 9 | 9 | 9 | 8 | 9 | 6 |
| **5p** | 5 | 6 | 6 | 6 | 4 | 7 | 7 | 7 | 5 | 3 | 8 | 8 | 7 | 6 | 8 | 2 | 2 | 4 | 6 | 8 | 7 | 8 | 7 | 6 | 6 | 6 | 5 | 6 | 5 | 9 | 7 | 1 | 3 | 5 | 7 | 7 | 8 | 8 | 7 | 6 | 6 | 6 |
| **5q** | 5 | 6 | 6 | 6 | 4 | 7 | 7 | 7 | 5 | 3 | 8 | 8 | 8 | 8 | 8 | 4 | 4 | 2 | 6 | 9 | 9 | 9 | 8 | 7 | 9 | 7 | 7 | 7 | 7 | 9 | 7 | 5 | 5 | 1 | 7 | 9 | 10 | 10 | 10 | 10 | 10 | 8 |
| **5r** | 5 | 5 | 6 | 4 | 4 | 4 | 5 | 3 | 3 | 5 | 3 | 3 | 5 | 4 | 4 | 6 | 6 | 6 | 2 | 4 | 4 | 3 | 2 | 1 | 5 | 2 | 3 | 3 | 3 | 5 | 3 | 7 | 7 | 7 | 3 | 5 | 5 | 5 | 5 | 4 | 5 | 2 |
| **5s** | 5 | 5 | 4 | 4 | 6 | 5 | 3 | 3 | 5 | 7 | 4 | 6 | 3 | 2 | 4 | 7 | 7 | 8 | 4 | 4 | 3 | 6 | 5 | 5 | 1 | 5 | 4 | 4 | 3 | 5 | 5 | 6 | 8 | 9 | 5 | 3 | 4 | 4 | 4 | 3 | 2 | 4 |
| **5t** | 5 | 5 | 4 | 6 | 6 | 6 | 3 | 3 | 5 | 7 | 7 | 7 | 3 | 2 | 4 | 8 | 8 | 8 | 4 | 8 | 8 | 7 | 6 | 5 | 3 | 5 | 5 | 4 | 3 | 5 | 5 | 9 | 9 | 9 | 5 | 1 | 8 | 8 | 8 | 8 | 6 | 6 |
| **6a** | 6 | 6 | 7 | 5 | 5 | 5 | 6 | 4 | 4 | 6 | 4 | 4 | 6 | 5 | 5 | 7 | 7 | 7 | 3 | 5 | 5 | 4 | 3 | 2 | 6 | 3 | 4 | 4 | 4 | 6 | 4 | 8 | 8 | 8 | 4 | 6 | 6 | 6 | 5 | 4 | 6 | 1 |
| **6b** | 6 | 6 | 7 | 5 | 5 | 5 | 6 | 4 | 4 | 6 | 4 | 4 | 6 | 5 | 5 | 7 | 7 | 7 | 3 | 5 | 5 | 4 | 3 | 2 | 6 | 3 | 4 | 4 | 4 | 6 | 4 | 8 | 8 | 8 | 4 | 6 | 6 | 6 | 5 | 4 | 6 | 1 |
| **6c** | 6 | 6 | 6 | 5 | 5 | 5 | 5 | 4 | 4 | 6 | 4 | 4 | 5 | 4 | 5 | 7 | 7 | 7 | 3 | 5 | 4 | 4 | 3 | 2 | 5 | 3 | 3 | 4 | 3 | 6 | 4 | 7 | 8 | 8 | 4 | 5 | 5 | 5 | 4 | 3 | 5 | 3 |
| **6d** | 6 | 6 | 5 | 5 | 5 | 5 | 4 | 4 | 4 | 6 | 4 | 4 | 4 | 2 | 5 | 7 | 7 | 7 | 3 | 4 | 3 | 4 | 3 | 2 | 4 | 3 | 2 | 3 | 2 | 6 | 4 | 6 | 8 | 8 | 4 | 4 | 4 | 4 | 3 | 2 | 4 | 1 |
| **6e** | 6 | 5 | 5 | 5 | 5 | 4 | 4 | 4 | 4 | 6 | 3 | 5 | 4 | 2 | 5 | 7 | 7 | 7 | 3 | 3 | 2 | 5 | 4 | 3 | 3 | 3 | 2 | 3 | 2 | 6 | 4 | 6 | 8 | 8 | 4 | 4 | 3 | 3 | 2 | 1 | 3 | 2 |
| **6f** | 6 | 6 | 5 | 6 | 6 | 5 | 4 | 4 | 5 | 7 | 4 | 6 | 4 | 3 | 5 | 6 | 6 | 8 | 4 | 4 | 3 | 6 | 5 | 4 | 2 | 4 | 3 | 4 | 3 | 6 | 5 | 5 | 7 | 9 | 5 | 4 | 4 | 4 | 4 | 4 | 2 | 3 |
| **6g** | 6 | 6 | 5 | 7 | 7 | 6 | 4 | 4 | 6 | 8 | 5 | 7 | 4 | 3 | 5 | 7 | 7 | 9 | 5 | 5 | 4 | 7 | 6 | 5 | 2 | 5 | 4 | 5 | 4 | 6 | 6 | 6 | 8 | 10 | 6 | 4 | 5 | 5 | 5 | 5 | 3 | 4 |
| **6h** | 6 | 6 | 5 | 7 | 7 | 7 | 4 | 4 | 6 | 8 | 5 | 7 | 4 | 3 | 5 | 8 | 8 | 9 | 5 | 5 | 4 | 7 | 6 | 6 | 2 | 6 | 5 | 5 | 4 | 6 | 6 | 7 | 9 | 10 | 6 | 4 | 5 | 5 | 5 | 5 | 3 | 5 |
| **6i** | 6 | 6 | 5 | 7 | 7 | 7 | 4 | 4 | 6 | 8 | 5 | 7 | 4 | 3 | 5 | 8 | 8 | 9 | 5 | 5 | 4 | 7 | 6 | 6 | 2 | 6 | 5 | 5 | 4 | 6 | 6 | 7 | 9 | 10 | 6 | 4 | 5 | 5 | 5 | 5 | 3 | 5 |
| **6j** | 6 | 6 | 5 | 7 | 7 | 7 | 4 | 4 | 6 | 8 | 5 | 7 | 4 | 3 | 5 | 8 | 8 | 9 | 5 | 5 | 4 | 7 | 6 | 6 | 2 | 6 | 5 | 5 | 4 | 6 | 6 | 7 | 9 | 10 | 6 | 4 | 5 | 5 | 5 | 5 | 3 | 5 |
| **6k** | 6 | 6 | 5 | 7 | 7 | 7 | 4 | 4 | 6 | 8 | 5 | 7 | 4 | 3 | 5 | 8 | 8 | 9 | 5 | 5 | 4 | 7 | 6 | 6 | 2 | 6 | 5 | 5 | 4 | 6 | 6 | 7 | 9 | 10 | 6 | 4 | 5 | 5 | 5 | 5 | 3 | 5 |
| **6L** | 6 | 6 | 5 | 7 | 7 | 7 | 4 | 4 | 6 | 8 | 8 | 8 | 4 | 3 | 5 | 9 | 9 | 9 | 5 | 9 | 8 | 8 | 7 | 6 | 4 | 6 | 6 | 5 | 4 | 6 | 6 | 10 | 10 | 10 | 6 | 2 | 9 | 9 | 9 | 9 | 7 | 7 |
| **6m** | 6 | 5 | 5 | 7 | 7 | 4 | 4 | 4 | 6 | 8 | 3 | 5 | 4 | 3 | 5 | 7 | 7 | 9 | 5 | 3 | 2 | 5 | 4 | 5 | 2 | 5 | 4 | 5 | 4 | 6 | 6 | 6 | 8 | 10 | 6 | 4 | 3 | 3 | 3 | 3 | 1 | 4 |
| **6n** | 6 | 5 | 6 | 7 | 7 | 4 | 6 | 6 | 7 | 8 | 3 | 5 | 6 | 5 | 7 | 9 | 9 | 9 | 6 | 3 | 2 | 5 | 4 | 5 | 4 | 6 | 6 | 7 | 6 | 8 | 7 | 8 | 10 | 10 | 7 | 6 | 3 | 3 | 3 | 3 | 1 | 6 |
| **6o** | 6 | 6 | 5 | 5 | 5 | 6 | 4 | 4 | 4 | 6 | 5 | 5 | 4 | 3 | 5 | 5 | 5 | 7 | 3 | 5 | 4 | 5 | 4 | 3 | 3 | 3 | 2 | 3 | 2 | 6 | 4 | 4 | 6 | 8 | 4 | 4 | 5 | 5 | 4 | 3 | 3 | 3 |
| **6p** | 6 | 6 | 5 | 5 | 5 | 6 | 4 | 4 | 4 | 6 | 5 | 6 | 4 | 3 | 5 | 7 | 7 | 7 | 3 | 5 | 4 | 5 | 4 | 3 | 4 | 3 | 2 | 2 | 2 | 6 | 4 | 6 | 8 | 8 | 4 | 4 | 5 | 5 | 4 | 3 | 5 | 3 |
| **6q** | 6 | 7 | 7 | 7 | 5 | 8 | 6 | 6 | 6 | 4 | 7 | 7 | 6 | 5 | 7 | 3 | 3 | 5 | 5 | 7 | 6 | 7 | 6 | 5 | 5 | 5 | 4 | 5 | 4 | 8 | 6 | 2 | 4 | 6 | 6 | 6 | 7 | 7 | 6 | 5 | 5 | 5 |
| **6r** | 6 | 7 | 6 | 5 | 5 | 8 | 5 | 4 | 4 | 6 | 7 | 6 | 5 | 4 | 5 | 7 | 7 | 7 | 3 | 7 | 6 | 6 | 5 | 4 | 5 | 4 | 3 | 2 | 3 | 4 | 4 | 8 | 8 | 8 | 4 | 5 | 7 | 7 | 6 | 5 | 7 | 5 |
| **6s** | 6 | 7 | 7 | 5 | 7 | 8 | 6 | 4 | 6 | 7 | 8 | 8 | 6 | 5 | 3 | 9 | 9 | 9 | 5 | 9 | 8 | 8 | 7 | 6 | 6 | 6 | 5 | 4 | 5 | 2 | 6 | 10 | 10 | 10 | 6 | 6 | 9 | 9 | 8 | 7 | 9 | 7 |
| **7a** | 7 | 6 | 7 | 8 | 8 | 5 | 7 | 7 | 8 | 9 | 4 | 6 | 7 | 6 | 8 | 10 | 10 | 10 | 7 | 4 | 3 | 6 | 5 | 6 | 5 | 7 | 7 | 8 | 7 | 9 | 8 | 9 | 11 | 11 | 8 | 7 | 4 | 4 | 4 | 4 | 2 | 7 |
| **7b** | 7 | 7 | 6 | 8 | 8 | 7 | 5 | 5 | 7 | 9 | 6 | 8 | 5 | 4 | 6 | 9 | 9 | 10 | 6 | 6 | 5 | 8 | 7 | 7 | 3 | 7 | 6 | 6 | 5 | 7 | 7 | 8 | 10 | 11 | 7 | 5 | 6 | 6 | 6 | 6 | 4 | 6 |
| **7c** | 7 | 8 | 7 | 6 | 6 | 9 | 6 | 5 | 5 | 7 | 7 | 7 | 6 | 5 | 6 | 8 | 8 | 8 | 4 | 8 | 7 | 7 | 6 | 5 | 6 | 5 | 4 | 3 | 4 | 3 | 5 | 9 | 9 | 9 | 5 | 6 | 8 | 8 | 7 | 6 | 8 | 6 |
| **7d** | 7 | 7 | 6 | 6 | 6 | 7 | 5 | 5 | 5 | 5 | 6 | 6 | 5 | 4 | 6 | 4 | 4 | 6 | 4 | 6 | 5 | 6 | 5 | 4 | 4 | 4 | 3 | 4 | 3 | 7 | 5 | 3 | 5 | 7 | 5 | 5 | 6 | 6 | 5 | 4 | 4 | 4 |
| **7e** | 7 | 7 | 6 | 8 | 8 | 7 | 5 | 5 | 7 | 9 | 6 | 8 | 5 | 4 | 6 | 8 | 8 | 10 | 6 | 6 | 5 | 8 | 7 | 6 | 3 | 6 | 5 | 6 | 5 | 7 | 7 | 7 | 9 | 11 | 7 | 5 | 6 | 6 | 6 | 6 | 4 | 5 |
| **7f** | 7 | 7 | 6 | 8 | 8 | 8 | 5 | 5 | 7 | 9 | 9 | 9 | 5 | 4 | 6 | 10 | 10 | 10 | 6 | 10 | 9 | 9 | 8 | 7 | 5 | 7 | 7 | 6 | 5 | 7 | 7 | 11 | 11 | 11 | 7 | 3 | 10 | 10 | 10 | 10 | 8 | 8 |
| **8a** | 8 | 7 | 8 | 9 | 9 | 6 | 8 | 8 | 9 | 10 | 5 | 7 | 8 | 7 | 9 | 11 | 11 | 11 | 8 | 5 | 4 | 7 | 6 | 7 | 6 | 8 | 8 | 9 | 8 | 10 | 9 | 10 | 12 | 12 | 9 | 8 | 5 | 5 | 5 | 5 | 3 | 8 |
| **8b** | 8 | 8 | 7 | 9 | 9 | 8 | 6 | 6 | 8 | 10 | 7 | 9 | 6 | 5 | 7 | 10 | 10 | 11 | 7 | 7 | 6 | 9 | 8 | 8 | 4 | 8 | 7 | 7 | 6 | 8 | 8 | 9 | 11 | 12 | 8 | 6 | 7 | 7 | 7 | 7 | 5 | 7 |
| **8c** | 8 | 8 | 7 | 9 | 9 | 8 | 6 | 6 | 8 | 10 | 7 | 9 | 6 | 5 | 7 | 9 | 9 | 11 | 7 | 7 | 6 | 9 | 8 | 7 | 4 | 7 | 6 | 7 | 6 | 8 | 8 | 8 | 10 | 12 | 8 | 6 | 7 | 7 | 7 | 7 | 5 | 6 |

*(continued)*

Palenque Palace Path Matrix Phase 5 (Cont.)

| | 5g | 5h | 5i | 5j | 5k | 5L | 5m | 5n | 5o | 5p | 5q | 5r | 5s | 5t | 6a | 6b | 6c | 6d | 6e | 6f | 6g | 6h | 6i | 6j | 6k | 6L | 6m | 6n | 6o | 6p | 6q | 6r | 6s | 7a | 7b | 7c | 7d | 7e | 7f | 8a | 8b | 8c |
|---|---|---|---|---|---|---|---|---|---|---|---|---|---|---|---|---|---|---|---|---|---|---|---|---|---|---|---|---|---|---|---|---|---|---|---|---|---|---|---|---|---|---|
| c | 5 | 5 | 5 | 5 | 5 | 5 | 5 | 5 | 5 | 5 | 5 | 5 | 5 | 5 | 6 | 6 | 6 | 6 | 6 | 6 | 6 | 6 | 6 | 6 | 6 | 6 | 6 | 6 | 6 | 6 | 6 | 6 | 6 | 7 | 7 | 7 | 7 | 7 | 7 | 8 | 8 | 8 |
| 1a | 4 | 4 | 4 | 4 | 5 | 6 | 5 | 6 | 6 | 6 | 6 | 5 | 5 | 5 | 6 | 6 | 6 | 6 | 5 | 6 | 6 | 6 | 6 | 6 | 6 | 6 | 5 | 5 | 6 | 6 | 7 | 7 | 7 | 6 | 7 | 8 | 7 | 7 | 7 | 7 | 8 | 8 |
| 1b | 5 | 5 | 5 | 5 | 5 | 5 | 4 | 5 | 6 | 6 | 6 | 6 | 4 | 4 | 7 | 7 | 6 | 5 | 5 | 5 | 5 | 5 | 5 | 5 | 5 | 5 | 5 | 6 | 5 | 5 | 7 | 6 | 7 | 7 | 6 | 7 | 6 | 6 | 6 | 8 | 7 | 7 |
| 1c | 5 | 5 | 5 | 5 | 4 | 4 | 4 | 4 | 4 | 6 | 6 | 4 | 4 | 6 | 5 | 5 | 5 | 5 | 5 | 6 | 7 | 7 | 7 | 7 | 7 | 7 | 7 | 7 | 5 | 5 | 7 | 5 | 5 | 8 | 8 | 6 | 6 | 8 | 8 | 9 | 9 | 9 |
| 1d | 5 | 5 | 5 | 5 | 4 | 4 | 4 | 4 | 6 | 4 | 4 | 4 | 6 | 6 | 5 | 5 | 5 | 5 | 5 | 6 | 7 | 7 | 7 | 7 | 7 | 7 | 7 | 7 | 5 | 5 | 5 | 5 | 7 | 8 | 8 | 6 | 6 | 8 | 8 | 9 | 9 | 9 |
| 2a | 3 | 3 | 3 | 3 | 4 | 7 | 5 | 7 | 7 | 7 | 7 | 4 | 5 | 6 | 5 | 5 | 5 | 5 | 4 | 5 | 6 | 7 | 7 | 7 | 7 | 7 | 4 | 4 | 6 | 6 | 8 | 8 | 8 | 5 | 7 | 9 | 7 | 7 | 8 | 6 | 8 | 8 |
| 2b | 6 | 6 | 6 | 6 | 4 | 4 | 3 | 4 | 5 | 7 | 7 | 5 | 3 | 3 | 6 | 6 | 5 | 4 | 4 | 4 | 4 | 4 | 4 | 4 | 4 | 4 | 4 | 6 | 4 | 4 | 6 | 5 | 6 | 7 | 5 | 6 | 5 | 5 | 5 | 8 | 6 | 6 |
| 2c | 4 | 4 | 4 | 4 | 3 | 3 | 3 | 3 | 3 | 7 | 7 | 3 | 3 | 3 | 4 | 4 | 4 | 4 | 4 | 4 | 4 | 4 | 4 | 4 | 4 | 4 | 4 | 6 | 4 | 4 | 6 | 4 | 4 | 7 | 5 | 5 | 5 | 5 | 5 | 8 | 6 | 6 |
| 2d | 4 | 4 | 4 | 4 | 3 | 3 | 3 | 3 | 5 | 5 | 5 | 3 | 5 | 5 | 4 | 4 | 4 | 4 | 4 | 5 | 6 | 6 | 6 | 6 | 6 | 6 | 6 | 7 | 4 | 4 | 6 | 4 | 6 | 8 | 7 | 5 | 5 | 7 | 7 | 9 | 8 | 8 |
| 2e | 6 | 6 | 6 | 6 | 5 | 5 | 5 | 5 | 7 | 3 | 3 | 5 | 7 | 7 | 6 | 6 | 6 | 6 | 6 | 7 | 8 | 8 | 8 | 8 | 8 | 8 | 8 | 8 | 6 | 6 | 4 | 6 | 7 | 9 | 9 | 7 | 5 | 9 | 9 | 10 | 10 | 10 |
| 3a | 2 | 2 | 2 | 2 | 3 | 5 | 4 | 5 | 7 | 8 | 8 | 3 | 4 | 7 | 4 | 4 | 4 | 4 | 3 | 4 | 5 | 5 | 5 | 5 | 5 | 8 | 3 | 3 | 5 | 5 | 7 | 7 | 8 | 4 | 6 | 7 | 6 | 6 | 9 | 5 | 7 | 7 |
| 3b | 2 | 2 | 2 | 2 | 3 | 5 | 4 | 5 | 6 | 8 | 8 | 3 | 6 | 7 | 4 | 4 | 4 | 4 | 5 | 6 | 7 | 7 | 7 | 7 | 7 | 8 | 5 | 5 | 5 | 6 | 7 | 6 | 8 | 6 | 8 | 7 | 6 | 8 | 9 | 7 | 9 | 9 |
| 3c | 6 | 6 | 6 | 6 | 4 | 4 | 3 | 4 | 5 | 7 | 8 | 5 | 3 | 3 | 6 | 6 | 5 | 4 | 4 | 4 | 4 | 4 | 4 | 4 | 4 | 4 | 4 | 6 | 4 | 4 | 6 | 5 | 6 | 7 | 5 | 6 | 5 | 5 | 5 | 8 | 6 | 6 |
| 3d | 5 | 5 | 5 | 5 | 3 | 3 | 2 | 3 | 4 | 6 | 8 | 4 | 2 | 2 | 5 | 5 | 4 | 2 | 2 | 3 | 3 | 3 | 3 | 3 | 3 | 3 | 3 | 5 | 3 | 3 | 5 | 4 | 5 | 6 | 4 | 5 | 4 | 4 | 4 | 7 | 5 | 5 |
| 3e | 5 | 5 | 5 | 5 | 4 | 4 | 4 | 4 | 2 | 8 | 8 | 4 | 4 | 4 | 5 | 5 | 5 | 5 | 5 | 5 | 5 | 5 | 5 | 5 | 5 | 5 | 5 | 7 | 5 | 5 | 7 | 5 | 3 | 8 | 6 | 6 | 6 | 6 | 6 | 9 | 7 | 7 |
| 3f | 7 | 7 | 7 | 7 | 6 | 6 | 6 | 6 | 8 | 2 | 4 | 6 | 7 | 8 | 7 | 7 | 7 | 7 | 7 | 6 | 7 | 8 | 8 | 8 | 8 | 9 | 7 | 9 | 5 | 7 | 3 | 7 | 9 | 10 | 9 | 8 | 4 | 8 | 10 | 11 | 10 | 9 |
| 3g | 7 | 7 | 7 | 7 | 6 | 6 | 6 | 6 | 8 | 2 | 4 | 6 | 7 | 8 | 7 | 7 | 7 | 7 | 7 | 6 | 7 | 8 | 8 | 8 | 8 | 9 | 7 | 9 | 5 | 7 | 3 | 7 | 9 | 10 | 9 | 8 | 4 | 8 | 10 | 11 | 10 | 9 |
| 3h | 7 | 7 | 7 | 7 | 6 | 6 | 6 | 6 | 8 | 4 | 2 | 6 | 8 | 8 | 7 | 7 | 7 | 7 | 7 | 8 | 9 | 9 | 9 | 9 | 9 | 9 | 9 | 9 | 7 | 7 | 5 | 7 | 9 | 10 | 10 | 8 | 6 | 10 | 10 | 11 | 11 | 11 |
| 3i | 3 | 3 | 3 | 3 | 2 | 2 | 2 | 2 | 4 | 6 | 6 | 2 | 4 | 4 | 3 | 3 | 3 | 3 | 3 | 4 | 5 | 5 | 5 | 5 | 5 | 5 | 5 | 6 | 3 | 3 | 5 | 3 | 5 | 7 | 6 | 4 | 4 | 6 | 6 | 8 | 7 | 7 |
| 4a | 3 | 3 | 3 | 3 | 4 | 5 | 4 | 6 | 8 | 8 | 9 | 4 | 4 | 8 | 5 | 5 | 5 | 4 | 3 | 4 | 5 | 5 | 5 | 5 | 5 | 9 | 3 | 3 | 5 | 5 | 7 | 7 | 9 | 4 | 6 | 8 | 6 | 6 | 10 | 5 | 7 | 7 |
| 4b | 3 | 3 | 3 | 3 | 4 | 4 | 3 | 5 | 8 | 7 | 9 | 4 | 3 | 8 | 5 | 5 | 4 | 3 | 2 | 3 | 4 | 4 | 4 | 4 | 4 | 8 | 2 | 2 | 4 | 4 | 6 | 6 | 8 | 3 | 5 | 7 | 5 | 5 | 9 | 4 | 6 | 6 |
| 4c | 2 | 2 | 2 | 2 | 3 | 5 | 4 | 5 | 7 | 8 | 9 | 3 | 6 | 7 | 4 | 4 | 4 | 4 | 5 | 6 | 7 | 7 | 7 | 7 | 7 | 8 | 5 | 5 | 5 | 5 | 7 | 6 | 8 | 6 | 8 | 7 | 6 | 8 | 9 | 7 | 9 | 9 |
| 4d | 1 | 1 | 1 | 1 | 2 | 4 | 3 | 4 | 6 | 7 | 8 | 2 | 5 | 6 | 3 | 3 | 3 | 3 | 4 | 5 | 6 | 6 | 6 | 6 | 6 | 7 | 4 | 4 | 4 | 4 | 6 | 5 | 7 | 5 | 7 | 6 | 5 | 7 | 8 | 6 | 8 | 8 |
| 4e | 2 | 2 | 2 | 2 | 1 | 3 | 2 | 3 | 5 | 6 | 7 | 1 | 5 | 5 | 2 | 2 | 2 | 2 | 3 | 4 | 5 | 6 | 6 | 6 | 6 | 6 | 5 | 5 | 3 | 3 | 5 | 4 | 6 | 6 | 7 | 5 | 4 | 6 | 7 | 7 | 8 | 7 |
| 4f | 6 | 6 | 6 | 6 | 4 | 4 | 3 | 4 | 5 | 6 | 9 | 5 | 1 | 3 | 6 | 6 | 5 | 4 | 3 | 2 | 2 | 2 | 2 | 2 | 2 | 4 | 2 | 4 | 3 | 4 | 5 | 5 | 6 | 5 | 3 | 6 | 4 | 3 | 5 | 6 | 4 | 4 |
| 4g | 3 | 3 | 3 | 3 | 2 | 2 | 2 | 3 | 5 | 6 | 7 | 2 | 5 | 5 | 3 | 3 | 3 | 3 | 3 | 4 | 5 | 6 | 6 | 6 | 6 | 6 | 5 | 6 | 3 | 3 | 5 | 4 | 6 | 7 | 7 | 5 | 4 | 6 | 7 | 8 | 8 | 7 |
| 4h | 4 | 4 | 4 | 4 | 2 | 1 | 1 | 2 | 5 | 5 | 7 | 3 | 4 | 5 | 4 | 4 | 3 | 2 | 2 | 3 | 4 | 5 | 5 | 5 | 5 | 6 | 4 | 6 | 2 | 2 | 4 | 3 | 5 | 7 | 6 | 4 | 3 | 5 | 7 | 8 | 7 | 6 |
| 4i | 4 | 4 | 4 | 4 | 3 | 2 | 2 | 1 | 5 | 6 | 7 | 3 | 4 | 4 | 4 | 4 | 4 | 3 | 3 | 4 | 5 | 5 | 5 | 5 | 5 | 5 | 5 | 7 | 3 | 2 | 5 | 2 | 4 | 8 | 6 | 3 | 4 | 6 | 6 | 9 | 7 | 7 |
| 4j | 4 | 4 | 4 | 4 | 2 | 2 | 1 | 2 | 5 | 5 | 7 | 3 | 3 | 3 | 4 | 4 | 3 | 2 | 2 | 3 | 4 | 4 | 4 | 4 | 4 | 4 | 4 | 6 | 2 | 2 | 4 | 3 | 5 | 7 | 5 | 4 | 3 | 5 | 5 | 8 | 6 | 6 |
| 4k | 6 | 6 | 6 | 6 | 5 | 5 | 5 | 5 | 1 | 9 | 9 | 5 | 5 | 5 | 6 | 6 | 6 | 6 | 6 | 6 | 6 | 6 | 6 | 6 | 6 | 6 | 6 | 8 | 6 | 6 | 8 | 4 | 2 | 9 | 7 | 3 | 7 | 7 | 7 | 10 | 8 | 8 |
| 4L | 4 | 4 | 4 | 4 | 3 | 3 | 3 | 3 | 5 | 7 | 7 | 3 | 5 | 5 | 4 | 4 | 4 | 4 | 4 | 5 | 6 | 6 | 6 | 6 | 6 | 6 | 6 | 7 | 4 | 4 | 6 | 4 | 6 | 8 | 7 | 5 | 5 | 7 | 7 | 9 | 8 | 8 |
| 4m | 8 | 8 | 8 | 8 | 6 | 6 | 5 | 7 | 9 | 1 | 5 | 7 | 6 | 9 | 8 | 8 | 7 | 6 | 6 | 5 | 6 | 7 | 7 | 7 | 7 | 10 | 6 | 8 | 4 | 6 | 2 | 8 | 10 | 9 | 8 | 9 | 3 | 7 | 11 | 10 | 9 | 8 |
| 4n | 8 | 8 | 8 | 8 | 7 | 7 | 7 | 7 | 9 | 3 | 5 | 7 | 8 | 9 | 8 | 8 | 8 | 8 | 8 | 7 | 8 | 9 | 9 | 9 | 9 | 10 | 8 | 10 | 6 | 8 | 4 | 8 | 10 | 11 | 10 | 9 | 5 | 9 | 11 | 12 | 11 | 10 |
| 4o | 8 | 8 | 8 | 8 | 7 | 7 | 7 | 7 | 9 | 5 | 1 | 7 | 9 | 9 | 8 | 8 | 8 | 8 | 8 | 9 | 10 | 10 | 10 | 10 | 10 | 10 | 10 | 10 | 8 | 8 | 6 | 8 | 10 | 11 | 11 | 9 | 7 | 11 | 11 | 12 | 12 | 12 |
| 4p | 4 | 4 | 4 | 4 | 3 | 3 | 3 | 3 | 5 | 7 | 7 | 3 | 5 | 5 | 4 | 4 | 4 | 4 | 4 | 5 | 6 | 6 | 6 | 6 | 6 | 6 | 6 | 7 | 4 | 4 | 6 | 4 | 6 | 8 | 7 | 5 | 5 | 7 | 7 | 9 | 8 | 8 |
| 4q | 6 | 6 | 6 | 6 | 4 | 4 | 3 | 4 | 5 | 7 | 9 | 5 | 3 | 1 | 6 | 6 | 5 | 4 | 4 | 4 | 4 | 4 | 4 | 4 | 4 | 2 | 4 | 6 | 4 | 4 | 6 | 5 | 6 | 7 | 5 | 6 | 5 | 5 | 3 | 8 | 6 | 6 |
| 5a | 4 | 4 | 4 | 4 | 5 | 5 | 4 | 6 | 9 | 8 | 10 | 5 | 4 | 8 | 6 | 6 | 5 | 4 | 3 | 4 | 5 | 5 | 5 | 5 | 5 | 9 | 3 | 3 | 5 | 5 | 7 | 7 | 9 | 4 | 6 | 8 | 6 | 6 | 10 | 5 | 7 | 7 |
| 5b | 4 | 4 | 4 | 4 | 5 | 5 | 4 | 6 | 9 | 8 | 10 | 5 | 4 | 8 | 6 | 6 | 5 | 4 | 3 | 4 | 5 | 5 | 5 | 5 | 5 | 9 | 3 | 3 | 5 | 5 | 7 | 7 | 9 | 4 | 6 | 8 | 6 | 6 | 10 | 5 | 7 | 7 |
| 5c | 4 | 4 | 4 | 4 | 4 | 4 | 3 | 5 | 9 | 7 | 10 | 5 | 4 | 8 | 5 | 5 | 4 | 3 | 2 | 4 | 5 | 5 | 5 | 5 | 5 | 9 | 3 | 3 | 4 | 4 | 6 | 6 | 8 | 4 | 6 | 7 | 5 | 6 | 10 | 5 | 7 | 7 |
| 5d | 4 | 4 | 4 | 4 | 3 | 3 | 2 | 4 | 8 | 6 | 10 | 4 | 3 | 8 | 4 | 4 | 3 | 2 | 1 | 4 | 5 | 5 | 5 | 5 | 5 | 9 | 3 | 3 | 3 | 3 | 5 | 5 | 7 | 4 | 6 | 6 | 4 | 6 | 10 | 5 | 7 | 7 |
| 5e | 4 | 4 | 4 | 4 | 5 | 5 | 4 | 6 | 9 | 6 | 10 | 5 | 2 | 6 | 6 | 6 | 5 | 4 | 3 | 2 | 3 | 3 | 3 | 3 | 3 | 7 | 1 | 1 | 3 | 5 | 5 | 7 | 9 | 2 | 4 | 8 | 4 | 4 | 8 | 3 | 5 | 5 |

| | | | | | | | | | | | | | | | | | | | | | | | | | | | | | | | | | | | | | | | | | | |
|---|---|---|---|---|---|---|---|---|---|---|---|---|---|---|---|---|---|---|---|---|---|---|---|---|---|---|---|---|---|---|---|---|---|---|---|---|---|---|---|---|---|---|
| **5f** | 3 | 3 | 3 | 3 | 2 | 3 | 2 | 4 | 6 | 6 | 8 | 2 | 4 | 6 | 1 | 1 | 3 | 1 | 2 | 3 | 4 | 5 | 5 | 5 | 5 | 7 | 4 | 6 | 3 | 3 | 5 | 5 | 7 | 7 | 6 | 6 | 4 | 5 | 8 | 8 | 7 | 6 |
| **5g** | x | 2 | 2 | 2 | 3 | 5 | 4 | 5 | 7 | 8 | 9 | 3 | 6 | 7 | 4 | 4 | 4 | 4 | 5 | 6 | 7 | 7 | 7 | 7 | 7 | 8 | 5 | 5 | 5 | 5 | 7 | 6 | 8 | 6 | 8 | 7 | 6 | 8 | 9 | 7 | 9 | 9 |
| **5h** | 2 | x | 2 | 2 | 3 | 5 | 4 | 5 | 7 | 8 | 9 | 3 | 6 | 7 | 4 | 4 | 4 | 4 | 5 | 6 | 7 | 7 | 7 | 7 | 7 | 8 | 5 | 5 | 5 | 5 | 7 | 6 | 8 | 6 | 8 | 7 | 6 | 8 | 9 | 7 | 9 | 9 |
| **5i** | 2 | 2 | x | 2 | 3 | 5 | 4 | 5 | 7 | 8 | 9 | 3 | 6 | 7 | 4 | 4 | 4 | 4 | 5 | 6 | 7 | 7 | 7 | 7 | 7 | 8 | 5 | 5 | 5 | 5 | 7 | 6 | 8 | 6 | 8 | 7 | 6 | 8 | 9 | 7 | 9 | 9 |
| **5j** | 2 | 2 | 2 | x | 3 | 5 | 4 | 5 | 7 | 8 | 9 | 3 | 6 | 7 | 4 | 4 | 4 | 4 | 5 | 6 | 7 | 7 | 7 | 7 | 7 | 8 | 5 | 5 | 5 | 5 | 7 | 6 | 8 | 6 | 8 | 7 | 6 | 8 | 9 | 7 | 9 | 9 |
| **5k** | 3 | 3 | 3 | 3 | x | 2 | 1 | 4 | 6 | 5 | 8 | 2 | 4 | 5 | 3 | 3 | 3 | 2 | 2 | 3 | 4 | 5 | 5 | 5 | 5 | 6 | 4 | 6 | 2 | 2 | 4 | 4 | 6 | 7 | 6 | 5 | 3 | 5 | 7 | 8 | 7 | 6 |
| **5L** | 5 | 5 | 5 | 5 | 2 | x | 1 | 3 | 6 | 5 | 8 | 4 | 4 | 6 | 4 | 4 | 3 | 2 | 2 | 3 | 4 | 5 | 5 | 5 | 5 | 6 | 4 | 6 | 2 | 2 | 4 | 4 | 6 | 7 | 6 | 5 | 3 | 5 | 7 | 8 | 7 | 6 |
| **5m** | 4 | 4 | 4 | 4 | 1 | 1 | x | 2 | 6 | 4 | 8 | 3 | 3 | 4 | 3 | 3 | 2 | 1 | 1 | 2 | 3 | 4 | 4 | 4 | 4 | 5 | 3 | 5 | 1 | 1 | 3 | 3 | 5 | 6 | 5 | 4 | 2 | 4 | 6 | 7 | 6 | 5 |
| **5n** | 5 | 5 | 5 | 5 | 4 | 3 | 2 | x | 4 | 6 | 8 | 5 | 5 | 5 | 5 | 5 | 4 | 3 | 3 | 4 | 5 | 6 | 6 | 6 | 6 | 6 | 5 | 7 | 3 | 1 | 5 | 1 | 3 | 8 | 7 | 2 | 4 | 6 | 7 | 9 | 8 | 7 |
| **5o** | 7 | 7 | 7 | 7 | 6 | 6 | 6 | 4 | x | 10 | 10 | 6 | 6 | 6 | 7 | 7 | 7 | 7 | 7 | 7 | 7 | 7 | 7 | 7 | 7 | 7 | 7 | 10 | 7 | 5 | 9 | 3 | 1 | 11 | 8 | 2 | 8 | 8 | 8 | 12 | 9 | 9 |
| **5p** | 8 | 8 | 8 | 8 | 5 | 5 | 4 | 6 | 10 | x | 6 | 7 | 5 | 8 | 7 | 7 | 6 | 5 | 5 | 4 | 5 | 6 | 6 | 6 | 6 | 9 | 5 | 6 | 3 | 5 | 1 | 7 | 9 | 7 | 7 | 8 | 2 | 6 | 10 | 8 | 8 | 7 |
| **5q** | 9 | 9 | 9 | 9 | 8 | 8 | 8 | 8 | 10 | 6 | x | 8 | 10 | 10 | 9 | 9 | 9 | 9 | 9 | 10 | 11 | 11 | 11 | 11 | 11 | 11 | 11 | 11 | 9 | 9 | 7 | 9 | 11 | 12 | 12 | 10 | 8 | 12 | 12 | 13 | 13 | 13 |
| **5r** | 3 | 3 | 3 | 3 | 2 | 4 | 3 | 5 | 6 | 7 | 8 | x | 5 | 6 | 3 | 3 | 1 | 2 | 3 | 5 | 6 | 7 | 7 | 7 | 7 | 7 | 6 | 6 | 4 | 4 | 6 | 6 | 7 | 7 | 8 | 7 | 5 | 7 | 8 | 8 | 9 | 8 |
| **5s** | 6 | 6 | 6 | 6 | 4 | 4 | 3 | 5 | 6 | 5 | 10 | 5 | x | 4 | 5 | 5 | 4 | 3 | 2 | 1 | 1 | 1 | 1 | 1 | 1 | 5 | 1 | 3 | 2 | 4 | 4 | 6 | 7 | 4 | 2 | 7 | 3 | 2 | 6 | 5 | 3 | 3 |
| **5t** | 7 | 7 | 7 | 7 | 5 | 6 | 4 | 5 | 6 | 8 | 10 | 6 | 4 | x | 7 | 7 | 6 | 5 | 5 | 5 | 5 | 5 | 5 | 5 | 5 | 1 | 5 | 7 | 5 | 5 | 7 | 6 | 7 | 8 | 6 | 7 | 6 | 6 | 2 | 9 | 7 | 7 |
| **6a** | 4 | 4 | 4 | 4 | 3 | 4 | 3 | 5 | 7 | 7 | 9 | 3 | 5 | 7 | x | 2 | 3 | 2 | 3 | 4 | 5 | 6 | 6 | 6 | 6 | 8 | 5 | 7 | 4 | 4 | 6 | 6 | 8 | 8 | 7 | 7 | 5 | 6 | 9 | 9 | 8 | 7 |
| **6b** | 4 | 4 | 4 | 4 | 3 | 4 | 3 | 5 | 7 | 7 | 9 | 3 | 5 | 7 | 2 | x | 3 | 2 | 3 | 4 | 5 | 6 | 6 | 6 | 6 | 8 | 5 | 7 | 4 | 4 | 6 | 6 | 8 | 8 | 7 | 7 | 5 | 6 | 9 | 9 | 8 | 7 |
| **6c** | 4 | 4 | 4 | 4 | 3 | 3 | 2 | 4 | 7 | 6 | 9 | 1 | 4 | 6 | 3 | 3 | x | 1 | 2 | 3 | 4 | 5 | 5 | 5 | 5 | 7 | 4 | 6 | 3 | 3 | 5 | 5 | 7 | 7 | 6 | 6 | 4 | 5 | 8 | 8 | 7 | 6 |
| **6d** | 4 | 4 | 4 | 4 | 2 | 2 | 1 | 3 | 7 | 5 | 9 | 2 | 3 | 5 | 2 | 2 | 1 | x | 1 | 2 | 3 | 4 | 4 | 4 | 4 | 6 | 3 | 5 | 2 | 2 | 4 | 4 | 6 | 6 | 5 | 5 | 3 | 4 | 7 | 7 | 6 | 5 |
| **6e** | 5 | 5 | 5 | 5 | 2 | 2 | 1 | 3 | 7 | 5 | 9 | 3 | 2 | 5 | 3 | 3 | 2 | 1 | x | 1 | 2 | 3 | 3 | 3 | 3 | 6 | 2 | 4 | 2 | 2 | 4 | 4 | 6 | 5 | 4 | 5 | 3 | 3 | 7 | 6 | 5 | 4 |
| **6f** | 6 | 6 | 6 | 6 | 3 | 3 | 2 | 4 | 7 | 4 | 10 | 5 | 1 | 5 | 4 | 4 | 3 | 2 | 1 | x | 1 | 2 | 2 | 2 | 2 | 6 | 1 | 3 | 1 | 3 | 3 | 5 | 7 | 4 | 3 | 6 | 2 | 2 | 7 | 5 | 4 | 3 |
| **6g** | 7 | 7 | 7 | 7 | 4 | 4 | 3 | 5 | 7 | 5 | 11 | 6 | 1 | 5 | 5 | 5 | 4 | 3 | 2 | 1 | x | 1 | 2 | 2 | 2 | 6 | 2 | 4 | 2 | 4 | 4 | 6 | 8 | 5 | 3 | 7 | 3 | 1 | 7 | 6 | 4 | 2 |
| **6h** | 7 | 7 | 7 | 7 | 5 | 5 | 4 | 6 | 7 | 6 | 11 | 7 | 1 | 5 | 6 | 6 | 5 | 4 | 3 | 2 | 1 | x | 2 | 2 | 2 | 6 | 2 | 4 | 3 | 5 | 5 | 7 | 8 | 5 | 3 | 8 | 4 | 2 | 7 | 6 | 4 | 3 |
| **6i** | 7 | 7 | 7 | 7 | 5 | 5 | 4 | 6 | 7 | 6 | 11 | 7 | 1 | 5 | 6 | 6 | 5 | 4 | 3 | 2 | 2 | 2 | x | 2 | 2 | 6 | 2 | 4 | 3 | 5 | 5 | 7 | 8 | 5 | 3 | 8 | 4 | 3 | 7 | 6 | 4 | 4 |
| **6j** | 7 | 7 | 7 | 7 | 5 | 5 | 4 | 6 | 7 | 6 | 11 | 7 | 1 | 5 | 6 | 6 | 5 | 4 | 3 | 2 | 2 | 2 | 2 | x | 2 | 6 | 2 | 4 | 3 | 5 | 5 | 7 | 8 | 5 | 3 | 8 | 4 | 3 | 7 | 6 | 4 | 4 |
| **6k** | 7 | 7 | 7 | 7 | 5 | 5 | 4 | 6 | 7 | 6 | 11 | 7 | 1 | 5 | 6 | 6 | 5 | 4 | 3 | 2 | 2 | 2 | 2 | 2 | x | 6 | 2 | 4 | 3 | 5 | 5 | 7 | 8 | 5 | 1 | 8 | 4 | 3 | 7 | 6 | 2 | 4 |
| **6L** | 8 | 8 | 8 | 8 | 6 | 6 | 5 | 6 | 7 | 9 | 11 | 7 | 5 | 1 | 8 | 8 | 7 | 6 | 6 | 6 | 6 | 6 | 6 | 6 | 6 | x | 6 | 8 | 6 | 6 | 8 | 7 | 8 | 9 | 7 | 8 | 7 | 7 | 1 | 10 | 8 | 8 |
| **6m** | 5 | 5 | 5 | 5 | 4 | 4 | 3 | 5 | 7 | 5 | 11 | 6 | 1 | 5 | 5 | 5 | 4 | 3 | 2 | 1 | 2 | 2 | 2 | 2 | 2 | 6 | x | 2 | 2 | 4 | 4 | 6 | 8 | 3 | 3 | 7 | 3 | 3 | 7 | 4 | 4 | 4 |
| **6n** | 5 | 5 | 5 | 5 | 6 | 6 | 5 | 7 | 10 | 6 | 11 | 6 | 3 | 7 | 7 | 7 | 6 | 5 | 4 | 3 | 4 | 4 | 4 | 4 | 4 | 8 | 2 | x | 4 | 6 | 6 | 8 | 10 | 1 | 5 | 9 | 5 | 5 | 9 | 2 | 6 | 6 |
| **6o** | 5 | 5 | 5 | 5 | 2 | 2 | 1 | 3 | 7 | 3 | 9 | 4 | 2 | 5 | 4 | 4 | 3 | 2 | 2 | 1 | 2 | 3 | 3 | 3 | 3 | 6 | 2 | 4 | x | 2 | 2 | 4 | 6 | 5 | 4 | 5 | 1 | 3 | 7 | 6 | 4 | 4 |
| **6p** | 5 | 5 | 5 | 5 | 2 | 2 | 1 | 1 | 5 | 5 | 9 | 4 | 4 | 5 | 4 | 4 | 3 | 2 | 2 | 3 | 4 | 5 | 5 | 5 | 5 | 6 | 4 | 6 | 2 | x | 4 | 2 | 4 | 7 | 6 | 3 | 3 | 4 | 7 | 8 | 7 | 6 |
| **6q** | 7 | 7 | 7 | 7 | 4 | 4 | 3 | 5 | 9 | 1 | 7 | 6 | 4 | 7 | 6 | 6 | 5 | 4 | 4 | 3 | 4 | 5 | 5 | 5 | 5 | 8 | 4 | 6 | 2 | 4 | x | 6 | 8 | 7 | 6 | 7 | 1 | 5 | 9 | 8 | 7 | 6 |
| **6r** | 6 | 6 | 6 | 6 | 4 | 4 | 3 | 1 | 3 | 7 | 9 | 6 | 6 | 6 | 6 | 6 | 5 | 4 | 4 | 5 | 6 | 7 | 7 | 7 | 7 | 7 | 6 | 8 | 4 | 2 | 6 | x | 2 | 9 | 8 | 1 | 5 | 7 | 8 | 10 | 9 | 8 |
| **6s** | 8 | 8 | 8 | 8 | 6 | 6 | 5 | 3 | 1 | 9 | 11 | 7 | 7 | 7 | 8 | 8 | 7 | 6 | 6 | 7 | 8 | 8 | 8 | 8 | 8 | 8 | 8 | 10 | 6 | 4 | 8 | 2 | x | 11 | 9 | 1 | 7 | 9 | 9 | 12 | 10 | 10 |
| **7a** | 6 | 6 | 6 | 6 | 7 | 7 | 6 | 8 | 11 | 7 | 12 | 7 | 4 | 8 | 8 | 8 | 7 | 6 | 5 | 4 | 5 | 5 | 5 | 5 | 5 | 9 | 3 | 1 | 5 | 7 | 7 | 9 | 11 | x | 6 | 10 | 6 | 6 | 10 | 1 | 7 | 7 |
| **7b** | 8 | 8 | 8 | 8 | 6 | 6 | 5 | 7 | 8 | 7 | 12 | 8 | 2 | 6 | 7 | 7 | 6 | 5 | 4 | 3 | 3 | 3 | 3 | 3 | 1 | 7 | 3 | 5 | 4 | 6 | 6 | 8 | 9 | 6 | x | 9 | 5 | 4 | 8 | 7 | 1 | 5 |
| **7c** | 7 | 7 | 7 | 7 | 5 | 5 | 4 | 2 | 2 | 8 | 10 | 7 | 7 | 7 | 7 | 7 | 6 | 5 | 5 | 6 | 7 | 8 | 8 | 8 | 8 | 8 | 7 | 9 | 5 | 3 | 7 | 1 | 1 | 10 | 9 | x | 6 | 8 | 9 | 11 | 10 | 9 |
| **7d** | 6 | 6 | 6 | 6 | 3 | 3 | 2 | 4 | 8 | 2 | 8 | 5 | 3 | 6 | 5 | 5 | 4 | 3 | 3 | 2 | 3 | 4 | 4 | 4 | 4 | 7 | 3 | 5 | 1 | 3 | 1 | 5 | 7 | 6 | 5 | 6 | x | 4 | 8 | 7 | 6 | 5 |
| **7e** | 8 | 8 | 8 | 8 | 5 | 5 | 4 | 6 | 8 | 6 | 12 | 7 | 2 | 6 | 6 | 6 | 5 | 4 | 3 | 2 | 1 | 2 | 3 | 3 | 3 | 7 | 3 | 5 | 3 | 4 | 5 | 7 | 9 | 6 | 4 | 8 | 4 | x | 8 | 7 | 5 | 1 |
| **7f** | 9 | 9 | 9 | 9 | 7 | 7 | 6 | 7 | 8 | 10 | 12 | 8 | 6 | 2 | 9 | 9 | 8 | 7 | 7 | 7 | 7 | 7 | 7 | 7 | 7 | 1 | 7 | 9 | 7 | 7 | 9 | 8 | 9 | 10 | 8 | 9 | 8 | 8 | x | 11 | 9 | 9 |
| **8a** | 7 | 7 | 7 | 7 | 8 | 8 | 7 | 9 | 12 | 8 | 13 | 8 | 5 | 9 | 9 | 9 | 8 | 7 | 6 | 5 | 6 | 6 | 6 | 6 | 6 | 10 | 4 | 2 | 6 | 8 | 8 | 10 | 12 | 1 | 7 | 11 | 7 | 7 | 11 | x | 8 | 8 |
| **8b** | 9 | 9 | 9 | 9 | 7 | 7 | 6 | 8 | 9 | 8 | 13 | 9 | 3 | 7 | 8 | 8 | 7 | 6 | 5 | 4 | 4 | 4 | 4 | 4 | 2 | 8 | 4 | 6 | 4 | 7 | 7 | 9 | 10 | 7 | 1 | 10 | 6 | 5 | 9 | 8 | x | 6 |
| **8c** | 9 | 9 | 9 | 9 | 6 | 6 | 5 | 7 | 9 | 7 | 13 | 8 | 3 | 7 | 7 | 7 | 6 | 5 | 4 | 3 | 2 | 3 | 4 | 4 | 4 | 8 | 4 | 6 | 4 | 6 | 6 | 8 | 10 | 7 | 5 | 9 | 5 | 1 | 9 | 8 | 6 | x |

Palenque Palace Path Matrix Phase 6

| | c | 1a | 1b | 1c | 1d | 1e | 1f | 1g | 2a | 2b | 2c | 2d | 2e | 2f | 2g | 3a | 3b | 3c | 3d | 3e | 3f | 3g | 3h | 3i | 3j | 3k | 3L | 3m | 3n | 4a | 4b | 4c | 4d | 4e | 4f | 4g | 4h | 4i | 4j | 4k | 4L | 4m | 4n | 4o | 4p |
|---|---|---|---|---|---|---|---|---|---|---|---|---|---|---|---|---|---|---|---|---|---|---|---|---|---|---|---|---|---|---|---|---|---|---|---|---|---|---|---|---|---|---|---|---|---|
| c | x | 1 | 1 | 1 | 1 | 1 | 1 | 1 | 2 | 2 | 2 | 2 | 2 | 2 | 2 | 3 | 3 | 3 | 3 | 3 | 3 | 3 | 3 | 3 | 3 | 3 | 3 | 3 | 3 | 4 | 4 | 4 | 4 | 4 | 4 | 4 | 4 | 4 | 4 | 4 | 4 | 4 | 4 | 4 | 4 |
| 1a | 1 | x | 1 | 2 | 2 | 2 | 2 | 2 | 1 | 2 | 3 | 3 | 3 | 3 | 3 | 2 | 3 | 3 | 4 | 4 | 4 | 4 | 4 | 4 | 4 | 4 | 4 | 2 | 4 | 3 | 3 | 3 | 4 | 5 | 5 | 5 | 5 | 5 | 5 | 5 | 4 | 5 | 5 | 5 | 5 |
| 1b | 1 | 1 | x | 2 | 2 | 2 | 2 | 2 | 2 | 1 | 3 | 3 | 3 | 3 | 3 | 3 | 2 | 2 | 3 | 4 | 4 | 4 | 4 | 4 | 4 | 4 | 4 | 3 | 4 | 4 | 4 | 4 | 5 | 5 | 5 | 5 | 5 | 5 | 5 | 5 | 3 | 5 | 4 | 5 | 5 |
| 1c | 1 | 2 | 2 | x | 2 | 2 | 2 | 2 | 3 | 3 | 3 | 3 | 3 | 3 | 3 | 4 | 4 | 4 | 4 | 4 | 4 | 4 | 4 | 4 | 4 | 4 | 4 | 4 | 4 | 5 | 5 | 5 | 5 | 5 | 5 | 5 | 5 | 5 | 5 | 5 | 5 | 5 | 5 | 5 | 5 |
| 1d | 1 | 2 | 2 | 2 | x | 2 | 2 | 2 | 3 | 3 | 1 | 1 | 1 | 3 | 3 | 4 | 3 | 4 | 2 | 2 | 2 | 2 | 2 | 2 | 4 | 4 | 4 | 4 | 2 | 5 | 5 | 5 | 3 | 3 | 3 | 3 | 3 | 3 | 3 | 3 | 4 | 3 | 3 | 3 | 3 |
| 1e | 1 | 2 | 2 | 2 | 2 | x | 2 | 2 | 3 | 3 | 3 | 3 | 3 | 3 | 3 | 4 | 4 | 4 | 4 | 4 | 4 | 4 | 4 | 4 | 4 | 4 | 4 | 4 | 4 | 5 | 5 | 5 | 5 | 5 | 5 | 5 | 5 | 5 | 5 | 5 | 5 | 5 | 5 | 5 | 5 |
| 1f | 1 | 2 | 2 | 2 | 2 | 2 | x | 2 | 3 | 3 | 3 | 3 | 3 | 3 | 3 | 4 | 4 | 4 | 4 | 4 | 4 | 4 | 4 | 4 | 4 | 4 | 4 | 4 | 4 | 5 | 5 | 5 | 5 | 5 | 5 | 5 | 5 | 5 | 5 | 5 | 5 | 5 | 5 | 5 | 5 |
| 1g | 1 | 2 | 2 | 2 | 2 | 2 | 2 | x | 3 | 3 | 3 | 3 | 3 | 1 | 1 | 4 | 4 | 4 | 4 | 4 | 4 | 4 | 4 | 2 | 2 | 2 | 2 | 4 | 4 | 5 | 5 | 4 | 3 | 3 | 3 | 3 | 3 | 3 | 3 | 3 | 6 | 5 | 5 | 5 | 5 |
| 2a | 2 | 1 | 2 | 3 | 3 | 3 | 3 | 3 | x | 3 | 4 | 4 | 4 | 4 | 4 | 1 | 4 | 4 | 5 | 5 | 5 | 5 | 5 | 5 | 5 | 5 | 5 | 1 | 5 | 2 | 2 | 2 | 3 | 4 | 5 | 5 | 5 | 5 | 5 | 5 | 5 | 6 | 6 | 6 | 6 |
| 2b | 2 | 2 | 1 | 3 | 3 | 3 | 3 | 3 | 3 | x | 4 | 4 | 4 | 4 | 4 | 4 | 1 | 1 | 2 | 4 | 5 | 5 | 5 | 5 | 5 | 5 | 5 | 4 | 5 | 5 | 5 | 5 | 6 | 6 | 6 | 6 | 6 | 6 | 6 | 5 | 2 | 5 | 3 | 4 | 6 |
| 2c | 2 | 3 | 3 | 3 | 1 | 3 | 3 | 3 | 4 | 4 | x | 2 | 2 | 4 | 4 | 5 | 2 | 4 | 1 | 1 | 3 | 3 | 3 | 3 | 5 | 5 | 5 | 5 | 3 | 6 | 6 | 5 | 4 | 4 | 4 | 4 | 4 | 4 | 4 | 4 | 3 | 2 | 2 | 2 | 4 |
| 2d | 2 | 3 | 3 | 3 | 1 | 3 | 3 | 3 | 4 | 4 | 2 | x | 2 | 4 | 4 | 5 | 4 | 5 | 3 | 3 | 1 | 1 | 1 | 3 | 5 | 5 | 5 | 5 | 3 | 6 | 6 | 5 | 4 | 4 | 4 | 4 | 4 | 4 | 4 | 4 | 3 | 4 | 4 | 2 | 2 |
| 2e | 2 | 3 | 3 | 3 | 1 | 3 | 3 | 3 | 4 | 4 | 2 | 2 | x | 4 | 4 | 4 | 4 | 5 | 3 | 3 | 3 | 3 | 3 | 1 | 3 | 5 | 5 | 4 | 1 | 6 | 6 | 3 | 2 | 2 | 2 | 2 | 2 | 2 | 2 | 2 | 5 | 4 | 4 | 4 | 4 |
| 2f | 2 | 3 | 3 | 3 | 3 | 3 | 3 | 1 | 4 | 4 | 4 | 4 | 4 | x | 2 | 5 | 4 | 4 | 4 | 4 | 4 | 4 | 4 | 3 | 1 | 1 | 1 | 5 | 5 | 6 | 6 | 5 | 4 | 4 | 4 | 4 | 4 | 4 | 4 | 4 | 6 | 6 | 6 | 6 | 6 |
| 2g | 2 | 3 | 3 | 3 | 3 | 3 | 3 | 1 | 4 | 4 | 4 | 4 | 4 | 2 | x | 4 | 5 | 5 | 5 | 5 | 5 | 5 | 5 | 1 | 1 | 3 | 3 | 4 | 3 | 6 | 6 | 3 | 2 | 2 | 2 | 2 | 2 | 2 | 2 | 2 | 6 | 6 | 6 | 6 | 6 |
| 3a | 3 | 2 | 3 | 4 | 4 | 4 | 4 | 4 | 1 | 4 | 5 | 5 | 4 | 5 | 4 | x | 5 | 5 | 6 | 6 | 6 | 6 | 6 | 6 | 5 | 6 | 6 | 2 | 5 | 1 | 1 | 1 | 2 | 3 | 4 | 4 | 4 | 4 | 4 | 4 | 6 | 7 | 7 | 7 | 7 |
| 3b | 3 | 3 | 2 | 4 | 3 | 4 | 4 | 4 | 4 | 1 | 2 | 4 | 4 | 4 | 5 | 5 | x | 2 | 1 | 3 | 5 | 5 | 5 | 5 | 6 | 6 | 6 | 5 | 5 | 6 | 6 | 6 | 5 | 6 | 6 | 6 | 6 | 6 | 6 | 4 | 1 | 4 | 2 | 3 | 6 |
| 3c | 3 | 3 | 2 | 4 | 4 | 4 | 4 | 4 | 4 | 1 | 4 | 5 | 5 | 4 | 5 | 5 | 2 | x | 3 | 5 | 6 | 6 | 6 | 6 | 6 | 6 | 6 | 5 | 6 | 6 | 6 | 6 | 7 | 7 | 7 | 7 | 7 | 7 | 7 | 6 | 3 | 6 | 4 | 5 | 7 |
| 3d | 3 | 4 | 3 | 4 | 2 | 4 | 4 | 4 | 5 | 2 | 1 | 3 | 3 | 4 | 5 | 6 | 1 | 3 | x | 2 | 4 | 4 | 4 | 4 | 6 | 6 | 6 | 6 | 4 | 7 | 7 | 6 | 5 | 6 | 7 | 7 | 7 | 7 | 7 | 5 | 2 | 3 | 1 | 3 | 5 |
| 3e | 3 | 4 | 4 | 4 | 2 | 4 | 4 | 4 | 5 | 4 | 1 | 3 | 3 | 4 | 5 | 6 | 3 | 5 | 2 | x | 2 | 4 | 4 | 4 | 6 | 6 | 6 | 6 | 4 | 7 | 7 | 6 | 5 | 5 | 5 | 5 | 5 | 5 | 5 | 5 | 3 | 1 | 3 | 1 | 5 |
| 3f | 3 | 4 | 4 | 4 | 2 | 4 | 4 | 4 | 5 | 5 | 3 | 1 | 3 | 4 | 5 | 6 | 5 | 6 | 4 | 2 | x | 2 | 2 | 4 | 6 | 6 | 6 | 6 | 4 | 7 | 7 | 6 | 5 | 5 | 5 | 5 | 5 | 5 | 5 | 5 | 3 | 3 | 5 | 1 | 3 |
| 3g | 3 | 4 | 4 | 4 | 2 | 4 | 4 | 4 | 5 | 5 | 3 | 1 | 3 | 4 | 5 | 6 | 5 | 6 | 4 | 4 | 2 | x | 1 | 4 | 6 | 6 | 6 | 6 | 2 | 7 | 7 | 6 | 5 | 5 | 5 | 5 | 5 | 5 | 5 | 5 | 5 | 5 | 5 | 3 | 1 |
| 3h | 3 | 4 | 4 | 4 | 2 | 4 | 4 | 4 | 5 | 5 | 3 | 1 | 3 | 4 | 5 | 6 | 5 | 6 | 4 | 4 | 2 | 1 | x | 4 | 6 | 6 | 6 | 6 | 3 | 7 | 7 | 6 | 5 | 5 | 5 | 5 | 5 | 5 | 5 | 5 | 5 | 5 | 5 | 3 | 1 |
| 3i | 3 | 4 | 4 | 4 | 2 | 4 | 4 | 2 | 5 | 5 | 3 | 3 | 1 | 3 | 1 | 6 | 5 | 6 | 4 | 4 | 4 | 4 | 4 | x | 2 | 4 | 4 | 3 | 2 | 4 | 4 | 2 | 1 | 1 | 1 | 1 | 1 | 1 | 1 | 1 | 4 | 5 | 5 | 5 | 5 |
| 3j | 3 | 4 | 4 | 4 | 4 | 4 | 4 | 2 | 5 | 5 | 5 | 5 | 3 | 1 | 1 | 5 | 6 | 6 | 6 | 6 | 6 | 6 | 6 | 2 | x | 2 | 2 | 5 | 4 | 6 | 6 | 4 | 3 | 3 | 3 | 3 | 3 | 3 | 3 | 3 | 6 | 7 | 7 | 7 | 7 |
| 3k | 3 | 4 | 4 | 4 | 4 | 4 | 4 | 2 | 5 | 5 | 5 | 5 | 5 | 1 | 3 | 6 | 6 | 6 | 6 | 6 | 6 | 6 | 6 | 4 | 2 | x | 2 | 6 | 6 | 7 | 7 | 6 | 5 | 5 | 5 | 5 | 5 | 5 | 5 | 5 | 7 | 7 | 7 | 7 | 7 |
| 3L | 3 | 4 | 4 | 4 | 4 | 4 | 4 | 2 | 5 | 5 | 5 | 5 | 5 | 1 | 3 | 6 | 6 | 6 | 6 | 6 | 6 | 6 | 6 | 4 | 2 | 2 | x | 6 | 6 | 7 | 7 | 6 | 5 | 5 | 5 | 5 | 5 | 5 | 5 | 5 | 7 | 7 | 7 | 7 | 7 |
| 3m | 3 | 2 | 3 | 4 | 4 | 4 | 4 | 4 | 1 | 4 | 5 | 5 | 4 | 5 | 4 | 2 | 5 | 5 | 6 | 6 | 6 | 6 | 6 | 3 | 5 | 6 | 6 | x | 6 | 3 | 3 | 1 | 2 | 3 | 4 | 4 | 4 | 4 | 4 | 4 | 6 | 7 | 7 | 7 | 7 |
| 3n | 3 | 4 | 4 | 4 | 2 | 4 | 4 | 4 | 5 | 5 | 3 | 3 | 1 | 5 | 3 | 5 | 5 | 6 | 4 | 4 | 4 | 2 | 3 | 2 | 4 | 6 | 6 | 6 | x | 6 | 6 | 4 | 3 | 3 | 3 | 3 | 3 | 3 | 3 | 3 | 6 | 5 | 5 | 5 | 3 |
| 4a | 4 | 3 | 4 | 5 | 5 | 5 | 5 | 5 | 2 | 5 | 6 | 6 | 6 | 6 | 6 | 1 | 6 | 6 | 7 | 7 | 7 | 7 | 7 | 4 | 6 | 7 | 7 | 3 | 6 | x | 1 | 2 | 3 | 4 | 5 | 5 | 5 | 5 | 5 | 5 | 7 | 8 | 8 | 8 | 8 |
| 4b | 4 | 3 | 4 | 5 | 5 | 5 | 5 | 5 | 2 | 5 | 6 | 6 | 6 | 6 | 6 | 1 | 6 | 6 | 7 | 7 | 7 | 7 | 7 | 4 | 6 | 7 | 7 | 3 | 6 | 1 | x | 2 | 3 | 4 | 5 | 5 | 5 | 5 | 5 | 5 | 7 | 8 | 8 | 8 | 8 |
| 4c | 4 | 3 | 4 | 5 | 5 | 5 | 5 | 4 | 2 | 5 | 5 | 5 | 3 | 5 | 3 | 1 | 6 | 6 | 6 | 6 | 6 | 6 | 6 | 2 | 4 | 6 | 6 | 1 | 4 | 2 | 2 | x | 1 | 2 | 3 | 3 | 3 | 3 | 3 | 3 | 5 | 7 | 7 | 6 | 7 |
| 4d | 4 | 4 | 5 | 5 | 3 | 5 | 5 | 3 | 3 | 6 | 4 | 4 | 2 | 4 | 2 | 2 | 5 | 7 | 5 | 5 | 5 | 5 | 5 | 1 | 3 | 5 | 5 | 2 | 3 | 3 | 3 | 1 | x | 1 | 2 | 2 | 2 | 2 | 2 | 2 | 4 | 6 | 6 | 5 | 6 |
| 4e | 4 | 5 | 5 | 5 | 3 | 5 | 5 | 3 | 4 | 6 | 4 | 4 | 2 | 4 | 2 | 3 | 6 | 7 | 6 | 5 | 5 | 5 | 5 | 1 | 3 | 5 | 5 | 3 | 3 | 4 | 4 | 2 | 1 | x | 1 | 2 | 2 | 2 | 2 | 2 | 5 | 6 | 6 | 5 | 6 |
| 4f | 4 | 5 | 5 | 5 | 3 | 5 | 5 | 3 | 5 | 6 | 4 | 4 | 2 | 4 | 2 | 4 | 6 | 7 | 7 | 5 | 5 | 5 | 5 | 1 | 3 | 5 | 5 | 4 | 3 | 5 | 5 | 3 | 2 | 1 | x | 2 | 2 | 2 | 2 | 2 | 5 | 6 | 6 | 5 | 6 |
| 4g | 4 | 5 | 5 | 5 | 3 | 5 | 5 | 3 | 5 | 6 | 4 | 4 | 2 | 4 | 2 | 4 | 6 | 7 | 7 | 5 | 5 | 5 | 5 | 1 | 3 | 5 | 5 | 4 | 3 | 5 | 5 | 3 | 2 | 2 | 2 | x | 1 | 2 | 2 | 2 | 5 | 6 | 6 | 5 | 6 |
| 4h | 4 | 5 | 5 | 5 | 3 | 5 | 5 | 3 | 5 | 6 | 4 | 4 | 2 | 4 | 2 | 4 | 6 | 7 | 7 | 5 | 5 | 5 | 5 | 1 | 3 | 5 | 5 | 4 | 3 | 5 | 5 | 3 | 2 | 2 | 2 | 1 | x | 1 | 2 | 2 | 5 | 6 | 6 | 5 | 6 |
| 4i | 4 | 5 | 5 | 5 | 3 | 5 | 5 | 3 | 5 | 6 | 4 | 4 | 2 | 4 | 2 | 4 | 6 | 7 | 7 | 5 | 5 | 5 | 5 | 1 | 3 | 5 | 5 | 4 | 3 | 5 | 5 | 3 | 2 | 2 | 2 | 2 | 1 | x | 1 | 2 | 5 | 6 | 6 | 5 | 6 |
| 4j | 4 | 5 | 5 | 5 | 3 | 5 | 5 | 3 | 5 | 6 | 4 | 4 | 2 | 4 | 2 | 4 | 6 | 7 | 7 | 5 | 5 | 5 | 5 | 1 | 3 | 5 | 5 | 4 | 3 | 5 | 5 | 3 | 2 | 2 | 2 | 2 | 2 | 1 | x | 2 | 5 | 6 | 6 | 5 | 6 |
| 4k | 4 | 5 | 5 | 5 | 3 | 5 | 5 | 3 | 5 | 5 | 4 | 4 | 2 | 4 | 2 | 4 | 4 | 6 | 5 | 5 | 5 | 5 | 5 | 1 | 3 | 5 | 5 | 4 | 3 | 5 | 5 | 3 | 2 | 2 | 2 | 2 | 2 | 2 | 2 | x | 5 | 6 | 6 | 3 | 6 |
| 4L | 4 | 4 | 3 | 5 | 4 | 5 | 5 | 6 | 5 | 2 | 3 | 3 | 5 | 6 | 6 | 6 | 1 | 3 | 2 | 3 | 3 | 5 | 5 | 4 | 6 | 7 | 7 | 6 | 6 | 7 | 7 | 5 | 4 | 5 | 5 | 5 | 5 | 5 | 5 | 5 | x | 4 | 3 | 2 | 6 |
| 4m | 4 | 5 | 5 | 5 | 3 | 5 | 5 | 5 | 6 | 5 | 2 | 4 | 4 | 6 | 6 | 7 | 4 | 6 | 3 | 1 | 3 | 5 | 5 | 5 | 7 | 7 | 7 | 7 | 5 | 8 | 8 | 7 | 6 | 6 | 6 | 6 | 6 | 6 | 6 | 6 | 4 | x | 4 | 2 | 6 |
| 4n | 4 | 5 | 4 | 5 | 3 | 5 | 5 | 5 | 6 | 3 | 2 | 4 | 4 | 6 | 6 | 7 | 2 | 4 | 1 | 3 | 5 | 5 | 5 | 5 | 7 | 7 | 7 | 7 | 5 | 8 | 8 | 7 | 6 | 6 | 6 | 6 | 6 | 6 | 6 | 6 | 3 | 4 | x | 4 | 6 |

| | | | | | | | | | | | | | | | | | | | | | | | | | | | | | | | | | | | | | | | | | | | | | |
|---|---|---|---|---|---|---|---|---|---|---|---|---|---|---|---|---|---|---|---|---|---|---|---|---|---|---|---|---|---|---|---|---|---|---|---|---|---|---|---|---|---|---|---|---|---|
| **4o** | 4 | 5 | 5 | 5 | 3 | 5 | 5 | 5 | 6 | 4 | 2 | 2 | 4 | 6 | 6 | 7 | 3 | 5 | 3 | 1 | 1 | 3 | 3 | 5 | 7 | 7 | 7 | 7 | 5 | 8 | 8 | 6 | 5 | 5 | 5 | 5 | 5 | 5 | 5 | 3 | 2 | 2 | 4 | x | 4 |
| **4p** | 4 | 5 | 5 | 5 | 3 | 5 | 5 | 5 | 6 | 6 | 4 | 2 | 4 | 6 | 6 | 7 | 6 | 7 | 5 | 5 | 3 | 1 | 1 | 5 | 7 | 7 | 7 | 7 | 3 | 8 | 8 | 7 | 6 | 6 | 6 | 6 | 6 | 6 | 6 | 6 | 6 | 6 | 6 | 4 | x |
| **4q** | 4 | 5 | 4 | 5 | 3 | 5 | 5 | 5 | 6 | 6 | 4 | 2 | 2 | 6 | 6 | 6 | 6 | 7 | 5 | 5 | 3 | 1 | 2 | 3 | 5 | 7 | 7 | 6 | 1 | 7 | 7 | 5 | 4 | 4 | 4 | 4 | 4 | 4 | 4 | 4 | 5 | 6 | 6 | 4 | 2 |
| **4r** | 4 | 5 | 5 | 5 | 5 | 5 | 5 | 3 | 6 | 6 | 6 | 6 | 6 | 2 | 4 | 7 | 7 | 7 | 7 | 7 | 7 | 7 | 7 | 5 | 3 | 3 | 1 | 7 | 7 | 8 | 8 | 7 | 6 | 6 | 6 | 6 | 6 | 6 | 6 | 6 | 8 | 8 | 8 | 8 | 8 |
| **4s** | 4 | 5 | 5 | 5 | 5 | 5 | 5 | 3 | 6 | 6 | 6 | 6 | 4 | 2 | 2 | 7 | 7 | 7 | 7 | 7 | 7 | 7 | 7 | 3 | 1 | 1 | 3 | 7 | 5 | 7 | 7 | 5 | 4 | 4 | 4 | 4 | 4 | 4 | 4 | 4 | 7 | 8 | 8 | 8 | 8 |
| **4t** | 4 | 5 | 5 | 5 | 5 | 5 | 5 | 3 | 6 | 6 | 6 | 6 | 4 | 2 | 2 | 6 | 7 | 7 | 7 | 7 | 7 | 7 | 7 | 3 | 1 | 1 | 3 | 7 | 5 | 7 | 7 | 5 | 4 | 4 | 4 | 4 | 4 | 4 | 4 | 4 | 7 | 8 | 8 | 8 | 8 |
| **4u** | 4 | 4 | 3 | 5 | 4 | 5 | 5 | 5 | 5 | 2 | 3 | 5 | 5 | 6 | 6 | 6 | 1 | 3 | 2 | 4 | 5 | 6 | 6 | 6 | 7 | 7 | 7 | 6 | 7 | 7 | 7 | 7 | 6 | 7 | 7 | 7 | 7 | 7 | 7 | 5 | 2 | 5 | 3 | 4 | 7 |
| **4v** | 4 | 4 | 3 | 5 | 4 | 5 | 5 | 5 | 5 | 2 | 3 | 5 | 5 | 6 | 6 | 6 | 1 | 3 | 2 | 4 | 5 | 6 | 6 | 6 | 7 | 7 | 7 | 6 | 7 | 7 | 7 | 7 | 6 | 7 | 7 | 7 | 7 | 7 | 7 | 5 | 2 | 5 | 3 | 4 | 7 |
| **4w** | 4 | 4 | 3 | 5 | 4 | 5 | 5 | 5 | 5 | 2 | 3 | 5 | 5 | 6 | 6 | 6 | 1 | 3 | 2 | 4 | 5 | 6 | 6 | 6 | 7 | 7 | 7 | 6 | 7 | 7 | 7 | 7 | 6 | 7 | 7 | 7 | 7 | 7 | 7 | 5 | 2 | 5 | 3 | 4 | 7 |
| **4x** | 4 | 4 | 3 | 5 | 4 | 5 | 5 | 5 | 5 | 2 | 3 | 5 | 5 | 6 | 6 | 6 | 1 | 3 | 2 | 4 | 5 | 6 | 6 | 5 | 7 | 7 | 7 | 6 | 7 | 6 | 7 | 5 | 4 | 5 | 6 | 6 | 6 | 6 | 6 | 5 | 2 | 5 | 3 | 4 | 7 |
| **4y** | 4 | 5 | 5 | 5 | 3 | 5 | 5 | 3 | 5 | 6 | 4 | 4 | 2 | 4 | 2 | 4 | 6 | 7 | 7 | 5 | 5 | 5 | 5 | 1 | 3 | 5 | 5 | 4 | 3 | 5 | 5 | 3 | 2 | 2 | 2 | 2 | 2 | 2 | 2 | 2 | 5 | 6 | 6 | 6 | 6 |
| **4z** | 4 | 3 | 4 | 5 | 5 | 5 | 5 | 5 | 2 | 5 | 6 | 6 | 4 | 6 | 4 | 2 | 6 | 6 | 7 | 7 | 7 | 7 | 7 | 3 | 5 | 7 | 7 | 1 | 5 | 3 | 3 | 1 | 2 | 3 | 4 | 4 | 4 | 4 | 4 | 4 | 7 | 8 | 8 | 7 | 8 |
| **4aa** | 4 | 4 | 3 | 5 | 4 | 5 | 5 | 5 | 5 | 2 | 3 | 5 | 5 | 6 | 6 | 6 | 1 | 3 | 2 | 4 | 5 | 6 | 6 | 6 | 7 | 7 | 7 | 6 | 7 | 7 | 7 | 7 | 6 | 7 | 7 | 7 | 7 | 7 | 7 | 5 | 2 | 5 | 3 | 4 | 7 |
| **5a** | 5 | 4 | 5 | 6 | 6 | 6 | 6 | 6 | 3 | 6 | 7 | 7 | 6 | 7 | 6 | 2 | 7 | 7 | 8 | 8 | 8 | 8 | 8 | 5 | 7 | 8 | 8 | 4 | 7 | 1 | 2 | 3 | 4 | 5 | 6 | 6 | 6 | 6 | 6 | 6 | 8 | 9 | 9 | 9 | 9 |
| **5b** | 5 | 4 | 5 | 6 | 6 | 6 | 6 | 6 | 3 | 6 | 7 | 7 | 6 | 7 | 6 | 2 | 7 | 7 | 8 | 8 | 8 | 8 | 8 | 5 | 7 | 8 | 8 | 4 | 7 | 1 | 2 | 3 | 4 | 5 | 6 | 6 | 6 | 6 | 6 | 6 | 8 | 9 | 9 | 9 | 9 |
| **5c** | 5 | 4 | 5 | 6 | 6 | 6 | 6 | 6 | 3 | 6 | 7 | 7 | 6 | 7 | 6 | 2 | 7 | 7 | 8 | 8 | 8 | 8 | 8 | 5 | 7 | 8 | 8 | 4 | 7 | 1 | 2 | 3 | 4 | 5 | 6 | 6 | 6 | 6 | 6 | 6 | 8 | 9 | 9 | 9 | 9 |
| **5d** | 5 | 4 | 5 | 6 | 6 | 6 | 6 | 6 | 3 | 6 | 7 | 7 | 6 | 7 | 6 | 2 | 7 | 7 | 8 | 8 | 8 | 8 | 8 | 5 | 7 | 8 | 8 | 4 | 7 | 1 | 2 | 3 | 4 | 5 | 6 | 6 | 6 | 6 | 6 | 6 | 8 | 9 | 9 | 9 | 9 |
| **5e** | 5 | 4 | 5 | 6 | 6 | 6 | 6 | 6 | 3 | 6 | 7 | 7 | 6 | 7 | 6 | 2 | 7 | 7 | 7 | 8 | 8 | 8 | 8 | 5 | 7 | 8 | 8 | 4 | 7 | 1 | 2 | 3 | 4 | 5 | 6 | 6 | 6 | 6 | 6 | 6 | 7 | 9 | 8 | 9 | 9 |
| **5f** | 5 | 5 | 4 | 6 | 5 | 6 | 6 | 6 | 6 | 3 | 4 | 6 | 6 | 7 | 7 | 7 | 2 | 4 | 3 | 5 | 6 | 7 | 7 | 7 | 8 | 8 | 8 | 7 | 8 | 8 | 8 | 8 | 7 | 8 | 8 | 8 | 8 | 8 | 8 | 8 | 3 | 6 | 4 | 5 | 8 |
| **5g** | 5 | 4 | 5 | 6 | 6 | 6 | 6 | 5 | 3 | 6 | 6 | 6 | 4 | 6 | 4 | 2 | 7 | 7 | 7 | 7 | 7 | 7 | 7 | 3 | 5 | 7 | 7 | 2 | 5 | 3 | 3 | 1 | 2 | 3 | 4 | 4 | 4 | 4 | 4 | 4 | 6 | 8 | 8 | 7 | 8 |
| **5h** | 5 | 4 | 5 | 6 | 6 | 6 | 6 | 5 | 3 | 6 | 6 | 6 | 4 | 6 | 4 | 2 | 7 | 7 | 7 | 7 | 7 | 7 | 7 | 3 | 5 | 7 | 7 | 2 | 5 | 3 | 3 | 1 | 2 | 3 | 4 | 4 | 4 | 4 | 4 | 4 | 6 | 8 | 8 | 7 | 8 |
| **5i** | 5 | 4 | 5 | 6 | 6 | 6 | 6 | 5 | 3 | 6 | 6 | 6 | 4 | 6 | 4 | 2 | 7 | 7 | 7 | 7 | 7 | 7 | 7 | 3 | 5 | 7 | 7 | 2 | 5 | 3 | 3 | 1 | 2 | 3 | 4 | 4 | 4 | 4 | 4 | 4 | 6 | 8 | 8 | 7 | 8 |
| **5j** | 5 | 5 | 6 | 6 | 4 | 6 | 6 | 4 | 4 | 7 | 5 | 5 | 3 | 5 | 3 | 3 | 6 | 8 | 6 | 6 | 6 | 6 | 6 | 2 | 4 | 6 | 6 | 3 | 4 | 4 | 4 | 2 | 1 | 2 | 3 | 3 | 3 | 3 | 3 | 3 | 5 | 7 | 7 | 6 | 7 |
| **5k** | 5 | 4 | 5 | 6 | 5 | 6 | 6 | 5 | 3 | 6 | 6 | 6 | 4 | 6 | 4 | 2 | 7 | 7 | 7 | 7 | 7 | 7 | 7 | 3 | 5 | 7 | 7 | 2 | 5 | 3 | 3 | 1 | 2 | 3 | 4 | 4 | 4 | 4 | 4 | 4 | 6 | 8 | 8 | 7 | 8 |
| **5L** | 5 | 5 | 6 | 6 | 4 | 6 | 6 | 4 | 4 | 6 | 5 | 5 | 2 | 5 | 3 | 3 | 5 | 7 | 6 | 6 | 6 | 6 | 6 | 2 | 4 | 6 | 6 | 3 | 4 | 4 | 4 | 2 | 1 | 2 | 3 | 3 | 3 | 3 | 3 | 3 | 4 | 7 | 7 | 6 | 7 |
| **5m** | 5 | 5 | 6 | 6 | 4 | 6 | 6 | 4 | 4 | 5 | 5 | 5 | 2 | 5 | 3 | 3 | 4 | 6 | 5 | 6 | 6 | 6 | 6 | 2 | 4 | 6 | 6 | 3 | 4 | 4 | 4 | 2 | 1 | 2 | 3 | 3 | 3 | 3 | 3 | 3 | 3 | 7 | 7 | 5 | 7 |
| **5n** | 5 | 5 | 4 | 6 | 5 | 6 | 6 | 6 | 6 | 3 | 4 | 6 | 5 | 7 | 5 | 5 | 2 | 4 | 3 | 5 | 6 | 7 | 7 | 4 | 6 | 6 | 8 | 5 | 6 | 5 | 6 | 4 | 3 | 4 | 5 | 5 | 5 | 5 | 5 | 5 | 3 | 6 | 4 | 5 | 8 |
| **5o** | 5 | 5 | 4 | 6 | 5 | 6 | 6 | 6 | 6 | 3 | 4 | 5 | 5 | 7 | 5 | 5 | 2 | 4 | 3 | 5 | 4 | 6 | 6 | 5 | 6 | 8 | 8 | 5 | 6 | 6 | 6 | 3 | 3 | 4 | 5 | 5 | 5 | 5 | 5 | 4 | 1 | 5 | 4 | 3 | 7 |
| **5p** | 5 | 6 | 6 | 6 | 4 | 6 | 6 | 4 | 6 | 4 | 5 | 4 | 3 | 5 | 3 | 5 | 3 | 5 | 4 | 6 | 3 | 5 | 5 | 2 | 4 | 6 | 6 | 5 | 4 | 6 | 6 | 4 | 3 | 3 | 3 | 3 | 3 | 3 | 3 | 1 | 2 | 4 | 5 | 2 | 6 |
| **5q** | 5 | 5 | 4 | 6 | 4 | 6 | 6 | 5 | 6 | 3 | 4 | 3 | 4 | 6 | 4 | 6 | 2 | 4 | 3 | 5 | 2 | 4 | 4 | 3 | 5 | 7 | 7 | 6 | 5 | 7 | 7 | 5 | 4 | 4 | 4 | 4 | 4 | 4 | 4 | 2 | 1 | 3 | 4 | 1 | 5 |
| **5r** | 5 | 6 | 6 | 6 | 4 | 6 | 6 | 6 | 7 | 7 | 5 | 3 | 5 | 7 | 7 | 8 | 7 | 8 | 6 | 6 | 4 | 2 | 2 | 6 | 8 | 8 | 8 | 8 | 4 | 9 | 9 | 8 | 7 | 7 | 7 | 7 | 7 | 7 | 7 | 7 | 7 | 7 | 7 | 5 | 1 |
| **5s** | 5 | 6 | 5 | 6 | 4 | 6 | 6 | 6 | 7 | 4 | 3 | 5 | 5 | 7 | 7 | 8 | 3 | 5 | 2 | 4 | 6 | 6 | 6 | 6 | 8 | 8 | 8 | 8 | 6 | 9 | 9 | 8 | 7 | 7 | 7 | 7 | 7 | 7 | 7 | 6 | 4 | 5 | 1 | 5 | 7 |
| **5t** | 5 | 6 | 6 | 6 | 6 | 6 | 6 | 4 | 7 | 7 | 7 | 7 | 5 | 3 | 3 | 7 | 6 | 8 | 7 | 8 | 8 | 8 | 8 | 4 | 2 | 2 | 4 | 7 | 6 | 8 | 8 | 5 | 5 | 5 | 5 | 5 | 5 | 5 | 5 | 5 | 7 | 9 | 8 | 9 | 9 |
| **5u** | 5 | 6 | 6 | 6 | 6 | 6 | 6 | 4 | 7 | 7 | 7 | 7 | 7 | 3 | 5 | 8 | 8 | 8 | 8 | 8 | 8 | 8 | 8 | 6 | 4 | 4 | 2 | 8 | 8 | 10 | 9 | 8 | 7 | 7 | 7 | 7 | 7 | 7 | 7 | 7 | 9 | 9 | 9 | 9 | 9 |
| **5v** | 5 | 4 | 5 | 6 | 6 | 6 | 6 | 6 | 3 | 6 | 7 | 7 | 6 | 7 | 6 | 2 | 7 | 7 | 8 | 8 | 8 | 8 | 8 | 5 | 7 | 8 | 8 | 8 | 7 | 1 | 2 | 3 | 3 | 5 | 6 | 6 | 6 | 6 | 6 | 6 | 8 | 9 | 9 | 9 | 9 |
| **5w** | 5 | 4 | 5 | 6 | 6 | 6 | 6 | 5 | 3 | 6 | 6 | 6 | 4 | 6 | 4 | 2 | 7 | 7 | 7 | 7 | 7 | 7 | 7 | 3 | 5 | 7 | 7 | 2 | 5 | 3 | 3 | 1 | 2 | 3 | 4 | 4 | 4 | 4 | 4 | 4 | 7 | 8 | 8 | 7 | 8 |
| **5x** | 5 | 5 | 4 | 6 | 5 | 6 | 6 | 6 | 6 | 3 | 4 | 6 | 6 | 7 | 7 | 7 | 2 | 4 | 3 | 5 | 6 | 7 | 7 | 7 | 8 | 8 | 8 | 7 | 8 | 8 | 8 | 8 | 7 | 8 | 8 | 8 | 8 | 8 | 8 | 6 | 3 | 6 | 4 | 5 | 8 |
| **6a** | 6 | 5 | 6 | 7 | 7 | 7 | 7 | 6 | 4 | 7 | 7 | 7 | 5 | 7 | 5 | 3 | 8 | 8 | 8 | 8 | 8 | 8 | 8 | 4 | 6 | 8 | 8 | 3 | 6 | 4 | 4 | 2 | 3 | 4 | 5 | 5 | 5 | 5 | 5 | 5 | 7 | 9 | 9 | 8 | 9 |
| **6b** | 6 | 6 | 7 | 7 | 5 | 7 | 7 | 5 | 5 | 8 | 6 | 6 | 4 | 6 | 4 | 4 | 7 | 9 | 7 | 7 | 7 | 7 | 7 | 3 | 5 | 7 | 7 | 4 | 5 | 5 | 5 | 3 | 2 | 3 | 4 | 4 | 4 | 4 | 4 | 4 | 6 | 8 | 8 | 7 | 8 |
| **6c** | 6 | 6 | 7 | 7 | 5 | 7 | 7 | 5 | 5 | 8 | 6 | 6 | 4 | 6 | 4 | 4 | 7 | 9 | 7 | 7 | 7 | 7 | 7 | 3 | 5 | 7 | 7 | 4 | 5 | 5 | 5 | 3 | 2 | 3 | 4 | 4 | 4 | 4 | 4 | 4 | 6 | 8 | 8 | 7 | 8 |
| **6d** | 6 | 6 | 7 | 7 | 5 | 7 | 7 | 5 | 5 | 8 | 6 | 6 | 4 | 6 | 4 | 4 | 7 | 9 | 7 | 7 | 7 | 7 | 7 | 3 | 5 | 7 | 7 | 4 | 5 | 5 | 5 | 3 | 2 | 3 | 4 | 4 | 4 | 4 | 4 | 4 | 6 | 8 | 8 | 7 | 8 |
| **6e** | 6 | 6 | 7 | 7 | 5 | 7 | 7 | 5 | 5 | 8 | 6 | 6 | 4 | 6 | 4 | 4 | 7 | 9 | 7 | 7 | 7 | 7 | 7 | 3 | 5 | 7 | 7 | 4 | 5 | 5 | 5 | 3 | 2 | 3 | 4 | 4 | 4 | 4 | 4 | 4 | 6 | 8 | 8 | 7 | 8 |
| **6f** | 6 | 6 | 7 | 7 | 5 | 7 | 7 | 5 | 5 | 8 | 6 | 6 | 4 | 6 | 4 | 4 | 7 | 9 | 7 | 7 | 7 | 7 | 7 | 3 | 5 | 7 | 7 | 4 | 5 | 5 | 5 | 3 | 2 | 3 | 4 | 4 | 4 | 4 | 4 | 4 | 6 | 8 | 8 | 7 | 8 |
| **6g** | 6 | 5 | 6 | 7 | 6 | 7 | 7 | 6 | 4 | 7 | 7 | 7 | 5 | 7 | 5 | 3 | 8 | 8 | 8 | 8 | 8 | 8 | 8 | 4 | 6 | 8 | 8 | 3 | 6 | 4 | 4 | 2 | 3 | 4 | 5 | 5 | 5 | 5 | 5 | 5 | 7 | 9 | 9 | 8 | 9 |
| **6h** | 6 | 6 | 7 | 7 | 5 | 7 | 7 | 5 | 5 | 8 | 6 | 6 | 4 | 6 | 4 | 4 | 7 | 9 | 7 | 7 | 7 | 7 | 7 | 3 | 5 | 7 | 7 | 4 | 5 | 5 | 5 | 3 | 2 | 3 | 4 | 4 | 4 | 4 | 4 | 4 | 6 | 8 | 8 | 7 | 8 |
| **6i** | 6 | 5 | 6 | 7 | 7 | 7 | 7 | 6 | 4 | 7 | 7 | 7 | 5 | 7 | 5 | 3 | 6 | 8 | 8 | 8 | 8 | 8 | 8 | 4 | 6 | 8 | 8 | 3 | 6 | 4 | 4 | 2 | 3 | 4 | 5 | 5 | 5 | 5 | 5 | 5 | 7 | 9 | 8 | 8 | 9 |

(*continued*)

Palenque Palace Path Matrix Phase 6 (Cont.)

| | c | 1a | 1b | 1c | 1d | 1e | 1f | 1g | 2a | 2b | 2c | 2d | 2e | 2f | 2g | 3a | 3b | 3c | 3d | 3e | 3f | 3g | 3h | 3i | 3j | 3k | 3L | 3m | 3n | 4a | 4b | 4c | 4d | 4e | 4f | 4g | 4h | 4i | 4j | 4k | 4L | 4m | 4n | 4o | 4p |
|---|---|---|---|---|---|---|---|---|---|---|---|---|---|---|---|---|---|---|---|---|---|---|---|---|---|---|---|---|---|---|---|---|---|---|---|---|---|---|---|---|---|---|---|---|---|
| **6j** | 6 | 6 | 7 | 7 | 5 | 7 | 7 | 5 | 5 | 7 | 6 | 6 | 4 | 6 | 4 | 4 | 5 | 7 | 6 | 7 | 7 | 7 | 7 | 3 | 5 | 7 | 7 | 4 | 5 | 5 | 5 | 3 | 2 | 3 | 4 | 4 | 4 | 4 | 4 | 4 | 5 | 8 | 7 | 6 | 8 |
| **6k** | 6 | 6 | 7 | 7 | 5 | 7 | 7 | 5 | 5 | 7 | 6 | 6 | 4 | 6 | 4 | 4 | 5 | 7 | 6 | 7 | 7 | 7 | 7 | 3 | 5 | 7 | 7 | 4 | 5 | 5 | 5 | 3 | 2 | 3 | 4 | 4 | 4 | 4 | 4 | 4 | 4 | 8 | 7 | 6 | 8 |
| **6L** | 6 | 6 | 7 | 7 | 5 | 7 | 7 | 5 | 5 | 7 | 6 | 6 | 4 | 6 | 4 | 4 | 5 | 7 | 6 | 7 | 7 | 7 | 7 | 3 | 5 | 7 | 7 | 4 | 5 | 5 | 5 | 3 | 2 | 3 | 4 | 4 | 4 | 4 | 4 | 4 | 4 | 8 | 7 | 6 | 8 |
| **6m** | 6 | 6 | 7 | 7 | 5 | 7 | 7 | 5 | 5 | 4 | 6 | 6 | 4 | 6 | 4 | 4 | 5 | 7 | 6 | 7 | 5 | 7 | 7 | 3 | 5 | 7 | 7 | 4 | 5 | 5 | 5 | 3 | 2 | 3 | 4 | 4 | 4 | 4 | 4 | 4 | 2 | 6 | 5 | 4 | 8 |
| **6n** | 6 | 6 | 5 | 7 | 6 | 7 | 7 | 7 | 7 | 4 | 5 | 6 | 6 | 8 | 6 | 6 | 3 | 5 | 4 | 6 | 5 | 7 | 7 | 6 | 8 | 9 | 9 | 6 | 7 | 7 | 7 | 4 | 4 | 5 | 6 | 6 | 6 | 6 | 6 | 5 | 2 | 6 | 5 | 4 | 8 |
| **6o** | 6 | 6 | 5 | 7 | 6 | 7 | 7 | 7 | 6 | 4 | 5 | 7 | 6 | 8 | 6 | 5 | 3 | 5 | 4 | 6 | 7 | 8 | 8 | 5 | 7 | 7 | 9 | 6 | 7 | 4 | 5 | 5 | 4 | 5 | 6 | 6 | 6 | 6 | 6 | 6 | 4 | 7 | 5 | 6 | 9 |
| **6p** | 6 | 6 | 5 | 7 | 6 | 7 | 7 | 6 | 6 | 4 | 5 | 7 | 5 | 7 | 5 | 5 | 3 | 5 | 4 | 6 | 7 | 8 | 8 | 4 | 6 | 6 | 8 | 5 | 6 | 5 | 6 | 4 | 3 | 4 | 5 | 5 | 5 | 5 | 5 | 5 | 4 | 7 | 5 | 6 | 9 |
| **6q** | 6 | 6 | 5 | 7 | 6 | 7 | 7 | 5 | 5 | 4 | 5 | 6 | 4 | 6 | 4 | 4 | 3 | 5 | 4 | 6 | 7 | 7 | 7 | 3 | 5 | 5 | 7 | 4 | 5 | 5 | 5 | 3 | 2 | 3 | 4 | 4 | 4 | 4 | 4 | 4 | 4 | 7 | 5 | 6 | 8 |
| **6r** | 6 | 6 | 5 | 7 | 6 | 7 | 7 | 7 | 7 | 4 | 5 | 7 | 6 | 8 | 6 | 6 | 3 | 5 | 4 | 6 | 7 | 8 | 8 | 5 | 7 | 7 | 9 | 6 | 7 | 6 | 7 | 5 | 4 | 5 | 6 | 6 | 6 | 6 | 6 | 6 | 4 | 7 | 5 | 6 | 9 |
| **6s** | 6 | 6 | 5 | 7 | 6 | 7 | 7 | 7 | 7 | 4 | 5 | 7 | 6 | 8 | 6 | 6 | 3 | 5 | 4 | 6 | 7 | 8 | 8 | 5 | 7 | 7 | 9 | 6 | 7 | 6 | 7 | 5 | 4 | 5 | 6 | 6 | 6 | 6 | 6 | 6 | 4 | 7 | 5 | 6 | 9 |
| **6t** | 6 | 6 | 5 | 7 | 6 | 7 | 7 | 7 | 7 | 4 | 5 | 7 | 6 | 8 | 6 | 6 | 3 | 5 | 4 | 6 | 7 | 8 | 8 | 5 | 7 | 7 | 9 | 6 | 7 | 6 | 7 | 5 | 4 | 5 | 6 | 6 | 6 | 6 | 6 | 6 | 4 | 7 | 5 | 6 | 9 |
| **6u** | 6 | 6 | 5 | 7 | 6 | 7 | 7 | 7 | 7 | 4 | 5 | 7 | 6 | 8 | 6 | 6 | 3 | 5 | 4 | 6 | 7 | 8 | 8 | 5 | 7 | 7 | 9 | 6 | 7 | 6 | 7 | 5 | 4 | 5 | 6 | 6 | 6 | 6 | 6 | 6 | 4 | 7 | 5 | 6 | 9 |
| **6v** | 6 | 6 | 5 | 7 | 6 | 7 | 7 | 7 | 7 | 4 | 5 | 7 | 7 | 8 | 8 | 8 | 3 | 5 | 4 | 6 | 7 | 8 | 8 | 8 | 9 | 9 | 9 | 8 | 9 | 9 | 9 | 9 | 8 | 9 | 9 | 9 | 9 | 9 | 9 | 7 | 4 | 7 | 5 | 6 | 9 |
| **6w** | 6 | 5 | 6 | 7 | 7 | 7 | 7 | 7 | 4 | 7 | 8 | 8 | 7 | 8 | 7 | 3 | 7 | 8 | 8 | 9 | 9 | 9 | 9 | 6 | 8 | 9 | 9 | 5 | 8 | 2 | 3 | 4 | 5 | 6 | 7 | 7 | 7 | 7 | 7 | 7 | 8 | 10 | 9 | 10 | 10 |
| **6x** | 6 | 5 | 6 | 7 | 7 | 7 | 7 | 7 | 4 | 6 | 7 | 8 | 7 | 8 | 8 | 3 | 5 | 7 | 6 | 8 | 9 | 9 | 9 | 6 | 8 | 9 | 9 | 5 | 8 | 2 | 3 | 4 | 5 | 6 | 7 | 7 | 7 | 7 | 7 | 7 | 6 | 9 | 7 | 8 | 10 |
| **6y** | 6 | 7 | 7 | 7 | 7 | 7 | 7 | 5 | 7 | 6 | 7 | 9 | 6 | 4 | 4 | 8 | 5 | 7 | 6 | 8 | 9 | 9 | 9 | 5 | 3 | 3 | 5 | 6 | 7 | 7 | 7 | 5 | 4 | 5 | 6 | 6 | 6 | 6 | 6 | 6 | 6 | 9 | 7 | 8 | 10 |
| **6z** | 6 | 7 | 6 | 7 | 5 | 7 | 7 | 5 | 7 | 5 | 6 | 5 | 4 | 6 | 4 | 6 | 4 | 6 | 5 | 7 | 4 | 6 | 6 | 3 | 5 | 7 | 7 | 6 | 5 | 7 | 7 | 5 | 4 | 4 | 4 | 4 | 4 | 4 | 4 | 2 | 3 | 5 | 5 | 3 | 7 |
| **6aa** | 6 | 5 | 6 | 7 | 7 | 7 | 7 | 7 | 4 | 7 | 8 | 8 | 7 | 8 | 7 | 3 | 6 | 8 | 7 | 9 | 9 | 9 | 9 | 6 | 8 | 9 | 9 | 7 | 7 | 2 | 3 | 4 | 5 | 6 | 7 | 7 | 7 | 7 | 7 | 7 | 7 | 10 | 8 | 9 | 10 |
| **6ab** | 6 | 7 | 6 | 7 | 5 | 7 | 7 | 7 | 8 | 5 | 4 | 6 | 6 | 8 | 7 | 9 | 4 | 6 | 3 | 5 | 7 | 7 | 7 | 6 | 8 | 9 | 9 | 9 | 7 | 10 | 10 | 8 | 7 | 7 | 7 | 7 | 7 | 7 | 7 | 5 | 5 | 6 | 2 | 6 | 8 |
| **7a** | 7 | 6 | 7 | 8 | 8 | 8 | 8 | 8 | 5 | 7 | 8 | 9 | 8 | 9 | 9 | 4 | 6 | 8 | 7 | 9 | 10 | 10 | 10 | 7 | 9 | 10 | 10 | 6 | 9 | 3 | 4 | 5 | 6 | 7 | 8 | 8 | 8 | 8 | 8 | 8 | 7 | 10 | 8 | 9 | 11 |
| **7b** | 7 | 6 | 6 | 8 | 7 | 8 | 8 | 8 | 5 | 5 | 6 | 8 | 7 | 9 | 7 | 4 | 4 | 6 | 5 | 7 | 8 | 9 | 9 | 6 | 8 | 8 | 10 | 6 | 8 | 3 | 4 | 5 | 5 | 6 | 7 | 7 | 7 | 7 | 7 | 7 | 5 | 9 | 6 | 7 | 10 |
| **7c** | 7 | 7 | 6 | 8 | 7 | 8 | 8 | 8 | 7 | 5 | 6 | 8 | 7 | 9 | 7 | 7 | 4 | 6 | 5 | 7 | 8 | 9 | 9 | 6 | 8 | 8 | 10 | 7 | 8 | 7 | 8 | 6 | 5 | 6 | 7 | 7 | 7 | 7 | 7 | 7 | 5 | 8 | 6 | 7 | 10 |
| **7d** | 7 | 6 | 7 | 8 | 8 | 8 | 8 | 7 | 5 | 6 | 7 | 8 | 6 | 8 | 6 | 4 | 5 | 7 | 6 | 8 | 9 | 10 | 9 | 5 | 7 | 8 | 9 | 4 | 7 | 3 | 4 | 3 | 4 | 5 | 6 | 6 | 6 | 6 | 6 | 6 | 6 | 9 | 7 | 8 | 10 |
| **7e** | 7 | 6 | 7 | 8 | 7 | 8 | 8 | 7 | 5 | 8 | 8 | 8 | 6 | 8 | 6 | 4 | 7 | 9 | 9 | 9 | 9 | 9 | 9 | 5 | 7 | 9 | 9 | 4 | 7 | 5 | 5 | 3 | 4 | 5 | 6 | 6 | 6 | 6 | 6 | 6 | 8 | 10 | 10 | 9 | 10 |
| **7f** | 7 | 7 | 8 | 8 | 6 | 8 | 8 | 6 | 6 | 7 | 7 | 7 | 5 | 7 | 5 | 5 | 6 | 8 | 7 | 8 | 8 | 8 | 8 | 4 | 6 | 8 | 8 | 5 | 6 | 4 | 6 | 4 | 3 | 4 | 5 | 5 | 5 | 5 | 5 | 5 | 6 | 9 | 8 | 7 | 9 |
| **7g** | 7 | 6 | 7 | 8 | 7 | 8 | 8 | 7 | 5 | 8 | 8 | 8 | 6 | 8 | 6 | 4 | 9 | 9 | 9 | 9 | 9 | 9 | 9 | 5 | 7 | 9 | 9 | 4 | 7 | 5 | 5 | 3 | 4 | 5 | 6 | 6 | 6 | 6 | 6 | 6 | 8 | 10 | 10 | 9 | 10 |
| **7h** | 7 | 7 | 6 | 8 | 7 | 8 | 8 | 7 | 6 | 5 | 6 | 8 | 6 | 8 | 6 | 6 | 4 | 6 | 5 | 7 | 8 | 9 | 9 | 5 | 7 | 7 | 7 | 6 | 7 | 4 | 7 | 4 | 4 | 5 | 6 | 6 | 6 | 6 | 6 | 6 | 5 | 8 | 6 | 7 | 10 |
| **7i** | 7 | 7 | 8 | 8 | 6 | 8 | 8 | 6 | 6 | 7 | 7 | 7 | 5 | 7 | 5 | 5 | 6 | 8 | 7 | 8 | 8 | 8 | 8 | 4 | 6 | 8 | 8 | 5 | 6 | 6 | 6 | 4 | 3 | 4 | 5 | 5 | 5 | 5 | 5 | 5 | 6 | 9 | 8 | 7 | 9 |
| **7j** | 7 | 7 | 6 | 8 | 7 | 8 | 8 | 6 | 6 | 5 | 6 | 7 | 5 | 7 | 5 | 5 | 4 | 6 | 5 | 7 | 8 | 8 | 8 | 4 | 6 | 6 | 8 | 5 | 6 | 6 | 6 | 4 | 3 | 4 | 5 | 5 | 5 | 5 | 5 | 5 | 5 | 8 | 6 | 7 | 9 |
| **7k** | 7 | 7 | 6 | 8 | 6 | 8 | 8 | 6 | 6 | 5 | 7 | 7 | 5 | 7 | 5 | 5 | 4 | 8 | 7 | 8 | 6 | 8 | 8 | 4 | 6 | 8 | 8 | 5 | 6 | 6 | 6 | 4 | 3 | 4 | 5 | 5 | 5 | 5 | 5 | 5 | 3 | 7 | 6 | 5 | 9 |
| **7L** | 7 | 8 | 6 | 8 | 7 | 8 | 8 | 6 | 6 | 5 | 6 | 8 | 7 | 5 | 5 | 9 | 4 | 6 | 5 | 7 | 8 | 9 | 9 | 4 | 4 | 4 | 6 | 5 | 6 | 76 | 6 | 4 | 3 | 4 | 5 | 5 | 5 | 5 | 5 | 5 | 5 | 8 | 6 | 7 | 9 |
| **7m** | 7 | 8 | 7 | 8 | 6 | 8 | 8 | 6 | 8 | 6 | 6 | 6 | 5 | 7 | 5 | 7 | 5 | 7 | 5 | 7 | 5 | 7 | 7 | 4 | 6 | 8 | 8 | 7 | 6 | 8 | 8 | 6 | 5 | 5 | 5 | 5 | 5 | 5 | 5 | 3 | 4 | 6 | 4 | 4 | 8 |
| **7n** | 7 | 7 | 6 | 8 | 7 | 8 | 8 | 6 | 6 | 5 | 6 | 7 | 5 | 7 | 5 | 5 | 4 | 6 | 5 | 7 | 8 | 8 | 8 | 4 | 6 | 6 | 8 | 5 | 6 | 6 | 6 | 4 | 3 | 4 | 5 | 5 | 5 | 5 | 5 | 5 | 5 | 8 | 5 | 7 | 9 |
| **7o** | 7 | 8 | 7 | 8 | 6 | 8 | 8 | 7 | 9 | 6 | 5 | 7 | 6 | 8 | 6 | 8 | 5 | 7 | 4 | 6 | 6 | 8 | 8 | 5 | 7 | 9 | 9 | 8 | 7 | 9 | 9 | 7 | 6 | 6 | 6 | 6 | 6 | 6 | 6 | 4 | 5 | 7 | 3 | 5 | 9 |
| **7p** | 7 | 7 | 6 | 8 | 7 | 8 | 8 | 8 | 8 | 5 | 6 | 8 | 8 | 9 | 9 | 9 | 4 | 6 | 5 | 7 | 8 | 9 | 9 | 9 | 10 | 10 | 10 | 9 | 10 | 10 | 10 | 10 | 9 | 10 | 10 | 10 | 10 | 10 | 10 | 8 | 5 | 8 | 6 | 7 | 10 |
| **8a** | 8 | 7 | 8 | 9 | 8 | 9 | 9 | 8 | 6 | 9 | 9 | 9 | 7 | 9 | 7 | 5 | 10 | 10 | 10 | 10 | 10 | 10 | 10 | 6 | 8 | 10 | 10 | 5 | 8 | 6 | 6 | 4 | 5 | 6 | 7 | 7 | 7 | 7 | 7 | 7 | 9 | 11 | 11 | 10 | 11 |
| **8b** | 8 | 7 | 8 | 9 | 9 | 9 | 9 | 9 | 6 | 8 | 9 | 10 | 10 | 10 | 10 | 5 | 7 | 9 | 8 | 10 | 11 | 11 | 11 | 8 | 10 | 11 | 11 | 7 | 10 | 4 | 5 | 6 | 7 | 8 | 9 | 9 | 9 | 9 | 9 | 9 | 8 | 11 | 9 | 10 | 12 |
| **8c** | 8 | 8 | 7 | 9 | 7 | 9 | 9 | 7 | 7 | 6 | 7 | 8 | 6 | 8 | 6 | 6 | 5 | 7 | 6 | 8 | 9 | 9 | 9 | 5 | 7 | 7 | 9 | 6 | 7 | 7 | 7 | 5 | 4 | 5 | 6 | 6 | 6 | 6 | 6 | 6 | 6 | 9 | 7 | 8 | 10 |
| **8d** | 8 | 7 | 8 | 9 | 9 | 9 | 9 | 8 | 6 | 7 | 8 | 9 | 7 | 9 | 7 | 5 | 6 | 8 | 7 | 9 | 10 | 11 | 10 | 6 | 8 | 9 | 10 | 5 | 8 | 4 | 5 | 4 | 5 | 6 | 7 | 7 | 7 | 7 | 7 | 7 | 7 | 10 | 8 | 9 | 11 |
| **8e** | 8 | 8 | 7 | 9 | 8 | 9 | 9 | 7 | 7 | 6 | 7 | 8 | 6 | 8 | 6 | 6 | 5 | 7 | 6 | 8 | 9 | 9 | 9 | 5 | 7 | 7 | 9 | 6 | 7 | 7 | 7 | 5 | 4 | 5 | 6 | 6 | 6 | 6 | 6 | 6 | 6 | 9 | 7 | 8 | 10 |
| **8f** | 8 | 8 | 7 | 9 | 8 | 9 | 9 | 7 | 7 | 6 | 7 | 8 | 6 | 8 | 6 | 6 | 5 | 7 | 6 | 8 | 9 | 9 | 9 | 5 | 7 | 7 | 9 | 6 | 7 | 7 | 7 | 5 | 4 | 5 | 6 | 6 | 6 | 6 | 6 | 6 | 6 | 9 | 7 | 8 | 10 |
| **8g** | 8 | 8 | 7 | 9 | 8 | 9 | 9 | 9 | 8 | 6 | 7 | 9 | 8 | 10 | 8 | 8 | 5 | 7 | 6 | 8 | 9 | 10 | 10 | 7 | 9 | 9 | 11 | 8 | 9 | 8 | 9 | 7 | 6 | 7 | 8 | 8 | 8 | 8 | 8 | 8 | 6 | 9 | 7 | 8 | 11 |
| **9a** | 9 | 8 | 9 | 10 | 10 | 10 | 10 | 10 | 7 | 9 | 10 | 11 | 11 | 11 | 11 | 6 | 8 | 10 | 9 | 11 | 12 | 12 | 12 | 9 | 11 | 12 | 12 | 8 | 11 | 5 | 6 | 7 | 8 | 9 | 10 | 10 | 10 | 10 | 10 | 10 | 9 | 12 | 10 | 11 | 13 |
| **9b** | 9 | 8 | 9 | 10 | 9 | 10 | 10 | 9 | 7 | 10 | 10 | 10 | 8 | 10 | 8 | 6 | 11 | 11 | 11 | 11 | 11 | 11 | 11 | 7 | 9 | 11 | 11 | 6 | 9 | 7 | 7 | 5 | 6 | 7 | 8 | 8 | 8 | 8 | 8 | 8 | 10 | 12 | 12 | 11 | 12 |
| **9c** | 9 | 9 | 8 | 10 | 9 | 10 | 10 | 8 | 8 | 7 | 8 | 9 | 7 | 9 | 7 | 7 | 6 | 8 | 7 | 9 | 10 | 10 | 10 | 6 | 8 | 8 | 10 | 7 | 8 | 8 | 8 | 6 | 5 | 6 | 7 | 7 | 7 | 7 | 7 | 7 | 7 | 10 | 8 | 9 | 11 |
| **10a** | 10 | 9 | 10 | 11 | 11 | 11 | 11 | 11 | 8 | 10 | 11 | 12 | 12 | 12 | 12 | 7 | 9 | 11 | 10 | 12 | 13 | 13 | 13 | 10 | 12 | 13 | 13 | 9 | 12 | 6 | 7 | 8 | 9 | 10 | 11 | 11 | 11 | 11 | 11 | 11 | 10 | 13 | 11 | 12 | 14 |

| | 4q | 4r | 4s | 4t | 4u | 4v | 4w | 4x | 4y | 4z | 4aa | 5a | 5b | 5c | 5d | 5e | 5f | 5g | 5h | 5i | 5j | 5k | 5L | 5m |
|---|---|---|---|---|---|---|---|---|---|---|---|---|---|---|---|---|---|---|---|---|---|---|---|---|
| **c** | 4 | 4 | 4 | 4 | 4 | 4 | 4 | 4 | 4 | 4 | 4 | 5 | 5 | 5 | 5 | 5 | 5 | 5 | 5 | 5 | 5 | 5 | 5 | 5 |
| **1a** | 5 | 5 | 5 | 5 | 4 | 4 | 4 | 4 | 5 | 3 | 4 | 4 | 4 | 4 | 4 | 4 | 5 | 4 | 4 | 4 | 5 | 4 | 5 | 5 |
| **1b** | 4 | 5 | 5 | 5 | 3 | 3 | 3 | 3 | 5 | 4 | 3 | 5 | 5 | 5 | 5 | 5 | 4 | 5 | 5 | 5 | 6 | 5 | 6 | 6 |
| **1c** | 5 | 5 | 5 | 5 | 5 | 5 | 5 | 5 | 5 | 5 | 5 | 6 | 6 | 6 | 6 | 6 | 6 | 6 | 6 | 6 | 6 | 6 | 6 | 6 |
| **1d** | 3 | 5 | 5 | 5 | 4 | 4 | 4 | 4 | 3 | 5 | 4 | 6 | 6 | 6 | 6 | 6 | 5 | 6 | 6 | 6 | 4 | 5 | 4 | 4 |
| **1e** | 5 | 5 | 5 | 5 | 5 | 5 | 5 | 5 | 5 | 5 | 5 | 6 | 6 | 6 | 6 | 6 | 6 | 6 | 6 | 6 | 6 | 6 | 6 | 6 |
| **1f** | 5 | 5 | 5 | 5 | 5 | 5 | 5 | 5 | 5 | 5 | 5 | 6 | 6 | 6 | 6 | 6 | 6 | 6 | 6 | 6 | 6 | 6 | 6 | 6 |
| **1g** | 5 | 3 | 3 | 3 | 5 | 5 | 5 | 5 | 3 | 5 | 5 | 6 | 6 | 6 | 6 | 6 | 6 | 5 | 5 | 5 | 4 | 5 | 4 | 4 |
| **2a** | 6 | 6 | 6 | 6 | 5 | 5 | 5 | 5 | 5 | 2 | 5 | 3 | 3 | 3 | 3 | 3 | 6 | 3 | 3 | 3 | 4 | 3 | 4 | 4 |
| **2b** | 6 | 6 | 6 | 6 | 2 | 2 | 2 | 2 | 6 | 5 | 2 | 6 | 6 | 6 | 6 | 6 | 3 | 6 | 6 | 6 | 7 | 6 | 6 | 5 |
| **2c** | 4 | 6 | 6 | 6 | 3 | 3 | 3 | 3 | 4 | 6 | 3 | 7 | 7 | 7 | 7 | 7 | 4 | 6 | 6 | 6 | 5 | 6 | 5 | 5 |
| **2d** | 2 | 6 | 6 | 6 | 5 | 5 | 5 | 5 | 4 | 6 | 5 | 7 | 7 | 7 | 7 | 7 | 6 | 6 | 6 | 6 | 5 | 6 | 5 | 5 |
| **2e** | 2 | 6 | 4 | 4 | 5 | 5 | 5 | 5 | 2 | 4 | 5 | 6 | 6 | 6 | 6 | 6 | 6 | 4 | 4 | 4 | 3 | 4 | 2 | 2 |
| **2f** | 6 | 2 | 2 | 2 | 6 | 6 | 6 | 6 | 4 | 6 | 6 | 7 | 7 | 7 | 7 | 7 | 7 | 6 | 6 | 6 | 5 | 6 | 5 | 5 |
| **2g** | 6 | 4 | 2 | 2 | 6 | 6 | 6 | 6 | 2 | 4 | 6 | 6 | 6 | 6 | 6 | 6 | 7 | 4 | 4 | 4 | 3 | 4 | 3 | 3 |
| **3a** | 6 | 7 | 7 | 6 | 6 | 6 | 6 | 6 | 4 | 2 | 6 | 2 | 2 | 2 | 2 | 2 | 7 | 2 | 2 | 2 | 3 | 2 | 3 | 3 |
| **3b** | 6 | 7 | 7 | 7 | 1 | 1 | 1 | 1 | 6 | 6 | 1 | 7 | 7 | 7 | 7 | 7 | 2 | 7 | 7 | 7 | 6 | 7 | 5 | 4 |
| **3c** | 7 | 7 | 7 | 7 | 3 | 3 | 3 | 3 | 7 | 6 | 3 | 7 | 7 | 7 | 7 | 7 | 4 | 7 | 7 | 7 | 8 | 7 | 7 | 6 |
| **3d** | 5 | 7 | 7 | 7 | 2 | 2 | 2 | 2 | 7 | 7 | 2 | 8 | 8 | 8 | 8 | 7 | 3 | 7 | 7 | 7 | 6 | 7 | 6 | 5 |
| **3e** | 5 | 7 | 7 | 7 | 4 | 4 | 4 | 4 | 5 | 7 | 4 | 8 | 8 | 8 | 8 | 8 | 5 | 7 | 7 | 7 | 6 | 7 | 6 | 6 |
| **3f** | 3 | 7 | 7 | 7 | 5 | 5 | 5 | 5 | 5 | 7 | 5 | 8 | 8 | 8 | 8 | 8 | 6 | 7 | 7 | 7 | 6 | 7 | 6 | 6 |
| **3g** | 1 | 7 | 7 | 7 | 6 | 6 | 6 | 6 | 5 | 7 | 6 | 8 | 8 | 8 | 8 | 8 | 7 | 7 | 7 | 7 | 6 | 7 | 6 | 6 |
| **3h** | 2 | 7 | 7 | 7 | 6 | 6 | 6 | 6 | 5 | 7 | 6 | 8 | 8 | 8 | 8 | 8 | 7 | 7 | 7 | 7 | 6 | 7 | 6 | 6 |
| **3i** | 3 | 5 | 3 | 3 | 6 | 6 | 6 | 5 | 1 | 3 | 6 | 5 | 5 | 5 | 5 | 5 | 7 | 3 | 3 | 3 | 2 | 3 | 2 | 2 |
| **3j** | 5 | 3 | 1 | 1 | 7 | 7 | 7 | 7 | 3 | 5 | 7 | 7 | 7 | 7 | 7 | 7 | 8 | 5 | 5 | 5 | 4 | 5 | 4 | 4 |
| **3k** | 7 | 3 | 1 | 1 | 7 | 7 | 7 | 7 | 5 | 7 | 7 | 8 | 8 | 8 | 8 | 8 | 8 | 7 | 7 | 7 | 6 | 7 | 6 | 6 |
| **3L** | 7 | 1 | 3 | 3 | 7 | 7 | 7 | 7 | 5 | 7 | 7 | 8 | 8 | 8 | 8 | 8 | 8 | 7 | 7 | 7 | 6 | 7 | 6 | 6 |
| **3m** | 6 | 7 | 7 | 7 | 6 | 6 | 6 | 6 | 4 | 1 | 6 | 4 | 4 | 4 | 4 | 4 | 7 | 2 | 2 | 2 | 3 | 2 | 3 | 3 |
| **3n** | 1 | 7 | 5 | 5 | 7 | 7 | 7 | 7 | 3 | 5 | 7 | 7 | 7 | 7 | 7 | 7 | 8 | 5 | 5 | 5 | 4 | 5 | 4 | 4 |
| **4a** | 7 | 8 | 7 | 7 | 7 | 7 | 7 | 6 | 5 | 3 | 7 | 1 | 1 | 1 | 1 | 1 | 8 | 3 | 3 | 3 | 4 | 3 | 4 | 4 |
| **4b** | 7 | 8 | 7 | 7 | 7 | 7 | 7 | 7 | 5 | 3 | 7 | 2 | 2 | 2 | 2 | 2 | 8 | 3 | 3 | 3 | 4 | 3 | 4 | 4 |
| **4c** | 5 | 7 | 5 | 5 | 7 | 7 | 7 | 5 | 3 | 1 | 7 | 3 | 3 | 3 | 3 | 3 | 8 | 1 | 1 | 1 | 2 | 1 | 2 | 2 |
| **4d** | 4 | 6 | 4 | 4 | 6 | 6 | 6 | 4 | 2 | 2 | 6 | 4 | 4 | 4 | 4 | 4 | 7 | 2 | 2 | 2 | 1 | 2 | 1 | 1 |
| **4e** | 4 | 6 | 4 | 4 | 7 | 7 | 7 | 5 | 2 | 3 | 7 | 5 | 5 | 5 | 5 | 5 | 8 | 3 | 3 | 3 | 2 | 3 | 2 | 2 |
| **4f** | 4 | 6 | 4 | 4 | 7 | 7 | 7 | 6 | 2 | 4 | 7 | 6 | 6 | 6 | 6 | 6 | 8 | 4 | 4 | 4 | 3 | 4 | 3 | 3 |
| **4g** | 4 | 6 | 4 | 4 | 7 | 7 | 7 | 6 | 2 | 4 | 7 | 6 | 6 | 6 | 6 | 6 | 8 | 4 | 4 | 4 | 3 | 4 | 3 | 3 |
| **4h** | 4 | 6 | 4 | 4 | 7 | 7 | 7 | 6 | 2 | 4 | 7 | 6 | 6 | 6 | 6 | 6 | 8 | 4 | 4 | 4 | 3 | 4 | 3 | 3 |
| **4i** | 4 | 6 | 4 | 4 | 7 | 7 | 7 | 6 | 2 | 4 | 7 | 6 | 6 | 6 | 6 | 6 | 8 | 4 | 4 | 4 | 3 | 4 | 3 | 3 |
| **4j** | 4 | 6 | 4 | 4 | 7 | 7 | 7 | 6 | 2 | 4 | 7 | 6 | 6 | 6 | 6 | 6 | 8 | 4 | 4 | 4 | 3 | 4 | 3 | 3 |
| **4k** | 4 | 6 | 4 | 4 | 5 | 5 | 5 | 5 | 2 | 4 | 5 | 6 | 6 | 6 | 6 | 6 | 8 | 4 | 4 | 4 | 3 | 4 | 3 | 3 |
| **4L** | 5 | 8 | 7 | 7 | 2 | 2 | 2 | 2 | 5 | 7 | 2 | 8 | 8 | 8 | 8 | 7 | 3 | 6 | 6 | 6 | 5 | 6 | 4 | 3 |
| **4m** | 6 | 8 | 8 | 8 | 5 | 5 | 5 | 5 | 6 | 8 | 5 | 9 | 9 | 9 | 9 | 9 | 6 | 8 | 8 | 8 | 7 | 8 | 7 | 7 |
| **4n** | 6 | 8 | 8 | 8 | 3 | 3 | 3 | 3 | 6 | 8 | 3 | 9 | 9 | 9 | 9 | 8 | 4 | 8 | 8 | 8 | 7 | 8 | 7 | 7 |

| | 5n | 5o | 5p | 5q | 5r | 5s | 5t | 5u | 5v | 5w | 5x | 6a | 6b | 6c | 6d | 6e | 6f | 6g | 6h | 6i | 6j |
|---|---|---|---|---|---|---|---|---|---|---|---|---|---|---|---|---|---|---|---|---|---|
| **c** | 5 | 5 | 5 | 5 | 5 | 5 | 5 | 5 | 5 | 5 | 5 | 6 | 6 | 6 | 6 | 6 | 6 | 6 | 6 | 6 | 6 |
| **1a** | 5 | 5 | 6 | 5 | 6 | 6 | 6 | 6 | 4 | 4 | 5 | 5 | 6 | 6 | 6 | 6 | 6 | 5 | 6 | 5 | 6 |
| **1b** | 4 | 4 | 6 | 4 | 6 | 5 | 6 | 6 | 5 | 5 | 4 | 6 | 7 | 7 | 7 | 7 | 7 | 6 | 7 | 6 | 7 |
| **1c** | 6 | 6 | 6 | 6 | 6 | 6 | 6 | 6 | 6 | 6 | 6 | 7 | 7 | 7 | 7 | 7 | 7 | 7 | 7 | 7 | 7 |
| **1d** | 5 | 5 | 4 | 4 | 4 | 4 | 6 | 6 | 6 | 6 | 5 | 7 | 5 | 5 | 5 | 5 | 5 | 6 | 5 | 7 | 5 |
| **1e** | 6 | 6 | 6 | 6 | 6 | 6 | 6 | 6 | 6 | 6 | 6 | 7 | 7 | 7 | 7 | 7 | 7 | 7 | 7 | 7 | 7 |
| **1f** | 6 | 6 | 6 | 6 | 6 | 6 | 6 | 6 | 6 | 6 | 6 | 7 | 7 | 7 | 7 | 7 | 7 | 7 | 7 | 7 | 7 |
| **1g** | 6 | 6 | 4 | 5 | 6 | 6 | 4 | 4 | 6 | 5 | 6 | 6 | 5 | 5 | 5 | 5 | 5 | 6 | 5 | 6 | 5 |
| **2a** | 6 | 6 | 6 | 6 | 7 | 7 | 7 | 7 | 3 | 3 | 6 | 4 | 5 | 5 | 5 | 5 | 5 | 4 | 5 | 4 | 5 |
| **2b** | 3 | 3 | 4 | 3 | 7 | 4 | 7 | 7 | 6 | 6 | 3 | 7 | 8 | 8 | 8 | 8 | 8 | 7 | 8 | 7 | 7 |
| **2c** | 4 | 4 | 5 | 4 | 5 | 3 | 7 | 7 | 7 | 6 | 4 | 7 | 6 | 6 | 6 | 6 | 6 | 7 | 6 | 7 | 6 |
| **2d** | 6 | 5 | 4 | 3 | 3 | 5 | 7 | 7 | 7 | 6 | 6 | 7 | 6 | 6 | 6 | 6 | 6 | 7 | 6 | 7 | 6 |
| **2e** | 5 | 5 | 3 | 4 | 5 | 5 | 5 | 7 | 6 | 4 | 6 | 5 | 4 | 4 | 4 | 4 | 4 | 5 | 4 | 5 | 4 |
| **2f** | 7 | 7 | 5 | 6 | 7 | 7 | 3 | 3 | 7 | 6 | 7 | 7 | 6 | 6 | 6 | 6 | 6 | 7 | 6 | 7 | 6 |
| **2g** | 5 | 5 | 3 | 4 | 7 | 7 | 3 | 5 | 6 | 4 | 7 | 5 | 4 | 4 | 4 | 4 | 4 | 5 | 4 | 5 | 4 |
| **3a** | 5 | 5 | 5 | 6 | 8 | 8 | 7 | 8 | 2 | 2 | 7 | 3 | 4 | 4 | 4 | 4 | 4 | 3 | 4 | 3 | 4 |
| **3b** | 2 | 2 | 3 | 2 | 7 | 3 | 6 | 8 | 7 | 7 | 2 | 8 | 7 | 7 | 7 | 7 | 7 | 8 | 7 | 6 | 5 |
| **3c** | 4 | 4 | 5 | 4 | 8 | 5 | 8 | 8 | 7 | 7 | 4 | 8 | 9 | 9 | 9 | 9 | 9 | 8 | 9 | 8 | 7 |
| **3d** | 3 | 3 | 4 | 3 | 6 | 2 | 7 | 8 | 8 | 7 | 3 | 8 | 7 | 7 | 7 | 7 | 7 | 8 | 7 | 8 | 6 |
| **3e** | 5 | 5 | 6 | 5 | 6 | 4 | 8 | 8 | 8 | 7 | 5 | 8 | 7 | 7 | 7 | 7 | 7 | 8 | 7 | 8 | 7 |
| **3f** | 6 | 4 | 3 | 2 | 4 | 6 | 8 | 8 | 8 | 7 | 6 | 8 | 7 | 7 | 7 | 7 | 7 | 8 | 7 | 8 | 7 |
| **3g** | 7 | 6 | 5 | 4 | 2 | 6 | 8 | 8 | 8 | 7 | 7 | 8 | 7 | 7 | 7 | 7 | 7 | 8 | 7 | 8 | 7 |
| **3h** | 7 | 6 | 5 | 4 | 2 | 6 | 8 | 8 | 8 | 7 | 7 | 8 | 7 | 7 | 7 | 7 | 7 | 8 | 7 | 8 | 7 |
| **3i** | 4 | 5 | 2 | 3 | 6 | 6 | 4 | 6 | 5 | 3 | 7 | 4 | 3 | 3 | 3 | 3 | 3 | 4 | 3 | 4 | 3 |
| **3j** | 6 | 6 | 4 | 5 | 8 | 8 | 2 | 4 | 7 | 5 | 8 | 6 | 5 | 5 | 5 | 5 | 5 | 6 | 5 | 6 | 5 |
| **3k** | 6 | 8 | 6 | 7 | 8 | 8 | 2 | 4 | 8 | 7 | 8 | 8 | 7 | 7 | 7 | 7 | 7 | 8 | 7 | 8 | 7 |
| **3L** | 8 | 8 | 6 | 7 | 8 | 8 | 4 | 2 | 8 | 7 | 8 | 8 | 7 | 7 | 7 | 7 | 7 | 8 | 7 | 8 | 7 |
| **3m** | 5 | 5 | 5 | 6 | 8 | 8 | 7 | 8 | 8 | 2 | 7 | 3 | 4 | 4 | 4 | 4 | 4 | 3 | 4 | 3 | 4 |
| **3n** | 6 | 6 | 4 | 5 | 4 | 6 | 6 | 8 | 7 | 5 | 8 | 6 | 5 | 5 | 5 | 5 | 5 | 6 | 5 | 6 | 5 |
| **4a** | 5 | 6 | 6 | 7 | 9 | 9 | 8 | 10 | 1 | 3 | 8 | 4 | 5 | 5 | 5 | 5 | 5 | 4 | 5 | 4 | 5 |
| **4b** | 6 | 6 | 6 | 7 | 9 | 9 | 8 | 9 | 2 | 3 | 8 | 4 | 5 | 5 | 5 | 5 | 5 | 4 | 5 | 4 | 5 |
| **4c** | 4 | 3 | 4 | 5 | 8 | 8 | 5 | 8 | 3 | 1 | 8 | 2 | 3 | 3 | 3 | 3 | 3 | 2 | 3 | 2 | 3 |
| **4d** | 3 | 3 | 3 | 4 | 7 | 7 | 5 | 7 | 3 | 2 | 7 | 3 | 2 | 2 | 2 | 2 | 2 | 3 | 2 | 3 | 2 |
| **4e** | 4 | 4 | 3 | 4 | 7 | 7 | 5 | 7 | 5 | 3 | 8 | 4 | 3 | 3 | 3 | 3 | 3 | 4 | 3 | 4 | 3 |
| **4f** | 5 | 5 | 3 | 4 | 7 | 7 | 5 | 7 | 6 | 4 | 8 | 5 | 4 | 4 | 4 | 4 | 4 | 5 | 4 | 5 | 4 |
| **4g** | 5 | 5 | 3 | 4 | 7 | 7 | 5 | 7 | 6 | 4 | 8 | 5 | 4 | 4 | 4 | 4 | 4 | 5 | 4 | 5 | 4 |
| **4h** | 5 | 5 | 3 | 4 | 7 | 7 | 5 | 7 | 6 | 4 | 8 | 5 | 4 | 4 | 4 | 4 | 4 | 5 | 4 | 5 | 4 |
| **4i** | 5 | 5 | 3 | 4 | 7 | 7 | 5 | 7 | 6 | 4 | 8 | 5 | 4 | 4 | 4 | 4 | 4 | 5 | 4 | 5 | 4 |
| **4j** | 5 | 5 | 3 | 4 | 7 | 7 | 5 | 7 | 6 | 4 | 8 | 5 | 4 | 4 | 4 | 4 | 4 | 5 | 4 | 5 | 4 |
| **4k** | 5 | 4 | 1 | 2 | 7 | 6 | 5 | 7 | 6 | 4 | 6 | 5 | 4 | 4 | 4 | 4 | 4 | 5 | 4 | 5 | 4 |
| **4L** | 3 | 1 | 2 | 1 | 7 | 4 | 7 | 9 | 8 | 7 | 3 | 7 | 6 | 6 | 6 | 6 | 6 | 7 | 6 | 7 | 5 |
| **4m** | 6 | 5 | 4 | 3 | 7 | 5 | 9 | 9 | 9 | 8 | 6 | 9 | 8 | 8 | 8 | 8 | 8 | 9 | 8 | 9 | 8 |
| **4n** | 4 | 4 | 5 | 4 | 7 | 1 | 8 | 9 | 9 | 8 | 4 | 9 | 8 | 8 | 8 | 8 | 8 | 9 | 8 | 8 | 7 |

*(continued)*

Palenque Palace Path Matrix Phase 6 (Cont.)

| | 4q | 4r | 4s | 4t | 4u | 4v | 4w | 4x | 4y | 4z | 4aa | 5a | 5b | 5c | 5d | 5e | 5f | 5g | 5h | 5i | 5j | 5k | 5L | 5m | 5n | 5o | 5p | 5q | 5r | 5s | 5t | 5u | 5v | 5w | 5x | 6a | 6b | 6c | 6d | 6e | 6f | 6g | 6h | 6i | 6j |
|---|---|---|---|---|---|---|---|---|---|---|---|---|---|---|---|---|---|---|---|---|---|---|---|---|---|---|---|---|---|---|---|---|---|---|---|---|---|---|---|---|---|---|---|---|---|
| **4o** | 4 | 8 | 8 | 8 | 4 | 4 | 4 | 4 | 6 | 7 | 4 | 9 | 9 | 9 | 9 | 9 | 5 | 7 | 7 | 7 | 6 | 7 | 6 | 5 | 5 | 3 | 2 | 1 | 5 | 5 | 9 | 9 | 9 | 7 | 5 | 8 | 7 | 7 | 7 | 7 | 7 | 8 | 7 | 8 | 6 |
| **4p** | 2 | 8 | 8 | 8 | 7 | 7 | 7 | 7 | 6 | 8 | 7 | 9 | 9 | 9 | 9 | 9 | 8 | 8 | 8 | 8 | 7 | 8 | 7 | 7 | 8 | 7 | 6 | 5 | 1 | 7 | 9 | 9 | 9 | 8 | 8 | 9 | 8 | 8 | 8 | 8 | 8 | 9 | 8 | 9 | 8 |
| **4q** | x | 8 | 6 | 6 | 7 | 7 | 7 | 7 | 4 | 6 | 7 | 8 | 8 | 8 | 8 | 8 | 8 | 6 | 6 | 6 | 5 | 6 | 5 | 5 | 7 | 7 | 5 | 5 | 3 | 7 | 7 | 9 | 8 | 6 | 8 | 7 | 6 | 6 | 6 | 6 | 6 | 7 | 6 | 7 | 6 |
| **4r** | 8 | x | 4 | 4 | 8 | 8 | 8 | 8 | 6 | 8 | 8 | 9 | 9 | 9 | 9 | 9 | 9 | 8 | 8 | 8 | 7 | 8 | 7 | 7 | 9 | 9 | 7 | 8 | 9 | 9 | 5 | 1 | 9 | 8 | 9 | 9 | 8 | 8 | 8 | 8 | 8 | 9 | 8 | 9 | 8 |
| **4s** | 6 | 4 | x | 2 | 8 | 8 | 8 | 6 | 4 | 6 | 8 | 8 | 8 | 8 | 8 | 8 | 9 | 6 | 6 | 6 | 5 | 6 | 5 | 5 | 5 | 7 | 5 | 6 | 9 | 9 | 1 | 5 | 8 | 6 | 9 | 7 | 6 | 6 | 6 | 6 | 6 | 7 | 6 | 7 | 6 |
| **4t** | 6 | 4 | 2 | x | 8 | 8 | 8 | 8 | 4 | 6 | 8 | 8 | 8 | 8 | 8 | 9 | 9 | 6 | 6 | 6 | 5 | 6 | 5 | 5 | 7 | 7 | 5 | 6 | 9 | 9 | 3 | 5 | 8 | 6 | 9 | 7 | 6 | 6 | 6 | 6 | 6 | 7 | 6 | 7 | 6 |
| **4u** | 7 | 8 | 8 | 8 | x | 2 | 2 | 2 | 7 | 7 | 2 | 8 | 8 | 8 | 8 | 7 | 1 | 8 | 8 | 8 | 7 | 8 | 6 | 5 | 3 | 3 | 4 | 3 | 8 | 4 | 7 | 9 | 8 | 8 | 3 | 9 | 8 | 8 | 8 | 8 | 8 | 9 | 8 | 7 | 6 |
| **4v** | 7 | 8 | 8 | 8 | 2 | x | 1 | 2 | 7 | 7 | 2 | 8 | 8 | 8 | 8 | 7 | 3 | 8 | 8 | 8 | 7 | 8 | 6 | 5 | 3 | 3 | 4 | 3 | 8 | 4 | 7 | 9 | 8 | 8 | 3 | 9 | 8 | 8 | 8 | 8 | 8 | 9 | 8 | 7 | 6 |
| **4w** | 7 | 8 | 8 | 8 | 2 | 1 | x | 2 | 7 | 7 | 2 | 8 | 8 | 8 | 8 | 7 | 3 | 8 | 8 | 8 | 7 | 8 | 6 | 5 | 3 | 3 | 4 | 3 | 8 | 4 | 7 | 9 | 8 | 8 | 3 | 9 | 8 | 8 | 8 | 8 | 8 | 9 | 8 | 7 | 6 |
| **4x** | 7 | 8 | 6 | 8 | 2 | 2 | 2 | x | 7 | 6 | 2 | 7 | 7 | 7 | 6 | 5 | 3 | 8 | 8 | 8 | 5 | 6 | 4 | 3 | 1 | 3 | 4 | 3 | 8 | 4 | 5 | 9 | 7 | 8 | 3 | 7 | 6 | 6 | 6 | 6 | 6 | 7 | 6 | 5 | 4 |
| **4y** | 4 | 6 | 4 | 4 | 7 | 7 | 7 | 7 | x | 4 | 7 | 6 | 6 | 6 | 6 | 6 | 8 | 4 | 4 | 4 | 3 | 4 | 3 | 3 | 5 | 5 | 3 | 4 | 7 | 7 | 5 | 7 | 6 | 4 | 8 | 5 | 4 | 4 | 4 | 4 | 4 | 5 | 4 | 5 | 4 |
| **4z** | 6 | 8 | 6 | 6 | 7 | 7 | 7 | 6 | 4 | x | 7 | 4 | 4 | 4 | 4 | 4 | 8 | 2 | 2 | 2 | 3 | 2 | 3 | 3 | 5 | 5 | 5 | 4 | 9 | 9 | 7 | 9 | 4 | 2 | 8 | 3 | 4 | 4 | 4 | 4 | 4 | 3 | 4 | 3 | 4 |
| **4aa** | 7 | 8 | 8 | 8 | 2 | 2 | 2 | 2 | 7 | 7 | x | 8 | 8 | 8 | 8 | 7 | 3 | 8 | 8 | 8 | 7 | 8 | 6 | 5 | 3 | 3 | 4 | 3 | 8 | 4 | 7 | 9 | 8 | 8 | 1 | 9 | 8 | 8 | 8 | 8 | 8 | 9 | 8 | 7 | 6 |
| **5a** | 8 | 9 | 8 | 8 | 8 | 8 | 8 | 7 | 6 | 4 | 8 | x | 2 | 2 | 2 | 2 | 9 | 4 | 4 | 4 | 5 | 4 | 5 | 5 | 6 | 7 | 7 | 8 | 10 | 10 | 9 | 10 | 2 | 4 | 9 | 5 | 6 | 6 | 6 | 6 | 6 | 5 | 6 | 5 | 6 |
| **5b** | 8 | 9 | 8 | 8 | 8 | 8 | 8 | 7 | 6 | 4 | 8 | 2 | x | 2 | 2 | 2 | 9 | 4 | 4 | 4 | 5 | 4 | 5 | 5 | 6 | 7 | 7 | 8 | 10 | 10 | 9 | 10 | 2 | 4 | 9 | 5 | 6 | 6 | 6 | 6 | 6 | 5 | 6 | 5 | 6 |
| **5c** | 8 | 9 | 8 | 8 | 8 | 8 | 8 | 7 | 6 | 4 | 8 | 2 | 2 | x | 2 | 2 | 9 | 4 | 4 | 4 | 5 | 4 | 5 | 5 | 6 | 7 | 7 | 8 | 10 | 10 | 9 | 10 | 2 | 4 | 9 | 5 | 6 | 6 | 6 | 6 | 6 | 5 | 6 | 5 | 6 |
| **5d** | 8 | 9 | 8 | 8 | 8 | 8 | 8 | 6 | 6 | 4 | 8 | 2 | 2 | 2 | x | 2 | 9 | 4 | 4 | 4 | 5 | 4 | 5 | 5 | 5 | 7 | 7 | 8 | 10 | 10 | 8 | 10 | 2 | 4 | 9 | 5 | 6 | 6 | 6 | 6 | 6 | 5 | 6 | 3 | 4 |
| **5e** | 8 | 9 | 8 | 9 | 7 | 7 | 7 | 5 | 6 | 4 | 7 | 2 | 2 | 2 | 2 | x | 8 | 4 | 4 | 4 | 5 | 4 | 5 | 5 | 4 | 7 | 7 | 8 | 10 | 9 | 8 | 10 | 2 | 4 | 8 | 5 | 6 | 6 | 6 | 6 | 6 | 5 | 6 | 5 | 6 |
| **5f** | 8 | 9 | 9 | 9 | 1 | 3 | 3 | 3 | 8 | 8 | 3 | 9 | 9 | 9 | 9 | 8 | x | 9 | 9 | 9 | 8 | 9 | 7 | 6 | 4 | 4 | 5 | 4 | 9 | 5 | 8 | 10 | 9 | 9 | 4 | 10 | 9 | 9 | 9 | 9 | 9 | 10 | 9 | 8 | 7 |
| **5g** | 6 | 8 | 6 | 6 | 8 | 8 | 8 | 8 | 4 | 2 | 8 | 4 | 4 | 4 | 4 | 4 | 9 | x | 2 | 2 | 3 | 2 | 3 | 3 | 5 | 5 | 5 | 6 | 9 | 9 | 7 | 9 | 4 | 2 | 9 | 3 | 4 | 4 | 4 | 4 | 4 | 3 | 4 | 3 | 4 |
| **5h** | 6 | 8 | 6 | 6 | 8 | 8 | 8 | 8 | 4 | 2 | 8 | 4 | 4 | 4 | 4 | 4 | 9 | 2 | x | 2 | 3 | 2 | 3 | 3 | 5 | 5 | 5 | 6 | 9 | 9 | 7 | 9 | 4 | 2 | 9 | 3 | 4 | 4 | 4 | 4 | 4 | 3 | 4 | 3 | 4 |
| **5i** | 6 | 8 | 6 | 6 | 8 | 8 | 8 | 8 | 4 | 2 | 8 | 4 | 4 | 4 | 4 | 4 | 9 | 2 | 2 | x | 3 | 2 | 3 | 3 | 5 | 5 | 5 | 6 | 9 | 9 | 7 | 9 | 4 | 2 | 9 | 3 | 4 | 4 | 4 | 4 | 4 | 3 | 4 | 3 | 4 |
| **5j** | 5 | 7 | 5 | 5 | 7 | 7 | 7 | 5 | 3 | 3 | 7 | 5 | 5 | 5 | 5 | 5 | 8 | 3 | 3 | 3 | x | 3 | 2 | 2 | 4 | 4 | 4 | 5 | 8 | 8 | 6 | 8 | 5 | 3 | 8 | 4 | 1 | 1 | 1 | 1 | 1 | 4 | 1 | 4 | 3 |
| **5k** | 6 | 8 | 6 | 6 | 8 | 8 | 8 | 6 | 4 | 2 | 8 | 4 | 4 | 4 | 4 | 4 | 9 | 2 | 2 | 2 | 3 | x | 3 | 3 | 5 | 5 | 5 | 6 | 9 | 10 | 7 | 9 | 4 | 2 | 9 | 1 | 4 | 4 | 4 | 4 | 4 | 1 | 4 | 1 | 4 |
| **5L** | 5 | 7 | 5 | 5 | 6 | 6 | 6 | 4 | 3 | 3 | 6 | 5 | 5 | 5 | 5 | 5 | 7 | 3 | 3 | 3 | 2 | 3 | x | 1 | 3 | 3 | 4 | 5 | 8 | 8 | 6 | 8 | 5 | 3 | 7 | 4 | 3 | 3 | 3 | 3 | 3 | 4 | 3 | 4 | 2 |
| **5m** | 5 | 7 | 5 | 5 | 5 | 5 | 5 | 3 | 3 | 3 | 5 | 5 | 5 | 5 | 5 | 5 | 6 | 3 | 3 | 3 | 2 | 3 | 1 | x | 2 | 2 | 4 | 4 | 8 | 7 | 4 | 8 | 5 | 3 | 6 | 4 | 3 | 3 | 3 | 3 | 3 | 4 | 3 | 4 | 1 |
| **5n** | 7 | 9 | 5 | 7 | 3 | 3 | 3 | 1 | 5 | 5 | 3 | 6 | 6 | 6 | 5 | 4 | 4 | 5 | 5 | 5 | 4 | 5 | 3 | 2 | x | 4 | 5 | 4 | 9 | 5 | 4 | 10 | 6 | 5 | 4 | 6 | 5 | 5 | 5 | 5 | 5 | 6 | 5 | 4 | 3 |
| **5o** | 7 | 9 | 7 | 7 | 3 | 3 | 3 | 3 | 5 | 5 | 3 | 7 | 7 | 7 | 7 | 7 | 4 | 5 | 5 | 5 | 4 | 5 | 3 | 2 | 4 | x | 3 | 2 | 8 | 5 | 6 | 10 | 7 | 5 | 4 | 6 | 5 | 5 | 5 | 5 | 5 | 6 | 5 | 5 | 3 |
| **5p** | 5 | 7 | 5 | 5 | 4 | 4 | 4 | 4 | 3 | 5 | 4 | 7 | 7 | 7 | 7 | 7 | 5 | 5 | 5 | 5 | 4 | 5 | 4 | 4 | 5 | 3 | x | 1 | 7 | 5 | 6 | 8 | 7 | 5 | 5 | 6 | 5 | 5 | 5 | 5 | 5 | 6 | 5 | 6 | 5 |
| **5q** | 5 | 8 | 6 | 6 | 3 | 3 | 3 | 3 | 4 | 4 | 3 | 8 | 8 | 8 | 8 | 8 | 4 | 6 | 6 | 6 | 5 | 6 | 5 | 4 | 4 | 2 | 1 | x | 6 | 5 | 7 | 9 | 8 | 6 | 4 | 7 | 6 | 6 | 6 | 6 | 6 | 7 | 6 | 7 | 5 |
| **5r** | 3 | 9 | 9 | 9 | 8 | 8 | 8 | 8 | 7 | 9 | 8 | 10 | 10 | 10 | 10 | 10 | 9 | 9 | 9 | 9 | 8 | 9 | 8 | 8 | 9 | 8 | 7 | 6 | x | 8 | 10 | 10 | 10 | 9 | 9 | 10 | 9 | 9 | 9 | 9 | 9 | 10 | 9 | 10 | 9 |
| **5s** | 7 | 9 | 9 | 9 | 4 | 4 | 4 | 4 | 7 | 9 | 4 | 10 | 10 | 10 | 10 | 9 | 5 | 9 | 9 | 9 | 8 | 10 | 8 | 7 | 5 | 5 | 5 | 5 | 8 | x | 9 | 10 | 10 | 9 | 5 | 11 | 9 | 9 | 9 | 9 | 9 | 11 | 9 | 9 | 8 |
| **5t** | 7 | 5 | 1 | 3 | 7 | 7 | 7 | 5 | 5 | 7 | 7 | 9 | 9 | 9 | 8 | 8 | 8 | 7 | 7 | 7 | 6 | 7 | 6 | 4 | 4 | 6 | 6 | 7 | 10 | 9 | x | 6 | 9 | 7 | 8 | 8 | 7 | 7 | 7 | 7 | 7 | 8 | 7 | 7 | 5 |
| **5u** | 9 | 1 | 5 | 5 | 9 | 9 | 9 | 9 | 7 | 9 | 9 | 10 | 10 | 10 | 10 | 10 | 10 | 9 | 9 | 9 | 8 | 9 | 8 | 8 | 10 | 10 | 8 | 9 | 10 | 10 | 6 | x | 10 | 9 | 10 | 10 | 9 | 9 | 9 | 9 | 9 | 10 | 9 | 10 | 9 |
| **5v** | 8 | 9 | 8 | 8 | 8 | 8 | 8 | 7 | 6 | 4 | 8 | 2 | 2 | 2 | 2 | 2 | 9 | 4 | 4 | 4 | 5 | 4 | 5 | 5 | 6 | 7 | 7 | 8 | 10 | 10 | 9 | 10 | x | 4 | 9 | 5 | 6 | 6 | 6 | 6 | 6 | 5 | 6 | 5 | 6 |
| **5w** | 6 | 8 | 6 | 6 | 8 | 8 | 8 | 8 | 4 | 2 | 8 | 4 | 4 | 4 | 4 | 4 | 9 | 2 | 2 | 2 | 3 | 2 | 3 | 3 | 5 | 5 | 5 | 6 | 9 | 9 | 7 | 9 | 4 | x | 9 | 3 | 4 | 4 | 4 | 4 | 4 | 3 | 4 | 3 | 4 |
| **5x** | 8 | 9 | 9 | 9 | 3 | 3 | 3 | 3 | 8 | 8 | 1 | 9 | 9 | 9 | 9 | 8 | 4 | 9 | 9 | 9 | 8 | 9 | 7 | 6 | 4 | 4 | 5 | 4 | 9 | 5 | 8 | 10 | 9 | 9 | x | 10 | 9 | 9 | 9 | 9 | 9 | 10 | 9 | 8 | 7 |
| **6a** | 7 | 9 | 7 | 7 | 9 | 9 | 9 | 7 | 5 | 3 | 9 | 5 | 5 | 5 | 5 | 5 | 10 | 3 | 3 | 3 | 4 | 1 | 4 | 4 | 6 | 6 | 6 | 7 | 10 | 11 | 8 | 10 | 5 | 3 | 10 | x | 5 | 5 | 5 | 5 | 5 | 2 | 3 | 2 | 5 |
| **6b** | 6 | 8 | 6 | 6 | 8 | 8 | 8 | 6 | 4 | 4 | 8 | 6 | 6 | 6 | 6 | 6 | 9 | 4 | 4 | 4 | 1 | 4 | 3 | 3 | 5 | 5 | 5 | 6 | 9 | 9 | 7 | 9 | 6 | 4 | 9 | 5 | x | 2 | 2 | 2 | 2 | 3 | 2 | 5 | 4 |
| **6c** | 6 | 8 | 6 | 6 | 8 | 8 | 8 | 6 | 4 | 4 | 8 | 6 | 6 | 6 | 6 | 6 | 9 | 4 | 4 | 4 | 1 | 4 | 3 | 3 | 5 | 5 | 5 | 6 | 9 | 9 | 7 | 9 | 6 | 4 | 9 | 5 | 2 | x | 2 | 2 | 2 | 3 | 2 | 5 | 4 |
| **6d** | 6 | 8 | 6 | 6 | 8 | 8 | 8 | 6 | 4 | 4 | 8 | 6 | 6 | 6 | 6 | 6 | 9 | 4 | 4 | 4 | 1 | 4 | 3 | 3 | 5 | 5 | 5 | 6 | 9 | 9 | 7 | 9 | 6 | 4 | 9 | 5 | 2 | 2 | x | 2 | 2 | 3 | 2 | 5 | 4 |
| **6e** | 6 | 8 | 6 | 6 | 8 | 8 | 8 | 6 | 4 | 4 | 8 | 6 | 6 | 6 | 6 | 6 | 9 | 4 | 4 | 4 | 1 | 4 | 3 | 3 | 5 | 5 | 5 | 6 | 9 | 9 | 7 | 9 | 6 | 4 | 9 | 5 | 2 | 2 | 2 | x | 2 | 3 | 2 | 5 | 4 |
| **6f** | 6 | 8 | 6 | 6 | 8 | 8 | 8 | 6 | 4 | 4 | 8 | 6 | 6 | 6 | 6 | 6 | 9 | 4 | 4 | 4 | 1 | 4 | 3 | 3 | 5 | 5 | 5 | 6 | 9 | 9 | 7 | 9 | 6 | 4 | 9 | 5 | 2 | 2 | 2 | 2 | x | 3 | 2 | 5 | 4 |
| **6g** | 7 | 9 | 7 | 7 | 9 | 9 | 9 | 7 | 5 | 3 | 9 | 5 | 5 | 5 | 5 | 5 | 10 | 3 | 3 | 3 | 4 | 1 | 4 | 4 | 6 | 6 | 6 | 7 | 10 | 11 | 8 | 10 | 5 | 3 | 10 | 2 | 3 | 3 | 3 | 3 | 3 | x | 3 | 2 | 3 |
| **6h** | 6 | 8 | 6 | 6 | 8 | 8 | 8 | 6 | 4 | 4 | 8 | 6 | 6 | 6 | 6 | 6 | 9 | 4 | 4 | 4 | 1 | 4 | 3 | 3 | 5 | 5 | 5 | 6 | 9 | 9 | 7 | 9 | 6 | 4 | 9 | 3 | 2 | 2 | 2 | 2 | 2 | 3 | x | 3 | 2 |

| | | | | | | | | | | | | | | | | | | | | | | | | | | | | | | | | | | | | | | | | | | | | | |
|---|---|---|---|---|---|---|---|---|---|---|---|---|---|---|---|---|---|---|---|---|---|---|---|---|---|---|---|---|---|---|---|---|---|---|---|---|---|---|---|---|---|---|---|---|---|
| **6i** | 7 | 9 | 7 | 7 | 7 | 7 | 7 | 5 | 5 | 3 | 7 | 5 | 5 | 5 | 3 | 5 | 8 | 3 | 3 | 3 | 4 | 1 | 4 | 4 | 4 | 5 | 6 | 7 | 10 | 9 | 7 | 10 | 5 | 3 | 8 | 2 | 5 | 5 | 5 | 5 | 5 | 2 | 3 | x | 3 |
| **6j** | 6 | 8 | 6 | 6 | 6 | 6 | 6 | 4 | 4 | 4 | 6 | 6 | 6 | 6 | 4 | 6 | 7 | 4 | 4 | 4 | 3 | 4 | 2 | 1 | 3 | 3 | 5 | 5 | 9 | 8 | 5 | 9 | 6 | 4 | 7 | 5 | 4 | 4 | 4 | 4 | 4 | 3 | 2 | 3 | x |
| **6k** | 6 | 8 | 6 | 6 | 6 | 6 | 6 | 4 | 4 | 4 | 6 | 6 | 6 | 6 | 5 | 6 | 7 | 4 | 4 | 4 | 3 | 4 | 2 | 1 | 3 | 3 | 5 | 5 | 9 | 8 | 5 | 9 | 6 | 4 | 7 | 5 | 4 | 4 | 4 | 4 | 4 | 4 | 3 | 4 | 1 |
| **6L** | 6 | 8 | 6 | 6 | 6 | 6 | 6 | 4 | 4 | 4 | 6 | 6 | 6 | 6 | 6 | 6 | 7 | 4 | 4 | 4 | 3 | 4 | 2 | 1 | 3 | 3 | 5 | 5 | 9 | 8 | 5 | 9 | 6 | 4 | 7 | 5 | 4 | 4 | 4 | 4 | 4 | 5 | 4 | 5 | 2 |
| **6m** | 6 | 8 | 6 | 6 | 4 | 4 | 4 | 4 | 4 | 4 | 4 | 6 | 6 | 6 | 6 | 6 | 5 | 4 | 4 | 4 | 3 | 4 | 2 | 1 | 3 | 1 | 4 | 3 | 9 | 6 | 5 | 9 | 6 | 4 | 5 | 5 | 4 | 4 | 4 | 4 | 4 | 5 | 4 | 5 | 2 |
| **6n** | 8 | 10 | 8 | 8 | 4 | 4 | 4 | 4 | 6 | 6 | 4 | 8 | 8 | 8 | 8 | 8 | 5 | 6 | 6 | 6 | 5 | 6 | 4 | 3 | 5 | 1 | 4 | 3 | 9 | 6 | 7 | 11 | 8 | 6 | 5 | 7 | 6 | 6 | 6 | 6 | 6 | 7 | 6 | 7 | 4 |
| **6o** | 8 | 10 | 6 | 8 | 4 | 4 | 4 | 2 | 6 | 6 | 4 | 5 | 5 | 5 | 5 | 3 | 5 | 6 | 6 | 6 | 5 | 5 | 4 | 3 | 1 | 5 | 7 | 5 | 10 | 6 | 5 | 11 | 5 | 6 | 5 | 6 | 6 | 6 | 6 | 6 | 6 | 6 | 6 | 4 | 4 |
| **6p** | 7 | 9 | 5 | 7 | 4 | 4 | 4 | 2 | 5 | 5 | 4 | 6 | 6 | 6 | 4 | 4 | 5 | 5 | 5 | 5 | 4 | 4 | 3 | 2 | 1 | 4 | 6 | 5 | 10 | 6 | 4 | 10 | 6 | 5 | 5 | 5 | 5 | 5 | 5 | 5 | 5 | 5 | 5 | 3 | 3 |
| **6q** | 6 | 8 | 4 | 6 | 4 | 4 | 4 | 2 | 4 | 4 | 4 | 7 | 7 | 7 | 5 | 5 | 5 | 4 | 4 | 4 | 3 | 4 | 2 | 1 | 1 | 3 | 5 | 5 | 9 | 6 | 3 | 9 | 7 | 4 | 5 | 5 | 4 | 4 | 4 | 4 | 4 | 5 | 4 | 4 | 2 |
| **6r** | 8 | 10 | 6 | 8 | 4 | 4 | 4 | 2 | 6 | 6 | 4 | 7 | 7 | 7 | 6 | 5 | 5 | 6 | 6 | 6 | 5 | 6 | 4 | 3 | 1 | 5 | 6 | 5 | 10 | 6 | 5 | 11 | 7 | 6 | 5 | 7 | 6 | 6 | 6 | 6 | 6 | 7 | 6 | 5 | 4 |
| **6s** | 8 | 10 | 6 | 8 | 4 | 4 | 4 | 2 | 6 | 6 | 4 | 7 | 7 | 7 | 6 | 5 | 5 | 6 | 6 | 6 | 5 | 6 | 4 | 3 | 1 | 5 | 6 | 5 | 10 | 6 | 5 | 11 | 7 | 6 | 5 | 7 | 6 | 6 | 6 | 6 | 6 | 7 | 6 | 5 | 4 |
| **6t** | 8 | 10 | 6 | 8 | 4 | 4 | 4 | 2 | 6 | 6 | 4 | 7 | 7 | 7 | 6 | 5 | 5 | 6 | 6 | 6 | 5 | 6 | 4 | 3 | 1 | 5 | 6 | 5 | 10 | 6 | 5 | 11 | 7 | 6 | 5 | 7 | 6 | 6 | 6 | 6 | 6 | 7 | 6 | 5 | 4 |
| **6u** | 8 | 10 | 6 | 8 | 4 | 4 | 4 | 2 | 6 | 6 | 4 | 7 | 7 | 7 | 6 | 5 | 5 | 6 | 6 | 6 | 5 | 6 | 4 | 3 | 1 | 5 | 6 | 5 | 10 | 6 | 5 | 11 | 7 | 6 | 5 | 7 | 6 | 6 | 6 | 6 | 6 | 7 | 6 | 5 | 4 |
| **6v** | 9 | 10 | 10 | 10 | 4 | 4 | 4 | 4 | 9 | 9 | 2 | 10 | 10 | 10 | 10 | 9 | 5 | 10 | 10 | 10 | 9 | 10 | 8 | 7 | 5 | 5 | 6 | 5 | 10 | 6 | 9 | 11 | 10 | 10 | 1 | 11 | 10 | 10 | 10 | 10 | 10 | 11 | 10 | 9 | 8 |
| **6w** | 9 | 10 | 9 | 10 | 8 | 8 | 8 | 6 | 7 | 5 | 8 | 3 | 3 | 3 | 3 | 1 | 9 | 5 | 5 | 5 | 6 | 5 | 6 | 6 | 5 | 8 | 8 | 9 | 11 | 10 | 9 | 11 | 3 | 5 | 9 | 6 | 7 | 7 | 7 | 7 | 7 | 6 | 7 | 6 | 7 |
| **6x** | 9 | 10 | 8 | 10 | 6 | 6 | 6 | 4 | 7 | 5 | 6 | 3 | 3 | 3 | 3 | 1 | 7 | 5 | 5 | 5 | 6 | 5 | 6 | 5 | 3 | 7 | 8 | 7 | 11 | 8 | 7 | 11 | 3 | 5 | 7 | 6 | 7 | 7 | 7 | 7 | 7 | 6 | 7 | 6 | 6 |
| **6y** | 8 | 6 | 2 | 4 | 6 | 6 | 6 | 4 | 6 | 6 | 6 | 9 | 9 | 9 | 7 | 7 | 7 | 6 | 6 | 6 | 5 | 6 | 5 | 3 | 3 | 7 | 7 | 8 | 11 | 8 | 1 | 7 | 9 | 6 | 7 | 7 | 6 | 6 | 6 | 6 | 6 | 7 | 6 | 6 | 4 |
| **6z** | 6 | 8 | 6 | 6 | 5 | 5 | 5 | 5 | 4 | 6 | 5 | 8 | 8 | 8 | 8 | 8 | 6 | 6 | 6 | 6 | 5 | 6 | 5 | 5 | 6 | 4 | 1 | 2 | 8 | 4 | 7 | 9 | 8 | 6 | 6 | 7 | 6 | 6 | 6 | 6 | 6 | 7 | 6 | 7 | 6 |
| **6aa** | 9 | 10 | 8 | 9 | 7 | 7 | 6 | 5 | 7 | 5 | 6 | 3 | 3 | 3 | 1 | 3 | 8 | 5 | 5 | 5 | 6 | 3 | 5 | 4 | 4 | 5 | 8 | 8 | 11 | 9 | 7 | 11 | 3 | 5 | 7 | 4 | 7 | 7 | 7 | 7 | 7 | 4 | 5 | 2 | 6 |
| **6ab** | 8 | 9 | 9 | 9 | 5 | 5 | 7 | 5 | 7 | 9 | 7 | 11 | 11 | 11 | 11 | 10 | 6 | 9 | 9 | 9 | 8 | 9 | 8 | 8 | 6 | 6 | 4 | 5 | 9 | 1 | 10 | 10 | 11 | 9 | 8 | 10 | 9 | 9 | 9 | 9 | 9 | 10 | 9 | 10 | 9 |
| **7a** | 10 | 11 | 9 | 11 | 7 | 7 | 7 | 5 | 8 | 6 | 7 | 4 | 4 | 4 | 4 | 2 | 8 | 6 | 6 | 6 | 7 | 6 | 7 | 6 | 4 | 8 | 9 | 8 | 12 | 9 | 8 | 12 | 4 | 6 | 8 | 7 | 8 | 8 | 8 | 8 | 8 | 7 | 8 | 7 | 7 |
| **7b** | 9 | 11 | 7 | 9 | 5 | 5 | 5 | 3 | 7 | 6 | 5 | 4 | 4 | 4 | 4 | 2 | 6 | 6 | 6 | 6 | 6 | 6 | 5 | 4 | 2 | 6 | 7 | 6 | 11 | 7 | 6 | 12 | 4 | 6 | 6 | 7 | 7 | 7 | 7 | 7 | 7 | 7 | 7 | 7 | 5 |
| **7c** | 9 | 11 | 7 | 9 | 5 | 5 | 5 | 3 | 7 | 7 | 5 | 8 | 8 | 8 | 7 | 6 | 6 | 7 | 7 | 7 | 6 | 7 | 5 | 4 | 2 | 6 | 7 | 6 | 11 | 7 | 6 | 12 | 8 | 7 | 6 | 8 | 7 | 7 | 7 | 7 | 7 | 8 | 7 | 6 | 5 |
| **7d** | 8 | 10 | 7 | 8 | 6 | 6 | 6 | 4 | 6 | 4 | 6 | 4 | 4 | 4 | 2 | 4 | 7 | 4 | 4 | 4 | 3 | 2 | 4 | 3 | 3 | 5 | 7 | 7 | 11 | 8 | 6 | 11 | 4 | 4 | 7 | 3 | 4 | 4 | 4 | 4 | 4 | 3 | 4 | 1 | 2 |
| **7e** | 8 | 10 | 8 | 8 | 8 | 8 | 8 | 6 | 6 | 4 | 8 | 6 | 6 | 6 | 4 | 6 | 9 | 4 | 4 | 4 | 5 | 2 | 4 | 3 | 5 | 5 | 7 | 7 | 11 | 10 | 7 | 11 | 6 | 4 | 9 | 3 | 6 | 6 | 6 | 6 | 6 | 1 | 4 | 3 | 2 |
| **7f** | 7 | 9 | 7 | 7 | 7 | 7 | 7 | 5 | 5 | 5 | 7 | 5 | 5 | 5 | 3 | 5 | 8 | 7 | 7 | 7 | 4 | 3 | 3 | 2 | 4 | 4 | 6 | 6 | 10 | 9 | 6 | 10 | 5 | 7 | 8 | 4 | 5 | 5 | 5 | 5 | 5 | 2 | 3 | 2 | 1 |
| **7g** | 8 | 10 | 8 | 8 | 10 | 10 | 10 | 8 | 6 | 4 | 10 | 6 | 6 | 6 | 6 | 6 | 11 | 4 | 4 | 4 | 5 | 2 | 5 | 5 | 7 | 7 | 7 | 8 | 11 | 11 | 9 | 11 | 6 | 4 | 11 | 3 | 6 | 6 | 6 | 6 | 6 | 1 | 4 | 3 | 4 |
| **7h** | 8 | 10 | 6 | 8 | 5 | 5 | 5 | 3 | 6 | 5 | 5 | 7 | 7 | 7 | 3 | 5 | 6 | 5 | 5 | 5 | 5 | 3 | 4 | 3 | 2 | 5 | 7 | 6 | 11 | 8 | 5 | 11 | 7 | 5 | 6 | 4 | 6 | 6 | 6 | 6 | 6 | 4 | 6 | 2 | 4 |
| **7i** | 7 | 9 | 7 | 7 | 7 | 7 | 7 | 5 | 5 | 5 | 7 | 7 | 7 | 7 | 5 | 7 | 8 | 5 | 5 | 5 | 2 | 5 | 3 | 2 | 4 | 4 | 6 | 6 | 10 | 9 | 6 | 10 | 7 | 5 | 8 | 6 | 3 | 3 | 3 | 3 | 3 | 4 | 1 | 4 | 1 |
| **7j** | 7 | 9 | 5 | 7 | 5 | 5 | 5 | 3 | 5 | 5 | 5 | 8 | 8 | 8 | 6 | 6 | 6 | 5 | 5 | 5 | 4 | 5 | 3 | 2 | 2 | 4 | 6 | 6 | 10 | 7 | 4 | 10 | 8 | 5 | 6 | 6 | 5 | 5 | 5 | 5 | 5 | 6 | 5 | 5 | 3 |
| **7k** | 7 | 9 | 7 | 7 | 5 | 5 | 5 | 5 | 5 | 5 | 5 | 7 | 7 | 7 | 7 | 7 | 6 | 5 | 5 | 5 | 4 | 5 | 3 | 2 | 4 | 2 | 5 | 4 | 10 | 7 | 6 | 10 | 7 | 5 | 6 | 6 | 5 | 5 | 5 | 5 | 5 | 6 | 5 | 6 | 3 |
| **7L** | 7 | 7 | 3 | 5 | 5 | 5 | 5 | 3 | 5 | 5 | 5 | 8 | 8 | 8 | 6 | 6 | 6 | 5 | 5 | 5 | 4 | 5 | 4 | 2 | 2 | 6 | 6 | 7 | 10 | 7 | 2 | 8 | 8 | 5 | 6 | 6 | 5 | 5 | 5 | 5 | 5 | 6 | 5 | 5 | 3 |
| **7m** | 7 | 9 | 7 | 7 | 6 | 6 | 6 | 6 | 5 | 7 | 6 | 9 | 9 | 9 | 9 | 9 | 7 | 7 | 7 | 7 | 6 | 7 | 6 | 6 | 7 | 5 | 2 | 3 | 9 | 3 | 8 | 10 | 9 | 7 | 7 | 8 | 7 | 7 | 7 | 7 | 7 | 8 | 7 | 8 | 7 |
| **7n** | 7 | 9 | 5 | 7 | 5 | 5 | 5 | 3 | 5 | 5 | 5 | 8 | 8 | 8 | 6 | 6 | 6 | 5 | 5 | 5 | 4 | 5 | 3 | 2 | 2 | 4 | 6 | 6 | 10 | 7 | 4 | 10 | 8 | 5 | 6 | 6 | 5 | 5 | 5 | 5 | 5 | 6 | 5 | 5 | 3 |
| **7o** | 8 | 10 | 8 | 8 | 6 | 6 | 6 | 6 | 6 | 8 | 6 | 10 | 10 | 10 | 10 | 10 | 7 | 8 | 8 | 8 | 7 | 8 | 7 | 7 | 7 | 6 | 3 | 4 | 10 | 2 | 9 | 11 | 10 | 8 | 7 | 9 | 8 | 8 | 8 | 8 | 8 | 9 | 8 | 9 | 8 |
| **7p** | 10 | 11 | 11 | 11 | 5 | 5 | 5 | 5 | 10 | 10 | 3 | 11 | 11 | 11 | 11 | 10 | 6 | 11 | 11 | 11 | 10 | 11 | 9 | 8 | 6 | 6 | 7 | 6 | 11 | 7 | 10 | 12 | 11 | 11 | 2 | 12 | 11 | 11 | 11 | 11 | 11 | 12 | 11 | 10 | 9 |
| **8a** | 9 | 11 | 9 | 9 | 11 | 11 | 11 | 9 | 7 | 5 | 11 | 7 | 7 | 7 | 7 | 7 | 12 | 5 | 5 | 5 | 6 | 3 | 6 | 6 | 8 | 8 | 8 | 9 | 12 | 12 | 10 | 12 | 7 | 5 | 12 | 4 | 7 | 7 | 7 | 7 | 7 | 2 | 5 | 4 | 5 |
| **8b** | 11 | 12 | 10 | 12 | 8 | 8 | 8 | 6 | 9 | 7 | 8 | 5 | 5 | 5 | 5 | 3 | 9 | 7 | 7 | 7 | 8 | 7 | 8 | 7 | 5 | 9 | 10 | 9 | 13 | 10 | 9 | 13 | 5 | 7 | 9 | 8 | 9 | 9 | 9 | 9 | 9 | 8 | 9 | 8 | 8 |
| **8c** | 8 | 10 | 6 | 8 | 6 | 6 | 6 | 4 | 6 | 6 | 6 | 9 | 9 | 9 | 7 | 7 | 7 | 6 | 6 | 6 | 5 | 6 | 4 | 3 | 3 | 5 | 7 | 7 | 11 | 8 | 5 | 11 | 9 | 6 | 7 | 7 | 6 | 6 | 6 | 6 | 6 | 7 | 6 | 6 | 4 |
| **8d** | 9 | 11 | 8 | 9 | 7 | 7 | 7 | 5 | 7 | 5 | 7 | 5 | 5 | 5 | 3 | 5 | 8 | 5 | 5 | 5 | 4 | 3 | 5 | 4 | 4 | 8 | 8 | 8 | 12 | 9 | 7 | 12 | 5 | 5 | 8 | 4 | 5 | 5 | 5 | 5 | 5 | 4 | 5 | 2 | 3 |
| **8e** | 8 | 10 | 6 | 8 | 6 | 6 | 6 | 4 | 6 | 6 | 6 | 9 | 9 | 9 | 7 | 7 | 7 | 6 | 6 | 6 | 5 | 6 | 4 | 3 | 3 | 5 | 7 | 7 | 11 | 8 | 5 | 11 | 9 | 6 | 7 | 7 | 6 | 6 | 6 | 6 | 6 | 7 | 6 | 6 | 4 |
| **8f** | 8 | 10 | 6 | 8 | 6 | 6 | 6 | 4 | 6 | 6 | 6 | 9 | 9 | 9 | 7 | 7 | 7 | 6 | 6 | 6 | 5 | 6 | 4 | 3 | 3 | 5 | 7 | 7 | 11 | 8 | 5 | 11 | 9 | 6 | 7 | 7 | 6 | 6 | 6 | 6 | 6 | 7 | 6 | 6 | 4 |
| **8g** | 10 | 12 | 8 | 10 | 6 | 6 | 6 | 4 | 8 | 8 | 6 | 9 | 9 | 9 | 8 | 7 | 7 | 8 | 8 | 8 | 7 | 8 | 6 | 5 | 3 | 7 | 8 | 7 | 12 | 8 | 7 | 13 | 9 | 8 | 7 | 9 | 8 | 8 | 8 | 8 | 8 | 9 | 8 | 7 | 6 |
| **9a** | 12 | 13 | 11 | 13 | 9 | 9 | 9 | 7 | 10 | 8 | 9 | 6 | 6 | 6 | 6 | 4 | 10 | 8 | 8 | 8 | 9 | 8 | 9 | 8 | 6 | 10 | 11 | 10 | 14 | 11 | 10 | 14 | 6 | 8 | 10 | 9 | 10 | 10 | 10 | 10 | 10 | 9 | 10 | 9 | 9 |
| **9b** | 10 | 12 | 10 | 10 | 12 | 12 | 12 | 10 | 8 | 6 | 12 | 8 | 8 | 8 | 8 | 8 | 13 | 6 | 6 | 6 | 7 | 4 | 7 | 7 | 9 | 9 | 9 | 10 | 13 | 13 | 11 | 13 | 8 | 6 | 13 | 5 | 8 | 8 | 8 | 8 | 8 | 3 | 6 | 5 | 6 |
| **9c** | 9 | 11 | 7 | 9 | 7 | 7 | 7 | 5 | 7 | 7 | 7 | 10 | 10 | 10 | 8 | 8 | 8 | 7 | 7 | 7 | 6 | 7 | 5 | 4 | 4 | 6 | 8 | 8 | 12 | 9 | 6 | 12 | 10 | 7 | 8 | 8 | 7 | 7 | 7 | 7 | 7 | 8 | 7 | 7 | 5 |
| **10a** | 13 | 14 | 12 | 14 | 10 | 10 | 10 | 8 | 11 | 9 | 10 | 7 | 7 | 7 | 7 | 5 | 11 | 9 | 9 | 9 | 10 | 9 | 10 | 9 | 7 | 11 | 12 | 11 | 15 | 12 | 11 | 15 | 7 | 9 | 11 | 10 | 11 | 11 | 11 | 11 | 11 | 10 | 11 | 10 | 10 |

Palenque Palace Path Matrix Phase 6 (Cont.)

| | 6k | 6L | 6m | 6n | 6o | 6p | 6q | 6r | 6s | 6t | 6u | 6v | 6w | 6x | 6y | 6z | 6aa | 6ab | 7a | 7b | 7c | 7d | 7e | 7f | 7g | 7h | 7i | 7j | 7k | 7L | 7m | 7n | 7o | 7p | 8a | 8b | 8c | 8d | 8e | 8f | 8g | 9a | 9b | 9c | 10a |
|---|---|---|---|---|---|---|---|---|---|---|---|---|---|---|---|---|---|---|---|---|---|---|---|---|---|---|---|---|---|---|---|---|---|---|---|---|---|---|---|---|---|---|---|---|---|
| c | 6 | 6 | 6 | 6 | 6 | 6 | 6 | 6 | 6 | 6 | 6 | 6 | 6 | 6 | 6 | 6 | 6 | 6 | 7 | 7 | 7 | 7 | 7 | 7 | 7 | 7 | 7 | 7 | 7 | 7 | 7 | 7 | 7 | 7 | 8 | 8 | 8 | 8 | 8 | 8 | 8 | 9 | 9 | 9 | 10 |
| 1a | 6 | 6 | 6 | 6 | 6 | 6 | 6 | 6 | 6 | 6 | 6 | 6 | 5 | 5 | 7 | 7 | 5 | 7 | 6 | 6 | 7 | 6 | 6 | 7 | 6 | 7 | 7 | 7 | 7 | 8 | 8 | 7 | 8 | 7 | 7 | 7 | 8 | 7 | 8 | 8 | 8 | 8 | 8 | 9 | 9 |
| 1b | 7 | 7 | 7 | 5 | 5 | 5 | 5 | 5 | 5 | 5 | 5 | 5 | 6 | 6 | 7 | 6 | 6 | 6 | 7 | 6 | 6 | 7 | 7 | 8 | 7 | 6 | 8 | 6 | 6 | 6 | 7 | 6 | 7 | 6 | 8 | 8 | 7 | 8 | 7 | 7 | 7 | 9 | 9 | 8 | 10 |
| 1c | 7 | 7 | 7 | 7 | 7 | 7 | 7 | 7 | 7 | 7 | 7 | 7 | 7 | 7 | 7 | 7 | 7 | 7 | 8 | 8 | 8 | 8 | 8 | 8 | 8 | 8 | 8 | 8 | 8 | 8 | 8 | 8 | 8 | 8 | 9 | 9 | 9 | 9 | 9 | 9 | 9 | 10 | 10 | 10 | 11 |
| 1d | 5 | 5 | 5 | 6 | 6 | 6 | 6 | 6 | 6 | 6 | 6 | 6 | 7 | 7 | 7 | 5 | 7 | 5 | 8 | 7 | 7 | 8 | 7 | 6 | 7 | 7 | 6 | 7 | 6 | 7 | 6 | 7 | 6 | 7 | 8 | 9 | 7 | 9 | 8 | 8 | 8 | 10 | 9 | 9 | 11 |
| 1e | 7 | 7 | 7 | 7 | 7 | 7 | 7 | 7 | 7 | 7 | 7 | 7 | 7 | 7 | 7 | 7 | 7 | 7 | 8 | 8 | 8 | 8 | 8 | 8 | 8 | 8 | 8 | 8 | 8 | 8 | 8 | 8 | 8 | 8 | 9 | 9 | 9 | 9 | 9 | 9 | 9 | 10 | 10 | 10 | 11 |
| 1f | 7 | 7 | 7 | 7 | 7 | 7 | 7 | 7 | 7 | 7 | 7 | 7 | 7 | 7 | 7 | 7 | 7 | 7 | 8 | 8 | 8 | 8 | 8 | 8 | 8 | 8 | 8 | 8 | 8 | 8 | 8 | 8 | 8 | 8 | 9 | 9 | 9 | 9 | 9 | 9 | 9 | 10 | 10 | 10 | 11 |
| 1g | 5 | 5 | 5 | 7 | 7 | 6 | 5 | 7 | 7 | 7 | 7 | 7 | 7 | 7 | 5 | 5 | 7 | 7 | 8 | 8 | 8 | 7 | 7 | 6 | 7 | 7 | 6 | 6 | 6 | 6 | 6 | 6 | 7 | 8 | 8 | 9 | 7 | 8 | 7 | 7 | 9 | 10 | 9 | 8 | 11 |
| 2a | 5 | 5 | 5 | 7 | 6 | 6 | 5 | 7 | 7 | 7 | 7 | 7 | 4 | 4 | 7 | 7 | 4 | 8 | 5 | 5 | 7 | 5 | 5 | 6 | 5 | 6 | 6 | 6 | 6 | 6 | 8 | 6 | 9 | 8 | 6 | 6 | 7 | 6 | 7 | 7 | 8 | 7 | 7 | 8 | 8 |
| 2b | 7 | 7 | 4 | 4 | 4 | 4 | 4 | 4 | 4 | 4 | 4 | 4 | 7 | 6 | 6 | 5 | 7 | 5 | 7 | 5 | 5 | 6 | 8 | 7 | 8 | 5 | 7 | 5 | 5 | 5 | 6 | 5 | 6 | 5 | 9 | 8 | 6 | 7 | 6 | 6 | 6 | 9 | 10 | 7 | 10 |
| 2c | 6 | 6 | 6 | 5 | 5 | 5 | 5 | 5 | 5 | 5 | 5 | 5 | 8 | 7 | 7 | 6 | 8 | 4 | 8 | 6 | 6 | 7 | 8 | 7 | 8 | 6 | 7 | 6 | 7 | 6 | 6 | 6 | 5 | 6 | 9 | 9 | 7 | 8 | 7 | 7 | 7 | 10 | 10 | 8 | 11 |
| 2d | 6 | 6 | 6 | 6 | 7 | 7 | 6 | 7 | 7 | 7 | 7 | 7 | 8 | 8 | 9 | 5 | 8 | 6 | 9 | 8 | 8 | 8 | 8 | 7 | 8 | 8 | 7 | 7 | 7 | 8 | 6 | 7 | 7 | 8 | 9 | 10 | 8 | 9 | 8 | 8 | 9 | 11 | 10 | 9 | 12 |
| 2e | 4 | 4 | 4 | 6 | 6 | 5 | 4 | 6 | 6 | 6 | 6 | 7 | 7 | 7 | 6 | 4 | 7 | 6 | 8 | 7 | 7 | 6 | 6 | 5 | 6 | 6 | 5 | 5 | 5 | 7 | 5 | 5 | 6 | 8 | 7 | 10 | 6 | 7 | 6 | 6 | 8 | 11 | 8 | 7 | 12 |
| 2f | 6 | 6 | 6 | 8 | 8 | 7 | 6 | 8 | 8 | 8 | 8 | 8 | 8 | 8 | 4 | 6 | 8 | 8 | 9 | 9 | 9 | 8 | 8 | 7 | 8 | 8 | 7 | 7 | 7 | 5 | 7 | 7 | 8 | 9 | 9 | 10 | 8 | 9 | 8 | 8 | 10 | 11 | 10 | 9 | 12 |
| 2g | 4 | 4 | 4 | 6 | 6 | 5 | 4 | 6 | 6 | 6 | 6 | 8 | 7 | 8 | 4 | 4 | 7 | 7 | 9 | 7 | 7 | 6 | 6 | 5 | 6 | 6 | 5 | 5 | 5 | 5 | 5 | 5 | 6 | 9 | 7 | 10 | 6 | 7 | 6 | 6 | 8 | 11 | 8 | 7 | 12 |
| 3a | 4 | 4 | 4 | 6 | 5 | 5 | 4 | 6 | 6 | 6 | 6 | 8 | 3 | 3 | 8 | 6 | 3 | 9 | 4 | 4 | 7 | 4 | 4 | 5 | 4 | 6 | 5 | 5 | 5 | 9 | 7 | 5 | 8 | 9 | 5 | 5 | 6 | 5 | 6 | 6 | 8 | 6 | 6 | 7 | 7 |
| 3b | 5 | 5 | 5 | 3 | 3 | 3 | 3 | 3 | 3 | 3 | 3 | 3 | 7 | 5 | 5 | 4 | 6 | 4 | 6 | 4 | 4 | 5 | 7 | 6 | 9 | 4 | 6 | 4 | 4 | 4 | 5 | 4 | 5 | 4 | 10 | 7 | 5 | 6 | 5 | 5 | 5 | 8 | 11 | 6 | 9 |
| 3c | 7 | 7 | 7 | 5 | 5 | 5 | 5 | 5 | 5 | 5 | 5 | 5 | 8 | 7 | 7 | 6 | 8 | 6 | 8 | 6 | 6 | 7 | 9 | 8 | 9 | 6 | 8 | 6 | 8 | 6 | 7 | 6 | 7 | 6 | 10 | 9 | 7 | 8 | 7 | 7 | 7 | 10 | 11 | 8 | 11 |
| 3d | 6 | 6 | 6 | 4 | 4 | 4 | 4 | 4 | 4 | 4 | 4 | 4 | 8 | 6 | 6 | 5 | 7 | 3 | 7 | 5 | 5 | 6 | 9 | 7 | 9 | 5 | 7 | 5 | 7 | 5 | 5 | 5 | 4 | 5 | 10 | 8 | 6 | 7 | 6 | 6 | 6 | 9 | 11 | 7 | 10 |
| 3e | 7 | 7 | 5 | 5 | 6 | 6 | 6 | 6 | 6 | 6 | 6 | 6 | 9 | 8 | 8 | 7 | 9 | 5 | 9 | 7 | 7 | 8 | 9 | 8 | 9 | 7 | 8 | 7 | 6 | 7 | 7 | 7 | 6 | 7 | 10 | 10 | 8 | 9 | 8 | 8 | 8 | 11 | 11 | 9 | 12 |
| 3f | 7 | 7 | 5 | 5 | 7 | 7 | 7 | 7 | 7 | 7 | 7 | 7 | 9 | 9 | 9 | 4 | 9 | 7 | 10 | 8 | 8 | 9 | 9 | 8 | 9 | 8 | 8 | 8 | 6 | 8 | 5 | 8 | 6 | 8 | 10 | 11 | 9 | 10 | 9 | 9 | 9 | 12 | 11 | 10 | 13 |
| 3g | 7 | 7 | 7 | 7 | 8 | 8 | 7 | 8 | 8 | 8 | 8 | 8 | 9 | 9 | 9 | 6 | 9 | 7 | 10 | 9 | 9 | 9 | 9 | 8 | 9 | 9 | 8 | 8 | 8 | 9 | 7 | 8 | 8 | 9 | 10 | 11 | 9 | 10 | 9 | 9 | 10 | 12 | 11 | 10 | 13 |
| 3h | 7 | 7 | 7 | 7 | 8 | 8 | 7 | 8 | 8 | 8 | 8 | 8 | 9 | 9 | 9 | 6 | 9 | 7 | 10 | 9 | 9 | 9 | 9 | 8 | 9 | 9 | 8 | 8 | 8 | 9 | 7 | 8 | 8 | 9 | 10 | 11 | 9 | 10 | 9 | 9 | 10 | 12 | 11 | 10 | 13 |
| 3i | 3 | 3 | 3 | 6 | 5 | 4 | 3 | 5 | 5 | 5 | 5 | 8 | 6 | 6 | 5 | 3 | 6 | 6 | 7 | 6 | 6 | 5 | 5 | 4 | 5 | 5 | 4 | 4 | 4 | 4 | 4 | 4 | 5 | 9 | 6 | 8 | 5 | 6 | 5 | 5 | 7 | 9 | 7 | 6 | 10 |
| 3j | 5 | 5 | 5 | 8 | 7 | 6 | 5 | 7 | 7 | 7 | 7 | 9 | 8 | 8 | 3 | 5 | 8 | 8 | 9 | 8 | 8 | 7 | 7 | 6 | 7 | 7 | 6 | 6 | 6 | 4 | 6 | 6 | 7 | 10 | 8 | 10 | 7 | 8 | 7 | 7 | 9 | 11 | 9 | 8 | 12 |
| 3k | 7 | 7 | 7 | 9 | 7 | 6 | 5 | 7 | 7 | 7 | 7 | 9 | 9 | 9 | 3 | 7 | 9 | 9 | 10 | 8 | 8 | 8 | 9 | 8 | 9 | 7 | 8 | 6 | 8 | 4 | 8 | 6 | 9 | 10 | 10 | 11 | 7 | 9 | 7 | 7 | 9 | 12 | 11 | 8 | 13 |
| 3L | 7 | 7 | 7 | 9 | 9 | 8 | 7 | 9 | 9 | 9 | 9 | 9 | 9 | 9 | 5 | 7 | 9 | 9 | 10 | 10 | 10 | 9 | 9 | 8 | 9 | 7 | 8 | 8 | 8 | 6 | 8 | 8 | 9 | 10 | 10 | 11 | 9 | 10 | 9 | 9 | 11 | 12 | 11 | 10 | 13 |
| 3m | 4 | 4 | 4 | 6 | 6 | 5 | 4 | 6 | 6 | 6 | 6 | 8 | 5 | 5 | 6 | 6 | 7 | 9 | 6 | 6 | 7 | 4 | 4 | 5 | 4 | 6 | 5 | 5 | 5 | 5 | 7 | 5 | 8 | 9 | 5 | 7 | 6 | 5 | 6 | 6 | 8 | 8 | 6 | 7 | 9 |
| 3n | 5 | 5 | 5 | 7 | 7 | 6 | 5 | 7 | 7 | 7 | 7 | 9 | 8 | 8 | 7 | 5 | 7 | 7 | 9 | 8 | 8 | 7 | 7 | 6 | 7 | 7 | 6 | 6 | 6 | 6 | 6 | 6 | 7 | 10 | 8 | 10 | 7 | 8 | 7 | 7 | 9 | 11 | 9 | 8 | 12 |
| 4a | 5 | 5 | 5 | 7 | 4 | 5 | 5 | 6 | 6 | 6 | 6 | 9 | 2 | 2 | 7 | 7 | 2 | 10 | 3 | 3 | 7 | 3 | 5 | 4 | 5 | 4 | 6 | 6 | 6 | 6 | 8 | 6 | 9 | 10 | 6 | 4 | 7 | 4 | 7 | 7 | 8 | 5 | 7 | 8 | 6 |
| 4b | 5 | 5 | 5 | 7 | 5 | 6 | 5 | 7 | 7 | 7 | 7 | 9 | 3 | 3 | 7 | 7 | 3 | 10 | 4 | 4 | 8 | 4 | 5 | 6 | 5 | 7 | 6 | 6 | 6 | 6 | 8 | 6 | 9 | 10 | 6 | 5 | 7 | 5 | 7 | 7 | 9 | 6 | 7 | 8 | 7 |
| 4c | 3 | 3 | 3 | 4 | 5 | 4 | 3 | 5 | 5 | 5 | 5 | 9 | 4 | 4 | 5 | 5 | 4 | 8 | 5 | 5 | 6 | 3 | 3 | 4 | 3 | 4 | 4 | 4 | 4 | 4 | 6 | 4 | 7 | 10 | 4 | 6 | 5 | 4 | 5 | 5 | 7 | 7 | 5 | 6 | 8 |
| 4d | 2 | 2 | 2 | 4 | 4 | 3 | 2 | 4 | 4 | 4 | 4 | 8 | 5 | 5 | 4 | 4 | 5 | 7 | 6 | 5 | 5 | 4 | 4 | 3 | 4 | 4 | 3 | 3 | 3 | 3 | 5 | 3 | 6 | 9 | 5 | 7 | 4 | 5 | 4 | 4 | 6 | 8 | 6 | 5 | 9 |
| 4e | 3 | 3 | 3 | 5 | 5 | 4 | 3 | 5 | 5 | 5 | 5 | 9 | 6 | 6 | 5 | 4 | 6 | 7 | 7 | 6 | 6 | 5 | 5 | 4 | 5 | 5 | 4 | 4 | 4 | 4 | 5 | 4 | 6 | 10 | 6 | 8 | 5 | 6 | 5 | 5 | 7 | 9 | 7 | 6 | 10 |
| 4f | 4 | 4 | 4 | 6 | 6 | 5 | 4 | 6 | 6 | 6 | 6 | 9 | 7 | 7 | 6 | 4 | 7 | 7 | 8 | 7 | 7 | 6 | 6 | 5 | 6 | 6 | 5 | 5 | 5 | 5 | 5 | 5 | 6 | 10 | 7 | 9 | 6 | 7 | 6 | 6 | 8 | 10 | 8 | 7 | 11 |
| 4g | 4 | 4 | 4 | 6 | 6 | 5 | 4 | 6 | 6 | 6 | 6 | 9 | 7 | 7 | 6 | 4 | 7 | 7 | 8 | 7 | 7 | 6 | 6 | 5 | 6 | 6 | 5 | 5 | 5 | 5 | 5 | 5 | 6 | 10 | 7 | 9 | 6 | 7 | 6 | 6 | 8 | 10 | 8 | 7 | 11 |
| 4h | 4 | 4 | 4 | 6 | 6 | 5 | 4 | 6 | 6 | 6 | 6 | 9 | 7 | 7 | 6 | 4 | 7 | 7 | 8 | 7 | 7 | 6 | 6 | 5 | 6 | 6 | 5 | 5 | 5 | 5 | 5 | 5 | 6 | 10 | 7 | 9 | 6 | 7 | 6 | 6 | 8 | 10 | 8 | 7 | 11 |
| 4i | 4 | 4 | 4 | 6 | 6 | 5 | 4 | 6 | 6 | 6 | 6 | 9 | 7 | 7 | 6 | 4 | 7 | 7 | 8 | 7 | 7 | 6 | 6 | 5 | 6 | 6 | 5 | 5 | 5 | 5 | 5 | 5 | 6 | 10 | 7 | 9 | 6 | 7 | 6 | 6 | 8 | 10 | 8 | 7 | 11 |
| 4j | 4 | 4 | 4 | 6 | 6 | 5 | 4 | 6 | 6 | 6 | 6 | 9 | 7 | 7 | 6 | 4 | 7 | 7 | 8 | 7 | 7 | 6 | 6 | 5 | 6 | 6 | 5 | 5 | 5 | 5 | 5 | 5 | 6 | 10 | 7 | 9 | 6 | 7 | 6 | 6 | 8 | 10 | 8 | 7 | 11 |
| 4k | 4 | 4 | 4 | 5 | 6 | 5 | 4 | 6 | 6 | 6 | 6 | 7 | 7 | 7 | 6 | 2 | 7 | 5 | 8 | 7 | 7 | 6 | 6 | 5 | 6 | 6 | 5 | 5 | 5 | 5 | 3 | 5 | 4 | 8 | 7 | 9 | 6 | 7 | 6 | 6 | 8 | 10 | 8 | 7 | 11 |
| 4L | 4 | 4 | 2 | 2 | 4 | 4 | 4 | 4 | 4 | 4 | 4 | 4 | 8 | 6 | 6 | 3 | 7 | 5 | 7 | 5 | 5 | 6 | 8 | 6 | 8 | 5 | 6 | 5 | 3 | 5 | 4 | 5 | 5 | 5 | 9 | 8 | 6 | 7 | 6 | 6 | 6 | 9 | 10 | 7 | 10 |
| 4m | 8 | 8 | 6 | 6 | 7 | 7 | 7 | 7 | 7 | 7 | 7 | 7 | 10 | 9 | 9 | 5 | 10 | 6 | 10 | 9 | 8 | 9 | 10 | 9 | 10 | 8 | 9 | 8 | 7 | 8 | 6 | 8 | 7 | 8 | 11 | 11 | 9 | 10 | 9 | 9 | 9 | 12 | 12 | 10 | 13 |
| 4n | 7 | 7 | 5 | 5 | 5 | 5 | 5 | 5 | 5 | 5 | 5 | 5 | 9 | 7 | 7 | 5 | 8 | 2 | 8 | 6 | 6 | 7 | 10 | 8 | 10 | 6 | 8 | 6 | 6 | 6 | 4 | 5 | 3 | 6 | 11 | 9 | 7 | 8 | 7 | 7 | 7 | 10 | 12 | 8 | 11 |

| | | | | | | | | | | | | | | | | | | | | | | | | | | | | | | | | | | | | | | | | | | | | | |
|---|---|---|---|---|---|---|---|---|---|---|---|---|---|---|---|---|---|---|---|---|---|---|---|---|---|---|---|---|---|---|---|---|---|---|---|---|---|---|---|---|---|---|---|---|---|
| **4o** | 6 | 6 | 4 | 4 | 6 | 6 | 6 | 6 | 6 | 6 | 6 | 6 | 10 | 8 | 8 | 3 | 9 | 6 | 9 | 7 | 7 | 8 | 9 | 7 | 9 | 7 | 7 | 7 | 5 | 7 | 4 | 7 | 5 | 7 | 10 | 10 | 8 | 9 | 8 | 8 | 8 | 11 | 11 | 9 | 12 |
| **4p** | 8 | 8 | 8 | 8 | 9 | 9 | 8 | 9 | 9 | 9 | 9 | 9 | 10 | 10 | 10 | 7 | 10 | 8 | 11 | 10 | 10 | 10 | 10 | 9 | 10 | 10 | 9 | 9 | 9 | 9 | 8 | 9 | 9 | 10 | 11 | 12 | 10 | 11 | 10 | 10 | 11 | 13 | 12 | 11 | 14 |
| **4q** | 6 | 6 | 6 | 8 | 8 | 7 | 6 | 8 | 8 | 8 | 8 | 9 | 9 | 9 | 8 | 6 | 9 | 8 | 10 | 9 | 9 | 8 | 8 | 7 | 8 | 8 | 7 | 7 | 7 | 7 | 7 | 7 | 8 | 10 | 9 | 11 | 8 | 9 | 8 | 8 | 10 | 12 | 10 | 9 | 13 |
| **4r** | 8 | 8 | 8 | 10 | 10 | 9 | 8 | 10 | 10 | 10 | 10 | 10 | 10 | 10 | 6 | 8 | 10 | 9 | 11 | 11 | 11 | 10 | 10 | 9 | 10 | 10 | 9 | 9 | 9 | 7 | 9 | 9 | 10 | 11 | 11 | 12 | 10 | 11 | 10 | 10 | 12 | 13 | 12 | 11 | 14 |
| **4s** | 6 | 6 | 6 | 8 | 6 | 5 | 4 | 6 | 6 | 6 | 6 | 10 | 9 | 8 | 2 | 6 | 8 | 9 | 9 | 7 | 7 | 7 | 8 | 7 | 8 | 6 | 7 | 5 | 7 | 3 | 7 | 5 | 8 | 11 | 9 | 10 | 6 | 8 | 6 | 6 | 8 | 11 | 10 | 7 | 12 |
| **4t** | 6 | 6 | 6 | 8 | 8 | 7 | 6 | 8 | 8 | 8 | 8 | 10 | 10 | 10 | 4 | 6 | 9 | 9 | 11 | 9 | 9 | 8 | 8 | 7 | 8 | 8 | 7 | 7 | 7 | 5 | 7 | 7 | 8 | 11 | 9 | 12 | 8 | 9 | 8 | 8 | 10 | 13 | 10 | 9 | 14 |
| **4u** | 6 | 6 | 4 | 4 | 4 | 4 | 4 | 4 | 4 | 4 | 4 | 4 | 8 | 6 | 6 | 5 | 7 | 5 | 7 | 5 | 5 | 6 | 8 | 7 | 10 | 5 | 7 | 5 | 5 | 5 | 6 | 5 | 6 | 5 | 11 | 8 | 6 | 7 | 6 | 6 | 6 | 9 | 12 | 7 | 10 |
| **4v** | 6 | 6 | 4 | 4 | 4 | 4 | 4 | 4 | 4 | 4 | 4 | 4 | 8 | 6 | 6 | 5 | 7 | 5 | 7 | 5 | 5 | 6 | 8 | 7 | 10 | 5 | 7 | 5 | 5 | 5 | 6 | 5 | 6 | 5 | 11 | 8 | 6 | 7 | 6 | 6 | 6 | 9 | 12 | 7 | 10 |
| **4w** | 6 | 6 | 4 | 4 | 4 | 4 | 4 | 4 | 4 | 4 | 4 | 4 | 8 | 6 | 6 | 5 | 6 | 7 | 7 | 5 | 5 | 6 | 8 | 7 | 10 | 5 | 7 | 5 | 5 | 5 | 6 | 5 | 6 | 5 | 11 | 8 | 6 | 7 | 6 | 6 | 6 | 9 | 12 | 7 | 10 |
| **4x** | 4 | 4 | 4 | 4 | 2 | 2 | 2 | 2 | 2 | 2 | 2 | 4 | 6 | 4 | 4 | 5 | 5 | 5 | 5 | 3 | 3 | 4 | 6 | 5 | 8 | 3 | 5 | 3 | 5 | 3 | 6 | 3 | 6 | 5 | 9 | 6 | 4 | 5 | 4 | 4 | 4 | 7 | 10 | 5 | 8 |
| **4y** | 4 | 4 | 4 | 6 | 6 | 5 | 4 | 6 | 6 | 6 | 6 | 9 | 7 | 7 | 6 | 4 | 7 | 7 | 8 | 7 | 7 | 6 | 6 | 5 | 6 | 6 | 5 | 5 | 5 | 5 | 5 | 5 | 6 | 10 | 7 | 9 | 6 | 7 | 6 | 6 | 8 | 10 | 8 | 7 | 11 |
| **4z** | 4 | 4 | 4 | 6 | 6 | 5 | 4 | 6 | 6 | 6 | 6 | 9 | 5 | 5 | 6 | 6 | 5 | 9 | 6 | 6 | 7 | 4 | 4 | 5 | 4 | 5 | 5 | 5 | 5 | 5 | 7 | 5 | 8 | 10 | 5 | 7 | 6 | 5 | 6 | 6 | 8 | 8 | 6 | 7 | 9 |
| **4aa** | 6 | 6 | 4 | 4 | 4 | 4 | 4 | 4 | 4 | 4 | 4 | 2 | 8 | 6 | 6 | 5 | 6 | 7 | 7 | 5 | 5 | 6 | 8 | 7 | 10 | 5 | 7 | 5 | 5 | 5 | 6 | 5 | 6 | 3 | 11 | 8 | 6 | 7 | 6 | 6 | 6 | 9 | 12 | 7 | 10 |
| **5a** | 6 | 6 | 6 | 8 | 5 | 6 | 7 | 7 | 7 | 7 | 7 | 10 | 3 | 3 | 9 | 8 | 3 | 11 | 4 | 4 | 8 | 4 | 6 | 5 | 6 | 7 | 7 | 8 | 7 | 8 | 9 | 8 | 10 | 11 | 7 | 5 | 9 | 5 | 9 | 9 | 9 | 6 | 8 | 10 | 7 |
| **5b** | 6 | 6 | 6 | 8 | 5 | 6 | 7 | 7 | 7 | 7 | 7 | 10 | 3 | 3 | 9 | 8 | 3 | 11 | 4 | 4 | 8 | 4 | 6 | 5 | 6 | 7 | 7 | 8 | 7 | 8 | 9 | 8 | 10 | 11 | 7 | 5 | 9 | 5 | 9 | 9 | 9 | 6 | 8 | 10 | 7 |
| **5c** | 6 | 6 | 6 | 8 | 5 | 6 | 7 | 7 | 7 | 7 | 7 | 10 | 3 | 3 | 9 | 8 | 3 | 11 | 4 | 4 | 8 | 4 | 6 | 5 | 6 | 7 | 7 | 8 | 7 | 8 | 9 | 8 | 10 | 11 | 7 | 5 | 9 | 5 | 9 | 9 | 9 | 6 | 8 | 10 | 7 |
| **5d** | 5 | 6 | 6 | 8 | 5 | 4 | 5 | 6 | 6 | 6 | 6 | 10 | 3 | 3 | 7 | 8 | 1 | 11 | 4 | 4 | 7 | 2 | 4 | 3 | 6 | 3 | 5 | 6 | 7 | 6 | 9 | 6 | 10 | 11 | 7 | 5 | 7 | 3 | 7 | 7 | 8 | 6 | 8 | 8 | 7 |
| **5e** | 6 | 6 | 6 | 8 | 3 | 4 | 5 | 5 | 5 | 5 | 5 | 9 | 1 | 1 | 7 | 8 | 3 | 10 | 2 | 2 | 6 | 4 | 6 | 5 | 6 | 5 | 7 | 6 | 7 | 6 | 9 | 6 | 10 | 10 | 7 | 3 | 7 | 5 | 7 | 7 | 7 | 4 | 8 | 8 | 5 |
| **5f** | 7 | 7 | 5 | 5 | 5 | 5 | 5 | 5 | 5 | 5 | 5 | 5 | 9 | 7 | 7 | 6 | 8 | 6 | 8 | 6 | 6 | 7 | 9 | 8 | 11 | 6 | 8 | 6 | 6 | 6 | 7 | 6 | 7 | 6 | 12 | 9 | 7 | 8 | 7 | 7 | 7 | 10 | 13 | 8 | 11 |
| **5g** | 4 | 4 | 4 | 6 | 6 | 5 | 4 | 6 | 6 | 6 | 6 | 10 | 5 | 5 | 6 | 6 | 5 | 9 | 6 | 6 | 7 | 4 | 4 | 7 | 4 | 5 | 5 | 5 | 5 | 5 | 7 | 5 | 8 | 11 | 5 | 7 | 6 | 5 | 6 | 6 | 8 | 8 | 6 | 7 | 9 |
| **5h** | 4 | 4 | 4 | 6 | 6 | 5 | 4 | 6 | 6 | 6 | 6 | 10 | 5 | 5 | 6 | 6 | 5 | 9 | 6 | 6 | 7 | 4 | 4 | 7 | 4 | 5 | 5 | 5 | 5 | 5 | 7 | 5 | 8 | 11 | 5 | 7 | 6 | 5 | 6 | 6 | 8 | 8 | 6 | 7 | 9 |
| **5i** | 4 | 4 | 4 | 6 | 6 | 5 | 4 | 6 | 6 | 6 | 6 | 10 | 5 | 5 | 6 | 6 | 5 | 9 | 6 | 6 | 7 | 4 | 4 | 7 | 4 | 5 | 5 | 5 | 5 | 5 | 7 | 5 | 8 | 11 | 5 | 7 | 6 | 5 | 6 | 6 | 8 | 8 | 6 | 7 | 9 |
| **5j** | 3 | 3 | 3 | 5 | 5 | 4 | 3 | 5 | 5 | 5 | 5 | 9 | 6 | 6 | 5 | 5 | 6 | 8 | 7 | 6 | 6 | 3 | 5 | 4 | 5 | 5 | 2 | 4 | 4 | 4 | 6 | 4 | 7 | 10 | 6 | 8 | 5 | 4 | 5 | 5 | 7 | 9 | 7 | 6 | 10 |
| **5k** | 4 | 4 | 4 | 6 | 5 | 4 | 4 | 6 | 6 | 6 | 6 | 10 | 5 | 5 | 6 | 6 | 3 | 9 | 6 | 6 | 7 | 2 | 2 | 3 | 2 | 3 | 5 | 5 | 5 | 5 | 7 | 5 | 8 | 11 | 3 | 7 | 6 | 3 | 6 | 6 | 8 | 8 | 4 | 7 | 9 |
| **5L** | 2 | 2 | 2 | 4 | 4 | 3 | 2 | 4 | 4 | 4 | 4 | 8 | 6 | 6 | 5 | 5 | 5 | 8 | 7 | 5 | 5 | 4 | 4 | 3 | 5 | 4 | 3 | 3 | 3 | 4 | 6 | 3 | 7 | 9 | 6 | 8 | 4 | 5 | 4 | 4 | 6 | 9 | 7 | 5 | 10 |
| **5m** | 1 | 1 | 1 | 3 | 3 | 2 | 1 | 3 | 3 | 3 | 3 | 7 | 6 | 5 | 3 | 5 | 4 | 8 | 6 | 4 | 4 | 3 | 3 | 2 | 5 | 3 | 2 | 2 | 2 | 2 | 6 | 2 | 7 | 8 | 6 | 7 | 3 | 4 | 3 | 3 | 5 | 8 | 7 | 4 | 9 |
| **5n** | 3 | 3 | 3 | 5 | 1 | 1 | 1 | 1 | 1 | 1 | 1 | 5 | 5 | 3 | 3 | 6 | 4 | 6 | 4 | 2 | 2 | 3 | 5 | 4 | 7 | 2 | 4 | 2 | 4 | 2 | 7 | 2 | 7 | 6 | 8 | 5 | 3 | 4 | 3 | 3 | 3 | 6 | 9 | 4 | 7 |
| **5o** | 3 | 3 | 1 | 1 | 5 | 4 | 3 | 5 | 5 | 5 | 5 | 5 | 8 | 7 | 7 | 4 | 5 | 6 | 8 | 6 | 6 | 5 | 5 | 4 | 7 | 5 | 4 | 4 | 2 | 6 | 5 | 4 | 6 | 6 | 8 | 9 | 5 | 8 | 5 | 5 | 7 | 10 | 9 | 6 | 11 |
| **5p** | 5 | 5 | 4 | 4 | 7 | 6 | 5 | 6 | 6 | 6 | 6 | 6 | 8 | 8 | 7 | 1 | 8 | 4 | 9 | 7 | 7 | 7 | 7 | 6 | 7 | 7 | 6 | 6 | 5 | 6 | 2 | 6 | 3 | 7 | 8 | 10 | 7 | 8 | 7 | 7 | 8 | 11 | 9 | 8 | 12 |
| **5q** | 5 | 5 | 3 | 3 | 5 | 5 | 5 | 5 | 5 | 5 | 5 | 5 | 9 | 7 | 8 | 2 | 8 | 5 | 8 | 6 | 6 | 7 | 7 | 6 | 8 | 6 | 6 | 6 | 4 | 7 | 3 | 6 | 4 | 6 | 9 | 9 | 7 | 8 | 7 | 7 | 7 | 10 | 10 | 8 | 11 |
| **5r** | 9 | 9 | 9 | 9 | 10 | 10 | 9 | 10 | 10 | 10 | 10 | 10 | 11 | 11 | 11 | 8 | 11 | 9 | 12 | 11 | 11 | 11 | 11 | 10 | 11 | 11 | 10 | 10 | 10 | 10 | 9 | 10 | 10 | 11 | 12 | 13 | 11 | 12 | 11 | 11 | 12 | 14 | 13 | 12 | 15 |
| **5s** | 8 | 8 | 6 | 6 | 6 | 6 | 6 | 6 | 6 | 6 | 6 | 6 | 10 | 8 | 8 | 4 | 9 | 1 | 9 | 7 | 7 | 8 | 10 | 9 | 11 | 8 | 9 | 7 | 7 | 7 | 3 | 7 | 2 | 7 | 12 | 10 | 8 | 9 | 8 | 8 | 8 | 11 | 13 | 9 | 12 |
| **5t** | 5 | 5 | 5 | 7 | 5 | 4 | 3 | 5 | 5 | 5 | 5 | 9 | 9 | 7 | 1 | 7 | 7 | 10 | 8 | 6 | 6 | 6 | 7 | 6 | 9 | 5 | 6 | 4 | 6 | 2 | 8 | 4 | 9 | 10 | 10 | 9 | 5 | 7 | 5 | 5 | 7 | 10 | 11 | 6 | 11 |
| **5u** | 9 | 9 | 9 | 11 | 11 | 10 | 9 | 11 | 11 | 11 | 11 | 11 | 11 | 11 | 7 | 9 | 11 | 10 | 12 | 12 | 12 | 11 | 11 | 10 | 11 | 11 | 10 | 10 | 10 | 8 | 10 | 10 | 11 | 12 | 12 | 13 | 11 | 12 | 11 | 11 | 13 | 14 | 13 | 12 | 15 |
| **5v** | 6 | 6 | 6 | 8 | 5 | 6 | 7 | 7 | 7 | 7 | 7 | 10 | 3 | 3 | 9 | 8 | 3 | 11 | 4 | 4 | 8 | 4 | 6 | 5 | 6 | 7 | 7 | 8 | 7 | 8 | 9 | 8 | 10 | 11 | 7 | 5 | 9 | 5 | 9 | 9 | 9 | 6 | 8 | 10 | 7 |
| **5w** | 4 | 4 | 4 | 6 | 6 | 5 | 4 | 6 | 6 | 6 | 6 | 10 | 5 | 5 | 6 | 6 | 5 | 9 | 6 | 6 | 7 | 4 | 4 | 7 | 4 | 5 | 5 | 5 | 5 | 5 | 7 | 5 | 8 | 11 | 5 | 7 | 6 | 5 | 6 | 6 | 8 | 8 | 6 | 7 | 9 |
| **5x** | 7 | 7 | 5 | 5 | 5 | 5 | 5 | 5 | 5 | 5 | 5 | 1 | 9 | 7 | 7 | 6 | 7 | 8 | 8 | 6 | 6 | 7 | 9 | 8 | 11 | 6 | 8 | 6 | 6 | 6 | 7 | 6 | 7 | 2 | 12 | 9 | 7 | 8 | 7 | 7 | 7 | 10 | 13 | 8 | 11 |
| **6a** | 5 | 5 | 5 | 7 | 6 | 5 | 5 | 7 | 7 | 7 | 7 | 11 | 6 | 6 | 7 | 7 | 4 | 10 | 7 | 7 | 8 | 3 | 3 | 4 | 3 | 4 | 6 | 6 | 6 | 6 | 8 | 6 | 9 | 12 | 4 | 8 | 7 | 4 | 7 | 7 | 9 | 9 | 5 | 8 | 10 |
| **6b** | 4 | 4 | 4 | 6 | 6 | 5 | 4 | 6 | 6 | 6 | 6 | 10 | 7 | 7 | 6 | 6 | 7 | 9 | 8 | 7 | 7 | 4 | 6 | 5 | 6 | 6 | 3 | 5 | 5 | 5 | 7 | 5 | 8 | 11 | 7 | 9 | 6 | 5 | 6 | 6 | 8 | 10 | 8 | 7 | 11 |
| **6c** | 4 | 4 | 4 | 6 | 6 | 5 | 4 | 6 | 6 | 6 | 6 | 10 | 7 | 7 | 6 | 6 | 7 | 9 | 8 | 7 | 7 | 4 | 6 | 5 | 6 | 6 | 3 | 5 | 5 | 5 | 7 | 5 | 8 | 11 | 7 | 9 | 6 | 5 | 6 | 6 | 8 | 10 | 8 | 7 | 11 |
| **6d** | 4 | 4 | 4 | 6 | 6 | 5 | 4 | 6 | 6 | 6 | 6 | 10 | 7 | 7 | 6 | 6 | 7 | 9 | 8 | 7 | 7 | 4 | 6 | 5 | 6 | 6 | 3 | 5 | 5 | 5 | 7 | 5 | 8 | 11 | 7 | 9 | 6 | 5 | 6 | 6 | 8 | 10 | 8 | 7 | 11 |
| **6e** | 4 | 4 | 4 | 6 | 6 | 5 | 4 | 6 | 6 | 6 | 6 | 10 | 7 | 7 | 6 | 6 | 7 | 9 | 8 | 7 | 7 | 4 | 6 | 5 | 6 | 6 | 3 | 5 | 5 | 5 | 7 | 5 | 8 | 11 | 7 | 9 | 6 | 5 | 6 | 6 | 8 | 10 | 8 | 7 | 11 |
| **6f** | 4 | 4 | 4 | 6 | 6 | 5 | 4 | 6 | 6 | 6 | 6 | 10 | 7 | 7 | 6 | 6 | 7 | 9 | 8 | 7 | 7 | 4 | 6 | 5 | 6 | 6 | 3 | 5 | 5 | 5 | 7 | 5 | 8 | 11 | 7 | 9 | 6 | 5 | 6 | 6 | 8 | 10 | 8 | 7 | 11 |
| **6g** | 4 | 5 | 5 | 7 | 6 | 5 | 5 | 7 | 7 | 7 | 7 | 11 | 6 | 6 | 7 | 7 | 4 | 10 | 7 | 7 | 8 | 3 | 1 | 2 | 1 | 4 | 4 | 6 | 6 | 6 | 8 | 6 | 9 | 12 | 2 | 8 | 7 | 4 | 7 | 7 | 9 | 9 | 3 | 8 | 10 |
| **6h** | 3 | 4 | 4 | 6 | 6 | 5 | 4 | 6 | 6 | 6 | 6 | 10 | 7 | 7 | 6 | 6 | 5 | 9 | 8 | 7 | 7 | 4 | 4 | 3 | 4 | 6 | 1 | 5 | 5 | 5 | 7 | 5 | 8 | 11 | 5 | 9 | 6 | 5 | 6 | 6 | 8 | 10 | 6 | 7 | 11 |
| **6i** | 4 | 5 | 5 | 7 | 4 | 3 | 4 | 5 | 5 | 5 | 5 | 9 | 6 | 6 | 6 | 7 | 2 | 10 | 7 | 7 | 6 | 1 | 3 | 2 | 3 | 2 | 4 | 5 | 6 | 5 | 8 | 5 | 9 | 10 | 4 | 8 | 6 | 2 | 6 | 6 | 7 | 9 | 5 | 7 | 10 |

*(continued)*

| | 6k | 6L | 6m | 6n | 6o | 6p | 6q | 6r | 6s | 6t | 6u | 6v | 6w | 6x | 6y | 6z | 6aa | 6ab | 7a | 7b | 7c | 7d | 7e | 7f | 7g | 7h | 7i | 7j | 7k | 7L | 7m | 7n | 7o | 7p | 8a | 8b | 8c | 8d | 8e | 8f | 8g | 9a | 9b | 9c | 10a |
|---|---|---|---|---|---|---|---|---|---|---|---|---|---|---|---|---|---|---|---|---|---|---|---|---|---|---|---|---|---|---|---|---|---|---|---|---|---|---|---|---|---|---|---|---|---|
| **6j** | 1 | 2 | 2 | 4 | 4 | 3 | 2 | 4 | 4 | 4 | 4 | 8 | 7 | 6 | 4 | 6 | 6 | 9 | 7 | 5 | 5 | 2 | 2 | 1 | 4 | 4 | 1 | 3 | 3 | 3 | 7 | 3 | 8 | 9 | 5 | 8 | 4 | 3 | 4 | 4 | 6 | 9 | 6 | 5 | 10 |
| **6k** | x | 2 | 2 | 4 | 4 | 3 | 2 | 4 | 4 | 4 | 4 | 8 | 7 | 6 | 4 | 6 | 4 | 9 | 7 | 5 | 5 | 3 | 3 | 2 | 5 | 4 | 2 | 3 | 3 | 3 | 7 | 3 | 8 | 9 | 6 | 8 | 4 | 4 | 4 | 4 | 6 | 9 | 7 | 5 | 10 |
| **6L** | 2 | x | 2 | 4 | 4 | 3 | 2 | 4 | 4 | 4 | 4 | 8 | 7 | 6 | 4 | 6 | 5 | 9 | 7 | 5 | 5 | 4 | 4 | 3 | 6 | 4 | 3 | 3 | 1 | 3 | 7 | 3 | 8 | 9 | 7 | 8 | 4 | 5 | 4 | 4 | 6 | 9 | 8 | 5 | 10 |
| **6m** | 2 | 2 | x | 2 | 4 | 3 | 2 | 4 | 4 | 4 | 4 | 6 | 7 | 6 | 4 | 6 | 5 | 7 | 7 | 5 | 5 | 4 | 4 | 3 | 6 | 4 | 3 | 3 | 1 | 3 | 7 | 3 | 8 | 7 | 7 | 8 | 4 | 5 | 4 | 4 | 6 | 9 | 8 | 5 | 10 |
| **6n** | 4 | 4 | 2 | x | 6 | 5 | 4 | 6 | 6 | 6 | 6 | 6 | 9 | 8 | 8 | 5 | 6 | 7 | 9 | 7 | 7 | 8 | 6 | 5 | 8 | 6 | 5 | 5 | 3 | 7 | 6 | 5 | 7 | 7 | 9 | 10 | 6 | 9 | 6 | 6 | 8 | 11 | 10 | 7 | 12 |
| **6o** | 4 | 4 | 4 | 6 | x | 1 | 2 | 2 | 2 | 2 | 2 | 6 | 4 | 2 | 4 | 7 | 4 | 7 | 3 | 1 | 3 | 3 | 6 | 5 | 7 | 2 | 5 | 3 | 5 | 3 | 8 | 3 | 8 | 7 | 8 | 4 | 4 | 4 | 4 | 4 | 4 | 5 | 9 | 5 | 6 |
| **6p** | 3 | 3 | 3 | 5 | 1 | x | 1 | 2 | 2 | 2 | 2 | 6 | 5 | 3 | 3 | 7 | 3 | 7 | 4 | 2 | 3 | 2 | 5 | 4 | 6 | 1 | 4 | 2 | 4 | 2 | 8 | 2 | 8 | 7 | 7 | 5 | 3 | 3 | 3 | 3 | 4 | 6 | 8 | 4 | 7 |
| **6q** | 2 | 2 | 2 | 4 | 2 | 1 | x | 2 | 2 | 2 | 2 | 6 | 6 | 4 | 2 | 6 | 4 | 7 | 5 | 3 | 3 | 3 | 4 | 3 | 6 | 2 | 3 | 1 | 3 | 1 | 7 | 1 | 8 | 7 | 7 | 6 | 2 | 4 | 2 | 2 | 4 | 7 | 8 | 3 | 8 |
| **6r** | 4 | 4 | 4 | 6 | 2 | 2 | 2 | x | 2 | 2 | 2 | 6 | 6 | 4 | 4 | 7 | 5 | 7 | 5 | 3 | 3 | 4 | 6 | 5 | 8 | 3 | 5 | 3 | 5 | 3 | 8 | 3 | 8 | 7 | 9 | 6 | 4 | 5 | 4 | 4 | 4 | 7 | 10 | 5 | 8 |
| **6s** | 4 | 4 | 4 | 6 | 2 | 2 | 2 | 2 | x | 2 | 2 | 6 | 6 | 4 | 4 | 7 | 5 | 7 | 5 | 3 | 3 | 4 | 6 | 5 | 8 | 3 | 5 | 3 | 5 | 3 | 8 | 3 | 8 | 7 | 9 | 6 | 4 | 5 | 4 | 4 | 4 | 7 | 10 | 5 | 8 |
| **6t** | 4 | 4 | 4 | 6 | 2 | 2 | 2 | 2 | 2 | x | 2 | 6 | 6 | 4 | 4 | 7 | 5 | 7 | 5 | 3 | 3 | 4 | 6 | 5 | 8 | 3 | 5 | 3 | 5 | 3 | 8 | 3 | 8 | 7 | 9 | 6 | 4 | 5 | 4 | 4 | 4 | 7 | 10 | 5 | 8 |
| **6u** | 4 | 4 | 4 | 6 | 2 | 2 | 2 | 2 | 2 | 2 | x | 6 | 6 | 4 | 4 | 7 | 5 | 7 | 5 | 3 | 1 | 4 | 6 | 5 | 8 | 3 | 5 | 3 | 5 | 3 | 8 | 3 | 8 | 7 | 9 | 6 | 4 | 5 | 4 | 4 | 2 | 7 | 10 | 5 | 8 |
| **6v** | 8 | 8 | 6 | 6 | 6 | 6 | 6 | 6 | 6 | 6 | 6 | x | 10 | 8 | 8 | 7 | 8 | 9 | 9 | 7 | 7 | 8 | 10 | 9 | 12 | 7 | 9 | 7 | 7 | 7 | 8 | 7 | 8 | 1 | 13 | 10 | 8 | 9 | 8 | 8 | 8 | 11 | 14 | 9 | 12 |
| **6w** | 7 | 7 | 7 | 9 | 4 | 5 | 6 | 6 | 6 | 6 | 6 | 10 | x | 2 | 8 | 9 | 4 | 11 | 3 | 3 | 7 | 5 | 7 | 6 | 7 | 6 | 8 | 7 | 8 | 7 | 10 | 7 | 11 | 11 | 8 | 4 | 8 | 6 | 8 | 8 | 8 | 5 | 9 | 9 | 6 |
| **6x** | 6 | 6 | 6 | 8 | 2 | 3 | 4 | 4 | 4 | 4 | 4 | 8 | 2 | x | 6 | 9 | 4 | 9 | 1 | 1 | 5 | 5 | 7 | 6 | 7 | 4 | 7 | 5 | 7 | 5 | 10 | 5 | 10 | 9 | 8 | 2 | 6 | 6 | 6 | 6 | 6 | 3 | 9 | 7 | 4 |
| **6y** | 4 | 4 | 4 | 8 | 4 | 3 | 2 | 4 | 4 | 4 | 4 | 8 | 8 | 6 | x | 8 | 6 | 9 | 7 | 5 | 5 | 5 | 6 | 5 | 8 | 4 | 5 | 3 | 5 | 1 | 9 | 3 | 10 | 9 | 9 | 8 | 4 | 6 | 4 | 4 | 6 | 9 | 10 | 5 | 10 |
| **6z** | 6 | 6 | 6 | 5 | 7 | 7 | 6 | 7 | 7 | 7 | 7 | 7 | 9 | 9 | 8 | x | 9 | 3 | 10 | 8 | 8 | 8 | 8 | 7 | 8 | 8 | 7 | 7 | 7 | 7 | 1 | 7 | 2 | 8 | 9 | 11 | 8 | 9 | 8 | 8 | 9 | 12 | 10 | 9 | 13 |
| **6aa** | 4 | 5 | 5 | 6 | 4 | 3 | 4 | 5 | 5 | 5 | 5 | 8 | 4 | 4 | 6 | 9 | x | 10 | 5 | 5 | 6 | 1 | 3 | 2 | 5 | 2 | 4 | 5 | 6 | 5 | 10 | 5 | 11 | 9 | 6 | 6 | 6 | 2 | 6 | 6 | 7 | 7 | 7 | 7 | 8 |
| **6ab** | 9 | 9 | 7 | 7 | 7 | 7 | 7 | 7 | 7 | 7 | 7 | 9 | 11 | 9 | 9 | 3 | 10 | x | 10 | 8 | 8 | 9 | 11 | 10 | 11 | 8 | 10 | 8 | 10 | 8 | 2 | 8 | 1 | 10 | 12 | 11 | 9 | 10 | 9 | 9 | 9 | 12 | 13 | 10 | 13 |
| **7a** | 7 | 7 | 7 | 9 | 3 | 4 | 5 | 5 | 5 | 5 | 5 | 9 | 3 | 1 | 7 | 10 | 5 | 10 | x | 2 | 6 | 6 | 8 | 7 | 8 | 5 | 8 | 6 | 8 | 6 | 11 | 6 | 11 | 10 | 9 | 1 | 7 | 7 | 7 | 7 | 7 | 2 | 10 | 8 | 3 |
| **7b** | 5 | 5 | 5 | 7 | 1 | 2 | 3 | 3 | 3 | 3 | 3 | 7 | 3 | 1 | 5 | 8 | 5 | 8 | 2 | x | 4 | 4 | 8 | 6 | 8 | 3 | 6 | 4 | 6 | 4 | 9 | 4 | 9 | 8 | 9 | 3 | 5 | 5 | 5 | 5 | 5 | 4 | 10 | 6 | 5 |
| **7c** | 5 | 5 | 5 | 7 | 3 | 3 | 3 | 3 | 3 | 3 | 1 | 7 | 7 | 5 | 5 | 8 | 6 | 8 | 6 | 4 | x | 5 | 7 | 6 | 9 | 4 | 6 | 4 | 6 | 4 | 9 | 4 | 9 | 8 | 10 | 7 | 5 | 6 | 5 | 5 | 1 | 8 | 11 | 6 | 9 |
| **7d** | 3 | 4 | 4 | 8 | 3 | 2 | 3 | 4 | 4 | 4 | 4 | 8 | 5 | 5 | 5 | 8 | 1 | 9 | 6 | 4 | 5 | x | 2 | 1 | 4 | 1 | 3 | 4 | 5 | 4 | 9 | 4 | 10 | 9 | 5 | 7 | 5 | 1 | 5 | 5 | 6 | 8 | 6 | 6 | 9 |
| **7e** | 3 | 4 | 4 | 6 | 6 | 5 | 4 | 6 | 6 | 6 | 6 | 10 | 7 | 7 | 6 | 8 | 3 | 11 | 8 | 8 | 7 | 2 | x | 1 | 2 | 5 | 3 | 5 | 5 | 5 | 9 | 5 | 10 | 11 | 3 | 9 | 6 | 3 | 6 | 6 | 8 | 10 | 4 | 7 | 11 |
| **7f** | 2 | 3 | 3 | 5 | 5 | 4 | 3 | 5 | 5 | 5 | 5 | 9 | 6 | 6 | 5 | 7 | 2 | 10 | 7 | 6 | 6 | 1 | 1 | x | 3 | 2 | 2 | 4 | 4 | 4 | 8 | 4 | 9 | 10 | 4 | 8 | 5 | 2 | 5 | 5 | 7 | 9 | 5 | 6 | 10 |
| **7g** | 5 | 6 | 6 | 8 | 7 | 6 | 6 | 8 | 8 | 8 | 8 | 12 | 7 | 7 | 8 | 8 | 5 | 11 | 8 | 8 | 9 | 4 | 2 | 3 | x | 5 | 5 | 7 | 7 | 7 | 9 | 7 | 10 | 13 | 1 | 9 | 8 | 5 | 8 | 8 | 10 | 10 | 2 | 9 | 11 |
| **7h** | 4 | 4 | 4 | 6 | 2 | 1 | 2 | 3 | 3 | 3 | 3 | 7 | 6 | 4 | 4 | 8 | 2 | 8 | 5 | 3 | 4 | 1 | 5 | 2 | 5 | x | 4 | 3 | 5 | 3 | 9 | 3 | 9 | 8 | 6 | 6 | 4 | 2 | 4 | 4 | 5 | 7 | 7 | 5 | 8 |
| **7i** | 2 | 3 | 3 | 5 | 5 | 4 | 3 | 5 | 5 | 5 | 5 | 9 | 8 | 7 | 5 | 7 | 4 | 10 | 8 | 6 | 6 | 3 | 3 | 2 | 5 | 4 | x | 4 | 4 | 4 | 8 | 4 | 9 | 10 | 6 | 9 | 5 | 4 | 5 | 5 | 7 | 10 | 7 | 6 | 11 |
| **7j** | 3 | 3 | 3 | 5 | 3 | 2 | 1 | 3 | 3 | 3 | 3 | 7 | 7 | 5 | 3 | 7 | 5 | 8 | 6 | 4 | 4 | 4 | 5 | 4 | 7 | 3 | 4 | x | 4 | 1 | 7 | 2 | 9 | 8 | 8 | 7 | 1 | 5 | 1 | 3 | 5 | 8 | 9 | 4 | 9 |
| **7k** | 3 | 1 | 1 | 3 | 5 | 4 | 3 | 5 | 5 | 5 | 5 | 7 | 8 | 7 | 5 | 7 | 6 | 10 | 8 | 6 | 6 | 5 | 5 | 4 | 7 | 5 | 4 | 4 | x | 4 | 8 | 4 | 9 | 8 | 8 | 9 | 5 | 6 | 5 | 5 | 7 | 10 | 9 | 6 | 11 |
| **7L** | 3 | 3 | 3 | 7 | 3 | 2 | 1 | 3 | 3 | 3 | 3 | 7 | 7 | 5 | 1 | 7 | 5 | 8 | 6 | 4 | 4 | 4 | 5 | 4 | 7 | 3 | 4 | 1 | 4 | x | 8 | 1 | 9 | 8 | 8 | 7 | 2 | 5 | 2 | 2 | 5 | 8 | 9 | 4 | 9 |
| **7m** | 7 | 7 | 7 | 6 | 8 | 8 | 7 | 8 | 8 | 8 | 8 | 8 | 10 | 10 | 9 | 1 | 10 | 2 | 11 | 9 | 9 | 9 | 9 | 8 | 9 | 9 | 8 | 7 | 8 | 8 | x | 7 | 1 | 9 | 10 | 12 | 8 | 10 | 8 | 8 | 10 | 13 | 11 | 9 | 14 |
| **7n** | 3 | 3 | 3 | 5 | 3 | 2 | 1 | 3 | 3 | 3 | 3 | 7 | 7 | 5 | 3 | 7 | 5 | 8 | 6 | 4 | 4 | 4 | 5 | 4 | 7 | 3 | 4 | 2 | 4 | 1 | 7 | x | 9 | 8 | 8 | 7 | 3 | 5 | 3 | 1 | 5 | 8 | 9 | 2 | 9 |
| **7o** | 8 | 8 | 8 | 7 | 8 | 8 | 8 | 8 | 8 | 8 | 8 | 8 | 11 | 10 | 10 | 2 | 11 | 1 | 11 | 9 | 9 | 10 | 10 | 9 | 10 | 9 | 9 | 9 | 9 | 9 | 1 | 9 | x | 9 | 11 | 12 | 10 | 11 | 10 | 10 | 10 | 13 | 12 | 11 | 14 |
| **7p** | 9 | 9 | 7 | 7 | 7 | 7 | 7 | 7 | 7 | 7 | 7 | 1 | 11 | 9 | 9 | 8 | 9 | 10 | 10 | 8 | 8 | 9 | 11 | 10 | 13 | 8 | 10 | 8 | 8 | 8 | 9 | 8 | 9 | x | 14 | 11 | 9 | 10 | 9 | 9 | 9 | 12 | 15 | 10 | 13 |
| **8a** | 6 | 7 | 7 | 9 | 8 | 7 | 7 | 9 | 9 | 9 | 9 | 13 | 8 | 8 | 9 | 9 | 6 | 12 | 9 | 9 | 10 | 5 | 3 | 4 | 1 | 6 | 6 | 8 | 8 | 8 | 10 | 8 | 11 | 14 | x | 10 | 9 | 6 | 9 | 9 | 11 | 11 | 1 | 10 | 12 |
| **8b** | 8 | 8 | 8 | 10 | 4 | 5 | 6 | 6 | 6 | 6 | 6 | 10 | 4 | 2 | 8 | 11 | 6 | 11 | 1 | 3 | 7 | 7 | 9 | 8 | 9 | 6 | 9 | 7 | 9 | 7 | 12 | 7 | 12 | 11 | 10 | x | 8 | 8 | 8 | 8 | 8 | 1 | 11 | 9 | 2 |
| **8c** | 4 | 4 | 4 | 6 | 4 | 3 | 2 | 4 | 4 | 4 | 4 | 8 | 8 | 6 | 4 | 8 | 6 | 9 | 7 | 5 | 5 | 5 | 6 | 5 | 8 | 4 | 5 | 1 | 5 | 2 | 8 | 3 | 10 | 9 | 9 | 8 | x | 6 | 2 | 4 | 6 | 9 | 10 | 5 | 10 |
| **8d** | 4 | 5 | 5 | 9 | 4 | 3 | 4 | 5 | 5 | 5 | 5 | 9 | 6 | 6 | 6 | 9 | 2 | 10 | 7 | 5 | 6 | 1 | 3 | 2 | 5 | 2 | 4 | 5 | 6 | 5 | 10 | 5 | 11 | 10 | 6 | 8 | 6 | x | 6 | 6 | 7 | 9 | 7 | 7 | 10 |
| **8e** | 4 | 4 | 4 | 6 | 4 | 3 | 2 | 4 | 4 | 4 | 4 | 8 | 8 | 6 | 4 | 8 | 6 | 9 | 7 | 5 | 5 | 5 | 6 | 5 | 8 | 4 | 5 | 1 | 5 | 2 | 8 | 3 | 10 | 9 | 9 | 8 | 2 | 6 | x | 4 | 6 | 9 | 10 | 5 | 10 |
| **8f** | 4 | 4 | 4 | 6 | 4 | 3 | 2 | 4 | 4 | 4 | 4 | 8 | 8 | 6 | 4 | 8 | 6 | 9 | 7 | 5 | 5 | 5 | 6 | 5 | 8 | 4 | 5 | 3 | 5 | 2 | 8 | 1 | 10 | 9 | 9 | 8 | 4 | 6 | 4 | x | 6 | 9 | 10 | 1 | 10 |
| **8g** | 6 | 6 | 6 | 8 | 4 | 4 | 4 | 4 | 4 | 4 | 2 | 8 | 8 | 6 | 6 | 9 | 7 | 9 | 7 | 5 | 1 | 6 | 8 | 7 | 10 | 5 | 7 | 5 | 7 | 5 | 10 | 5 | 10 | 9 | 11 | 8 | 6 | 7 | 6 | 6 | x | 9 | 12 | 7 | 10 |
| **9a** | 9 | 9 | 9 | 11 | 5 | 6 | 7 | 7 | 7 | 7 | 7 | 11 | 5 | 3 | 9 | 12 | 7 | 12 | 2 | 4 | 8 | 8 | 10 | 9 | 10 | 7 | 10 | 8 | 10 | 8 | 13 | 8 | 13 | 12 | 11 | 1 | 9 | 9 | 9 | 9 | 9 | x | 12 | 10 | 1 |
| **9b** | 7 | 8 | 8 | 10 | 9 | 8 | 8 | 10 | 10 | 10 | 10 | 14 | 9 | 9 | 10 | 10 | 7 | 13 | 10 | 10 | 11 | 6 | 4 | 5 | 2 | 7 | 7 | 9 | 9 | 9 | 11 | 9 | 12 | 15 | 1 | 11 | 10 | 7 | 10 | 10 | 12 | 12 | x | 11 | 13 |
| **9c** | 5 | 5 | 5 | 7 | 5 | 4 | 3 | 5 | 5 | 5 | 5 | 9 | 9 | 7 | 5 | 9 | 7 | 10 | 8 | 6 | 6 | 6 | 7 | 6 | 9 | 5 | 6 | 4 | 6 | 4 | 9 | 2 | 11 | 10 | 10 | 9 | 5 | 7 | 5 | 1 | 7 | 10 | 11 | x | 11 |
| **10a** | 10 | 10 | 10 | 12 | 6 | 7 | 8 | 8 | 8 | 8 | 8 | 12 | 6 | 4 | 10 | 13 | 8 | 13 | 3 | 5 | 9 | 9 | 11 | 10 | 11 | 8 | 11 | 9 | 11 | 9 | 14 | 9 | 14 | 13 | 12 | 2 | 10 | 10 | 10 | 10 | 10 | 1 | 13 | 11 | x |

# Bibliography


Altman, I. 1975. *The Environment and Social Behavior: Privacy, Personal Space, Territory, Crowding*. Belmont, California, Wadsworth Publishing Company.

Anawalt, P. 1981. *Indian Clothing before Cortes: Mesoamerican Costumes from the Codices*. Norman, University of Oklahoma Press.

Andrews, G. F. 1974. *Archaeological Survey of Palenque: Introduction and Miscellaneous Notes*. Andrews Papers, The Alexander Architectural Archive, The University of Texas Libraries. Austin, The University of Texas.

Andrews, G. F. 1975. *Maya Cities: Placemaking and Urbanization*. Norman, University of Oklahoma Press.

Arnold, D. 2002. *Reading Archaeological History*. London and New York, Routledge.

Ashmore, W. 1986. Peten Cosmology in the Maya Southeast: An Analysis of Architecture and Settlement Patterns at Classic Quirigua. In *The Southeast Maya Periphery*, edited by P. A. Urban: 35–49. Austin, University of Texas Press.

Ashmore, W. 1992. Deciphering Maya Site Plans. In *New Theories on the Ancient Maya*, edited by E. Danian and R. Sharer: 173–184. Philadelphia, University of Pennsylvania.

Ashmore, W. and J. A. Sabloff. 2002. Spatial Orders in Maya Civic Plans. *Latin American Antiquity*, 13(2): 201–215.

Ashmore, W. and J. A. Sabloff. 2003. Interpreting Ancient Maya Civic Plans: Reply to Smith. *Latin American Antiquity*, 14(2): 229–236.

Aveni, A., and H. Hartung. 1986. Maya City Planning and the Calendar. *Transactions of the American Philosophical Society*, 76(7): 1–71.

Baddeley, O. 1983. The Relationship of Ancient American Writing Systems to the Visual Arts. In *Text and Image in Pre-Columbian Art: Essays on the Interrelationship of the Verbal and Visual Arts*, edited by J. C. Berlo: 55–79. Oxford, B. A. R International Series.

Banning, E. B. 1996. Houses, Compounds and Mansions in the Prehistoric Near East. In *People Who Lived in Big Houses, Archaeological Perspectives on Large Domestic Structures*, edited by G. Coupland and E. Banning: 165–185. Madison, Wisconsin, Prehistory Press.

Barnhart, E. L. 2001. *The Palenque Mapping Project: Settlement and Urbanism at an Ancient Maya City*. Unpublished PhD dissertation. Austin, University of Texas.

Barnhart, E. 2002. Urbanism at Palenque. A Paper Presented at the Society for American Archaeology 67th Annual Meeting Denver, Colorado, March 20–24. *The PARI Journal*, 4(1): 10–16.

Baudez, C. F. 1986. Iconography and History of Copan. In *The South East Maya Periphery*, edited by P. A. Urban and E. M. Schortman: 17–26. Austin, University of Texas Press.

Bavin, L. J. 1996. *The Punishment Administered: Archaeology and Penal Institutions in the Swan River Colony, Western Australia. Unpublished PhD dissertation*. Perth, University of Western Australia.

Bell, P. A. and J. D. Fisher. 1978. *Environmental Psychology*. Philadelphia, Pennsylvania, W. B. Saunders Company.

Benson, E. P. 1978. Ritual Cloth and Palenque Kings. In *Segunda Mesa Redonda de Palenque, Part III*, edited by M. G. Robertson: 45–58. Pebble Beach, California, Pre-Columbian Art Research Center, Robert Louis Stevenson School.

Berlin, H. 1965. *Neue Funde zu alten Zeichnungen*. Ethnos 30, Ethnographical Museum, Stockholm.

Berlo, J. C. 1983. Conceptual Categories for the Study of Text and Images in Mesoamerica. In *Text and Image in Pre-Columbian Art: Essays on the Interrelationship of the Verbal and the Visual Arts*, edited by J. C. Berlo: 1–39. Oxford, B.A.R. International Series.

Beyer, H. 1937. *Studies on the Inscriptions of Chichen Itza*, Carnegie Institution of Washington, Publication no. 483, pp. 29–175 (*Contributions to American Archaeology, Volume IV, no. 21.*) Washington, D.C.

Blankholm, H. P. 1991. *Intrasite Spatial Analysis in Theory and Practice*. Denmark, Aarhus University Press.

Blanton, R. E. 1994. *House and Households: A Comparative Study*. New York, Plenum Press.

Blom, F. 1923. *Las Ruinas De Palenque, Xupá y Finca Encanto*. Mexico, D. F., Instituto Nacional de Anthropologia e Historia.

Boone, E. H. 1994. Introduction: Writing and Recording Knowledge. In *Writing Without Words: Alternative Literacies in Mesoamerica and the Andes*, edited by E. H. Boone and W. D. Mignolo: 3–26. Durham and London, Duke University Press.

Brown, C. 1991. Hieroglyphic Literacy in Ancient Mayaland: Inferences From Linguistic Data. *Current Anthropology*, 32(4): 489–496.

Bullard, W. R. 1960. Maya Settlement Patterns in Northern Peten, Guatemala. *American Antiquity*, 25(3): 355–372.

Calkins, R. G. 1985. *Monuments of Medieval Art*. Ithaca and London, Cornell University Press.

Chase, A. F. 1992a. Elites and the Changing Organization of Classic Maya Society. In *Mesoamerican Elites, An Archaeological Assessment*, edited by D. Z. Chase and A. F. Chase: 30–50. Norman, University of Oklahoma Press.

Chase, A. F. and D. Z. Chase. 1992a. Mesoamerican Elites: Assumptions, Definitions, and Models. In *Mesoamerican Elites, An Archaeological Assessment*, edited by D. Z. Chase and A. F. Chase: 3–17. Norman, University of Oklahoma Press.

Chase, A. F. and D. Z. Chase. 1992b. An Archaeological Assessment of Mesoamerican Elites. In *Mesoamerican Elites, An Archaeological Assessment*, edited by D. Z. Chase and A. F. Chase: 301–316. Norman, University of Oklahoma Press.

Chase, D. Z. 1992b. Postclassic Maya Elites: Ethnohistory and Archaeology. In *Mesoamerican Elites, An Archaeological Assessment*, edited by D. Z. Chase and A. F. Chase: 118–135. Norman, University of Oklahoma Press.

Chermayeff, S. and C. Alexander. 1963. *Community and Privacy*. New York, Anchor Books.

Christie, J. J. 2003a. Conclusions. *Maya Palaces and Elite Residences: An Interdisciplinary Approach*, edited by J. Joyce Christie: 315–336. Austin, University of Texas Press.

Christie, J. J. 2003b. Introduction. In *Maya Palaces and Elite Residences: An Interdisciplinary Approach*, edited by J. J. Christie: 1–12. Austin, University of Texas Press.

Clancy, F. 1986. Text and Image in the Tablets of the Cross Group at Palenque. *Peabody Museum of American Archaeology and Ethnology*, XI: 17–32.

Clark, P. J. 1956. Grouping of Spatial Distributions. *Science*, 123: 373–374.

Clarkson, P. B. 1979. *Classic Maya Attire as Indicators of Status, Role and Function: Or You Can Dress Them Up but You Can't Take Them Anywhere*. Unpublished MA thesis. Department of Archaeology, University of Calgary.

Coe, M. D. 1965. *The Jaguar's Children: Pre-Classic Central Mexico*. The Museum of Primitive Art, New York.

Coe, M. D. 2000. *The Maya*. London, Thames & Hudson.

Coe, M. D. and M. Van Stone. 2001. *Reading Maya Glyphs*. London, Thames & Hudson.

Cohodas, M. 1973. Some Unusual Aspects of Cross Group Symbolism. In *Proceedings of the Tercera Mesa Redonda de Palenque Part II, A Conference on the Art, Iconography and Dynastic History of Palenque*, edited by M. Greene Robertson: 215–232. Pebble Beach, California, Pre-Columbian Art Research Center, Robert Louis Stevenson School.

Cohodas, M. 1991. Ballgame Imagery of the Maya Lowlands: History and Iconography. In *The Mesoamerican Ballgame*, edited by V. L. Scarborough and D. R. Wilcox: 251–288. Tucson, University of Arizona Press.

Cordry, D. and D. Cordry. 1968. *Mexican Indian Costumes*. Austin, University of Texas Press.

Cortés, H. 1986. *Letters from Mexico*, edited and translated by A. Pagden. New Haven, Connecticut, Yale University Press.

Culbert, P. 1991. Maya Political History and Elite Interaction. In *Classic Maya Political History*, edited by P. Culbert: 311–347. Norman, University of Oklahoma Press.

de Landa, D. 1978. *Relación de las cosas de Yucatán*, 1566, translated by William Gates as Friar Diego de Landa – Yucatan Before and After the Conquest. Publication No. 20, the Maya Society, Baltimore.

Dev Sharma, K. 1991. *The Roots of Privacy: Understanding Private and Public Behaviour*. Prakashan, Popular.

Ebert, J. I. 1992. *Distributional Archaeology*. Albuquerque, New Mexico, University of New Mexico Press.

Ekholm, S. M. 1991. Ceramic Figurines and the Mesoamerican Ballgame. In *The Mesoamerican Ballgame*, edited by V. L. Scarborough and D. R. Wilcox: 243–249. Tucson, University of Arizona Press.

Fairclough, G. 1992. Meaningful Constructions – Spatial and Functional Analysis of Medieval Buildings. *Antiquity*, 66: 348–366.

Fash, B. 1992. Late Classic Architectural Sculpture Themes in Copan. *Ancient Mesoamerica*, 3: 89–104.

Fash, B., W. Fash, et al. 1992. *He at the Head of the Mat: Investigations of a Classic Maya Council House at Copan, Honduras*. Paper delivered at the 90th Annual Meeting of the American Anthropological Association, Chicago, Illinois, November 20–24.

Fash, W. 1983. *Maya State Formation: A Case Study and Its Implications*. PhD dissertation. Department of Archaeology, Harvard University.

Fash, W. L. 1991. *Scribes, Warriors and Kings: The City of Copán and the Ancient Maya*. London, Thames & Hudson.

Fash, W. L. 1998. Dynastic Architectural Programs: Intention and Design in Classic Maya Buildings at Copán and Other Sites. In *Function and Meaning in Classic Maya Architecture: A Symposium at Dumbarton Oaks 7th and 8th October 1994*, edited by S. Houston: 223–267. Washington, DC, Dumbarton Oaks.

Ferguson T. J. 1996. Historic Zuni Architecture and Society: Archaeological Application of Space Syntax. *Anthropological Papers*, 60: 1–176.

Foster, L. 2002. *Handbook to Life in the Ancient Maya World*. New York, Oxford University Press.

Foster, S. 1989a. Analysis of Spatial Patterns in Buildings (Access Analysis) as an Insight into Social Structure: Examples from the Scottish Atlantic Iron Age. *Antiquity*, 63: 40–50.

Foster, S. 1989b. Transformations in Social Space – the Iron Age in Orkney and Caithness. *Scottish Archaeological Review*, 6: 34–55.

Freidel, D. and L. Schele. 1988. Kingship in the Late Preclassic Maya Lowlands: The Instruments and Places of Ritual Power. *American Anthropologist*, 90(3): 547–568.

Freidel, D. and L. Schele. 1993. *Maya Cosmos: Three Thousand Years of the Shaman's Path.* New York, William Morrow and Company.

Fuson, R. H. 1969. The Orientation of Maya City Centers. *Annals of the Association of American Geographers*, 59: 494–511.

Giddens, A. 1979. *Central Problems in Social Theory: Action, Structure and Contradiction in Social Analysis.* London, MacMillan.

Giddens, A. 1981. *A Contemporary Critique of Historical Materialism.* London, MacMillan.

Giddens, A. 1984. *The Constitution of Society.* Cambridge, UK, Polity Press.

Gilchrist, R. 1994. *Gender and Material Culture: The Archaeology of Religious Women.* New York, Routledge.

Girouard, M. 1978. *Life in the English Country House.* New Haven, Connecticut, Yale University Press.

Gould, S. 1999. Planning, Development and Social Archaeology. The Familiar Past: Archaeologies of Later Historical Britain. In *Planning, Development and Social Archaeology*, edited by S. Tarlow and S. West: 140–154. London and New York, Routledge.

Grahame, M. 1995. *The House Of Pompeii: Space and Social Interaction.* PhD dissertation. Southampton, UK, Department of Archaeology, University of Southampton.

Grahame, M. 1997. Public and Private in the Roman House: The Spatial Order of the Casa del Fauno. In *Domestic Space in the Roman World: Pompeii and Beyond: Supplementary Series Number 22*, edited by R. Laurence and A. Wallace-Hadrill: 137–165. Oxford, Oxbow Books.

Greene Robertson, M. 1979. A Sequence for Palenque Painting Techniques. In *Maya Archaeology and Ethnohistory*, edited by N. Hammond and G. R. Willey: 149–171. Austin, University of Texas Press.

Greene Robertson, M. 1984. *The Sculpture of Palenque: The Temple of the Inscriptions.* Princeton, Princeton University Press.

Greene Robertson, M. 1985a. *The Sculpture of Palenque: The Early Buildings of the Palace and the Wall Paintings.* Princeton, Princeton University Press.

Greene Robertson, M. 1985b. *The Sculpture of Palenque: The Late Buildings of the Palace.* Princeton, Princeton University Press.

Greene Robertson, M. 1991. *The Sculpture of Palenque: The Cross Group, the North Group, the Oldvidado, and Other Pieces.* Princeton, Princeton University Press.

Greene Robertson, M. 1994. The Iconography of "Isolated Art Styles", that are "Group Supported" and "Individual Supported" Occurring at Chichen Itza and Uxmal. In *Hidden Among the Hills, Maya Archaeology of the Northwest Yucatan Peninsula*, edited by H. J. Prem: 197–212. Mockmuhl, Germany, Verlag Von Flemming.

Grube, N. 1992. Classic Maya Dance: Evidence from Hieroglyphs and Iconography. *Ancient Mesoamerica*, 3(2): 201–218.

Guderjan, T. H., R. J. Lichtenstein, et al. 2003. Elite Residences at Blue Creek, Belize. In *Maya Palaces and Elite Residences: An Interdisciplinary Approach*, edited by J. Joyce Christie: 13–45. Austin, University of Texas Press.

Hall, E. T. 1959. *The Silent Language.* New York, Doubleday.

Hall, E. T. 1966. *The Hidden Dimension.* New York, Doubleday.

Hallowell, I. A. 1977. Cultural Factors in Spatial Orientation: A Reader in the Study of Symbols and Meanings. In *Symbolic Anthropology*, edited by J. L. Dolgin, D. S. Kemnitzer, and D. M. Schneider: 131–150. New York, Columbia University Press.

Hammond, N. 1975. *Lubaantun: A Classic Maya Realm.* Peabody Museum Monographs, No. 2. Cambridge, Massachusetts.

Hammond, N. 1982. *Ancient Maya Civilization.* New Brunswick, Rutgers University Press.

Hanks, W. F. 1990. *Referential Practice: Language and Lived Space among the Maya.* Chicago, University of Chicago Press.

Hanson, J. and B. Hillier. 1982. Domestic Space Organization: Two Contemporary Space-Codes Compared. *Architecture and Behavior*, 2: 5–25.

Hanson, R. 1998. Continuity and Disjunction: The Pre-Classic Antecedents of Classic Maya Architecture. In *Function and Meaning in Classic Maya Architecture: A Symposium at Dumbarton Oaks 7th and 8th October 1994*, edited by S. Houston: 49–122. Washington, DC, Dumbarton Oaks.

Harrison, P. 1970. *The Central Acropolis, Tikal, Guatemala: A Preliminary Study of Functions and its Structural Components during the Late Classic Period.* PhD dissertation. Department of Archaeology, University of Pennsylvania.

Haviland, W. A. and H. Moholy-Nagy. 1992. Distinguishing the High and Mighty from the Hoi Polloi at Tikal, Guatemala. In *Mesoamerican Elites, An Archeological Assessment*, edited by D. Z. Chase and A. F. Chase: 50–60. Norman, University of Oklahoma Press.

Hillier, B. 1985. The Nature of the Artificial: The Contingent and the Necessary in Spatial Form in Architecture. *Geoforum*, 16(2): 163–178.

Hillier, B. and J. Hanson. 1984. *The Social Logic of Space.* Cambridge, University of Cambridge Press.

Hodder, I. 1972. Locational Models and the Study of Romano-British Settlement. In *Models in Archaeology*, edited by D. L. Clarke: 887–909. London, Methuen.

Hodder, I. and C. Orton. 1976. *Spatial Analysis in Archaeology.* London, Cambridge University Press.

Holmes, W. H. 1897. *Archaeological Studies Among the Ancient Cities of Mexico: Part II, Monuments of Chiapas, Oaxaca and the Valley of Mexico.* Chicago, Field Columbian Museum.

Houston, S. 1994. Literacy among the Pre-Columbian Maya: A Comparative Perspective. In *Writing Without Words*, edited by E. H. Boone and W. D. Mignolo: 27–50. Durham, Duke University Press.

Houston, S. 1998. Finding Function and Meaning in Classic Maya Architecture. In *Function and Meaning in Classic Maya Architecture: A Symposium at Dumbarton Oaks 7th and 8th October 1994*, edited by S. Houston: 519–539. Washington, DC, Dumbarton Oaks.

Houston, S. and D. Stuart. 1992. On Maya Hieroglyphic Literacy. *Current Anthropology*, 33(5): 589–593.

Houston, S. and D. Stuart. 2001. Peopling the Maya Court. In *Royal Courts of the Ancient Maya: Theory, Comparison and Synthesis, Part I*, edited by T. Inomata and S. Houston: 55–83. Oxford, Westview Press.

Jaffé, A. 1966. Symbolism in the Visual Arts. In *Man and His Symbols*, edited by C. G. Jung: 230–272. London, Aldus Books.

Jones, C., W. Ashmore, et al. 1983. The Quirigua Project: 1977 Season. In *Quirigua Reports 2*, edited by P. Schortman and A. Urban: 1–38. Philadelphia, University of Pennsylvania.

Justeson, J. S. 1989. The Representational Conventions of Mayan Hieroglyphic Writing. In *Text and Image in Maya Culture: Explorations in Language, Writing, and Representation*, edited by D. S. Rice and W. F. Hanks: 25–39. Salt Lake City, University of Utah Press.

Kostof, S. 1991. *The City Shaped: Urban Patterns and Meanings Through History*. London, Thames & Hudson.

Kowalski, J. K. 1998. Uxmal and the Puuc Zone: Monumental Architecture, Sculptured Facades and Political Power in the Terminal Classic Period. In *Maya*, edited by P. Schmidt: 400–425. New York, St Martin's Press.

Kowalski, J. K. 2003. Evidence for the Functions and Meanings of some Northern Maya Palaces. In *Maya Palaces and Elite Residences: An Interdisciplinary Approach*, edited by J. Joyce Christie: 204–253. Austin, University of Texas Press.

Kubler, G. 1985. The Design of Space in Maya Architecture. In *Studies in Ancient American and European Art: The Collected Essays of George Kubler*, edited by T. R. Reese: 242–255. New Haven, Connecticut, Yale University Press.

Lang, J. 1987. *Creating Archaeological Theory: The Role of the Behavioral Sciences in Environmental Design*. New York, Van Nostrand Reinhold Company.

Liendo Stuardo, R. L. 2003. Access Patterns in Maya Royal Precincts. *Maya Palaces and Elite Residences: An Interdisciplinary Approach*, edited by J. J. Christie: 184–203. Austin, University of Texas Press.

Lincoln, C. E. 1986. Chronology of Chichen Itza: A Review of the Literature. In *Late Lowland Maya Civilization: Classic to Postclassic*, edited by J. A. Sabloff and E. W. Andrews V: 141–199. Albuquerque, University of New Mexico Press.

Marcus, J. 1992a. *Mesoamerican Writing Systems: Propaganda, Myth, and History in Four Ancient Civilizations*. Princeton, New Jersey, Princeton University Press.

Marcus, J. 1992b. Royal Families, Royal Texts: Examples from the Zapotec and Maya. In *Mesoamerican Elites, An Archaeological Assessment*, edited by D. Z. Chase and A. F. Chase: 221–241. Norman, University of Oklahoma Press.

Marquina, I. 1964. *Arquitectura Prehispanica*. Mexico City, Instituto Nacional de Antrepologia e Historia.

Martin, S. 2003. *The Proceedings of the Maya Hieroglyphic Weekend: Maya Warfare and Politics, New Data and Perspectives, Maya Hieroglyphic Weekend October 4–5*, edited and transcribed by P. Wanyerka, Cleveland State University.

Martin, S. and N. Grube. 2000. *Chronicle of the Maya Kings and Queens: Deciphering the Dynasties of the Ancient Maya*. London, Thames & Hudson.

Mathews, D. 2004. *The Organization of Public and Private Space in the Late Classic Central Acropolis of Tikal and Palace of Palenque*. Honours thesis. Department of Archaeology, University of Calgary.

Mathews, P. 1996. *Notebook for the Inaugural Maya Hieroglyphic Workshop at Calgary. Department of Archaeology*. Calgary, University of Calgary.

Mathews, P. 2004. *Maya Hieroglyphic Weekend October 2–3, 2004: All in the Family, The Royal Genealogy of Palenque*. Department of Archaeology, Cleveland State University.

Maudslay, A. P. 1889–1902a. *Biologia Centrali-Americana: Archaeology. Volumes III–IV*. London, R. H. Porter and Dulau and Co.

Maudslay, A. P. 1889–1902b. *Biologia Centrali-Americana: Archaeology. Volume V*. London, R. H. Porter and Dulau and Co.

Maudslay, A. P. 1889–1902c. *Biologia Centrali-Americana: Archaeology. Volume VI*. London, R. H. Porter and Dulau and Co.

McAnany, P. and S. Plank. 2001. Perspectives on Actors, Gender Roles, and Architecture at Classic Maya Courts and Households. In *Royal Courts of the Ancient Maya: Theory Comparison and Synthesis Vol. 1*, edited by T. Inomata and S. Houston: 85–125. Oxford, Westview Press.

Miller, A. G. 1989. Comparing Maya Image and Text. In *Word and Image in the Maya Culture*, edited by W. F. Hanks and D. S. Rice: 176–188. Salt Lake City, University of Utah Press.

Miller, M. 1986. *The Bonampak Murals*. Princeton, New Jersey, Princeton University Press.

Miller, M. 1988. The Meaning and Function of the Main Acropolis, Copan. In *The Southeast Classic Maya Zone*, edited by E. H. Boone and G. R. Willey: 149–195. Washington, DC, Dumbarton Oaks.

Miller, M. 1998. A Design for Meaning in Maya Architecture. *Function and Meaning in Classic Maya Architecture: A Symposium at Dumbarton Oaks 7th and 8th October 1994*, edited by S. D. Houston: 187–223. Washington, DC, Dumbarton Oaks.

Miller, M. 1999. *Maya Art and Architecture*. London, Thames & Hudson.

Miller, M. and K. Taube. 1993. *The Gods and Symbols of Ancient Mexico and the Maya: Illustrated Dictionary of the Mesoamerican Region*. London, Thames & Hudson.

Miller, M. and S. D. Houston. 1987. The Classic Maya Ballgame and its Architectural Setting: A Study in the Relations Between Text and Image. *Anthropology and Aesthetics*, 14: 47–66.

Moore, J. D. 1992. Pattern and Meaning in Prehistoric Peruvian Architecture: The Architecture of Social Control in the Chinu State. *Latin American Antiquity*, 3(2): 95–113.

Moore, J. D. 1996. The Archaeology of Plazas and the Proxemics of Ritual: Three Andean Traditions. *American Anthropologist*, 98(3): 789–802.

Nanda, S. 1994. *Cultural Anthropology*. Belmont, California, Wadsworth.

Newman, O. 1972. *Defensible Space: People and Design in the Violent City*. London, Architectural Press.

Parmington, A. 2000. *Classic Maya Costume and the Subsidiary Office of "Sahal"*. Honours dissertation. Department of Archaeology, La Trobe University, Melbourne.

Parmington, A. 2003. Classic Maya Status and the Subsidiary "Office" of Sajal: A Comparative Study of Status as Represented in Costume and Composition in the Iconography of Monuments. *Mexicon*, XXV(2): 46–53.

Pearson, M. and C. Richards. 1994a. Architecture and Order: Spatial Representation and Archaeology. In *Architecture and Order: Spatial Representation and Archaeology, Approaches to Social Space*, edited by M. Pearson and C. Richards: 38–73. London, Routledge.

Pearson, M. and C. Richards. 1994b. Ordering the World: Perceptions of Architecture, Space and Time. In *Architecture and Order: Spatial Representation and Archaeology, Approaches to Social Space*, edited by M. Pearson and C. Richards: 1–37. London, Routledge.

Pendergast, D. M. 1992. Noblesse Oblige: The Elites of Altun Ha and Lamani, Belize. In *Mesoamerican Elites: An Archaeological Assessment*, edited by D. Z. Chase and A. F. Chase: 61–79. Norman, University of Oklahoma Press.

Peterson, K. K. 1983. Observations on the Ik Windows at Palenque. In *The Fifth Palenque Round Table Vol. VII*, edited by M. Greene Robertson: 19–29. San Francisco, Pre-Columbian Art Research Institute.

Pollock, H. E. D. 1965. Architecture of the Maya Lowlands. In *Handbook of Middle American Indians Vol. 2*, edited by R. Wauchope and G. R. Willey: 378–440. Austin, University of Texas Press.

Proskouriakoff, T. 1950. *A Study of Classic Maya Sculpture*. Washington, DC, Carnegie Institute of Washington.

Rands, R. 1974. A Chronological Framework for Palenque. In *Primera Mesa Redonda de Palenque (First Palenque Round Table), Part I*, edited by M. G. Robertson. Electronic version: 1–5. Pebble Beach, California, Robert Louis Stevenson School, Pre-Columbian Art Research.

Rands, R. 1985. The Ceramic Patterns and Traditions in the Palenque Area. In *Maya Ceramics – Papers from the 1985 Maya Ceramic Conference*, edited by P. M. Rice and R. J. Sharer: 203–241. Oxford, B. A. R International Series.

Rands, R. and B. Rands. 1957. The Ceramic Position of Palenque, Chiapas. *American Antiquity*, 23(2): 140–150.

Rands, R. and B. Rands. 1959. The Incensario Complex of Palenque, Chiapas. *American Antiquity*, 25(2): 225–236.

Rapoport, A. 1976. Social Aspect of Man-Environment Studies. In *The Mutual Interaction of People in Their Built Environment*, edited by A. Rapoport: 7–37. Paris, Mouton Publishers.

Rapoport, A. 1977. *Human Aspects of Urban Form: Towards a Man-Environment Approach to Urban Form and Design*. New York, Pergamon Press.

Reade, J. 1979. Ideology and Propaganda in Assyrian Art. In *Power and Propaganda: A Symposium on Ancient Empires, MCSA 7*, edited by M. Trolle Larsen: 329–343. Copenhagen, Akademisk Forlag.

Reents-Budet, D. 1989. Narrative in Classic Maya Art. In *Word and Image in Maya Culture: Explorations in Language, Writing and Representation*, edited by W. F. Hanks and D. S. Rice: 189–198. Salt Lake City, University of Utah Press.

Reents-Budet, D. 1994. *Painting the Maya Universe: Royal Ceramics of the Classic Period*. Durham, Duke University Press.

Rice, D. and P. Culbert, 1990. Historical Contexts for Population Reconstruction in the Maya Lowlands. In *Precolumbian Population History in the Maya Lowlands*, edited by T. P. Culbert and D. S. Rice. Albuquerque, University of New Mexico Press.

Ruz, L. A. 1958. Exploraciones Arqueológicas en Palenque: 1953. *Anales del Instituto Nacional de Antropología e Historia X*: 69–116. Mexico City, Instituto Nacional de Antropología e Historia.

Ruz, L. A. 1962. Exploraciones Arqueológicas en Palenque: 1957. *Anales del Instituto Nacional de Antropología e Historia XIV*: 35–90. Mexico City, Instituto Nacional de Antropología e Historia.

Samson, R. 1990. *Social Archaeology of Houses*. Edinburgh: Edinburgh University Press.

Sanders, W. T. 1992. Ranking and Stratification in Prehispanic Mesoamerica. In *Mesoamerican Elites, An Archaeological Assessment*, edited by D. Z. Chase and A. F. Chase: 278–291. Norman, University of Oklahoma Press.

Satterthwaite, L. 1935. *Palace Structures J-2 and J-6*. Piedras Negras Preliminary Papers 3. Philadelphia, University Museum, University of Pennsylvania.

Schele, L. 1984. Human Sacrifice among the Classic Maya. In *Ritual Human Sacrifice in Mesoamerica*, edited by H. Boone: 7–49. Washington, DC, Dumbarton Oaks.

Schele, L. 1985. Colour on Classic Architecture and Monumental Sculpture of the Southern Maya Lowlands. In *Painted Architecture and Polychrome Monumental Sculpture in Mesoamerica*, edited by E. H. Boone: 31–49. Washington, DC, Dumbarton Oaks.

Schele, L. 1986. Architectural Development and Political History of Palenque. In *City-States of the Maya: Art and Architecture: A Conference*, edited by E. P. Benson: 110–138. Denver, Colorado, Rocky Mountain Institute for Pre-Columbian Studies.

Schele, L. 1998. The Iconography of Maya Architectural Facades during the Late Classic Period. In *Function and Meaning in Classic Maya Architecture: A Symposium at Dumbarton Oaks 7th and 8th October 1994*, edited by S. Houston: 479–519. Washington, DC, Dumbarton Oaks.

Schele, L. and D. Freidel. 1990. *A Forest of Kings: Untold Stories of the Ancient Maya*. New York, William Morrow and Co.

Schele, L. and P. Mathews. 1979. *The Bodega of Palenque Chiapas, Mexico*. Washington, DC, Dumbarton Oaks.

Schele, L. and P. Mathews. 1998. *The Code of Kings: The Language of Seven Sacred Temples and Tombs*. New York, Scribner.

Schele, L. and M. Miller. 1986. *The Blood of Kings: Dynasty and Ritual in Maya Art*. New York, George Braziller, Inc., in association with Kimbell Art Museum, Fort Worth.

Seler, E. 1915. *Gesammelte Abhandlungen zur Amerikanischen Sprach- und Alterthumskunde*. Berlin, A. Ascher & Co.

Seler, E. 1976. *Observations and Studies in the Ruins of Palenque 1915*. Pebble Beach, California, Pre-Columbian Art Research Center, Robert Louis Stevenson School.

Sharer, R. 1989. The Preclassic Origins of Maya Writing: A Highland Perspective. In *Word and Image in Classic Maya Culture: Explorations in Language, Writing, and Representation*, edited by W. F. Hanks and D. S. Rice: 165–176. Salt Lake City, University of Utah Press.

Sharer, R. J. 1994. *The Ancient Maya* [Fifth Edition]. Stanford, California, Stanford University Press.

Sharer, R, J. with L. P. Traxler. 2006. *The Ancient Maya* [Sixth Edition]. Stanford, California, Stanford University Press.

Stuart, D. 2003. *A Cosmological Throne at Palenque*. Mesoweb: www.mesoweb.com/Stuart /notes/Throne.pdf.

Stuart, D. 2006. *The Palenque Mythology: Inscriptions and Interpretations of the Cross Group. Sourcebook for the 30th Maya Meetings, March 14–19, 2006*. 85–194. Austin, The University of Texas.

Tate, C. 1992. *Yaxchilan: The Design Of a Maya Ceremonial City*. Austin, University of Texas Press.

Taube, K. 1992. *The Major Gods of Ancient Yucatan. Studies in Pre-Columbian Art and Archaeology, 32*. Washington, DC, Dumbarton Oaks Research Libraries and Collections.

Taylor, D. 1983. *Classic Maya Costume: Regional Types of Dress*. Unpublished PhD dissertation. Department of Archaeology, Yale University.

Thompson, J. E. S. 1950. *Maya Hieroglyphic Writing: An Introduction*. University of Oklahoma Press, Norman.

Thompson, J. E. S. 1954. *The Rise and Fall of the Ancient Maya Civilization*. Norman, University of Oklahoma Press.

Tilley, C. 1994. *A Phenomenology of Landscape: Place, Paths and Monuments*. Oxford, UK, Berg.

Tozzer, A. M. 1941. *Landa's Relacion de las Cosas de Yucatan: A Translation*. Papers of the Peabody Museum of American Archaeology and Ethnology 28. Cambridge, Massachusetts, Harvard University.

Velecky, L. C. 1978. The Concept of Privacy. In *Privacy*, edited by J. B. Joung: 13–33. New York, John Wiley & Sons.

Viollet-le-Duc, E. E. 1987. *Lectures in Architecture*. New York, Dover Publications Inc.

von Meiss, P. 1986. *De la forme au lieu*. Lausanne, Presses polytechniques et universitaires romandes.

von Meiss, P. 1990. *Elements of Architecture: From Form to Place*. New York, Van Nostrand Reinhold.

Wallace-Hadrill, A. 1988. The Social Structure of the Roman House. *Papers of the British School at Rome*, 56: 43–97.

Walmsley, D. J. 1988. *Urban Living: The Individual in the City*. London, Longarm.

Washburn, D. K. 1974. Spatial Analysis of Occupation Floors II: The Application of Nearest Neighbour Analysis. *American Antiquity*, 39: 16–34.

Watson, M. 1970. *Proxemic Behaviour: A Cross-Cultural Study*. Paris, Mouton.

Webster, D. 1992. Maya Elites: The Perspective from Copan. In *Mesoamerican Elites, An Archaeological Assessment*, edited by D. Z. Chase and A. F. Chase: 135–168. Norman, University of Oklahoma Press.

Webster, D. 1998. Classic Maya Architecture: Implications and Comparisons. In *Function and Meaning in Classic Maya Architecture: A Symposium at Dumbarton Oaks 7th and 8th October 1994*, edited by S. Houston: 5–49. Washington, DC, Dumbarton Oaks.

Weitzmann, K. 1970. *Illustration in Roll and Codex: A Study of the Origin and Method of Text Illustration*. Princeton, Princeton University Press.

West, S. 1999. Social Space and the English Country House. In *The Familiar Past: Archaeologies of Later Historical Britain*, edited by S. Tarlow and S. West: 103–122. London and New York, Routledge.

Westin, A. F. 1967. *Privacy and Freedom*. New York, Atheneum.

Willey, G. R. 1980. Towards a Holistic View of Ancient Maya Civilisation. *Man*, 15: 249–266.

Young, J. B. 1978. Introduction: A Look at Privacy. In *Privacy*, edited by J. B. Young: 1–57. New York, John Wiley & Sons.

### *Web sites*

Foundation for the Advancement of Mesoamerican Studies, Inc. (FAMSI), www.famsi.org.

# Index